Advances in Rock Dynamics and Applications

Advances in Rock Dynamics and Applications

Edited by

Yingxin Zhou
Defence Science & Technology Agency (DSTA), Singapore

Jian Zhao
École Polytechnique Fédérale de Lausanne (EPFL), Switzerland

CRC Press
Taylor & Francis Group
Boca Raton London New York

CRC Press is an imprint of the
Taylor & Francis Group, an **informa** business

A BALKEMA BOOK

Photos on the cover: Fracture and failure of a rock material specimen under dynamic compression using split Hopkinson pressure bar. The duration between the first and last images is 400 microseconds. Photo courtesy of Xibing Li, Central South University, China.

CRC Press
Taylor & Francis Group
6000 Broken Sound Parkway NW, Suite 300
Boca Raton, FL 33487-2742

First issued in paperback 2018

CRC Press/Balkema is an imprint of the Taylor & Francis Group,
an informa business

ISBN-13: 978-0-415-61351-4 (hbk)
ISBN-13: 978-1-138-07296-1 (pbk)

Typeset by MPS Limited, a Macmillan Company, Chennai, India

Library of Congress Cataloging-in-Publication Data

Advances in rock dynamics and applications / edited by Yingxin Zhou,
Jian Zhao.
    p. cm.
    Includes bibliographical references and index.
    ISBN 978-0-415-61351-4 (hardback) — ISBN 978-0-203-09320-7 (eBook)
    1. Rock mechanics. I. Zhou, Yingxin. II. Zhao, Jian, 1960–
    TA706.A38   2011
    624.1'5132—dc22
                                                          2011013714

Published by: CRC Press/Balkema
                P.O. Box 447, 2300 AK Leiden, The Netherlands
                e-mail: Pub.NL@taylorandfrancis.com
                www.crcpress.com – www.taylorandfrancis.co.uk – www.balkema.nl

Visit the Taylor & Francis Web site at
http://www.taylorandfrancis.com

and the CRC Press Web site at
http://www.crcpress.com

Table of Contents

Contributing Authors

Carlo ALBERTINI, albertini.carlo@alice.it
Dynalab Impact Technologies, via Buonarroti 41, 21014-Laveno Mombello (Va), ITALY

Mehdi AMINI, amini_chermahini@yahoo.com
Tehran University, Department of Mining Engineering, Tehran, IRAN

Ömer AYDAN, aydan@scc.u-tokai.ac.jp
Tokai University, Department of Marine Civil Engineering, Orido 3-20-1, Shimizu-ku, Shizuoka, 424-8610, JAPAN

Tore BØRVIK, tore.borvik@ntnu.no
Department of Structural Engineering, Norwegian University of Science and Technology, NO-7491 Trondheim, NORWAY

Vladimir BRATOV, vladimir@bratov.com
Institute for Problems in Mechanical Engineering, Bolshoy pr. V.O., 61, 199178, St.-Petersburg.
St.-Petersburg State University, Faculty of Mathematics and Mechanics, Universitetskaya nab. 7-9, 199034, St.-Petersburg, RUSSIA

Jungang CAI, caijg@tritech.com.sg
Tritech Group Limited, 2 Kaki Bukit Place, Tritech Building, SINGAPORE 416180

Ezio CADONI, ezio.cadoni@supsi.ch
University of Applied Sciences of Southern Switzerland (SUPSI), DynaMat Laboratory, CAMPUS SUPSI-Trevano, C.P. 105, CH-6952 Canobbio, SWITZERLAND

Rong CHEN, cr27251345@gmail.com
National University of Defense Technology, College of Science, Changsha 410073, CHINA

Feng DAI, feng.dai@utoronto.ca
University of Toronto, Impact and Fracture Laboratory (IFL) at the Department of Civil Engineering and Lassonde Institute, 35 St. George St., Toronto, Ontario, CANADA M5S 1A4

Mitsuo DAIDO
Tokai University, Graduate School of Marine Science and Technology, Orido 3-20-1, Shimizu-ku, Shizuoka, 424-8610, JAPAN

Evgeny DOLMATOV
St.-Petersburg State University, Faculty of Mathematics and Mechanics, Universitet-skaya nab. 7-9, 199034, St.-Petersburg, RUSSIA

Melih GENIS, genis@karaelmas.edu.tr
Zonguldak Karaelmas University, Dept. of Mining Engineering, 67100, Zonguldak, TURKEY

Yossef H. HATZOR, hatzor@bgu.ac.il
Ben-Gurion University of the Negev, Department of Geological and Environmental Sciences, Beer-Sheva 84105, ISRAEL

Lei HE, helei@ntu.edu.sg
School of Civil and Environmental Engineering, Nanyang Technological University, Nanyang Avenue, SINGAPORE 639798

Takashi ITO, tak@toyota-ct.ac.jp
Toyota National College of Technology, Department of Civil Engineering, 2-1 Eisei-cho, Toyota, Aichi, 471-8525, JAPAN

Tohid KAZERANI, tohid.kazerani@epfl.ch
École Polytechnique Fédérale de Lausanne (EPFL), School of Architecture, Civil and Environmental Engineering, Laboratory for Rock Mechanics (LMR), CH-1015 Lausanne, SWITZERLAND

Halil KUMSAR, hkumsar@pau.edu.tr
Pamukkale University, Department of Geological Engineering, 20020, Denizli, TURKEY

Charlie Chunlin LI, charlie.c.li@ntnu.no
Norwegian University of Science and Technology (NTNU), NO-7491 Trondheim, NORWAY

Haibo LI, hbli@whrsm.ac.cn
State Key Laboratory of Geomechanics and Geotechnical Engineering, Institute of Rock and Soil Mechanics, Chinese Academy of Sciences, Wuhan 430071, CHINA

Jianchun LI, jcli@whrsm.ac.cn
State Key Laboratory of Geomechanics and Geotechnical Engineering, Institute of Rock and Soil Mechanics, Chinese Academy of Sciences, Wuhan 430071, CHINA

Junru LI, jrli@whrsm.ac.cn
State Key Laboratory of Geomechanics and Geotechnical Engineering, Institute of Rock and Soil Mechanics, Chinese Academy of Sciences, Wuhan 430071, CHINA

Xibing LI, xbli@mail.csu.edu.cn
School of Resources and Safety Engineering, Central South University, Changsha
410083, CHINA

Deshun LIU, deshunliu@hotmail.com
Hunan University of Science and Technology, Xiangtan, Hunan Province, CHINA

Guowei MA, ma@civil.uwa.edu.au
School of Civil and Resource Engineering, University of Western Australia, Perth, WA
6009, AUSTRALIA

Yoshimi OHTA, ota-yoshimi@jnes.go.jp
Japan Nuclear Energy Safety Organization (JNES), 17-1, Toranomon 3-chome,
Minato-ku, Tokyo 105-0001, JAPAN

Tso-Chien PAN, cpan@ntu.edu.sg
Protective Technology Research Centre, School of Civil and Environmental Engineering, Nanyang Technological University, Nanyang Avenue, SINGAPORE 639798

Yuri PETROV
Institute for Problems in Mechanical Engineering, Bolshoy pr. V.O., 61, 199178,
St.-Petersburg.
St.-Petersburg State University, Faculty of Mathematics and Mechanics, Universitet-
skaya nab. 7-9, 199034, St.-Petersburg, RUSSIA

Svein REMSETH, svein.remseth@ntnu.no
Department of Structural Engineering, Norwegian University of Science and Technology, NO-7491 Trondheim, NORWAY

Chong Chiang SEAH, schongch@dsta.gov.sg
Defence Science & Technology Agency, 1 Depot Road Defence Technology Tower A,
SINGAPORE 109679

Gen-Hua SHI, sghua@aol.com
DDA Company, 1746 Terrace Drive, Belmont, CA 94002, USA

Chun'an TANG, catang@mechsoft.cn
School of Civil Engineering, Dalian University of Technology, Dalian 116024, CHINA

Naohiko TOKASHIKI, tokasiki@tec.u-ryukyu.ac.jp
Ryukyu University, Department of Civil Engineering, 1 Senbaru, Nishihara, Okinawa,
903-0213 JAPAN

Grigory VOLKOV
Institute for Problems in Mechanical Engineering, Bolshoy pr. V.O., 61, 199178,
St.-Petersburg, RUSSIA

Xuejun WANG, xjwang@eagle.org
School of Civil and Environmental Engineering, Nanyang Technological University, Nanyang Avenue, SINGAPORE 639798

Kaiwen XIA, kaiwen.xia@utoronto.ca
University of Toronto, Impact and Fracture Laboratory (IFL) at the Department of Civil Engineering and Lassonde Institute, 35 St. George St., Toronto, Ontario, CANADA M5S 1A4

Gony YAGODA-BIRAN, yagoda@bgu.ac.il
Ben-Gurion University of the Negev, Department of Geological and Environmental Sciences, Beer-Sheva 84105, ISRAEL

Yuefeng YANG, yang.yue.feng@foxmail.com
School of Civil Engineering, Dalian University of Technology, Dalian 116024, CHINA

Tubing YIN
School of Resources and Safety Engineering, Central South University, Changsha 410083, CHINA

Gaofeng ZHAO, gaofeng.zhao@unsw.edu.au
School of Civil & Environmental Engineering, University of New South Wales (UNSW), Sydney, NSW 2052, AUSTRALIA

Jian ZHAO, Jian.zhao@epfl.ch
École Polytechnique Fédérale de Lausanne (EPFL), School of Architecture, Civil and Environmental Engineering, Laboratory for Rock Mechanics (LMR), CH-1015 Lausanne, SWITZERLAND

Xiaobao ZHAO, xbzhao@nju.edu.cn
Institute for Underground Space and Geoenvironment, School of Earth Sciences and Engineering, Nanjing University, Nanjing 210093, CHINA

Yingxin ZHOU, zyingxin@dsta.gov.sg
Defence Science & Technology Agency, 1 Depot Road Defence Technology Tower A, SINGAPORE 109679

Zilong ZHOU, zlzhou@mail.csu.edu.cn
School of Resources and Safety Engineering, Central South University, Changsha 410083, CHINA

Jianbo ZHU, Jianbo.zhu@epfl.ch
École Polytechnique Fédérale de Lausanne (EPFL), School of Architecture, Civil and Environmental Engineering, Laboratory for Rock Mechanics (LMR), CH-1015 Lausanne, SWITZERLAND

Yang ZOU
School of Resources and Safety Engineering, Central South University, Changsha 410083, CHINA

Preface

Dynamics has been an important part of mechanics in various disciplines. Rock dynamics, too, is an important part of rock mechanics, where an increased rate of loading induces a change in the mechanical behaviour of the rock materials and rock masses.

The study of rock dynamics is important because many rock mechanics and rock engineering problems involve dynamic loading ranging from earthquakes to vibrations to explosions, and rock failure under those dynamic loads as well as dynamic failure under static loads. However, due to the additional "4th" dimension of time, dynamics has been a more challenging topic to understand and to apply. It remains, at least in the discipline of rock mechanics, a relatively uncultivated territory, where research and knowledge are limited.

In 2008, the Commission on Rock Dynamics was set up within the International Society for Rock Mechanics (ISRM). One of the aims of the Commission is to share and exchange knowledge in rock dynamics research and to produce documents on the study and engineering applications of rock dynamics.

In the summer of 2009, the ISRM Commission on Rock Dynamics organised its first workshop in Lausanne, Switzerland. It was at this workshop that participants felt that there was a lack of a comprehensive knowledge base and the Commission should organise researchers to prepare a document summarising the state-of-the-art. This edited book is a direct result of that discussion.

The book aims to provide a summary of the current knowledge of rock dynamics for researchers and engineers. It consists of 18 chapters contributed by individual authors. The topics chosen are wide-ranging, covering fundamental theories of fracture dynamics and wave propagation, rock dynamic properties and testing methods, numerical modelling of rock dynamic failure, engineering applications in earthquakes, explosion loading and tunnel response, as well as dynamic rock support.

The editors would like to thank all the contributing authors. The editing effort by Ms Haiying Bian is greatly appreciated. The CRC Press team, particularly Mr Janjaap Blom and Mr Richard Gundel, also provided publishing support.

Yingxin Zhou and Jian Zhao
March 2011

Chapter 1

Introduction

Yingxin Zhou and Jian Zhao

1.1 SCOPE OF ROCK DYNAMICS

Rock dynamics, as a branch of rock mechanics, deals with the responses of rock (materials and masses) under dynamic stress fields, where an increased rate of loading (or impulsive loading) induces a change in the mechanical behaviour of the rock materials and rock masses. Figure 1.1 is an example showing the different failure behaviours for a rock material under static and dynamic loads.

Differing from static mechanics, dynamic stresses are in the forms of stress waves propagating in the loaded medium with time, and therefore the response of rock is influenced by, and interacts with, the stresses in motion. Rock dynamics deals not only with the end effects of the forces, but also the processes of the forces acting on the rocks. In these processes, both forces and objects are in motion. Rock dynamics specifically examines the processes of dynamic motions of both the forces and the rocks, at different scales varying from micro particles to rock blocks.

Rock dynamics as a science subject covers a wide scope related to forces, and responses of rock, in the time domain. It deals with the distribution of stress fields,

Figure 1.1 Rock specimens after failure under static (left) and dynamic (right) loads.

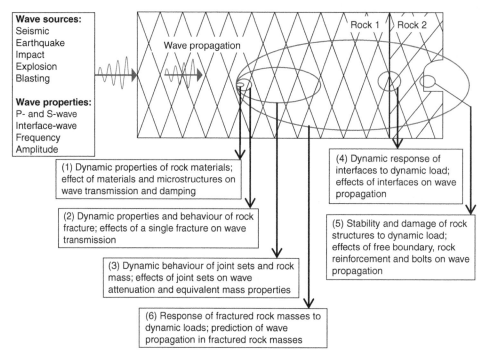

Figure 1.2 Typical rock dynamic problems in tunnels and caverns (after Zhao *et al.,* 1999).

responses and properties of rocks, and dynamic behaviour coupled with the physical environment.

Sources of dynamic loads include explosion, impact, and seismic events. These loads are typically given in the form of time histories of particle acceleration, velocity, or displacement.

The distribution of a dynamic stress field is in the form of a stress wave moving in the loaded medium, including the propagation behaviour of the stress wave. Stress wave propagation in rock masses is governed by wave transmission and transformation across the discontinuities (rock joints) in the rock masses.

The response of rock materials and rock masses under dynamic stress field includes displacements of rock at particle scale, material fracturing and failure, and large movements at discontinuities. Rock fracturing, for example, is a dynamic micro-scale process leading to macro-scale deformation and failure.

Rock dynamic behaviour is often coupled with, and frequently induced by, the physical environment, e.g. water and temperature. Changes of physical environment may alter the stress fields as well as the properties of the rock materials and rock masses, hence leading to dynamic responses of the rocks.

Rock dynamics has applications in mining, energy, environmental and civil engineering, when dynamic loads and behaviours are encountered. Figure 1.2 illustrates typical rock dynamics issues related to the construction and utilisation of a storage cavern. Some of the applications are summarised in, but not limited to, the items below.

a) Construction: rock excavation and fragmentation by blasting and by mechanical means, stability of rock mass and rock support under various dynamic loads, protection of rock falls, use of seismic waves for ground exploration;
b) Energy and mining: rock burst and support in deep mines, fracturing of hot rock in geothermal fields, effects of water injection and induced seismic events; and
c) Environment: earthquake effects on slopes and landslides, hazard and risk control due to explosion and blast, effects of blasting vibrations on existing structures, seismic damage to structures in and on rocks.

1.2 ISRM COMMISSION ON ROCK DYNAMICS

Understanding the effects of dynamic loading on rock and built structures (e.g. tunnels and caverns with their associated reinforcement and support) is essential in dealing with the various rock dynamics problems such as dynamic support design and safety assessment. However, guidance and standards in dynamic analysis and design are generally lacking, and much of the research work done on rock dynamics for military purposes has not been easily available for the general public. For example, there are no existing standard methods for rock dynamic testing, and in rock engineering practice, guidelines and design methodologies for dealing with dynamic problems are generally lacking.

It was against this background that the International Society for Rock Mechanics (ISRM) established a Commission on Rock Dynamics in January 2008. The aim of the Commission (ISRM 2010) is to:

i) Provide a forum for the sharing and exchange of knowledge in rock dynamics research and engineering applications, including organising commission meetings, workshops, seminars and short courses;
ii) Co-ordinate rock dynamic research activities within the ISRM community as well as with other research and professional organizations; and
iii) Produce reports and guidelines on the study and engineering applications of rock dynamics covering fundamental theories, dynamic properties of rock and rock mass, testing methods, tunnel response, and support design.

Specifically, the Commission's work scope (ISRM 2010) covers:

i) Characterisation of dynamic loading sources,
ii) Rock dynamic properties and their determination,
iii) Propagation of dynamic stress waves in geological media,
iv) Rock damage criteria and damage assessment, and
v) Dynamic rock support design.

Under the work plan of the Commission, and resulting from its first workshop held in Lausanne, Switzerland in June 2009, Suggested Methods for determining the dynamic strength parameters (uniaxial compression and the Brazilian tension) and fracture toughness of rock materials have been drafted, all based on the split Hopkinson pressure bar (SHPB) techniques. In addition, a thorough literature review was conducted by members of the Commission, and formed the basis for the workshop discussions and content of this book.

1.3 ABOUT THIS BOOK

This book is partially a result of the activities of the ISRM Commission on Rock Dynamics. Several contributions are made by non-members. The book is intended to present some recent advances in rock dynamics and engineering applications. It is to be used as a reference for research.

This book consists of 18 chapters representing rock dynamics research and applications. Efforts have been made to be as consistent as possible, in terms of uses of symbols, style and references.

While each chapter is independently prepared by individual authors, the 18 edited chapters have been organised into roughly five sections. Chapter 1 provides an introduction to the topic and background of the ISRM Commission on Rock Dynamics and this edited book, while Chapter 2 provides an overview of the state of the art in rock dynamics research. Chapters 3 to 7 discuss various testing techniques for determining the dynamic properties of rock material. Chapters 8 to 10 focus on some fundamental theories related to rock fracturing under dynamic loads and wave propagation in geological media. Chapters 11 to 14 deal with numerical modelling using some of the most advanced numerical techniques of both continuum and discontinuum methods focusing on micromechanics modelling of rock dynamics problems. Finally, Chapters 15 to 18 present some applications in interpretation of seismic effects, tunnel responses under explosion loading and dynamic rock support.

REFERENCES

ISRM, website of the International Society for Rock Mechanics (ISRM), Commission on Rock Dynamics, http://www.isrm.net (2010).

Zhao, J., Zhou, Y.X., Hefny, A.M., Cai, J.G., Chen, S.G., Li, H.B., Liu, J.F., Jain, M., Foo, S.T. and Seah, C.C.: Rock dynamics research related to cavern development for ammunition storage. *Tunnelling and Underground Space Technology* 14(4) (1999), pp.513–526.

Chapter 2

An overview of some recent progress in rock dynamics research

Jian Zhao

2.1 INTRODUCTION

Dynamics, as a branch of mechanics, deals with dynamic load (stress), deformation (strain) and failure (fracturing) in relation to time. Hence rock dynamics covers a wide scope, ranging from the initiation of dynamic loads, forms of dynamic loads, transmission and attenuation of dynamic loads, rock fracturing and damage under dynamic loading, to support of rock under dynamic conditions.

This chapter provides a summary of recent progress in some areas of rock dynamics. It covers stress wave propagation and attenuation, loading rate effects on rock strength and discontinuous micromechanics modelling of dynamic fracturing.

2.2 STRESS WAVE PROPAGATION AND ATTENUATION

Dynamic loads are generally presented in the form of stress waves. Stress waves, similar to other physical waves, attenuate during propagation, particularly at discontinuities. Since the rock masses are generally discontinuous, containing joint sets, stress wave attenuation at joints is the dominating cause of overall wave attenuation in rock masses. Current researches on wave propagation in rock masses have been focused on wave transmission and transformation across joints.

2.2.1 Dynamic loads and stress waves

Dynamic loads are generally the loads applied in a short duration, including impact, cyclic, explosion, and earthquake. For example, impact load, perhaps the most common dynamic load, is the load generated by knocking/hitting of one object onto another object, with very short time duration.

As distinct from static loads, which are generally treated as constant without change in time, dynamic loads change with time. An impact load typically rises quickly from zero to peak and ends in zero, within a very short loading duration. Therefore, they are in the form of waves. Typical forms of dynamic loads are illustrated in Figure 2.1.

The dynamic loading is applied at a point/plane in stress wave forms, and the stress moves further and applies to the next points/planes. The wave propagates at a speed that is governed by the medium in which the wave travels. This speed is

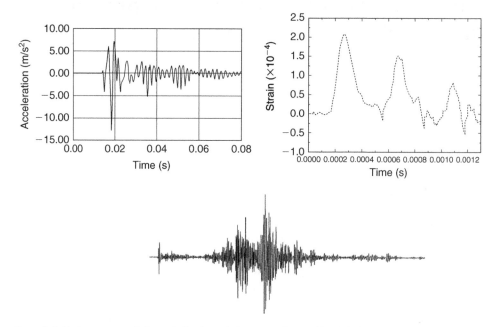

Figure 2.1 Various types of dynamic loads and their waveforms. Top left: blast wave measured from an explosive test (Zhao *et al.*, 1999b). Top right: impact wave measured from a SHPB test (Li, Ma and Huang, 2010). Bottom: ground acceleration of the Turkmenistan earthquake measured on 6 December 2000 (Landes, Ritter and Wedeken, 2009).

Table 2.1 Typical compressional wave velocities of various rocks.

Igneous Rock	P-Wave Velocity (m/s)	Sedimentary Rock	P-Wave Velocity (m/s)	Metamorphic Rock	P-Wave Velocity (m/s)
Granite	4500–6500	Conglomerate	1500–4500	Gneiss	5000–7000
Diorite	4500–6700	Sandstone	1500–5000	Schist	4500–6500
Gabbro	4500–7000	Shale	2000–4600	Phyllite	4500–6000
Rhyolite	4500–6000	Mudstone	2000–4600	Slate	3500–4500
Andesite	4500–6500	Dolomite	3500–6000	Marble	5000–6000
Basalt	5000–7000	Limestone	3500–6000	Quartzite	5000–7000

generally known as seismic velocity, and is the speed of the wave passing through the medium. For a specific medium, seismic velocity is a constant, unless the medium becomes discontinuous. The two most common seismic waves are the compressional (P) wave and the shear (S) wave. Typical values of P wave velocities in rocks are given in Table 2.1.

When a stress wave travels in a medium (solid or fluid), stress is applied to particles of the medium. The particles are accelerated to oscillate around their original positions. The speed of particle movement is termed the particle velocity, and it is the physical speed of particles moving back and forth in the direction the stress passing through.

Particle velocity should not be confused with the seismic velocity, as the latter has a much larger value. The particle velocity is governed by the magnitude and speed of the load. A high particle velocity is generally produced by a high amplitude of the stress wave. Peak particle velocity is often used as a key parameter assessing the failure and stability of rock masses and engineering structures in and on the rocks.

When a stress wave propagates across a rock mass, its amplitude is mainly attenuated at the presence of joints, due to the discontinuity in particle movements. Wave attenuation at joints accounts for a great deal of wave attenuation in a rock mass.

2.2.2 Theoretical approaches for wave propagation

There are mainly four models for studying the influences of joints on elastic wave propagation. They are the layered medium model (LMM), the displacement discontinuity model (DDM), the wave scattering model (WSM), and the equivalent medium model (EMM), as summarised in Table 2.2.

With the LMM, which is also termed as the perfect bonded interface model or the displacement continuity model by some researchers, both the stresses and displacements across the joint are continuous (Ewing, Jardetzky and Press, 1957; Brekhovskikh, 1980). There are two kinds of treatment of joints within the LMM. The joint can be modelled as a perfectly bonded interface, or as a layer of the filled weak medium sandwiched between two fully-bonded interfaces

The DDM treats each joint as a non-welded interface of zero thickness. It was originally developed by Mindlin (1960) and applied to seismic wave propagation by Schoenberg (1980). The basic assumption of this method is that, as a wave propagates through a joint, the particle displacements are discontinuous. The displacement discontinuity is equal to the stress divided by the specific joint stiffness. When the joint specific stiffness approaches infinity, the interface becomes a perfectly welded boundary, which can also be modelled with the LMM. When the joint specific stiffness approaches zero, the interface becomes a free surface. For joints with viscoelastic deformational behaviour, the particle velocities as well as the particle displacements are discontinuous (Pyrak-Nolte, Myer and Cook, 1990a). When the joint is filled with viscoelastic material, e.g., saturated sand or clay, due to the existence of the initial mass of the filled joint, besides the particle displacements and velocities, the stresses across the joint are also discontinuous.

The WSM treats the joint as a plane boundary with a distribution of small cracks and voids (Achenbach and Kitahara, 1986; Hudson, 1981; Hudson, Liu and Crampin, 1996). The wave reflection and transmission across a joint is the result of wave scattering through all cracks. According to this model, the stress waves propagating through the joint are considered to be uniformly scattered by the cracks, provided that the crack size is small compared with the wavelength. Apparent wave attenuation due to the scattering of energy at cracks is considered as the principal attenuation mechanism. The wave propagation is determined by crack geometry, crack distribution, crack density, saturation and other parameters. If cracks are filled with liquid, intrinsic attenuation can be taken into account based on the viscous dissipation by the filling liquid.

The EMM (White, 1983; Schoenberg and Muir, 1989; Schoenberg and Sayers, 1995; Li, Ma and Zhao, 2010) treats problems from the viewpoint of entirety. From the EMM, a material and the contained joints together are approximated by an equivalent

Table 2.2 Theoretical approaches for wave propagation in discontinuous medium.

Models	Boundary equations to describe the joint	Relations	Applications	Advantages and Disadvantages
LMM Layered medium model	$\Delta u_i = \dfrac{1}{f(d, M_r, M_f)}\sigma_{i3}$ (For filled joint) $\Delta u_i = 0$ (For perfectly bonded joint)	k and η can be obtained from M_r and M_f, or G_c, M_r and M_c; M_f can be obtained from G_c and M_c.	Filled joint; perfectly bonded joint.	A: accurate. D: very complex.
DDM Displacement discontinuity model	$\Delta u_i = \dfrac{1}{f(k, \eta)}\sigma_{i3}$		Non-perfectly bonded joint with thickness much smaller than wavelength.	A: simple. D: valid only when joint thickness is much smaller than wavelength.
WSM Wave scattering model	$\Delta u_i = \dfrac{1}{f(G_c, M_r, M_c)}\sigma_{i3}$		Joint containing a great number of cracks.	A: accurate. D: geometry and distribution of cracks are difficult to obtain.
EMM Equivalent medium model		Changes of equivalent moduli due to the presence of joints are a function of parameters used in boundary equations of LMM, DDM or WSM.	Estimate the overall influence of joints on wave transmission	A: convenient in engineering applications. D: loss of joint discreteness and accuracy.

Note: d is the joint thickness, M_r is the mechanical properties of the rock material, M_f is the mechanical properties of the filled medium, k is the joint specific stiffness, η is the joint specific viscosity, G_c is the geometric and distribution properties of the cracks, M_c is the mechanical properties of the cracks.

continuous, homogeneous and isotropic medium. Thus, stress waves propagate as if the jointed medium is continuous, homogeneous and isotropic. The effect of joints is lumped into effective moduli of the equivalent medium. The methods for calculating the effective moduli are mainly based on the geometry, structures, distributions of the joints, and the filling contained in the joints.

Wave propagation across a single joint has been extensively studied. However, joints in nature are in parallel form as joint sets. Multiple wave reflections among joint sets have great effect on wave propagation (Schoenberger and Levin, 1974; Cai and Zhao, 2000). The overall reflected and transmitted waves are the result of the superposition of reflected and transmitted waves arriving at different times. A simplified method was proposed by ignoring multiple wave reflections as a short-wavelength

Table 2.3 Different methods applicable for studying wave propagation in jointed rock masses.

Methods	Domain application	Dimension	Analyticity	Material damping
MC	Time	1D	Semi-analytical	Not considered
SMM	Frequency and time	1D and 2D	Semi-analytical	Considered
VWSM	Frequency and time	1D and 2D	Semi-analytical	Considered
SAM	Frequency and time	1D	Analytical	Considered

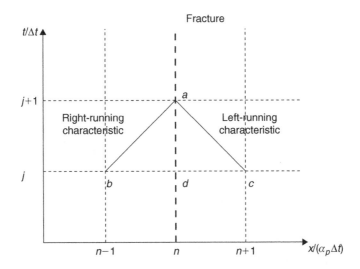

Figure 2.2 Characteristics in the nondimensional *x-t* plane (Cai and Zhao, 2000).

approximation (Pyrak-Nolte, Myer and Cook, 1990b; Myer *et al.*, 1995). The transmission coefficient across one joint set is calculated as the product of transmission coefficients of individual joints. However, laboratory experiments (Hopkins, Myer and Cook, 1988; Pyrak-Nolte, Myer and Cook, 1990b; Myer *et al.*, 1995) found that the simplified method was valid only when the first arriving wave was not contaminated by multiple wave reflections. When the incident wavelength is comparable to or larger than the joint spacing, the simplified method is not applicable. So far, there are four methods which take into account multiple wave reflections among joints, i.e. the method of characteristics (MC), the scattering matrix method (SMM), the virtual wave source method (VWSM), and the superposed analytical method (SAM). The characteristics of the methods are summarised in Table 2.3 and discussed in detail in Chapters 9 and 10.

The MC (Achenbach, 1973) is a mathematical tool for studying wave propagation across different layers, where multiple wave reflections are taken into account. Based on a one-dimensional wave equation, relations between particle velocity and stress along right- and left-running characteristics can be built (Fig. 2.2). Combined with the DDM, Cai and Zhao (2000) introduced the MC to study wave propagation across parallel joints with linear elastic deformational behaviours. Joints with nonlinear and

Coulomb slip behaviour (Zhao and Cai, 2001; Zhao, Zhao and Cai, 2006) were also studied with the MC.

The SMM, which is also termed the propagation matrix method, was originally used to study electromagnetic wave propagation (Collin, 1992), and adopted to study wave propagation across rock joints (Aki and Richards, 2002; Perino, Barla and Orta, 2010). When an elastic wave impinges on a discontinuity, a scattering phenomenon takes place and can be described by a scattering matrix. When more parallel joints are present, the scattering matrices of each one are combined according to a standard algorithm in order to describe the behaviour of the complete structure, with due consideration of all multiply-reflected waves. The global scattering matrix contains the global reflection and transmission coefficients of a set of parallel discontinuities.

The VWSM, combined with the EMM, is introduced initially for studying normally incident wave propagation across one joint set, where multiple wave reflections among the joints were considered (Li, Ma and Zhao, 2010). The VWS exists at each joint surface and produces a new wave, which is equal to the reflected wave, at each time when an incident wave propagates across the VWS. The VWSM is extended to study the effects of discretely jointed rock masses combined with the DDM (Zhu *et al.*, 2011). With the DDM, VWS exists at the joint position and represents the mechanical properties of the joint. It produces one reflected wave and one transmitted wave each time a normally incident wave arrives at the joint, two reflected waves and two transmitted waves each time an obliquely incident wave arrives at the joint.

Solutions for the MC, the SMM and the VWSM are not explicitly expressed and can be regarded as semi-analytical. The Superposed Analytical Method (SAM) is a new and explicitly expressed analytical method, where multiple wave reflections among joints are superimposed in the analytical solutions (Zhu, 2011). Assuming, but not limiting, that the background rock media of the opposite sides of each joint are identical, the mechanical properties are the same for every joint, and joints are equally spaced, the reflection and transmission coefficients across 2^n joints, which are considered as basic solutions, can be expressed as a function of the reflection and transmission coefficients across 2^{n-1} joints. Detailed description of the analytical solutions can be found in Zhu (2011).

$$R_{2^n} = R_{2^{n-1}} + \frac{T_{2^{n-1}}{}^2 R_{2^{n-1}} e^{i4\pi\xi}}{1 - R_{2^{n-1}}{}^2 e^{i4\pi\xi}}, \tag{2.1}$$

$$T_{2^n} = \frac{T_{2^{n-1}}{}^2 e^{i2\pi\xi}}{1 - R_{2^{n-1}}{}^2 e^{i4\pi\xi}} \tag{2.2}$$

where R and T are reflection and transmission coefficients, respectively, and ξ is the nondimensional joint spacing, which is defined as the ratio of joint spacing to the wavelength.

In the SAM, the reflection and transmission coefficients across other numbers of joints can be derived through basic solutions. This analytical method can be applied to joints described by different models only if the reflection and transmission coefficients across a single joint are available. It should be noted that this method can also be used to study the general cases where the background rock media of the opposite sides of each joint are different, the mechanical properties are different for every joint, and joints are not equally spaced. Besides, the method can be extended to study obliquely incident wave propagation across one joint set by using a matrix.

Some of the 1D and 2D analytical methods to take into account multiple wave reflections among joints, which are currently available and applicable to studying wave propagation across rock joints and rock masses, are summarised in Table 2.3. Depending on the problem to be solved, a specific model or method can be chosen and adopted.

The study on dynamic stress wave propagation across joints at present is limited to the assumption that the joint may deform (linear and non-linear) but is not damaged. This is often not true in reality. A joint could be crushed and sheared when the stress wave is imposed on it. Damage to the joint contact interface will consume energy and reduce further the wave transmission. Such interaction between wave transmission and joint damage has not been considered so far in the studies. It is envisaged that further research will explore this interaction by combining the works on wave propagation across joints and material fracturing/failure at joint surfaces.

2.2.3   Numerical modelling of wave propagation

Compared with theoretical and experimental studies, numerical modelling provides a convenient and economical approach to study wave propagation across a jointed rock mass, especially for complicated cases where theoretical solutions are impossible to obtain and experiments are difficult to conduct.

The representation of joints is a key difficulty in numerical modelling for wave propagation across jointed rock masses. In the finite element method (FEM), joints are often treated as individual elements called joint elements (Goodman, Taylor and Brekke, 1968; Ghaboussi, Wilson and Isenberg, 1973). Boundary interfaces are often used to model joints with the FEM and boundary element method (BEM) (Beer, 1986) or between BEMs (Crotty and Wardle, 1985; Pande, Beer and Williams, 1990). Joints are treated as slide lines in the finite difference method (FDM) (Schwer and Lindberg, 1992). In the discrete element method (DEM), a rock mass is represented as an assembly of discrete blocks and joints as interfaces between the blocks (Cundall, 1971; Shi, 1988).

The finite boundary of the computational model will cause elastic waves to be reflected and mixed with the original wave, which will make analysis of the modelling results more difficult. To solve these problems, an artificial boundary condition that can simulate a computational model without any finite boundaries is needed. This kind of boundary condition is also called a non-reflection boundary condition, which can eliminate the spurious reflections induced by the finite boundary. A number of non-reflection boundary conditions have been proposed in the past. For example, vicous boundary element (Lysmer and Kuhlemeyer, 1969), strip element (Liu and Achenbach, 1994) and infinite element (Gratkowski, Pichon and Razek, 1995) are implemented in FEM and DEM to realize non-reflection boundary.

The universal distinct element code (UDEC), a 2D DEM numerical program, has been widely adopted to study wave propagation across jointed rock masses. Lemos (1987) performed a study on S-wave attenuation across a single joint with Coulomb slip behaviour using UDEC. Brady *et al.* (1990) performed UDEC modelling on the slip of a single joint under an explosive line source. Chen (1999) verified the capability of UDEC to model the responses of jointed rock masses under explosion loading. Zhao *et al.* (2008) carried out numerical studies of P-wave propagation across multiple nonlinearly deformable joints with UDEC.

The Distinct Lattice Spring Model (DLSM) can also be used to study wave propagation across jointed rock masses (Zhu *et al.*, 2010). DLSM is a microstructure-based numerical model, which is meshless and has advantages in modelling dynamic problems including stress wave propagation.

2.2.4 Laboratory and field investigation

Pyrak-Nolte, Myer and Cook (1990a, 1990b) conducted experiments on wave propagation across one single joint and one joint set. It was found that joints had significant effects on wave propagation. The joint functioned as a high-frequency filter, i.e., only waves with low frequency can transmit across the joint. However, the multiple wave reflections among the joints were not studied in their research. Zhao *et al.* (2006a) carried out a series of laboratory tests to study wave propagation across one joint set. The transmitted pulses across joints are captured and compared with the results computed with the method of characteristics (MC). Generally, experimental results agree well with those obtained by the MC.

Wave propagation across a filled joint is also performed, where the incident wave is generated through a modified SHPB (Li and Ma, 2009). It is found that the joint width and water content have significant effect on wave transmission through a filled joint.

A two-dimensional physical model to investigate an elastic plane stress wave propagating across joints is established at EPFL (Wu *et al.*, 2011). Different from previous tests, this experimental apparatus can produce plane wave in 2D plates. It can also be used to study obliquely incident wave propagation across a joint set and multiple joint sets.

Cross-hole techniques have been used in a variety of geomechanical exploration and monitoring applications (Auld, 1977; McKenzie, Stacey and Gladwin, 1982; McCann and Baria, 1982; King, Myer and Rezowalli, 1986). The cross-hole method has been found to provide a particularly promising in situ test for studying wave propagation across jointed rock masses and the geomechanical characteristics of jointed rock masses. It was found that the propagation of stress waves in a rock mass containing joints is strongly influenced by the state of stress, changes in temperature, and degree of water saturation. Watanabe and Sassa (1996) performed site geological observation to detect the joints.

There are many criteria to relate stress wave and the performance of rock masses (e.g., Dowding, 1984, 1985, 1996). Among them, the PPV is used as a main stability criterion for engineering structures in and on rocks. Zhao *et al.* (1999b), Chong *et al.* (2002), and Zhou (2011) reported in situ experiments in jointed rock masses to investigate the rock joint effects on wave propagation. It was found that the PPV attenuates with the increase of distance from the charge centre, and the increase of incident angle between the joint strike and the wave propagation path.

2.2.5 Wave across multiple joint sets

Wave propagation across multiple joint sets will be further complicated due to the inter-secting of joint sets. With the EMM, Schoenberg and Muir (1989) and Schoenberg and Sayers (1995) incorporated the effects of multiple sets of parallel fractures by representing them as group elements. However, EMM have two limitations: loss of discreteness of wave attenuation and intrinsic frequency-dependent properties at individual fractures.

Due to the complexity of wave propagation across multiple joint sets, analytical solutions are difficult to obtain. Hence, numerical modelling and experimental tests are more suitable for studying wave propagation across multiple joint sets with the consideration of joint spacing, number of joint sets and joint sets intersecting angles. While research continues with numerical and physical modelling to obtain wave trans-mission and transformation across joints and joint sets, future work should also be directed to using numerical methods to simulate multiple jointed rock masses to obtain equivalent parameters for wave propagation, by considering joint frequency and distri-bution, joint shear and normal stiffness. The requirement for engineering applications is to be able to predict wave attenuation in a rock mass with known common rock mechanics characteristics.

2.3 LOADING RATE EFFECTS ON ROCK STRENGTH

Dynamic loads are usually associated with high amplitude and short duration stress pulse or a high loading rate. Mechanical properties of rock materials, including com-pressive strength, tensile strength, shear strength and fracture toughness, are affected by the loading rate. A proper understanding of the effect of loading rate on rock strength is important in the analysis of mechanical behaviour. Rate effect has been stud-ied experimentally by many researchers (e.g. Abbott, Cornish and Weil., 1964; Stowe and Ainsworth, 1968; Lindholm, Yeakley and Nagy, 1974; Goldsmith, Sackman and Ewerts., 1976; Grady et al., 1977; Li et al., 2001; Zhang et al., 2001; Lok et al., 2002; Backers et al., 2003; Zhang and Hao, 2003; Li et al., 2004; Fuenkajorn and Kenkhunthod, 2010; Liang et al., 2011). All these dynamic tests exhibit a general trend of increase in strength with increasing loading rate. However, the test results are rather scattered because of the complexity of rock types and rock properties.

This section will focus on the observations of rate effects on rock material strength from experiments and the studies of rate dependent mechanisms.

2.3.1 Dynamic tests on rock strengths

A fundamental difference between dynamic tests and quasi-static tests is that inertia and wave propagation effects become more pronounced at higher strain rates. Some excellent reviews about the testing methods of strain rate effect on many engineering materials such as concrete, ceramics, rock, silicon carbide and composite materials etc., are presented by Field et al. (2004), Gama et al. (2004) and Ramesh (2008), and also in Chapters 3, 4, 5 and 6 of this book. Ramesh (2008) classified the common impact tests into four categories according to the objective of the experiment, high-strain-rate

Table 2.4 Dynamic strength tests and apparatus.

Strain Rate (s^{-1})	Test Apparatus	Testing Principle	Applicability
$\leq 10^2$	Specialized hydraulic servo-controlled machines	Dynamic load applied by movement of a piston hydraulically driven by gas or oil	Uniaxial compression (e.g. Green and Perkins, 1968; Zhao et al., 1999a); dynamic triaxial compression (e.g. Li, Zhao and Li, 1999)
			Direct tension (e.g. Yan and Lin, 2006; Asprone et al., 2009); dynamic Brazilian indirect tension (e.g. Zhao and Li, 2000)
			Punch shear test (e.g. Zhao, Li and Zhao, 1998)
			Shear of rock joints (e.g. Barbero, Barla and Zaninetti, 1996; Kana et al., 1996)
$10^0 \sim 10^3$	Drop-weight machines	Gravitational potential energy	Flexural loading (e.g. Banthia et al., 1989)
			Impact and fragmentations (e.g. Whittles et al., 2006)
$10^1 \sim 10^3$	Hopkinson pressure bar	One-dimensional stress wave propagation theory	Uniaxial compression (e.g. Li et al., 2000; Li, Lok and Zhao, 2005; Cai et al., 2007; Zhou et al., 2010)
			Triaxial compression (e.g. Christensen, Swanson and Brown, 1972; Li et al., 2008; Frew et al., 2010)
			Direct tension (e.g. Cadoni, 2010; Huang, Chen and Xia, 2010a)
			Brazilian indirect tension (e.g. Wang, Li and Song, 2006; Cai et al., 2007; Dai and Xia, 2010)
			Flattened Brazilian disk (FBD) tension (e.g. Wang, Li and Xie, 2009)
			Semi-circular bend (SCB) test (e.g. Dai, Xia and Luo, 2008)
			One-point impact test (e.g. Belenky and Rittel, in press)
			Spalling test (e.g. Erzar and Forquin, 2010)
$> 10^3$	Gas gun	High-pressure gas driven projectile	Equations of state (e.g. Shang, Shen and Zhao, 2000)

experiments, wave-propagation experiments, dynamic failure experiments and direct impact experiments. Experimental techniques to obtain the strength of rock materials under dynamic loading are summarised in Table 2.4.

Ordinary hydraulic servo-controlled testing machines can load specimens at strain rates up to $10^{-3}\,s^{-1}$, but some specialized hydraulic servo-controlled machines such as those developed by Green and Perkins (1968), Logan and Handin (1970), Perkins, Green and Friedman (1970), Zhao et al. (1999a), Yan and Lin (2006), Asprone et al. (2009) and Cadoni (2010), can achieve strain rates up to $10^2\,s^{-1}$. However, the medium strain rate range (between 10^0 and $10^2\,\mathrm{s}^{-1}$) is very difficult to investigate. The primary approach to testing in this range uses drop-weight machines (Charlie *et al.*, 1993), but great care must be taken in interpreting the data because of the coupling between machine vibrations and wave propagation. The classical experimental technique in the high strain rate range of $10^1\sim10^4\,\mathrm{s}^{-1}$ is the Hopkinson pressure bar tests for the measurement of rock mechanical properties (Kumar, 1968; Li *et al.*, 2000; Frew, Forrestal and Chen, 2001; Li, Lok and Zhao, 2005; Cai *et al.*, 2007; Xia *et al.*, 2008; Dai *et al.*, 2010). At higher strain rates (i.e. exceeding $10^3\,\mathrm{s}^{-1}$), light gas guns have been successfully deployed to test the mechanical properties of rock materials (Shockey *et al.*, 1974; Shang, Shen and Zhao, 2000).

2.3.2 Loading rate effects on rock material strengths

Changes of rock strength with loading rate are primarily reported through laboratory tests. There have been many attempts to derive empirical equations to express the relationship between loading rate (or strain rate) and rock material strength.

Based on uniaxial compression tests with strain rate of 10^{-6}–$10^4\,\mathrm{s}^{-1}$ on limestone, Lankford (1981) proposed that:

$$\sigma_{dc} \propto \begin{cases} \dot{\varepsilon}^{1/(1+n_c)} & \dot{\varepsilon} < 10^2\,\mathrm{s}^{-1} \\ \dot{\varepsilon}^{1/n} & \dot{\varepsilon} > 10^2\,\mathrm{s}^{-1} \end{cases} \tag{2.3}$$

where σ_{dc} is the uniaxial dynamic compression strength, $\dot{\varepsilon}$ is the strain rate, n and n_c are material constants, and are equal to 0.3 and 130, respectively in his experiments. Lankford concluded that there exists a critical strain rate for a certain material. When the strain rate is smaller than the critical value, the compressive strength slightly increases with the strain rate. However, when the strain rate is larger than the critical value, the compressive strength switches to rapidly increase with the strain rate.

Olsson (1991) studied the uniaxial compressive strength of a tuff with a strain rate in the range 10^{-6} to $10^3\,\mathrm{s}^{-1}$. In his experiment, he also found a critical strain rate of $76\,\mathrm{s}^{-1}$, and gave the similar relationship,

$$\sigma_{dc} \propto \begin{cases} \dot{\varepsilon}^{0.007} & \dot{\varepsilon} < 76\,\mathrm{s}^{-1} \\ \dot{\varepsilon}^{0.35} & \dot{\varepsilon} > 76\,\mathrm{s}^{-1} \end{cases} \tag{2.4}$$

In addition, similar conclusions are drawn by Chong and Boresi (1990), and Lajtai, Duncan and Carter (1991).

Based on tests on a granite at strain rate of 10^{-4} to $10^0\,\mathrm{s}^{-1}$, Masuda, Mizutani and Yamada (1987) noted that the dynamic compressive strength increases with the strain rate, following the relationship given as:

$$\sigma_{dc} = C\log(\dot{\varepsilon}) + \sigma_c \tag{2.5}$$

where σ_c is the static uniaxial compressive strength, and C is a constant for the rock material.

Based on tests on a granite with strain rate between 10^{-4} and 10^0 s^{-1}, Zhao (2000) suggested the relationship can be unified and expressed as:

$$\sigma_{dc} = RSC_d \log(\dot{\sigma}_{dc}/\dot{\sigma}_{sc}) + \sigma_{sc} \tag{2.6}$$

where $\dot{\sigma}_{dc}$ is the dynamic loading rate; $\dot{\sigma}_{sc}$ is the quasi-static loading rate, σ_{sc} is the uniaxial compressive strength at quasi-static loading rate (0.5~1 MPa/s according to ISRM suggested methods), and RSC_d is the dynamic rock strength constant for the rock material.

Logan and Handin (1970) conducted quasi-dynamic triaxial compression tests of the Westerly granite at confining pressures up to 700 MPa, and found the failure strength increases proportionally with increasing loading rate. The rate of increase rises with increasing confining pressure. Green and Perkins (1968) and Masuda, Mizutani and Yamada (1987) also found that at a low confining pressure the effect of loading rate on the strength of a granite is smaller than that at a high confining pressure. However, Yang and Li (1994) reported that the loading rate sensitivity seems to decrease with increasing confining pressure on a marble. Dynamic triaxial compression tests on a granite (Li, Zhao and Li, 1999) showed that the increments of compressive strength with increasing loading rate are different under various confining pressures. The maximum rising rate is 86%, with the strain rates increasing from 10^{-4} to 10^0 s^{-1} under the confining pressure of 20 MPa. Zhao (2000) suggested the confining pressure effects can generally account for the effect on strength following the Hoek-Brown strength criterion.

Changes of dynamic tensile strength of rock materials with loading rate have also been reported extensively, mostly with the Brazilian tests (e.g. Price and Knill, 1966; Zhao and Li, 2000; Wang, Li and Song, 2006; Cai et al., 2007; Dai and Xia, 2010; Chen et al., 2009; Cho, Ogata and Kaneko, 2003; Erzar and Forquin, 2010; Asprone et al., 2009; Cadoni, 2010; Huang, Chen and Xia, 2010a). Results all showed that tensile strength increases with loading rate, with similar equations to those of compressive strength proposed based on the experiments.

Dynamic shear tests on rock materials done by Zhao, Li and Zhao (1998) and Fukui, Okubo and Ogawa (2004) concluded that rock material shear strength is also rate-dependent. When the loading rate increases by one order of magnitude, the shear strength increases by approximately 10%. Zhao (2000) further suggested that, based on the results of compression, tension and shear tests, the change of shear strength with loading rate is primarily the change of the cohesion but not the friction angle.

2.3.3   Fracture dynamics and strain rate mechanisms

Efforts have been made to study the mechanism governing the rate-dependent behaviour of rock materials (e.g. Kumar, 1968; Qi, Wang and Qian, 2009; Chong et al., 1980; Blanton, 1981; Chong and Boresi, 1990; Morozov and Petrov, 2000 and Ou, Duan and Huang, 2010).

Rock is typically a brittle and inhomogeneous material, containing initial defects such as grain boundaries, micro-cracks and pores. There have recently been increasing studies of inhomogeneity effects on the failure mechanism of rock materials. Some researchers (e.g. Cho, Ogata and Kaneko, 2003; Cho and Kaneko, 2004; Zhu and Tang, 2006; Zhou and Hao, 2008; Zhu, 2008) incorporated the rock inhomogeneity into numerical methods, and successfully simulated progressive failure of rock materials under both static and dynamic loading conditions. These analyses revealed that the differences are due to the stress concentrations and redistribution mechanisms in the rock. The rock inhomogeneity also contributes to the difference between the dynamic and static tensile strengths. In addition, Cho and Kaneko (2004) used the same method to investigate the influence of applied pressure waveforms on dynamic fracture processes in rocks.

Observation from the experiments showed that at high loading rates, rock materials fail with more fractures and fragments are of smaller size. This observation is often related to the strength increase. Since more fractures are generated at high loading rates, more energy is consumed hence leading to higher loads and higher strengths. There is certainly a connection between high density of fracturing and high strength. However, the reasons for more fracturing are still under investigation, and are believed to be micromechanics based (Kazerani and Zhao, 2010).

Micromechanics-based crack models have been investigated (e.g. Zhang, Wong and Davis, 1990; Wong, 1990; Wong et al., 2006; Brace and Bombolakis, 1963; Nemat-Nasser and Horii, 1982; Ashby and Hallam, 1986; Deng and Nemat-Nasser, 1992, 1994; Nemat-Nasser and Deng, 1994; Ravichandran and Subhash, 1995; Huang, Subhash and Vitton, 2002; Huang and Subhash, 2003; Zhou et al., 2004; Zhou and Yang, 2007; Li, Zhao and Li, 2000; Xie and Sanderson, 1995; Alves, 2005; Saksala, 2010; Wang, Sluys and de Borst, 1997; Ambrosio and Tortorelli, 1990; Bourdin, Larsen and Richardson, 2010; Larsen, Ortner and Süli, 2010). Paliwal and Ramesh (2008) developed an interacting micro-crack damage model based on sliding of pre-existing cracks for the estimation of the strain rate dependent constitutive behaviour of brittle materials, which shows a good agreement with experiments (Paliwal and Ramesh, 2008; Kimberley, Ramesh and Barnouin, 2010). In order to evaluate the variability of the mesoscale strain rate dependent constitutive behaviour in brittle materials, Graham-Brady (2010) improved on the interacting micro-crack damage model by incorporating statistical characterization of mesoscale random cracks.

Zuo et al. (2006) presented a rate-dependent damage model, the Dominant Crack Algorithm (DCA), for the damage of brittle materials based on the dominant crack. Zuo, Disilvestro and Richter (2010) recently proposed a rate-dependent crack mechanics based model by incorporating plastic deformation into the DCA model for damage and plasticity of brittle materials under dynamic loading.

Kazerani (2011) and Kazerani and Zhao (2010, 2011) studied rock fracturing with microscopic discrete element modelling and revealed that the rate dependency observed in the experiments may be due to several causes: the intrinsic rate-dependent properties of the microstructure, the structural rate dependent properties of the rock material composition, and the testing conditions. The structural rate dependent properties of the rock material composition are related to the mineral grain structure and homogeneity. The intrinsic rate dependent properties of the microstructure are on the cohesion of the bond between microelements. Testing conditions, such as end frictions at loading and supporting points/planes also contribute to the rate effects.

2.3.4 Rock dynamic strength criteria

Based on dynamic experimental data of the Bukit Timah granite (Li, Zhao and Li, 1999; Zhao *et al.*, 1999a; Zhao and Li, 2000), Zhao (2000) examined the applicability of the Mohr-Coulomb and the Hoek-Brown criteria to rock material strength in the dynamic range.

The Mohr-Coulomb criterion is only applicable to dynamic triaxial strength in the low confining pressure range. It appears that the change in the strength with loading rate is primarily due to the change of cohesion, and the internal friction angle seems unaffected by loading rate. The dynamic triaxial strength can be estimated as

$$c_d = \sigma_{dc}(1 - \sin \phi)/2 \cos \phi \tag{2.7}$$

$$\sigma_{d1} = \sigma_{dc} + \sigma_3(1 + \sin \phi)/(1 - \sin \phi) \tag{2.8}$$

where c_d is the dynamic cohesion, and ϕ is the friction angle.

The dynamic triaxial strength can be represented by the Hoek-Brown criterion at low and high confining pressure ranges for the loading rate range examined. It may be assumed that the parameter m (a constant in the Hoek-Brown criterion) is not affected by the loading rate. Hence, the dynamic triaxial strength can be estimated from

$$\sigma_{d1} = \sigma_3 + \sigma_{dc}(m\sigma_3/\sigma_{dc} + 1.0)^{0.5} \tag{2.9}$$

Additional testing data will provide further verification of the above conclusions for other rocks and for the wide range of loading rates.

2.4 NUMERICAL MODELLING OF ROCK DYNAMIC FRACTURING

Dynamic fracturing of rock governs the strength and failure mode, and is one of the most important research issues in rock dynamics. However, the real mechanism of the rate-dependency for fracturing pattern and mechanical properties of rock under dynamic loading is still not clear. Facing this problem, both the experimental method and the analytical method are limited. With the rapid advancement of computing technology, numerical methods provide powerful tools. The combination of numerical and physical modelling methods can be the best applicable solution to provide the insight of rock fracturing dynamics. In this section, numerical methods used for rock fracturing dynamics are briefly reviewed. Detailed reviews on the corresponding classical numerical methods and the newly developed numerical methods can be found in Chapters 13 and 14 of this book.

2.4.1 Numerical methods for fracturing modelling

Generally, numerical methods used in rock mechanics are classified into continuum based method, discontinuum based method and coupled continuum/discontinuum method (Jing, 2003). The continuum based methods are methods which based on continuum assumption, examples are the Finite Element Method (FEM) (Clough, 1960), the Finite Difference Method (FDM) (Malvern, 1969), and the Smoothed Particle Hydrodynamics (SPH) (Monaghan, 1988). The merits of continuum methods are directly inputting macro mechanical parameters which can be obtained from experiments and precisely modelling the stress state of pre-failure stage. Moreover, computer codes for continuum-based methods are also relatively mature, e.g., LS-DYNA (LSTC, 2010), ABQUS (SIMULIA, 2010), FLAC (ITASCACG, 2010) and RFPA (MECHSOFT, 2010) are commercial computer codes which can be used to model dynamic fracturing problems. However, the continuum assumptions in these continuum-based methods make them unsuitable for dealing with complete detachment and large-scale fracture opening problems. It is also difficult to apply continuum-based methods to solve problems which involve complex discontinuity, such as jointed rock masses and rock in post-failure state.

Discontinuum-based methods treat rock material or rock mass as an assembled model of blocks, particles or bars, e.g., the Distinct Element Method (DEM) (Cundall, 1971), Discontinuous Deformation Analysis (DDA) (Shi, 1988) and Distinct Lattice Spring Model (DLSM) (Zhao, 2010). In these methods, the fracturing process of rock is represented by the breakage of inter-block contacts or inter-particle bonds. Discontinuum-based methods can reproduce realistic rock failure processes especially the post failure stage. However, they are not best suited for stress state analysis of pre-failure rock. Available commercial computer codes based on DEM are UDEC/3DEC and PFC (ITASC, 2010) and DDA (Shi, 1988). There also exist some research codes, for example, DLSM (Zhao, 2010).

In order to overcome the limitations of both continuum and discrete methods, coupled methods have been developed in recent years. For example, the Numerical Manifold Method (NMM) (Shi, 1991) was developed to integrate DDA and FEM, the FEM/DEM method (Munjiza, 2004) is designed to couple FEM with DEM, and the Particle-based Manifold Method (PMM) (Zhao, 2009) was proposed to combine DLSM and NMM. The coupled method is capable of capturing both the pre-failure and the post-failure behaviour of rock materials. However, its implementation is difficult and no commercial codes are available now. There only exist some research codes, e.g., NMM (Shi, 1991), Y2D (Munjiza, 2004), m-DLSM (Zhao, 2010). In Table 2.5, a summary on these numerical methods and corresponding computer codes are listed.

Table 2.5 Numerical methods for rock dynamic problems.

Numerical Methods	Typical Software/Code	General Applicability
Continuum based: FEM, FDM, BEM, SPH	LS-DYNA, ABQUS, FLAC, RFPA	Displacement without element detachment
Discontinuum based: DEM, DLSM	UDEC/3DEC, PFC, DDA, DLSM	Element detachment, rock fracturing, rock block movement
Coupled/hybrid based: combined methods	FEMDEM, NMM, Y2D, m-DLSM, PMM	Multiscale, displacement, fracturing, and block movement combined

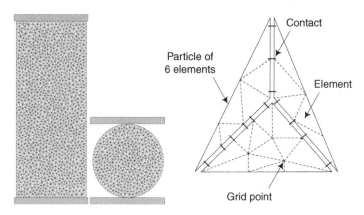

Figure 2.3   UDEC model for dynamic fracturing simulation (left), and basic unit (right).

2.4.2   Micromechanics modelling of rock dynamic fracturing using UDEC

Dynamic fracturing of heterogeneous materials such as rock and concrete cannot be modelled realistically without appealing to their microstructures. This requires that a successful numerical method must be capable of considering the formulation and evolution of micro discontinuities. Recently, the micro dynamic fracturing of rock is modeled by using UDEC through implementing a triangulation pre-processor and a rate-dependent cohesive law (Kazerani and Zhao, 2010). The basic scheme is shown in Figure 2.3, in which the material is represented as an assembly of distinct particles/bodies interacting at their boundaries. The interface between these particles is viewed as a contact which in fact represents grain-interface or grain cementation properties for igneous or sedimentary rocks, respectively. In order to model the dynamic fracturing of rock materials, a full rate-dependent cohesive law was proposed (Kazerani, 2011; Kazerani and Zhao, 2010). The model was used to model the tensile and compressive failure of rock materials, and compared well with experimental results (Kazerani and Zhao, 2011). It is also used for simulating the dynamic fracture toughness test of rock materials, dynamic crack propagation of PMMA plate (Kazerani and Zhao, 2011) and dynamic failure of joints under shear force (Kazerani, Zhao and Yang, 2010).

2.4.3   Particle-based Manifold Method (PMM) for multiscale rock dynamics modelling

Particle-based Manifold Method (PMM) is a new particle-based multi-scale numerical method and corresponding computer code, currently under development by EPFL-LMR (Zhao, 2009; Sun, Zhao and Zhao, 2011). PMM introduces the microscopic particle concept into the numerical manifold method (NMM) and rebuilds a particle manifold method (Fig. 2.4). It unifies continuum-discontinuum models at micro scale.

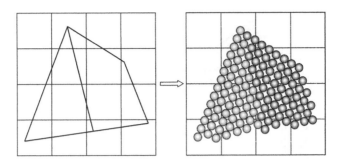

Figure 2.4 NMM model (left) of two blocks with rectangular mathematical meshes, and PMM model (right), where the blocks are approximately replaced by particles and the mathematical meshes remain.

Further, PMM can be incorporated with NMM for multi-scale modelling. In summary, PMM provides the following new features:

i) PMM is a dynamic model. Motion of discontinuum can be accurately described by inertial equations. Static simulation is also available when the velocity is ignored.

ii) PMM is a fully implicit model. All unknowns are solved by a global mathematical equation. The contact behaviours are described by the penalty method and the open-close iteration is inherited from manifold method to make contact state convergent.

iii) PMM extends NMM to micro scale simulation. By importing proper failure mechanisms, PMM could simulate explicit processes with implicit modelling.

iv) PMM is capable of presenting material nonlinearity and inhomogeneity. The separation of mathematical mesh and material mesh frees the description of physical domain without the limitation of drawing meshes. Inhomogeity is described at the micro scale.

v) An analytical sphere simplex integration is given to guarantee the accuracy of integration on physical domain.

vi) PMM has mobility of contact mechanism and failure model. PMM overcomes the difficulty of 3D implementation of NMM by replacing the polyhedron-to-polyhedron contact by the sphere-to-sphere contact.

The advantages of PMM, including unified implicit computational format, accurate dynamic simulation, and microscale and manifold features, make the model a suitable tool for analysing rock dynamics, especially when dealing with dynamic fracturing.

Multi-scale modelling is regarded as an exciting and promising methodology due to its ability to solve problems which cannot be handled directly by microscopic methods due to the limitation of computing capacitance (Guidault *et al.*, 2007; Hettich, Hund and Ramm, 2008; Xiao and Belytschko, 2003). The most direct way to build a

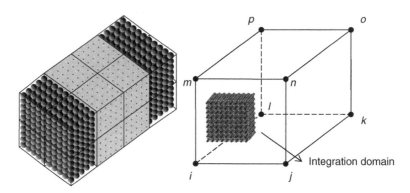

Figure 2.5 Particle based Manifold Method (PMM) model (left) and basic element (right).

multi-scale numerical model is to combine two different scale methods. This methodology has been widely used, for example, in the coupling of MD with continuum mechanics models (Mullins and Dokainish, 1982; Hasnaoui, van Swygenhoven and Derlet, 2003). The PMM is to couple with the DLSM (Distinct Lattice Spring Model) (Zhao, 2010; Zhao, Fang and Zhao, 2010) and the NMM. The computational model of PMM is shown in Figure 2.5. The PMM element is realized by replacing the physical domain of the manifold element in NMM by the particle-based DLSM model. The implementation details of this method are given by Zhao (2010). As a newly developed numerical method, only a few examples are given (e.g., Zhao, 2010; Zhu *et al.*, 2011; Kazerani, Zhao and Zhao, 2010). The implicit PMM and GPU based high performance PMM code is still under the development (Sun, Zhao and Zhao, 2011). The new computer code will provide useful solvers for rock dynamic problems at multiscale.

2.5 PROSPECTS OF ROCK DYNAMICS RESEARCH

Rock dynamics research is not limited to the aspects discussed in the previous sections. It has a much wider scope, with topics ranging from wave propagation, to response of rock material and rock mass, to engineering applications, dealing with microscopic fracturing of rock material to dynamic behaviour of rock masses (Zhao *et al.*, 2006b). There are indeed many issues yet to be covered in rock dynamics. Some of the important aspects requiring investigations are discussed below.

a) Wave propagation in rock joints

Further studies in this field need to be focused on the coupling of wave attenuation and joint geometrical properties, such as spacing, frequency, aperture, roughness and filling. Typical information on rock joints includes orientation, aperture and filling, surface roughness, spacing and frequency, which can be generally measured. Spacing, frequency and orientation can remain as geometrical parameters and can be the input for either analytical solutions or numerical modelling. Aperture and roughness can be correlated to mechanical properties such as joint normal stiffness and shear strength.

Therefore, it is possible to incorporate those rock joint parameters in the wave propagation analysis, particularly in numerical modelling, to estimate wave attenuation in the jointed rock masses.

To deal with filled joints, mechanical properties of filling materials (e.g., sand or clay) can be determined and incorporated into the wave propagation analytical solutions by treating the filling as a viscous material.

b) Wave propagation in rock masses

Studies along this line are to develop equivalent medium wave propagation parameters for jointed rock masses, by incorporating rock mass parameters. Statistic approached may be adopted to represent the geometrical distribution of joints and of the joint properties for rock masses. This can be achieved by performing a parametrical study using numerical modelling to generate a representative rock mass and then to obtain a wave attenuation coefficient for that rock mass.

c) Interaction of wave transmission and joint damage

Joint damage associates with energy consumption, and complicates the wave propagation equation. For analytical solutions, one must consider the energy balance at the failure of the joint surface asperities under compression and shearing.

There are possibilities for exploring the interaction between wave transmission and joint damage by physical and numerical modelling. For numerical modelling, the challenges will be the simulation of rock joint surface damage under dynamic loads. Micromechanical discrete element modelling is likely to be required in such cases in order to model the fracture and failure of rock joint surfaces.

d) Rock fracture induced seismic energy and wave

When a highly stressed (or strained) rock (material or joint) fractures, the stored strain energy is released at the facture plane. If the energy released is sufficiently large, it can cause induced seismic events. Physical experiment may offer direct observation on energy release patterns (amplitude and form), with good monitoring devices. Chapter 15 addresses this issue.

Numerical modelling, particularly micromechanics-based discrete element methods, will be good tools to capture the phases of statically-strained rock materials, sudden fracturing, and released and propagation of dynamic stress.

e) Mechanics of rock fracturing and rate effects

While it is clear that at high loading rate, rock material strengths increase and rock material fails with more fractures, it is not clear yet what is the cause of high density of fracturing. There are indeed many opportunities within this field to explore the mechanical and physical cause of rate effects on rock strength and failure pattern. For example, rate effects on fracture branching, rate effects on multiple fracture initiation, and rate effects on crack propagation velocity.

Further study also needs to be conducted on the shear strength of rock joints under dynamic loads, to understand the rate effects on shear strength and dilation.

f) Micromechanics modelling of rock fracturing and failure

As already mentioned in (e), numerical modelling of rock fracture and failure need to be micromechanics- and discrete-based. There are two aspects which need to be addressed.

One is to incorporate micromechanical constitutive laws and input parameters into the existing codes, such as UDEC and DDA. The second is to develop new microscale numerical codes with specific focus on modelling fracture initiation, propagation and branching. The need for correlation with physical modelling will also advocate experimental progress in terms of high-speed, high-resolution micromechanics monitoring and observation.

The other question that micromechanics modelling should address is the effect of element size. It is argued that if the elements are sufficiently small, the contact force between the elements will be sufficiently simple and non-rate dependent (Zhao, Wang and Tang, 2008). It needs to be verified and also determined with the element size.

g) Static-dynamic interaction

Rock dynamics also covers the dynamic failure processes under existing static loading conditions, as reflected by rock burst and spalling. Rock burst mechanism, failure pattern, energy release, fracture propagation velocity and distance are likely to be affected by static strain energy (in situ stress) and triggering mechanism (e.g. stress re-distribution due to excavation). The process may involve static-dynamic transition and interaction. Such a study will require multiscale and multimechanics approaches.

Other areas which can be explored are the interaction between rock fracturing and groundwater, gas and pore pressure and temperature, which may extend the approaches to multiphysics.

h) Rock and earthquake engineering applications

In parallel with the fundamental studies outlined above, rock dynamics will be continuously applied to engineering and construction. Stability of slopes and tunnels under various dynamic conditions (earthquake and explosion), reinforcement and support of rock slope and tunnels for dynamic loads, use of explosives and blast damage control, seismic and vibration hazard control, are some typical examples of engineering applications needing to be addressed.

REFERENCES

Abbott, B.W., Cornish, R.H. and Weil, N.A.: Techniques for studying strain rate effects in brittle materials. *Journal of Applied Polymer Science* 8(1) (1964), pp.151–167.

Achenbach, J.D.: *Wave Propagation in Elastic Solids*, North-Holland, Amsterdam, 1973.

Achenbach, J.G. and Kitahara, M.: Reflection and transmission of an obliquely incident wave by an array of spherical cavities. *Journal of the Acoustical Society of America* 80(4) (1986), pp.1209–1214.

Aki, K. and Richards P.G.: *Quantitative Seismology*, University Science Books, California, 2002.

Alves, L.M.: Fractal geometry concerned with stable and dynamic fracture mechanics, *Theoretical and Applied Fracture Mechanics* 44(1) (2005), pp.44–57.

Ambrosio, L. and Tortorelli, V.M.: Approximation of functionals depending on jumps by elliptic functionals via Γ-convergence. *Communications on Pure and Applied Mathematics* 43(8) (1990), pp.999–1036.

Ashby, M.F. and Hallam, S.D.: The failure of brittle solids containing small cracks under compressive stress states. *Acta Metallurgica* 34(3) (1986), pp.497–510.

Asprone, D., Cadoni, E., Prota, A. and Manfredi, G.: Dynamic behavior of a mediterranean natural stone under tensile loading. *International Journal of Rock Mechanics and Mining Sciences* 46(3) (2009), pp.514–520.

Auld, B.: Cross-hole and down-hole v_s by mechanical impulse. *Journal of Geotechnical Engineering Division, American Society Civil Engineers* 103, GT 12 (1977), pp.1381–1398.

Backers, T., Fardin, N., Dresen, G., and Stephansson, O.: Effect of loading rate on mode I fracture toughness, roughness and micromechanics of sandstone. *International Journal of Rock Mechanics and Mining Sciences*, 40(3) (2003), pp.425–433.

Banthia, N., Mindess, S., Bentur, A., and Pigeon, M.: Impact testing of concrete using a drop-weight impact machine. *Experimental Mechanics* 29(1) (1989), pp.63–69.

Barbero, M., Barla, G., and Zaninetti, A.: Dynamic shear strength of rock joints subjected to impulse loading. *International Journal of Rock Mechanics and Mining Science & Geomechanics Abstracts* 33(2) (1996), pp.141–151.

Beer G.: Implementation of combined boundary element-finite element analysis with applications in geomechanics. In: Banerjee P.K. and Watson J.O. (eds.), *Developments in Boundary Element Methods*. London: Applied Science, 1986, pp.191–226.

Belenky, A. and Rittel, D.A.: simple methodology to measure the dynamic flexural strength of brittle materials. *Experimental Mechanics*, In press, 2010

Bieniawski, Z.T. and Hawkes, I.: Suggested methods for determining tensile strength of rock materials. *International Journal of Rock Mechanics and Mining Sciences & Geomechanics Abstracts* 15(3) (1978), pp.99–103.

Blanton, T.L.: Effect of strain rates from 10-2 to 10 sec-1 in triaxial compression tests on three rocks. *International Journal of Rock Mechanics and Mining Sciences & Geomechanics Abstracts* 18(1) (1981), pp.47–62.

Bourdin, B., Larsen, C. and Richardson, C.: A time-discrete model for dynamic fracture based on crack regularization. *International Journal of Fracture* (2010), pp.1–11.

Brace, W.F. and Bombolakis, E.G.: A note on brittle crack growth in compression. *Journal of Geophysical Research* 68(12) (1963), pp.3709–3713.

Brady B.H., Hsiung S.H., Chowdhury A.H. and Philip J.: Verification studies on the UDEC computational model of jointed rock. *Mechanics Jointed Faulted Rock* (1990), pp.551–558.

Brekhovskikh, L.M.: *Waves in Layered Media*. Academic, San Diego, 1980.

Cadoni, E.: Dynamic characterization of orthogneiss rock subjected to intermediate and high strain rates in tension. *Rock Mechanics and Rock Engineering* 43(6) (2010), pp.667–676.

Cai, J.G. and J. Zhao: Effects of multiple parallel fractures on apparent attenuation of stress waves in rock masses. *International Journal of Rock Mechanics and Mining Sciences* 37(4) (2000), 661–682.

Cai, M., Kaiser, P.K., Suorineni, F. and Su, K.: A study on the dynamic behavior of the meuse/haute-marne argillite. *Physics and Chemistry of the Earth, Parts A/B/C*, 32(8–14) (2007), pp.907–916.

Charlie, W., Ross, C., Skinner, M. and Burleigh, J.: Evaluation of drop bar test for concrete and rock quality. *Geotechnical Testing Journal* 16(3) (1993), pp.350–364.

Chen S.G.: *Discrete Element Modelling of Jointed Rock Mass under Dynamic Loading*. Ph.D. Thesis. Nanyang Technological University, Singapore; 1999.

Chen, E.P.: Non-local effects on dynamic damage accumulation in brittle solids. *International Journal for Numerical and Analytical Methods in Geomechanics* 23(1) (1999), pp.1–21.

Chen, R., Xia, K., Dai, F., Lu, F. and Luo, S.N.: Determination of dynamic fracture parameters using a semi-circular bend technique in split Hopkinson pressure bar testing. *Engineering Fracture Mechanics*, 76(9) (2009), pp.1268–1276.

Cho, S.H., and Kaneko, K.: Influence of the applied pressure waveform on the dynamic fracture processes in rock. *International Journal of Rock Mechanics and Mining Sciences* 41(5) (2004), pp.771–784.

Cho, S.H., Ogata, Y., and Kaneko, K.: Strain-rate dependency of the dynamic tensile strength of rock. *International Journal of Rock Mechanics and Mining Sciences* 40(5) (2003), pp.763–777.

Chong, K., Zhou, Y., Seah, C.C. and Lim, H.S.: Large-scale tests – Airblast, Ground Shock, and Debris. *Proceedings of the International Symposium on Defence Construction*, Singapore, 17–18 April 2002.

Chong, K.P., and Boresi, A.P.: Strain rate dependent mechanical properties of new albany reference shale. *International Journal of Rock Mechanics and Mining Science & Geomechanics Abstracts* 27(3) (1990), 199–205.

Chong, K.P., Hoyt, P.M., Smith, J.W. and Paulsen, B.Y.: Effects of strain rate on oil shale fracturing. *International Journal of Rock Mechanics and Mining Sciences & Geomechanics Abstracts* 17(1) (1980), pp.35–43.

Christensen, R., Swanson, S. and Brown, W.: Split-hopkinson-bar tests on rock under confining pressure. *Experimental Mechanics* 12(11) (1972), pp.508–513.

Clough, R.W.: The finite element method in plane stress analysis. *Proceedings of the Second ASCE Conference on Electronic Computation*, Pittsburg, PA, 1960.

Collin, R.: *Foundations for Microwave Engineering*, McGraw-Hill, New York, 1992.

Crotty J.M. and Wardle L.J.: Boundary integral analysis of piecewise homogeneous media with structural discontinuities. *International Journal of Rock Mechanics and Mining Sciences & Geomechanics Abstracts* 22 (1985), pp.419–427.

Cundall P.A.: A computer model for simulating progressive large scale movements in blocky rock systems. *Proceedings of the Symposium of ISRM*, Nancy, France, 1971, Vol.1, pp.11–18.

Dai, F. and Xia, K.: Loading rate dependence of tensile strength anisotropy of barre granite. *Pure and Applied Geophysics*, 2010, pp.1–14.

Dai, F., Xia, K. and Luo, S.N.: Semicircular bend testing with split Hopkinson pressure bar for measuring dynamic tensile strength of brittle solids. *Review of Scientific Instruments*, 79(12) (2008), pp.123903-6.

Dai, F., Huang, S., Xia, K., and Tan, Z.: Some fundamental issues in dynamic compression and tension tests of rocks using split Hopkinson pressure bar. *Rock Mechanics and Rock Engineering* (2010), pp.1–10.

Deng, H. and Nemat-Nasser, S.: Dynamic damage evolution in brittle solids. *Mechanics of Materials* 14(2) (1992), pp.83–103.

Deng, H. and Nemat-Nasser, S.: Dynamic damage evolution of solids in compression: Microcracking, plastic flow, and brittle-ductile transition. *Journal of Engineering Materials and Technology* 116(3) (1994), pp.286–289.

Dowding, C.H.: Estimating earthquake damage from explosion testing of full-scale tunnels. *Adv. Tunnel Technology and Subsurface Use* 4(3) (1984), pp 113–117.

Dowding, C.H.: *Blasting Vibration Monitoring and Control*. Prentice-Hall, NJ, USA, 1985.

Dowding C.H.: *Construction Vibrations*. New Jersey, Prentice-Hall, 1996.

Erzar, B. and Forquin, P.: An experimental method to determine the tensile strength of concrete at high rates of strain. *Experimental Mechanics* 50(7) (2010), pp.941–955.

Ewing, W.M., Jardetzky, W.S. and Press, F.: *Elastic Waves in Layered Media*. McGraw-Hill, New York, 1957.

Field, J.E., Walley, S.M., Proud, W.G., Goldrein, H.T. and Siviour, C.R.: Review of experimental techniques for high rate deformation and shock studies. *International Journal of Impact Engineering* 30(7) (2004), pp.725–775.

Frew, D., Forrestal, M. and Chen, W.: A split Hopkinson pressure bar technique to determine compressive stress-strain data for rock materials. *Experimental Mechanics* 41(1) (2001), pp.40–46.

Frew, D.J., *et al.*: Development of a dynamic triaxial kolsky bar. *Measurement Science and Technology* 21(10) (2010), p.105704.

Fuenkajorn, K. and Kenkhunthod, N.: Influence of loading rate on deformability and compressive strength of three thai sandstones. *Geotechnical and Geological Engineering* 28(5) (2010), pp.707–715.

Fukui, K., Okubo, S. and Ogawa, A.: Some aspects of loading-rate dependency of sanjome andesite strengths. *International Journal of Rock Mechanics and Mining Sciences*, 41(7) (2004), pp.1215–1219.

Gama, B.A., Lopatnikov, S.L. and Gillespie, J.W.Jr.: Hopkinson bar experimental technique: A critical review. *Applied Mechanics Reviews* 57(4) (2004), pp.223–250.

Ghaboussi, J., Wilson, E.L. and Isenberg J.: Finite elements for rock joints and interfaces. *Journal of Soil Mechanics Foundation Division ASCE* 99 (1973), pp.833–848.

Goldsmith, W., Sackman, J.L., and Ewerts, C.: Static and dynamic fracture strength of barre granite. *International Journal of Rock Mechanics and Mining Sciences & Geomechanics Abstracts* 13(11) (1976), pp.303–309.

Goodman, R.E., Taylor, R.L. and Brekke, T.: A model for the mechanics of jointed rock. *Journal of Soil Mechanics Foundation Division ASCE* 94 (1968), pp.637–659.

Grady, D., Hollenbach, R., Schuler, K. and Callender, J.: Strain rate dependence in dolomite inferred from impact and static compression studies. *Journal of Geophysical Research* 82(8) (1977), pp.1325–1333.

Graham-Brady, L.: (2010), Statistical characterization of meso-scale uniaxial compressive strength in brittle materials with randomly occurring flaws. *International Journal of Solids and Structures* 47(18–19) (2010), pp.2398–2413.

Gratkowski, S., Pichon, L. and Razek, A.: Infinite elements for 2D unbounded wave problems. *International Journal of Computers and Mathematics in Electronic Engineering* 14 (1995), pp.65–69.

Green, S.J. and Perkins, R.D.: Uniaxial compression tests at varying strain rates on three geologic materials. *The 10th U.S. Symposium on Rock Mechanics (USRMS), May 20–22, 1968, Austin, TX*, pp. 68-0035.

Guidault, P.A., Allix, O., Champaney, L. and Navarro, J.P.: A two-scale approach with homogenization for the computation of cracked structures. *Computers and Structures*, 85 (2007), pp.1360–1371.

Hasnaoui, A., van Swygenhoven, H. and Derlet, P.M.: Dimples on nanocrystalline fracture surfaces as evidence for shear plane formation. *Science* 300 (2003), pp.1550–1552.

Hettich, T., Hund, A. and Ramm, E.: Modeling of failure in composites by X-FEM and level sets within a multiscale framework. *Computational Methods in Applied Mechanical Engineering*, 197 (2008), pp.414–424.

Hopkins, D.L., Myer, L.R., and Cook, N.G.W.: Seismic wave attenuation across parallel joints as a function of joint stiffness and spacing. *EOS Transactions of American Geophysical Union* 68 (1988), pp.1427–1436.

Huang, C. and Subhash, G.: Influence of lateral confinement on dynamic damage evolution during uniaxial compressive response of brittle solids. *Journal of the Mechanics and Physics of Solids* 51(6) (2003), pp.1089–1105.

Huang, S., Chen, R. and Xia, K.W.: Quantification of dynamic tensile parameters of rocks using a modified kolsky tension bar apparatus. *Journal of Rock Mechanics and Geotechnical Engineering* 2 (2010), pp.162–168.

Huang, C., Subhash, G. and Vitton, S.J.: A dynamic damage growth model for uniaxial compressive response of rock aggregates. *Mechanics of Materials* 34(5) (2002), pp.267–277.

Hudson, J.A.: Wave speeds and attenuation of elastic waves in material containing cracks. *Geophysical Journal of the Royal Astronomic Society* 64(1) (1981), pp.133–150.

Hudson, J.A., Liu, E.R. and Crampin S.: Transmission properties of a plane fault. *Geophysical Journal International* 125(2) (1996), pp.559–566.

ITASCA: *http://www.itascacg.com/*, 2010.

Jing, L.: A review of techniques, advances and outstanding issues in numerical modelling for rock mechanics and rock engineering. *International Journal of Rock Mechanics and Mining Sciences* 40 (2003), pp.283–353.

Johnson, G.R. and Holmquist, T.J.: An improved computational constitutive model for brittle materials. *High Pressure Science and Technology* (1994), pp.981–984.

Kana, D.D., Fox, D.J. and Hsiung, S.M.: Interlock/friction model for dynamic shear response in natural jointed rock. *International Journal of Rock Mechanics and Mining Sciences & Geomechanics Abstracts* 33(4) (1996), pp.371–386.

Kazerani, T. *Micromechanical Study of Rock Fracture and Fragmentation under Dynamic Loads Using Discrete Element Method.* Ph.D. thesis, École Polytechnique Fédérale de Lausanne (EPFL), Lausanne, Switzerland, 2011.

Kazerani, T. and Zhao J.: Micromechanical parameters in bonded particle method for modelling of brittle material failure. *International Journal of Numerical and Analytical Methods in Geomechanics* doi.org/10.1002/nag.884. (2010).

Kazerani, T. and Zhao, J.: Simulation of dynamic fracturing in brittle materials using discrete element method and a full rate dependent logic for cohesive contact. *Engineering Fracture Mechanics* (2011).

Kazerani, T., Zhao, J. and Yang, Z.Y.: Investigation of failure mode and shear strength of rock joints using discrete element method. *Proc. EUROCK 2010*, Lausanne, Switzerland, 2010, pp.235–238.

Kazerani, T., Zhao, G.F. and Zhao, J.: Dynamic fracturing simulation of brittle material using the Distinct Lattice Spring Model (DLSM) with a full rate-dependent cohesive law. *Rock Mechanics and Rock Engineering*; doi: 10.1007/s00603-010-0099-0 (2010).

Kimberley, J., Ramesh, K.T. and Barnouin, O.S.: Visualization of the failure of quartz under quasi-static and dynamic compression. *Journal of Geophysical Research* 115(B8) (2010), B08207.

King, M.S., Myer, L.R. and Rezowalli, J.J.: Experimental studies of elastic-wave propagation in a columnar-jointed rock mass. *Geophysical Prospecting* 34(8) (1986), pp.1185–1199.

Kumar, A.: The effect of stress rate and temperature on the strength of basalt and granite. *Geophysics*, 33(3) (1968), pp.501–510.

Lajtai, E.Z., Duncan, E.J.S. and Carter, B.J.: The effect of strain rate on rock strength. *Rock Mechanics and Rock Engineering* 24(2) (1991), pp.99–109.

Lambert, D.E. and Ross, A.C.: Strain rate effects on dynamic fracture and strength. *International Journal of Impact Engineering* 24(10) (2000), pp.985–998.

Landes, M., Ritter, J.R.R. and Wedeken, U.: Weighing earthquake waves. *Measurement* 42 (2009), pp.13–17.

Lankford, J.: The role of tensile microfracture in the strain rate dependence of compressive strength of fine-grained limestone-analogy with strong ceramics. *International Journal of Rock Mechanics and Mining Science & Geomechanics Abstracts* 18 (1981), pp.173–175.

Larsen, C.J., Ortner, C. and Süli, E.: Existence of solutions to a regularized model of dynamic fracture. *Mathematical Models and Methods in Applied Sciences* 20(7) (2010), pp.1021–1048.

Lemos, J.V.: A Distinct Element Model for Dynamic Analysis of Jointed Rock with Application to Dam Foundation and Fault Motion. *PhD Thesis*, University of Minnesota, Minneapolis, USA, 1987.

Li, H.B., Zhao, J. and Li, T.J.: Triaxial compression tests on a granite at different strain rates and confining pressures. *International Journal of Rock Mechanics and Mining Sciences* 36(8) (1999), pp.1057–1063.

Li, H.B., Zhao, J. and Li, T.J.: Micromechanical modelling of the mechanical properties of a granite under dynamic uniaxial compressive loads. *International Journal of Rock Mechanics and Mining Sciences* 37(6) (2000), pp.923–935.

Li, H.B., Zhao, J., Li, T.J. and Yuan, J.X.: Analytical simulation of the dynamic compressive strength of a granite using the sliding crack model. *International Journal for Numerical and Analytical Methods in Geomechanics* 25(9) (2001), pp.853–869.

Li, H.B., Zhao, J., Li, J.R., Liu, Y.Q. and Zhou, Q.C.: Experimental studies on the strength of different rock types under dynamic compression. *International Journal of Rock Mechanics and Mining Sciences* 41(Supplement 1) (2004), pp.68–73.

Li, J.C. and Ma, G.W. Experimental study of stress wave propagation across a filled rock joint, *Int J Rock Mech Min Sci* 46(3) (2009), pp. 471–478.

Li J.C., Ma G.W. and Huang X.: Analysis of wave propagation through a filled rock joint. *Rock Mechanics and Rock Engineering* (2010) DOI: 10.1007/s00603-009-0033-5.

Li, J.C., Ma, G.W. and Zhao, J.: An equivalent viscoelastic model for rock mass with parallel joints. *Journal of Geophysical Research*, 115 (2010), B03305, doi:10.1029/2008JB006241.

Li, X.B., Lok, T.S. and Zhao, J.: Dynamic characteristics of granite subjected to intermediate loading rate. *Rock Mechanics and Rock Engineering* 38(1) (2005), pp.21–39.

Li, X.B., Lok, T.S., Zhao, J. and Zhao, P.J.: Oscillation elimination in the Hopkinson bar apparatus and resultant complete dynamic stress-strain curves for rocks. *International Journal of Rock Mechanics and Mining Sciences* 37(7) (2000), pp.1055–1060.

Li, X.B., Zhou, Z.L., Lok, T.S., Hong, L. and Yin, T.B.: Innovative testing technique of rock subjected to coupled static and dynamic loads. *International Journal of Rock Mechanics and Mining Sciences* 45(5) (2008), pp.739–748.

Liang, W.G., Zhao, Y.S., Xu, S.G. and Dusseault, M.B.: Effect of strain rate on the mechanical properties of salt rock. *International Journal of Rock Mechanics and Mining Sciences* 48(1) (2011), pp.161–167.

Lindholm, U.S., Yeakley, L.M. and Nagy, A.: The dynamic strength and fracture properties of dresser basalt. *International Journal of Rock Mechanics and Mining Sciences & Geomechanics Abstracts* 11(5) (1974), pp.181–191.

Liu, G.R. and Achenbach, J.D. A strip element method for stress analysis of anisotropic linearly elastic solids. *Journal of Applied Mechanics* 61 (1994), pp.270–277.

Logan, J.M. and Handin, J.: Triaxial compression testing at intermediate strain rates. *The 12th U.S. Symposium on Rock Mechanics (USRMS)*, November 16–18, 1970, Rolla, MO, 70-0167.

Lok, T.S., Li, X.B., Liu, D. and Zhao, P.J.:Testing and response of large diameter brittle materials subjected to high strain rate. *Journal of Materials in Civil Engineering* 14(3) (2002), pp.262–269.

LSTC: *http://www.lstc.com/lsdyna.htm*, 2010.

Lysmer, J., Kuhlemeyer, R.L.: Finite dynamic model for infinite media. *Journal of Engineering Mechanics Div ASCE* 95 (1969), pp.859–877.

Malvern, L.E.: Introduction. In: *Mechanics of a Continuous Medium*. Englewood Cliffs, New Jersey: Prentice Hall, 1969.

Masuda, K., Mizutani, H. and Yamada, I.: Experimental study of strain-rate dependence and pressure dependence of failure properties of granite. *Journal of Physical Earth* 35 (1987), pp.37–66.

McCann, D.M. and Baria, R.: Inter-borehole seismic measurements and their application in rock mass assessment. In: *Goephysical Investigations in Connection with Geological Disposal of Radioactive Waste, Proceedings of a Nuclear Energy Agency OECD Workshop*, Ottawa, 1982, pp.123–135.

McKenzie, C.L., Stacey, G.P. and Gladwin, M.T.: Ultrasonic characteristics of a rock mass. *International Journal of Rock Mechanics and Mining Sciences and Geomechanical Abstracts* 19 (1982), pp.25–30.

MECHSOFT: *http://www.mechsoft.cn/*, 2010.

Mindlin, R.D.: Waves and vibrations in isotropic, elastic plates. In: Goodier, J. N. and Hoff, J.J. (eds): *Structural Mechanics*, Pergamon, New York, 1960, pp.199–232.

Monaghan, J.J.: An introduction to SPH. *Comput Phys Com* 48 (1988), pp.89–96.

Morozov, N.F. and Petrov, Y.V.: *Dynamic of Fracture*, Springer-Verlag, 2000.

Mullins, M. and Dokainish, M.A.: Simulation of the (001) Plane crack in alpha-iron employing a new boundary scheme. *Philos Mag A* 46 (1982), pp.771–787.

Munjiza, A.: *The combined finite-discrete element method*. John Wiley&Sons, Ltd, University of London, 2004.

Myer, L.R., Hopkins, D.L., Peterson, J.E. and Cook, N.G.W.: Seismic wave propagation across multiple fractures. *Fractured and jointed rock masses* (1995), pp.105–110.

Nemat-Nasser, S. and Deng, H.: Strain-rate effect on brittle failure in compression. *Acta Metallurgica et Materialia* 42(3) (1994), pp.1013–1024.

Nemat-Nasser, S. and Horii, H.: Compression-induced nonplanar crack extension with application to splitting, exfoliation, and rockburst. *Journal of Geophysical Research* 87(B8) (1982), pp.6805–6821.

Olsson, W.A.: The compressive strength of tuff as a function of strain rate from 10-6 to 103/sec. *International Journal of Rock Mechanics and Mining Sciences & Geomechanics Abstracts* 28(1) (1991), pp.115–118.

Ou, Z.C., Duan, Z.P. and Huang, F.L.: Analytical approach to the strain rate effect on the dynamic tensile strength of brittle materials. *International Journal of Impact Engineering* 37(8) (2010), pp.942–945.

Paliwal, B. and Ramesh, K.T.: An interacting micro-crack damage model for failure of brittle materials under compression. *Journal of the Mechanics and Physics of Solids* 56(3) (2008), pp.896–923.

Pande, G.N., Beer, G. and Williams, J.R.: *Numerical Methods in Rock Mechanics*. Wiley, New York, 1990.

Perino, A., Barla, G. and Orta, R.: Wave propagation in discontinuous media. *European Rock Mechanics Symposium 2010*, Int Soc Rock Mech, Lausanne, Switzerland, 2010.

Perino A., Zhu J.B., Li J.C., Barla G., Zhao J.: Theoretical methods for wave propagation across jointed rock masses. *Rock Mechanics and Rock Engineering* 43 (2010), pp.799–809.

Perkins, R.D., Green, S.J. and Friedman, M.: Uniaxial stress behavior of porphyritic tonalite at strain rates to 103/second. *International Journal of Rock Mechanics and Mining Sciences & Geomechanics Abstracts* 7(5) (1970), pp.527–528, IN5-IN6, pp.529–535.

Price, D.G. and Knill, J.L.: A study of the tensile strength of isotropic rocks. *Proceedings of the 1st Congress of International Society of Rock Mechanics*, Lisbon, 1966, 439–442.

Pyrak-Nolte, L.J., Myer, L.R. and Cook, N.G.W.: Transmission of seismic waves across single natural fractures. *Journal of Geophysical Research* 95 (1990a), pp.8617–8638.

Pyrak-Nolte, L.J., Myer, L.R. and Cook, N.G.W.: Anisotropy in seismic velocities and amplitudes from multiple parallel fractures. *Journal of Geophysical Research* 95 (1990b), pp.11345–11358.

Qi, C., Wang, M. and Qian, Q.: Strain-rate effects on the strength and fragmentation size of rocks. *International Journal of Impact Engineering* 36(12) (2009), pp.1355–1364.

Ramesh, K.T.: High rates and impact experiments. In: Sharpe, W.N. (ed): *Springer Handbook of Experimental Solid Mechanics*, Springer US, 2008, pp.929–960.

Ravichandran, G. and Subhash, G.: A micromechanical model for high strain rate behavior of ceramics. *International Journal of Solids and Structures* 32(17–18) (1995), pp.2627–2646.

Saksala, T.: Damage–viscoplastic consistency model with a parabolic cap for rocks with brittle and ductile behavior under low-velocity impact loading. *International Journal for Numerical and Analytical Methods in Geomechanics* 34(13) (2010), pp.1362–1386.

Schoenberg, M.: Elastic wave behavior across liner slip interfaces. *Journal of Acoustic Society of America* 68 (1980), pp.1516–1521.

Schoenberger, M. and Levin, F. K.: Apparent attenuation due to interbed multiples. *Geophysics* 39(3) (1974), pp.278–291.

Schoenberg, M. and Muir, F.: A calculus for finely layered anisotropic media. *Geophysics* 54 (1989), pp.581–589.

Schoenberg, M. and Sayers, C.M.: Seismic anisotropy of fractured rock. *Geophysics* 60 (1995), pp.204–211.

Schwer, L.E. and Lindberg, H.E.: Application brief: a finite element slideline approach for calculating tunnel response in jointed rock. *International Journal of Numerical and Analytical Methods in Geomechanics* 16 (1992), pp.529–540.

Shang, J.L., Shen, L.T. and Zhao, J.: Hugoniot equation of state of the bukit timah granite. *International Journal of Rock Mechanics and Mining Sciences* 37 (2000), pp.705–713.

Shi, G.H.: Discontinuous Deformation Analysis, A New Numerical Model for the Statics and Dynamics of Block Systems. *PhD thesis*, Univ. of California, Berkeley, Berkeley, Calif, 1988.

Shi, G.H.: Manifold method of material analysis. *Transactions of the 9th Army Conference on Applied Mathematics and Computations,* U.S. Army Research Office, Minneapolis, MN, (1991), pp.57–76.

Shockey, D.A., Curran, D.R., Seaman, L., Rosenberg, J.T. and Petersen, C.F.: Fragmentation of rock under dynamic loads. *International Journal of Rock Mechanics and Mining Sciences & Geomechanics Abstracts* 11(8) (1974), pp.303–317.

SIMULIA, *http://www.simulia.com/products/product_index.html*, 2010.

Stowe, R.L. and Ainsworth, D.L.: Effect of rate of loading on strength and young's modulus of elasticity of rock. *The 10th U.S. Symposium on Rock Mechanics (USRMS)*, May 20–22, 1968, Austin, TX, 68-0003.

Sun L., Zhao G.F,. Zhao J.: An introduction to practical manifold method for continuum-discontinuum modelling. In: *Proceedings of ICADD10*, Hawaii, 2011.

Wang, Q.Z., Li, W. and Song, X.L.: A method for testing dynamic tensile strength and elastic modulus of rock materials using SHPB. In: Dresen, G., Zang, A., and Stephansson, O. (eds.): *Rock Damage and Fluid Transport, Part I: Pageoph Topical Volumes*, Birkhäuser Basel, (2006), pp.1091–1100.

Wang, Q.Z., Li, W. and Xie, H.P.: Dynamic split tensile test of flattened brazilian disc of rock with SHPB setup. *Mechanics of Materials* 41(3) (2009), pp.252–260.

Wang, W.M., Sluys, L.J. and de Borst, R.: Viscoplasticity for instabilities due to strain softening and strain-rate softening. *International Journal for Numerical Methods in Engineering* 40(20) (1997), pp.3839–3864.

Watanabe, T. and Sassa, K.: Seismic attenuation tomography and its application to rock mass evaluation. *International Journal of Rock Mechanics and Mining Sciences and Geomechanical Abstracts* 33(5) (1996), pp.467–477,

White, J.E.: *Underground Sound*. Elsevier, New York, 1983.

Whittles, D.N., Kingman, S., Lowndes, I. and Jackson, K.: Laboratory and numerical investigation into the characteristics of rock fragmentation. *Minerals Engineering* 19(14) (2006), pp.1418–1429.

Wong, T.F.: A note on the propagation behavior of a crack nucleated by a dislocation pileup. *Journal of Geophysical Research* 95(B6) (1990), pp.8639–8646.

Wong, T.F., Wong, R.H.C., Chau, K.T. and Tang, C.A.: Microcrack statistics, weibull distribution and micromechanical modeling of compressive failure in rock. *Mechanics of Materials* 38(7) (2006), pp.664–681.

Wu, W., Zhu, J.B., Zhang, Q.B., Mathier, J.F. and Zhao, J.: Plane stress wave transmission and attenuation across single rock fracture, *Proceedings of International Rock Mechanics Congress*, Beijing, 2011.

Xia, K., Nasseri, M.H.B., Mohanty, B., Lu, F., Chen, R. and Luo, S.N.: Effects of microstructures on dynamic compression of barre granite. *International Journal of Rock Mechanics and Mining Sciences* 45(6) (2008), pp.879–887.

Xiao, S.P. and Belytschko, T.: A bridging domain method for coupling continua with molecular dynamics. *Computational Methods in Applied Mechanical Engineering* 193 (2003), pp.1645–1669.

Xie, H. and Sanderson, D.J.: Fractal kinematics of crack propagation in geomaterials. *Engineering Fracture Mechanics* 50(4) (1995), pp.529–536.

Yan, D., and Lin, G.: Dynamic properties of concrete in direct tension. *Cement and Concrete Research* 36(7) (2006), pp.1371–1378.

Yang, C.H. and Li, T.J.: The strain rate-dependent mechanical properties of marble and its constitutive relation. In: *Proceedings of International Conference of Computational Methods in Structural and Geotechnical Engineering*, Hong Kong, 1994.

Zhang, J.X., Wong, T.F. and Davis, D.M.: Micromechanics of pressure-induced grain crushing in porous rocks, *Journal of Geophysical Research* 95(B1) (1990), pp.341–352.

Zhang, Y.Q. and Hao, H.: Dynamic fracture in brittle solids at high rates of loading. *Journal of Applied Mechanics* 70(3) (2003), pp.454–457.

Zhang, Z.X., Yu, J., Kou, S.Q. and Lindqvist, P.A.: On study of influences of loading rate on fractal dimensions of fracture surfaces in gabbro. *Rock Mechanics and Rock Engineering* 34(3) (2001), pp.235–242.

Zhao, G.F.: *Development of Micro-macro Continuum-discontinuum Coupled Numerical Method*. PhD thesis. École Polytechnique Fédérale de Lausanne (EPFL), Lausanne, Switzerland (2010).

Zhao, G.F., Fang, J. and Zhao, J.: A 3D distinct lattice spring model for elasticity and dynamic failure. *International Journal of Numerical and Analytical Methods in Geomechnics* DOI: 10.1002/nag.930 (2010).

Zhao, J.: Applicability of mohr-coulomb and hoek-brown strength criteria to the dynamic strength of brittle rock. *International Journal of Rock Mechanics and Mining Sciences*, 37(7) (2000), pp.1115–1121.

Zhao, J.: *Development of a numerical modelling code adopting Particle based Manifold Method (PMM)*, Swiss National Science Foundation proposal 200021_127178 (2009).

Zhao, J. and Cai, J.G.: Transmission of elastic P-waves across single fractures with a nonlinear normal deformational behaviour. *Rock Mechanics and Rock Engineering* 34 (2001), pp.3–22.

Zhao, J. and Li, H.B.: Experimental determination of dynamic tensile properties of a granite. *International Journal of Rock Mechanics and Mining Sciences* 37(5) (2000), pp.861–866.

Zhao, J., Li, H.B. and Zhao, Y.H.: *Dynamics Strength Tests of the Bukit Timah Granite*. Geotechnical Research Report NTU/GT/98-2, Nanyang Technological University, Singapore, 1998, 106pp.

Zhao, J., Wang, S.J. and Tang, C.A.: Personal discussions and communications on micromechanics of dynamic failure of rock materials (2008).

Zhao, J., Li, H.B., Wu, M.B. and Li, T.J.: Dynamic uniaxial compression tests on a granite. *International Journal of Rock Mechanics and Mining Sciences* 36(2) (1999a), pp.273–277.

Zhao, J., Zhou, Y. Hefny, A.M., Cai, J.G., Chen, S.G., Li, H.B., Liu, J.F., Jain, M., Foo, S.T. and Seah, C.C.: Rock dynamics research related to cavern development for ammunition storage. *Tunnelling Underground Space Technology* 14 (1999b), pp.513–526.

Zhao, J., Cai, J.G., Zhao, X.B. and Li H.B.: Experimental study of ultrasonic wave attenuation across parallel fractures. *Geomechanics and Geoengineering* 1(2) (2006a), pp.87–103.

Zhao, J., Zhao, X.B., Ma, G.W., Li, H.B. and Zhou, Y.X.: Keynote: Recent studies of rock dynamics for underground development. *Geo-Shanghai 2006, ASCE Geotechnical Special Publication No.155* (2006b), pp.1–25.

Zhao, X.B., Zhao, J. and Cai, J.G.: P-wave transmission across fractures with nonlinear deformational behaviour. *International Journal of Numerical and Analytical Methods in Geomechanics* 30 (2006), pp.1097–1112.

Zhao, X.B., Zhao, J., Cai, J.G. and Hefny, A.M.: UDEC modelling on wave propagation across fractured rock masses. *Computers and Geotechnics*, 35 (2008), pp.97–104.

Zhou, X., Ha, Q., Zhang, Y. and Zhu, K.: Analysis of deformation localization and the complete stress-strain relation for brittle rock subjected to dynamic compressive loads. *International Journal of Rock Mechanics and Mining Sciences* 41(2) (2004), pp.311–319.

Zhou, X.P., and Yang, H.Q.: Micromechanical modeling of dynamic compressive responses of mesoscopic heterogenous brittle rock. *Theoretical and Applied Fracture Mechanics* 48(1) (2007), pp.1–20.

Zhou, X.Q. and Hao, H.: Modelling of compressive behaviour of concrete-like materials at high strain rate. *International Journal of Solids and Structures* 45(17) (2008), pp.4648–4661.

Zhou, Y.X.: Explosion loading and tunnel response. In: Zhou, Y.X. and Zhao, J. (eds.): *Advances in Rock Dynamics and Applications*, Taylor & Francis (2011).

Zhou, Z., Li, X., Ye, Z. and Liu, K.: Obtaining constitutive relationship for rate-dependent rock in SHPB tests. *Rock Mechanics and Rock Engineering* (2010), pp.1–10.

Zhu, J.B.: *Theoretical and numerical analyses of wave propagation in jointed rock masses*. PhD thesis. École Polytechnique Fédérale de Lausanne (EPFL), Lausanne, Switzerland (2011).

Zhu, J.B., Zhao, G.F., Zhao, X.B. and Zhao, J.: Validation study of the distinct lattice spring model (DLSM) on P-wave propagation across multiple parallel joints. *Computers and Geotechnics* 38(2) (2010), pp.298–304.

Zhu, J.B., Zhao, X.B., Li, J.C., Zhao, G.F. and Zhao, J.: Normally incident wave propagation across one joint set with Virtual Wave Source Method. *Journal of Applied Geophysics* 73(3) (2011), pp.283–288.

Zhu, W.C.: Numerical modelling of the effect of rock heterogeneity on dynamic tensile strength. *Rock Mechanics and Rock Engineering* 41(5) (2008), pp.771–779.

Zhu, W.C. and Tang, C.A.: Numerical simulation of brazilian disk rock failure under static and dynamic loading. *International Journal of Rock Mechanics and Mining Sciences* 43(2) (2006), pp.236–252.

Zuo, Q.H., Addessio, F.L., Dienes, J.K. and Lewis, M.W.: A rate-dependent damage model for brittle materials based on the dominant crack. *International Journal of Solids and Structures* 43(11–12) (2006), pp.3350–3380.

Zuo, Q.H., Disilvestro, D. and Richter, J.D.: A crack-mechanics based model for damage and plasticity of brittle materials under dynamic loading. *International Journal of Solids and Structures* 47(20) (2010), pp.2790–2798.

Chapter 3

Split Hopkinson pressure bar tests of rocks: Advances in experimental techniques and applications to rock strength and fracture

Kaiwen Xia, Feng Dai and Rong Chen

3.1 INTRODUCTION

The accurate measurement of rock dynamic mechanical properties has always been a very important task for a variety of rock engineering and geophysical applications, which include quarrying, drilling, rock bursts, blasts, earthquakes, and projectile penetrations. In these applications, the rock materials are subjected to dynamic loading over a wide range of loading rates. Therefore, accurate determination of dynamic strength and toughness properties of rocks over a wide range of loading rates is crucial. However, in sharp contrast to the static rock testing methods, no recommended methods have been suggested by the International Society of Rock Mechanics (ISRM). In addition, the existing dynamic testing results with different methods and instrumentations are so scattered that cross-referencing of others' results is unfeasible. It is thus necessary and urgent for the rock mechanics community to develop reliable suggested methods to standardize the mechanical testing of rocks under high loading rates.

To test dynamic mechanical properties of rocks, we need a reliable testing device. For testing rock materials under high strain rates ($10^2 \sim 10^3$ s^{-1}), split Hopkinson pressure bar (SHPB) is an ideal dynamic testing machine. As a widely used device to quantify the dynamic compressive response of various metallic materials at high loading or strain rates, SHPB was invented in 1949 by Kolsky (Kolsky, 1949; Kolsky, 1953). Shortly after that, SHPB was attempted by researchers to test brittle materials such as concretes (Ross, Thompson and Tedesco, 1989; Ross, Tedesco and Kuennen, 1995), ceramics (Chen and Ravichandran, 1996; Chen and Ravichandran, 2000) and rocks (Christensen, Swanson and Brown, 1972; Dai, Xia and Tang, 2010). However, some major limitations of using SHPB for brittle materials were not fully explored until two decades ago (Subhash, Ravichandran and Gray, 2000).

Unlike ductile metals, brittle materials have small failure strains ($<1\%$) and hence if the loading is too fast, as in a conventional SHPB test, the specimen may fail in a non-uniform manner (i.e., the front portion of the sample may be shattered while the back portion of the sample remains intact.). To achieve accurate measurements in SHPB tests, one has to make sure that the dynamic loading is slow enough so that the specimen is experiencing an essentially quasi-static load, and thus the deformation of the specimen is uniform. As a rule of thumb, it takes the loading stress wave to travel in the specimen 3–4 rounds for the stress to achieve such an equilibrium state. The pulse-shaping technique was proposed to slow down the loading rate and thus to minimize the so-called inertial effect associated with the stress wave loading (Frew, Forrestal and Chen, 2001). Another problem in conventional SHPB tests is that the specimen is subjected to multiple loading due to the reflection of the wave at the impact end of the incident bar. A momentum-trap technique was proposed to ensure single pulse loading and thus enable valid post-mortem analysis of the recovered specimen (Nemat-Nasser, Isaacs and Starrett, 1991). Other advancements in SHPB can be found in a recent review (Field *et al.*, 2004).

Using these new techniques in SHPB, we systematically measured the dynamic mechanical properties of rocks. A few new testing methods were developed to accurately measure the dynamic compressive strength and response, the dynamic tensile strength, and dynamic fracture parameters of rocks. For all these tests, we used core-based rock specimens to facilitate sample preparation. In the rock dynamic compression, we addressed the issue of the length to diameter ratio of the cylindrical rock specimen. In the static uniaxial compressive strength (UCS) tests, the length to diameter ratio is required to be 2 or more to minimize the end frictional effect; in SHPB tests, the friction is dynamic and thus the frictional effect is presumably smaller. Shorter specimen favors dynamic stress equilibrium but has worse frictional effect. An optimal length to diameter ratio was sought. The dynamic tensile strength measurements using SHPB were conducted using the Brazilian disc (BD) method. This method was fully validated on the dynamic force balance and quasi-static data reduction with the aid of high speed photography. We proposed the fracture onset detection to determine the correct value of the far-field load at failure for calculating the rock tensile strength. There are two methods used to measure the dynamic fracture toughness of rocks: the notched semi-circular bend (SCB) method and the cracked chevron-notched Brazilian disc (CCNBD) method. Using a special optical technique to monitor the crack surface opening distance (CSOD), we observed the stable fracture to unstable fracture transition in dynamic CCNBD tests. We also showed that using our optical device, the dynamic fracture energy and fracture velocity of rocks can be estimated.

The chapter is organized as follows. The principles of SHPB and the new testing techniques are covered in Section 2. The application of SHPB to dynamic compressive tests, dynamic tensile tests and dynamic fracture tests of rocks are discussed in Section 3, Section 4, and Section 5 respectively. Section 6 concludes the materials presented in the entire chapter.

3.2   PRINCIPLES OF SPLIT HOPKINSON PRESSURE BAR AND NEW TECHNIQUES

3.2.1   The split Hopkinson pressure bar system

SHPB is composed of three bars: a striker bar, an incident bar, and a transmitted bar (Gray, 2000). The impact of the striker bar on the free end of the incident bar induces a longitudinal compressive wave propagating in both directions. The left-propagating wave is fully released at the free end of the striker bar and forms the trailing end of the incident compressive pulse $-\varepsilon_i$ (Fig. 3.1). Upon reaching the bar-specimen interface, part of the incident wave is reflected as the reflected wave $-\varepsilon_r$ and the remainder passes through the specimen to the transmitted bar as the transmitted wave $-\varepsilon_t$.

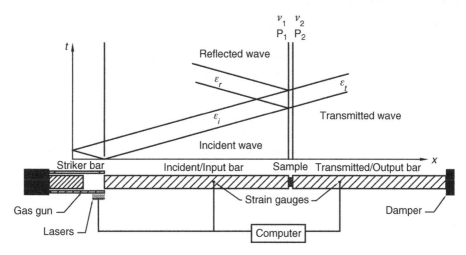

Figure 3.1 Schematics of a split Hopkinson pressure bar (SHPB) system and the *x-t* diagram of stress waves propagation in SHPB.

A 25 mm diameter SHPB system is used in this study. The length of the striker bar is 200 mm. The incident bar is 1500 mm long and the strain gauge station is 733 mm from the impact end of the bar. The transmitted bar is 1200 mm long and the stain gauge station is 655 mm away from the sample. An infrared detector system is used together with a two-channel TDS1021 digital oscilloscope to measure the velocity of the striker bar. An eight-channel Sigma digital oscilloscope by Nicolet is used to record and store the strain signals collected from the Wheatstone bridge circuits after amplification. Because the bar diameter is relatively small, it is suitable for testing fine- to medium-grained rocks. However, the methodologies developed can be applied to general dynamic rock testing, given that an SHPB system with appropriate diameter is chosen for a specific rock.

3.2.2   Standard analysis of SHPB

Based on the one dimensional stress wave theory, the dynamic forces on the incident end (P_1) and the transmitted end (P_2) of the specimen are (Kolsky, 1949; Kolsky, 1953):

$$P_1 = AE(\varepsilon_i + \varepsilon_r), \quad P_2 = AE\varepsilon_t \tag{3.1}$$

The velocities at the incident bar end (v_1) and the transmitted bar end (v_2) are:

$$v_1 = C(\varepsilon_i - \varepsilon_r), \quad v_2 = C\varepsilon_t \tag{3.2}$$

In the above equations, E is the Young's Modulus of the bar, A is the cross-sectional area of the bar, and C is the one dimensional longitudinal stress wave velocity of the bar.

3.2.3 The pulse-shaping technique

The loading pulse of the conventional SHPB system for materials testing at high strain rates has an approximately trapezoidal shape companied by high level of oscillations. The oscillations induced by the sharp rising portion of the incident wave results in much difficulty in achieving dynamic stress equilibrium state in the sample. However, the stress equilibrium is a prerequisite for valid SHPB tests.

In a review paper by Franz *et al.* discussing the incident pulse shaping for SHPB experiments with metal samples (Frantz, Follansbee and Wright, 1984), the authors emphasized that a slowly rising incident pulse is a preferred loading pulse in order to minimize the effects of dispersion and inertia, and thus facilitate dynamic stress equilibrium of the sample. Frantz, Follansbee and Wright (1984) presented experimental results to show a properly shaped loading pulse can not only provide stress equilibrium in the sample, but also generate a nearly constant strain rate in the sample. Gray and Blumenthal also discussed these issues in their recent review paper (Gray, 2000).

To shape the incident pulse, one way is to modify the geometry of the striker. For example, Christensen, Swanson and Brown (1972) used striker bars with a truncated-cone on the impact end in an attempt to produce ramp pulses, Frantz, Follansbee and Wright (1984) used a machined striker bar with a large radius on the impact face to generate a slowly rising incident pulse for the tests, Li *et al.* (2000) used tapered striker to generate an approximate half-sine loading waveform. Another way, maybe a more convenient way is to place a small, thin disc made of soft materials between the striker and the incident bar. The disc is called the pulse shaper and can be made of paper, aluminum, brass or stainless steel, with 0.1–2.0 mm in thickness. During tests, the striker impacts on the pulse shaper before the incident bar, thus generating a non-dispersive ramp pulse propagating into the incident bar. This incident pulse with slow-rising front facilitates the dynamic force balance the specimen (Frew, Forrestal and Chen, 2001; 2002). One example of waves with and without shaper is shown in Figure 3.2.

A wide variety of incident pulses can be produced by varying the geometry of the pulse shaper (Fig. 3.3). Depending on the materials of testing, different loading pulses are needed and can be achieved with proper shaper design.

The pulse-shaping technique in SHPB is especially useful for investigating dynamic response of brittle materials such as rocks (Frew, Forrestal and Chen, 2001; 2002). Without proper pulse-shaping, it is difficult to achieve dynamic stress equilibrium in such materials because the sample may fail immediately from its end in contact with the incident bar upon the arrival of the incident wave. In our SHPB tests, we use the C11000 copper as the main shaper to modify the incident wave from a rectangular shape to a ramped shape. In addition, a small rubber disc is placed in front of the copper shaper to further reduce the slope of rising portion of the pulse to a desired value.

3.2.4 The momentum-trap system

To ensure single pulse loading, the momentum-trap technique is adopted in our Hopkinson bar setup as shown in Figure 3.4. Figure 3.4a is the photograph of the momentum-trap system, which is composed of a momentum transfer flange that is

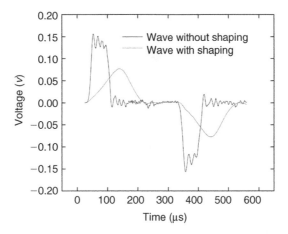

Figure 3.2 Stress waves from the incident bar with and without pulse shaper (the incident pulses are fully reflected because only the incident bar is used in these attempts).

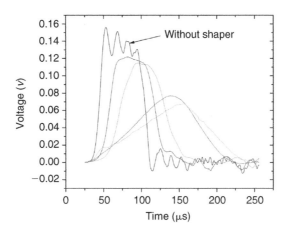

Figure 3.3 Different loading pulses produced by pulse-shaping with shaper.

attached to the impact end of the input/incident bar and a rigid mass that is attached to the supporting I-beam for the bar system.

Denoting the length of the incident bar by l, it takes $t_0 = 2l/C$ for the reflected wave to arrive at the impact end of the incident bar. The reflection wave is then reflected and changes from the tensile wave to compression wave at the input end. As a result, it will exert dynamic compression on the sample for a second time. This way, the sample in a conventional SHPB will thus experience multiple compressive loading. This kind of multi-loading complicates the post-mortem examination of tested samples (Nemat-Nasser, Isaacs and Starrett, 1991). A momentum-trap system similar to that proposed by Song and Chen (2004) is adopted here. The main idea of this method is to absorb the first reflection by a big mass that can be considered as rigid because of its large

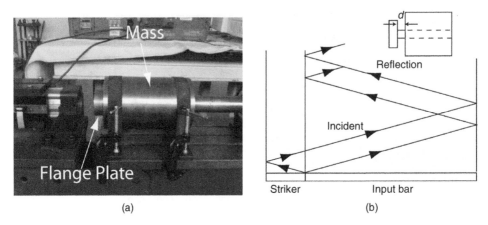

(a)   (b)

Figure 3.4 The momentum-trap system: (a) photograph and (b) *x-t* diagram showing its working principle.

impedance (which is equal to ρCA, where ρ is density) compared to the bar. As showed in the inset of Figure 3.4b, there is a gap between the flange and the rigid mass. The distance of the gap d is determined by the velocity of the striker v_0, the length of the input bar l and the shape of the input pulse. It is required that when the reflection wave arrives at the front end of the incident bar, the flange is in contact with the big mass. As a result, the reflected compressive wave will be changed to tension due to the interaction between the incident bar and the big mass through the flange. This requirement is expressed as:

$$d = C \int_0^{t_0} \varepsilon_i(t)dt \qquad (3.3)$$

If there is no pulse-shaper between the striker and the input bar, the particle velocity of the input bar after impact is $1/2\,v_0$ for the case where the striker and input bar are made of the same material. Denote the length of the striker by l_s, the total duration of the loading pulse is $t_1 = 2l_s/C$, which is usually much smaller than $t_0 = 2l/C$. The total displacement of the end of the incident bar (flange), which is equal to the gap between the flange and the rigid mass that we need to set is then $d = C\int_0^{t_0} \varepsilon_i(t)dt = \int_0^{t_1} v_0/2dt = v_0 l_s/C$. If there is a pulse-shaper between the striker and the incident bar, we should use the measured incidence pulse to determine the size of the gap using Equation (3.3).

As an example shown in Figure 3.5, the second compression is indeed reduced substantially by the momentum-trap so that the sample will experience essentially a single pulse loading. The second "loading" pulse is composed of a low amplitude compressive portion followed by a tensile portion. The tensile portion of the pulse will separate the incident bar from the sample, resulting in soft-recovery of the sample for valid post-mortem examination.

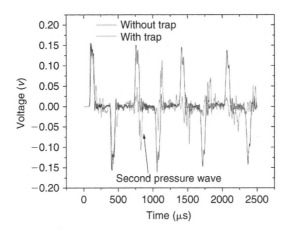

Figure 3.5 Comparison of stress waves from the incident bar with and without momentum trap.

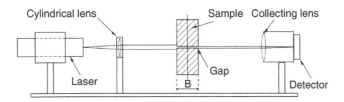

Figure 3.6 Schematics of the laser gap gauge (LGG) system.

3.2.5 The laser gap gauge (LGG)

In the fracture tests conducted on SHPB system, we developed an LGG system to monitor the opening of the notch to deduce the opening velocity of the cracked fragments (Chen *et al.*, 2009).

As shown schematically in Figure 3.6, the system consists of two major components: the collimated line laser source and the light detector. LGG is mounted perpendicular to the bar axis and the laser passes through the notch in the center of the specimen. During the test, as the notch opens up, the amount of light passing through the sample increases, leading to higher voltage output from the detector. The voltage is linearly proportional to the gap width and thus the crack surface displacement distance can be measured.

3.3 DYNAMIC COMPRESSIVE TEST

3.3.1 Introduction

The compressive tests of SHPB are based on two fundamental assumptions: 1) one dimensional (1D) elastic wave propagation in the bars and 2) homogeneous deformation of the sample (Kolsky, 1953). The assumption of 1D stress wave propagation is ensured by using long bars, and the elasticity of the bar deformation is guaranteed throughout the test by limiting the impacting velocity of the striker. The homogeneity of the sample deformation is affected by two factors: inertial effects (i.e. the axial inertial effect and the radial inertial effect) and the end frictional effect between the sample and the bars.

The ideal slenderness ratio (i.e., the length to diameter ratio) of the sample has long been studied because it plays a major role on the inertial effects during the dynamic SHPB test. Based on the synthetic analysis of both axial and radial inertial effects, Davies and Hunter (1963) suggested an optimal slenderness ratio of $L/D = \sqrt{3}v/2$, where L and D are the length and diameter of the cylindrical sample respectively, and v is the Poisson's ratio of the testing material. This slenderness ratio has been frequently used to design the sample geometry for metals (Meng and Li, 2003). To limit the inertial effects associated with stress wave loading, the slenderness ratio of samples can not be too large. When SHPB is first introduced to the dynamic testing community, the incident wave is in rectangular shape with a sharp rising edge and high oscillation, it is harder to minimize the axial inertial effects because it takes longer time for the sample to reach stress equilibrium as compared to the rising edge of the loading pulse. However, with recent developed pulse shaping techniques (Frew, Forrestal and Chen, 2001; 2002), even a relatively long compressive sample can easily obtain stress equilibrium, thus reducing the axial inertial effects to a negligible amount. In this case, the suggested ratio of $L/D = \sqrt{3}v/2$ by Davies and Hunter (1963) may be too conservative under current application of the pulse-shaping technique. For example, the slenderness ratio for an incompressible material (with v equal to 0.5) is determined to be 0.433 following the Davies and Hunter's formula. As will be discussed later, for short samples like this, friction at boundary may markedly affect the inner stress state, and thus the homogeneity of the sample deformation.

Frictional effect is another major concern in the SHPB test. As early as SHPB was first introduced as a useful dynamic testing tool, it was realized that the interfacial friction on both ends of the sample may affect the testing results (Kolsky, 1949; Kolsky, 1953). When the sample is loaded by the compressive stress wave in the SHPB test, it expands radially due to the Poisson's effect. If the sample/bar interfaces are not sufficiently lubricated, the resulting interfacial friction force can be significant. This friction force influences the accuracy of the testing results by applying a dynamic confinement to the compressive specimen, whose stress state should be one dimensional stress by assumption. This additional sample stress can yield pseudo rate effects of the material (Schey, Venner and Takomana, 1982). For example, Hauser *et al.* (1960) mistakenly concluded that the Aluminum alloy was a rate sensitive material because they glued the sample on the bars during their tests. In addition, the sample is no longer deformed uniformly because of this dynamic confinement (Narayanasamy and Pandey, 1997), whose effect is the largest on the ends and diminishes toward the centre of the specimen. Bell (1966) examined the distribution of stress and strain in the SHPB tests and found that there exists marked discrepancy between the measured strain from SHPB data reduction and the strain directly measured from the sample surface. With a finite difference method, Bertholf and Karnes (1975) simulated SHPB tests on samples with three types of slenderness ratios and interfacial friction conditions to investigate both inertial effects and interface frictional effects. They arrived at the same conclusion that without enough lubrication at the boundary interfaces, the stress state

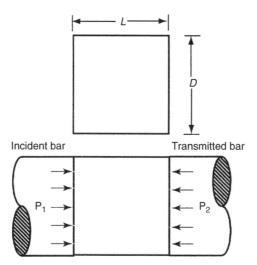

Figure 3.7 Close-view of the sample in the compression test using SHPB.

in the sample is inhomogeneous and big deviation of measurement occurs inevitably. Malinouski and Klepaczko (1986) presented a united analytic and numerical approach to investigate inertia and frictional effects in SHPB tests on an annealed Aluminum through the consideration of energy balance. They concluded that proper treatment of frictional effects, along with inertia is crucial for an exact determination of the material response during plastic deformation. Meng and Li (2003) recently revisited the combined effects of slenderness ratio and the interface friction numerically. To limit the frictional effect, the slenderness ratio of a compressive sample should be large enough. This can be manifested from a recommended slenderness ratio of 2 or larger for static compressive tests of rocks by ISRM (Bieniawski and Bernede, 1979). On the other hand, the slenderness ratio should be short enough to limit the inertial effects. Thus, an optimal slenderness ratio is thus needed to consider both the inertial effect and the frictional effect.

3.3.2 Data reduction

The histories of strain rate $\dot{\varepsilon}(t)$, strain $\varepsilon(t)$ and stress $\sigma(t)$ within the sample in the dynamic compression tests (Fig. 3.7) can be calculated as:

$$\begin{cases} \dot{\varepsilon}(t) = \dfrac{C}{L}(\varepsilon_i - \varepsilon_r - \varepsilon_t) \\ \varepsilon(t) = \dfrac{C}{L}\int_0^t (\varepsilon_i - \varepsilon_r - \varepsilon_t)dt \\ \sigma(t) = \dfrac{A}{2A_0}E(\varepsilon_i + \varepsilon_r + \varepsilon_t) \end{cases} \tag{3.4}$$

where L is the length of the sample and A_0 is the initial area of the sample.

Assuming the stress equilibrium or uniform deformation prevails during dynamic loading (i.e., $\varepsilon_i + \varepsilon_r = \varepsilon_t$), the commonly used formulas are obtained:

$$\begin{cases} \dot{\varepsilon}(t) = -\dfrac{2C}{L}\varepsilon_r \\ \varepsilon(t) = -\dfrac{2C}{L}\int_0^t \varepsilon_r dt \\ \sigma(t) = \dfrac{A}{A_0}E\varepsilon_t \end{cases} \tag{3.5}$$

3.3.3 Sample preparation

An isotropic fine-grained granitic rock, Laurentian granite (LG) is chosen for this research, whose mineralogical and mechanical characteristics are well documented (Nasseri, Mohanty and Robin, 2005). Laurentian granite is taken from the Laurentian region of Grenville province of the Precambrian Canadian Shield, north of St. Lawrence and north-west of Quebec City, Canada. The mineral grain size of LG varies from 0.2 to 2 mm with the average quartz grain size of 0.5 mm and the average feldspar grain size of 0.4 mm, with feldspar being the dominant mineral (60%) followed by quartz (33%). Biotite grain size is of the order of 0.3 mm and constitutes 3–5% of this rock. Rock cores with nominal diameters of 25 mm and 40 mm are first drilled from the rock blocks. For the 25 mm in diameter cores, we directly slice them to obtain cylindrical samples with varying slenderness ratios of 0.5, 1.0, 1.5 and 2.0. These cylindrical samples are prepared for dynamic compressive tests. For the dynamic BD test, we slice the 40 mm in diameter core into discs with nominal thickness of 20 mm. All the disc samples are polished afterwards resulting in surface roughness of less than 0.5% of the sample thickness.

3.3.4 Slenderness ratio

The choice of a proper slenderness ratio has been a fundamental issue in dynamic compression tests with SHPB because it has a major influence on the axial inertial effect: the higher the slenderness ratio, the larger the axial inertial effect and the smaller the relative radial inertial effect. In conventional SHPB tests, a rectangular incident wave is generated by a direct impact of the striker to the free end of the incident bar. This incident wave features a very sharp rising part and significant oscillations. For brittle solids like rocks with small failure strain, the sample may fail immediately from its end in contact with the incident bar by such incident pulses.

Recently, the pulse-shaping technique has been widely utilized for SHPB testing on engineering materials and it is especially useful for investigating dynamic response of brittle materials such as rocks (Frew, Forrestal and Chen, 2001; 2002). During tests, the striker impacts the pulse shapers right before the incident bar, generating a non-dispersive ramp pulse propagating into the incident bar and thus facilitating the dynamic stress equilibrium in the specimen (Frew, Forrestal and Chen, 2001; 2002). Under stress equilibrium, the stress gradient vanishes, and the inertial effects induced by stress wave propagation are minimized.

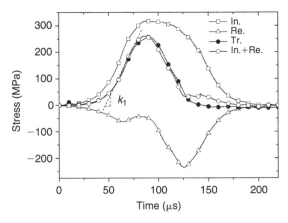

Figure 3.8 Dynamic stresses on both ends of disc specimen tested using a modified SHPB with careful pulse-shaping.

In the modified SHPB test, we use the C11000 copper disc in combination with a small rubber disc together as the shaper to transform the incident wave from a rectangular shape to a ramped shape (Xia *et al.*, 2008). Figure 3.8 shows the dynamic stress in a typical dynamic compressive test. Based on the 1D stress wave theory, the time zeros of the incident and reflection stress waves in Figure 3.8 are shifted to the sample-incident bar interface and the sample-transmitted bar interface, respectively. From Equation (3.1), the dynamic force on one side of the specimen P_1 is the sum of the incident and reflected waves (In.+Re.), and the dynamic force on the other side of the specimen P_2 is the transmitted wave (Tr.). It is shown that the time-varying stresses on both sides of the samples match with each other before the peak point is reached during the dynamic loading. The sample is thus in a state of dynamic stress equilibrium. It is also noted that the resulting stress on either side of the sample also features a linear portion before the peak, thus facilitating a constant loading rate via $\dot{\sigma} = k_1 A/A_0$. The parameter k_1 is illustrated in Figure 3.8.

We conducted compressive tests on cylindrical rock samples with varying slenderness ratio from 0.5, 1.0, 1.5 to 2.0. For all tests, we achieved dynamic stress equilibrium. Since there is no stress gradient in the sample, the axial inertial effect is thus negligible. In addition, to minimize the disturbance from the boundary frictions on the measured strength, the bar-sample interfaces for all samples are fully lubricated with vacuum grease. The measured compressive strengths with corresponding loading rates are shown in Figure 3.9. There are no significant differences of strengths from samples with selected slenderness ratios. For dynamic compressive tests on rocks, we conclude that with bar-sample interfaces fully lubricated and thus with axial inertial effects minimized, the slenderness ratio has little influence on the testing results within the range of 0.5 to 2. We thus suggest using samples with a slenderness ratio of 1 for dynamic compressive tests of rocks for convenience because it is difficult to hold shorter samples during sample fabrication.

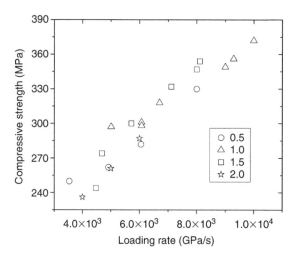

Figure 3.9 Dynamic compressive strengths with loading rates measured from rock samples with varying slenderness ratio of 0.5, 1.0, 1.5 and 2.0.

3.3.5 Frictional effect

To manifest the frictional effect, we conducted tests on samples with three different frictional boundaries on the bar-sample interfaces for sample slenderness ratio of 1: lubricated, dry and bonded. The bonded bar-sample interface completely restricts the motion of the rock surfaces on the bar and is believed to provide the maximum dynamic confinement to the sample. Through proper pulse shaping, we guarantee the dynamic stress balance on both ends of the sample (Fig. 3.8) and a constant loading rate has been achieved for all tests. Figure 3.10 illustrates the trend of the rate effects of compressive strength under these three boundary friction conditions. Samples with bar-sample interfaces fully lubricated yield the lowest measured compressive strength while the samples with bonded interfaces own the highest. The frictional effect in dynamic compressive tests on rocks is significant. The measured compressive strength increases with increasing friction involved in the tests. To obtain the actual dynamic compressive response of rocks, the bar-sample interfaces should be sufficiently lubricated.

Figure 3.11 shows the recovered samples with a) bonded b) dry c) lubricated bar/sample interfaces, coming from the tests with the data points of A, B and C in Figure 3.10, respectively. For these three typical tests, we load the sample with approximately the identical incident wave. However the damage levels of them are quite different.

With bar-sample interfaces fully lubricated, the samples are completely fragmented into small pieces (Fig. 3.11c), featuring a typical axial splitting failure mode. This failure mode confirms one dimensional stress state during the dynamic tests. In contrast, with friction at the boundary interfaces, the splitting is constrained significantly and the recovered samples feature a shear cone as shown in Figures 3.11a and b. With approximated similar loading rates, the strength values (at point A) measured for the sample with bonded interfaces with the maximum induced friction, the strength is the maximum (Fig. 3.11a); while fully lubricated sample yields the lowest value

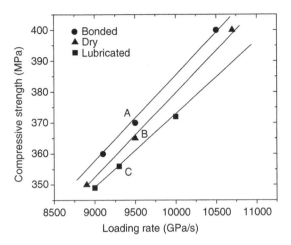

Figure 3.10 Dynamic compressive strengths with loading rates measured from rock samples (L/D = 1.0) with three interfacial friction boundaries.

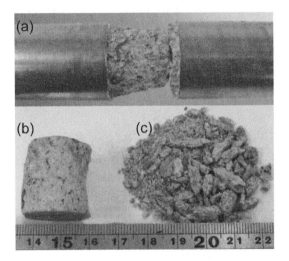

Figure 3.11 Photograph of recovered samples with (a) bonded (b) dry (c) lubricated bar/sample interfaces showing the damaged level.

(at point C). The sample with dry friction interfaces has an intermediate strength level (at point B). We can conclude here that without proper lubrication, the measurements strength values will be over-estimated.

3.4 DYNAMIC BRAZILIAN DISC TEST

3.4.1 Introduction

SHPB has also been adopted to conduct indirect tension tests for measuring the tensile strength of brittle solids like rocks. For examples, conventional SHPB tests were conducted using BD (Brazilian disc) method on marbles (Wang, Li and Song, 2006) and argillites (Cai *et al.*, 2007). These attempts followed the pioneer work on dynamic BD tests of concretes using SHPB (Ross, Thompson and Tedesco, 1989; Ross, Tedesco and Kuennen, 1995). Semi-circular bend samples were used also in SHPB to measure the flexural tensile strength of Laurentian granite (Dai, Xia and Tang, 2010). BD method has been suggested by ISRM as a recommended method for tensile strength measurement of rocks (Bieniawski and Hawkes, 1978). Using BD method, Zhao and Li (2000) measured the dynamic tensile properties of granite using a hydraulic loading system. For quasi-static and low speed BD tests, it is reasonable to use the standard static equation to calculate the tensile strength. However, for dynamic BD test conducted with SHPB featuring stress wave loading, the application of the quasi-static equation to the data reduction has not yet been rigorously checked (Wang, Li and Song, 2006; Cai *et al.*, 2007). In this work, we will investigate the conditions under which the static analysis is valid and address the importance of the fracture onset detection on the accurate data reduction.

3.4.2 Data reduction

A close-view of the dynamic BD sample in the SHPB system is schematically shown in Figure 3.12, where the disc sample is sandwiched between the incident bar and the transmitted bar. The principle of BD test comes from the fact that rocks are much weaker in tension than in compression and thus the diametrically loaded rock disc sample fails due to the tension along the loading diameter near the centre. The calculation equation of tensile strength is based on the 2D elastic analysis as:

$$\sigma_t = \frac{2P_f}{\pi DB} \tag{3.6}$$

where P_f is the load when the failure occurs, σ_t is the tensile strength, D and B are the diameter and the thickness of the disc, respectively.

A strain gauge is glued on the disc surface with 2 mm away from the centre of the disc (Fig. 3.12). The cracking of the disc centre emits elastic release waves upon cracking, and this wave causes sudden strain drop in the recorded strain gauge signal (Jiang *et al.*, 2004). The peak point of the strain gauge signal right before the sudden drop corresponds to the arrival of the release wave due to crack initiation. It is noted that the original strain gauge signal should be corrected accordingly considering the time the elastic wave propagates from the disc centre to the strain gauge.

3.4.3 Test without pulse-shaping

3.4.3.1 Dynamic force and failure sequence

Traditionally, by the direct impact of the striker on the free end of the incident bar in an SHPB test, the generated incident wave is a square compressive stress wave with a very sharp arising portion, which inevitably introduces high frequency oscillations. As a result, the dynamic forces on both ends of the sample vary significantly. Figure 3.13 depicts a large oscillation of dynamic force occurring on the incident side and a sizeable distinction between P_1 and P_2.

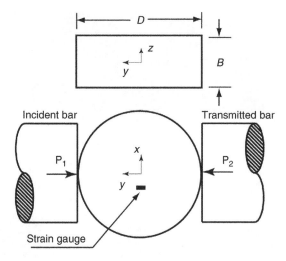

Figure 3.12   Close-view of the disc in a dynamic BD test using SHPB.

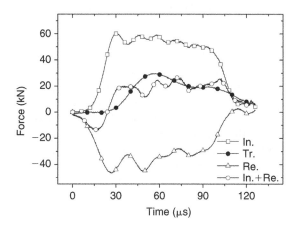

Figure 3.13   Dynamic forces on both ends of the disc specimen tested using a traditional SHPB without pulse-shaping.

3.4.3.2   Failure sequence from high speed camera

For a valid BD test, the disc sample should break first along the loading direction somewhere near the centre of the disc (Mellor and Hawkes, 1971; Hudson, Rummel and Brown, 1972). To verify this condition, we used a Photron Fastcam SA5 high speed camera to monitor the fracture processes of the BD test. The high speed camera is placed perpendicular to the sample surface with images taken at an inter frame interval of 3.8 μs. The failure process of this test without shaping the loading incident wave has been shown as top four images in Figure 3.14.

The time zero corresponds to the moment when the incident pulse arrives at the incident bar-sample interface. It can be seen that the first breakage emanates from the

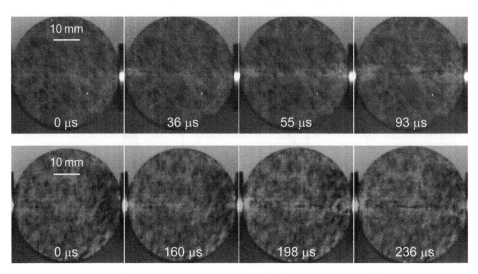

Figure 3.14 High-speed video images of two typical dynamic BD tests; Top row: BD test without pulse-shaping; Bottom row: BD test with careful pulse-shaping.

incident side of the sample at around 36 μs after the incident wave arrives at the bar-sample interface. Soon after that, damages also occur from the transmitted side of the sample (image at 55 μs). Thus, the splitting of the disc (image at 93 μs) is triggered by the damages at the loading points and then expands to the centre of the disc. We thus conclude that in this case, the working principle of BD test is violated. The rectangular incident loading wave with a sharp rising edge (Fig. 3.13) seems to affect the failure mode of the testing sample significantly. Since the cracking of the BD initiates from the loading ends, not from somewhere near the centre of the disc, the standard equation is invalid for reducing the tensile strength from the tensile stress history at the disc centre.

3.4.4 Test with pulse-shaping

3.4.4.1 Dynamic force and failure sequence

Figure 3.15 illustrates the time-varying forces in a typical test with careful pulse-shaping. The incident wave is shaped to a ramp pulse with a rising time of 180 μs, and a total pulse width of 300 μs. It is evident that the time-varying forces on both sides of the samples are almost identical before the peak point is reached during the dynamic loading. The resulting forces on either side of the sample also feature a linear portion before the peak, thus facilitating a constant loading rate via $\dot{\sigma} = 2k_2/(\pi DB)$, where the parameter k_2 is illustrated in Figure 3.15.

High speed camera has also been utilized to capture the failure sequences of the BD sample. Bottom four images in Figure 3.14 presents the key frames with representative features in the dynamic test. In sharp contrast to the images from the BD test without pulse-shaping, this disc cracks near the centre and the primary crack occurs at around

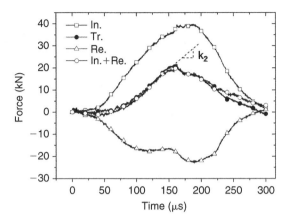

Figure 3.15 Dynamic forces on both ends of disc specimen tested using a modified SHPB with careful pulse-shaping.

$160\,\mu s$, and then propagates bilaterally to the loading ends. The next two frames illustrate the splitting trajectory of the sample; and the disc specimen is split completely into two fragments approximately along the centre line of the sample (Fig. 3.14). We also note that after the initiation of the primary crack, one secondary crack is visible near the loading ends at time instant $236\,\mu s$. Thus, since the splitting of the disc initiates near the centre, not from the loading ends, the tensile strength can be determined as long as we can accurately determine the tensile stress of the disc at failure.

3.4.4.2 *Validation of the quasi-static data analysis*

For a conventional dynamic compression tests with SHPB or direct tension tests with split Hopkinson tension bar (SHTB), the samples are cylindrical and thus the force balance on the ends ensures the stress equilibrium throughout the sample. However, the disc is two dimensional (2D), and force balance on the boundaries (Fig. 3.15) does not necessarily ensure dynamic equilibrium within the entire sample. A further comparison of the stress history at a point of interest from full dynamic analysis with that from quasi-static analysis is necessary. The dynamic finite element analysis represents the accurate stress history. The commercial finite element software ANSYS is employed for the analysis and the disc sample is meshed with quadrilateral eight-node element PLANE82, with total 4,800 elements and 14,561 nodes (Fig. 3.16).

Assuming linear elasticity, this analysis solves the following equation of motion with the Newmark time integration technique:

$$\nabla \cdot \boldsymbol{\sigma} = \rho \ddot{\mathbf{u}} \tag{3.7}$$

where $\boldsymbol{\sigma}$ is the stress tensor, ρ denotes density, and $\ddot{\mathbf{u}}$ is the second time derivative of the displacement vector u. The input loads in the finite element model are taken as the dynamic loading forces exerted on the incident and transmitted side of the specimen calculated using Equation (3.1) with measured waves.

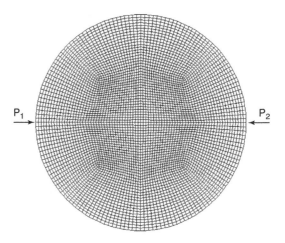

Figure 3.16 Mesh of the disc for the finite element analysis.

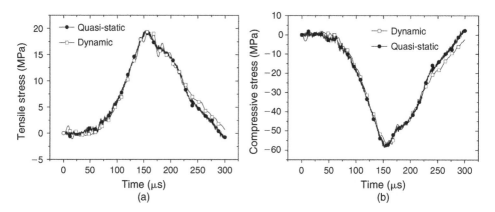

Figure 3.17 (a) Tensile stress σ_x and (b) compressive stress σ_y histories at the disc centre from both dynamic and quasi-static finite element analyses in a typical dynamic BD test with pulse shaping.

The transient dynamic stress history at the disc centre (potential failure spot) is calculated and compared with that from a quasi-static analysis employing Equation (3.6). The histories of the stress components σ_x (in tension) and σ_y (in compression) for dynamic and quasi-static finite element analyses are compared in Figures 3.17a and b respectively. The stress states at the disc centre for both quasi-static and dynamic data reductions match with each other. Thus, provided force balance on the sample ends, the quasi-static analysis with the far-field loading measured as input can accurately represent the stress history in the sample.

3.4.4.3 Determination of the fracture onset and dynamic tensile strength

Figure 3.18 shows the signal of the strain gauge mounted on the sample, compared with the transmitted force. Only one peak (A) of the signal is registered by the stain

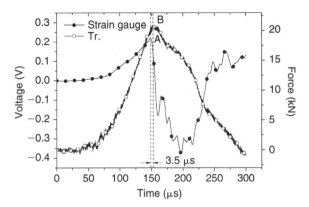

Figure 3.18 Comparison of strain gage signal with the transmitted force for a dynamic Brazilian test using a modified SHPB with careful pulse shaping.

gauge, occurring at time $149\,\mu s$. Thus, the breakage initiation time is designated by the unique trough A at the time of $149\,\mu s$. Because the peak transmitted force occurs at time $152.5\,\mu s$, it is thus delayed only by $3.5\,\mu s$ after the measured onset of breakage. We conclude that in this case, the peak far-field load matches with the breakage onset with a negligibly small time difference. The small time difference of $3.5\,\mu s$ can be interpreted as follows. The release waves travel at the sound speed of the rock material (around $5\,km/s$) and the distance between the fracture location and the supporting pin is $20\,mm$. Assume the fracture starts at the center of the disc, the distance from the center to the strain gauge is about $10\,mm$. The time difference between the two wave trajectories is thus about $2\,\mu s$. In other words, the theoretical time delay should be $2\,\mu s$ if the fracture onset matches the peak of the load exactly.

Due to the interaction between the release wave and the pins, the load on the transmitted side decreases (Fig. 3.12). Thus, the peak of the transmitted force can be regarded as synchronous with the single peak of the strain gauge signal (the rupture onset) for this specific test where the loading rate is not that fast. For this test, the dynamic tensile strength can be reduced from Equation (3.6) where the stress state is quasi-static, the tensile strength is calculated to be $18.9\,MPa$ at the loading rate of $233\,GPa/s$.

It has been long realized from quasi-static tests that the BD sample may fracture before the peak-load is reached (Mellor and Hawkes, 1971; Hudson, Rummel and Brown, 1972). This is also the case for the typical test shown in Figure 3.18. The reason for this phenomenon lies in the sample testing configuration. For an ideal BD test, the fracture will initiate in the center of the sample along the loading axis. At the fracture onset, the sample is still in contact with the two loading platens. The load can thus still increase until the sample is completely split into two halves. From part of the strain energy release from the fracture, the two halves will get transverse velocities to separate from each other. The two halves may rotate and lose contact with the platen during the separation process and this will lead to the unloading.

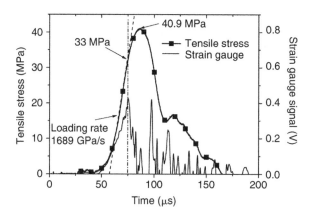

Figure 3.19 Tensile stress history with the strain gauge signal for detecting failure onset.

In quasi-static BD tests, the difference between the peak-load and the failure load may be smaller if a servo-controlled material testing machine is used and the transverse expansion of the disc is used as the controlling variable (Hudson, Rummel and Brwon., 1972). In dynamic BD tests using SHPB, there is no way to control the load using a feedback system, the mismatch of the measured peak-load and the failure load can be significant. The low loading rate test presented in Figure 3.18 features a reasonable match of the load, it is necessary though to address the load mismatch for higher rate loading tests where the foregoing discussed load mismatch may be significant.

Figure 3.19 shows the signal of the strain gauge mounted on the sample, compared with the transmitted force for a test featuring higher loading rate as compared to the test shown in Figure 3.18. It can be seen that the tensile stress at the fracture onset (i.e., tensile strength) is much lower than the peak stress determined from the far-field loading. One will overestimate the tensile strength by 20% if the peak-stress is used by mistake, which is significant. The accurate dynamic tensile strength measured using the dynamic BD test for this case is 33 MPa.

3.5 DYNAMIC FRACTURE TEST

3.5.1 Introduction

Dynamic fracture is frequently encountered in geophysical processes and engineering applications (e.g., earthquakes, airplane crashes, projectile penetrations, rock bursts and blasts). These processes are governed by material dynamic fracture parameters, such as initiation fracture toughness, fracture energy, fracture propagation toughness, and average fracture velocity. Therefore, accurate determination of these fracture parameters is crucial for understanding mechanisms of dynamic fracture and is also beneficial for hazard prevention and mitigation.

Most of the existing studies on material fracture have focused on the fracture initiation toughness measurement, mainly under quasi-static loading conditions. Fracture initiation toughness depicts the material resistance to crack reactivation. For brittle materials such as rocks, one can not simply use the standard methods of fracture tests developed for metals. Special sample geometries have been developed for fracture toughness measurements for brittle solids like ceramics and rocks. For example, ISRM recommended two suggested methods with three types of core-based specimens for determining the fracture toughness of rocks: chevron bend (CB) and short rod (SR) specimens in 1988 method (Ouchterlony, 1988) and cracked chevron notched Brazilian disc (CCNBD) specimen in 1995 method (Fowell et al., 1995).

Limited attempts have been made to measure the dynamic initiation fracture toughness of brittle solids, primarily due to the difficulties in experimentation and subsequent data interpretation. As reported in the pioneering work by Böhme and Kalthoff (1982), high loading rate test features significant inertial effect due to stress wave loading and this inertial effect complicates the data reduction. They demonstrated the inertial effect using a three point bending configuration loaded by a drop weight. They showed that the measured crack tip stress intensity factor (SIF) history using the shadow optical method of caustics did not synchronize with the load histories at supports.

Tang and Xu (1990) tried to measure dynamic fracture toughness of rocks using three point impact with a single Hopkinson bar, and Zhang et al. (1999, 2000) employed the split Hopkinson pressure bar (SHPB) technique to measure the dynamic fracture toughness of rocks with SR specimen. In these attempts, the evolution of SIF and the fracture toughness were calculated using quasi-static formulas without careful consideration of the inertial effect. To minimize the error induced by inertial effects, the pulse-shaping technique was employed to conduct dynamic fracture tests with the SHPB (Weerasooriya et al., 2006; Jiang and Vecchio, 2007). The pulse-shaping technique (Frew, Forrestal and Wright, 2001; 2002) facilitates dynamic force equilibrium and thus minimizes inertial effect. The fracture sample is therefore in a quasi-static state of deformation. Indeed, as it was observed by Owen, Zhuang and Rosakis (1998), the SIF value obtained by directly measuring the crack tip opening is consistent with that calculated with the quasi-static equation if the dynamic force balance is roughly achieved in split Hopkinson tension bar tests.

The dynamic fracture energy and the fracture propagation toughness of materials are directly related to the energy consumption during dynamic failures. For transparent polymers or polished metals, those properties could be readily measured with optical methods (Owen, Zhuang and Rosakis, 1998; Xia, Chalivendra and Rosakis, 2006). For rocks, the measurements on these fracture properties are rarely reported in the literature, albeit their direct relevance to the energy consumption during dynamic fracture. To our best knowledge, there is only one report on the dynamic propagation toughness of rocks (Bertram and Kalthoff, 2003).

Recently, a semi-circular bend (SCB) technique in SHPB tests to measure dynamic fracture parameters of rocks was proposed (Chen et al., 2009). Provided that the force balance is achieved with pulse shaping, the initiation fracture toughness can be obtained from the peak load by virtue of static analysis. A laser gap gauge system was developed to measure the crack surface opening displacement history. From this history and the stress wave measurements in the bars, the fracture energy, propagation fracture toughness, and fracture velocity can be determined (Chen et al., 2009).

A fundamental prerequisite for fracture testing via this semi-circular bend specimen is the fabrication of a sharp crack. A 1 mm notch was first made in the semi-circular rock disc (with 40 mm in diameter) and then sharpened with a diamond wire saw to achieve a tip radius of 0.25 mm. For rock with average grain size 0.5 mm or larger, the radius of the tip is smaller than the thickness of naturally formed cracks. This will result in valid fracture toughness measurement. This argument was supported by Lim *et al.* (1994). However, it will be very difficult to make a sharp enough crack tip for fine-grained rocks. To overcome this problem, a convenient way in the fracture test is to employ one of the V shaped (or chevron) notch specimens as suggested by ISRM (Ouchterlony, 1988; Fowell *et al.*, 1995). The V shaped ligament facilitates crack initiation emanating from the notch tip and thus avoids pre-cracking in the brittle solids. Subsequently, the crack propagates in a stable fashion until it reaches the critical crack length where the crack transitions to unstable growth. If the load is static, the load reaches its maximum at this critical crack length while the corresponding SIF has a minimum value. The V notched specimen has been conducted in the SHPB fracture test for rocks (Zhang *et al.*, 1999; 2000), and ceramics (Weerasooriya *et al.*, 2006). Zhang *et al.* (1999; 2000) conducted dynamic SHPB wedge tests using SR rock samples. The quasi-static equation proposed in the ISRM 1988 method was employed to determine the fracture toughness without evaluating the stress state in the sample. Weerasooriya *et al.* (2006) employed a V notched four point bend specimen to measure dynamic initiation toughness of ceramics. They applied the pulse-shaping technique to achieve force balance in the SHPB tests. The time-varying forces on both ends of the sample is almost the same during the loading. They thus concluded that the sample is in a quasi-static loading condition and a quasi-static data reduction is valid. We noticed that in these attempts on the dynamic initiation toughness measurements employing V notched samples (Zhang *et al.*, 1999; 2000; Weerasooriya *et al.*, 2006), no detailed evaluation has been conducted on the measurement principles. In addition, key fracture parameters such as dynamic fracture energy and the fracture propagation toughness were not measured. A thorough investigation on the dynamic fracture test employing the sample with a chevron notch is thus desirable. Among three standard ISRM specimens (Ouchterlony, 1988; Fowell *et al.*, 1995), the CCNBD specimen owns special merits such as: much higher failure load, fewer restrictions on the testing apparatus, larger tolerance on the specimen machining error, simpler testing procedure and lower scatter of test results (Fowell *et al.*, 1995). This CCNBD method has thus been widely used (Dwivedi *et al.*, 2000; Iqbal and Mohanty, 2007).

3.5.2 Semi-circular bend (SCB) method

3.5.2.1 Methodology

Figure 3.20 shows the notched SCB fracture sample sandwiched in the spit Hopkinson pressure bar (SHPB) system and the laser gap gauge (LGG) system.

Based on the ASTM standard E399-06e2 for rectangular three-point bending sample (2002), we propose a similar equation for calculating the stress intensity factor for mode-I fracture in current SCB specimen:

$$K_I(t) = \frac{P(t)S}{BR^{3/2}} \cdot Y\left(\frac{a}{R}\right) \qquad (3.8)$$

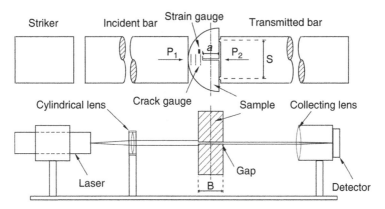

Figure 3.20 Schematics of the straight-through notched SCB specimen in the spit Hopkinson pressure bar (SHPB) system with laser gap gauge (LGG) system.

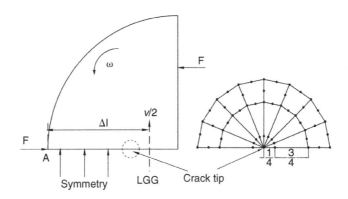

Figure 3.21 Schematic of the configuration for finite element analysis, and the quarter point element on the crack tip.

where $P(t)$ is the time-varying loading force, $Y(a/R)$ is a dimensionless geometry factor, which can be calculated with a standard finite element software package (e.g., ANSYS).

With the pulse-shaping technique, the forces applied on both sides of the sample during our SHPB tests are identical. The inertial effects are eliminated because there is no global force difference in the specimen to induce inertial forces (Weerasooriya *et al.*, 2006). We conduct finite element analysis to relate the far-field loading to the local stress intensity factor at the crack tip for a given specimen geometry. This process is called numerical calibration. Taking advantage of the symmetry of the problem, half-crack model is used to construct the finite element model. PLANE82 (eight-node) element is used in the analysis. To better simulate the stress singularity near the crack tip (r is the radius to the crack tip), 1/4 nodal element (singular element) (Barsoum, 1977) is applied to the vicinity of the crack tip in meshing the finite element model (Fig. 3.21).

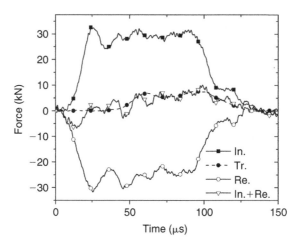

Figure 3.22 Dynamic forces on both ends of the notched SCB specimen tested using a conventional SHPB.

The Young's modulus is 92 GPa and the Poisson's ratio is 0.21 for Laurentian granite used in our simulation (Iqbal and Mohanty, 2006). The load is set as the boundary stresses at the left and right edge of the model plate while the lower edge of the model has the symmetric boundary condition. The resulting loading at the main crack is mode I. For a given load P, K_I can be obtained from the finite element analysis. The geometry factor $Y(a/R)$ for a given sample geometry follows Equation (3.8) as:

$$Y\left(\frac{a}{R}\right) = \frac{K_I B R^{3/2}}{PS} \tag{3.9}$$

After $Y(a/R)$ is numerically calibrated, $K_I(t)$ is directly calculated from Equation (3.8) for the loading history $P(t)$. The dynamic initiation fracture toughness K_{ID} corresponds to the peak point of the loading $P_{\max}$. There is an approximately linear region in $K_I(t)$, and its slope is taken as the fracture loading rate.

In the forgoing discussion, we have assumed that with the dynamic force balance achieved using the pulse-shaping technique, the quasi-static stress analysis is valid and the maximum load corresponding to the failure load. This has to be validated and the validation is covered in the following two sections.

3.5.2.2 SCB test without pulse-shaping

In a conventional SHPB tests, impact of the striker on the incident bar generates a square incident stress wave. The rising portion of the incident wave is too sharp that high frequency oscillations are inevitably introduced. Figure 3.22 shows the forces on both ends of the sample.

From Equation (3.1), the dynamic force on one side of the specimen P_1 is the sum of the incident (In.) and reflected (Re.) waves, and the dynamic force on the other side of the specimen P_2 is the transmitted wave (Tr.). It is evident from Figure 3.22 that a

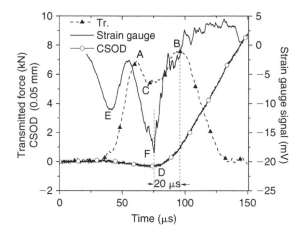

Figure 3.23 Comparison of CSOD and strain gage signal with the transmitted force of the notched SCB specimen tested using a conventional SHPB (the unit for CSOD is 0.05 mm).

large fluctuation of dynamic force occurs on the incident side and a sizeable distinction exists between forces on the two ends of the specimen.

The measured CSOD of the notched SCB specimen by LGG and the transmitted force in a conventional SHPB test are illustrated in Figure 3.23. Two force peaks A and B are identified in the transmitted force signal, occurring at time $62\,\mu s$ and $94\,\mu s$ respectively. More interestingly, over a rather long time period, upon $85\,\mu s$ after the incident stress wave arrives at the sample, the measured CSOD is negative. This means that the crack surface at the measuring site closes rather than opens, which is a manifestation of load inertia effect. The closing of the crack surface may lead to "loss of contact" between the transmitted side of the sample and the two pins (Böhme and Kalthoff, 1982). This explains why after the first peak A of the transmitted force, an obvious unloading is observed (Fig. 3.23). This unloading ends at trough C and then the load continuously rises until the second peak B. From the CSOD signal, we can see that the trough D almost synchronizes with C, indicating the completion of the unloading and the restart of the loading phase.

The signal of the strain gauge mounted on the sample surface is also depicted in Figure 3.23. Two troughs E and F are visible from the strain gauge signal. The first trough E occurs at time $39\,\mu s$ and the second trough F occurs at time $76\,\mu s$. The second trough F is lower and believed to coincide with the fracture initiation time at $76\,\mu s$. Because the peak transmitted force occurs at time $96\,\mu s$, the fracture initiation of the notched SCB sample is thus $20\,\mu s$ ahead of the peak transmitted load. These observations show that due to the inertial effect, the far-field loads on the sample boundary do not synchronize with the local load at the sample crack-tip. This kind of loading inertial effect is similar to what was observed by Böhme and Kalthoff in a different testing configuration (Böhme and Kalthoff, 1982).

Figure 3.24 shows the evolution of SIF from both quasi-static and dynamic data reductions. The static analysis is carried out using the transmitted force on both ends of

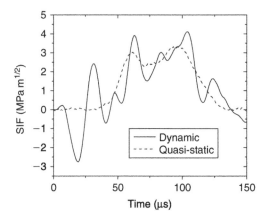

Figure 3.24 The evolution of SIF of the notched SCB specimen tested using a conventional SHPB with both quasi-static analysis and dynamic analysis.

the sample. The overall trends of the two curves match with each other but the dynamic SIF features huge fluctuation. Furthermore, the dynamic SIF is far from linear and therefore it is difficult to achieve a constant loading rate. Consequently, the SIF from the quasi-static data reduction with the far-field load recorded from the transmitted bar cannot reflect the transient SIF history in the notched SCB sample. The usage of the far-field loads such as the transmitted force to obtain the fracture toughness with a quasi-static analysis will lead to tremendous error in the results. The quasi-static equation is not valid for determining fracture toughness in a conventional SHPB test.

3.5.2.3 SCB test with pulse-shaping

A composite pulse shaper (a combination of a C11000 copper and a thin rubber shim) is utilized to shape the loading pulse. In a test with the same speed of striker as the previous case, the incident wave is shaped to a ramp pulse with a rising time of $150\,\mu s$, and a total pulse width of $300\,\mu s$ (Fig. 3.25).

Also shown in Figure 3.25 are the forces on both ends of the specimen. In contrast to Figure 3.22, the forces on the two ends of the specimen exhibit no fluctuation and they are almost identical before the maximum value is reached. The balance of dynamic forces on both ends of the sample is clearly achieved.

Figure 3.26 illustrates the measured CSOD of the notched SCB specimen by LGG and the transmitted force in a modified SHPB test. The measured CSOD is always positive and there is a single peak A in the transmitted force (Fig. 3.26), occurring at time $164\,\mu s$. The phenomenon of crack closing due to inertial effects vanishes completely in this case. Figure 3.26 also shows strain gauge signal mounted on the sample. Only one trough B signal is registered by the stain gauge, occurring at time $160\,\mu s$. Thus, the fracture initiation time is designated by the unique trough B at time $160\,\mu s$. Because the peak transmitted force occurs at time $164\,\mu s$, it is thus only $4\,\mu s$ after the measured fracture onset. We can conclude that in this case, the peak far-field load matches with the fracture onset with negligibly small time difference. The small time

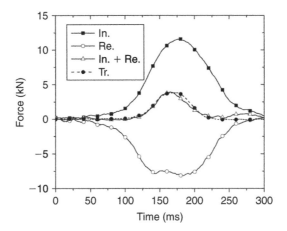

Figure 3.25 Dynamic forces on both ends of the notched SCB specimen tested using a modified SHPB.

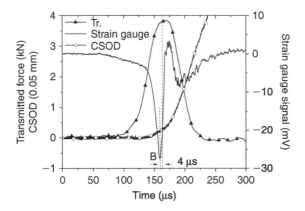

Figure 3.26 Comparison of CSOD and strain gauge signal with the transmitted force of the notched SCB specimen tested using a modified SHPB test (the unit for CSOD is 0.05 mm).

difference between them can be partially interpreted as follows. The load on the specimen increases with the incident pulse before it reaches the peak. At the fracture onset, release waves are emitted from the crack tip at the sound speed of the rock material. The distance between the crack tip and the supporting pin is 12 mm and it thus takes around 2.4 μs for the first release wave to reach the supporting pins. Due to the interaction between the release wave and the pins, the load on the transmitted side decreases (Fig. 3.26). In addition, between 160 μs and 164 μs, the curve of transmitted force is almost flat (Fig. 3.26). The 4 μs time difference will thus lead to negligibly small error in the final result of fracture toughness.

By carefully shaping the loading wave, the dynamic force balance on the boundary of the sample is achieved (Fig. 3.25). However, with a 2D geometric configuration, the force balance on the boundary does not necessarily guarantee the dynamic stress equilibrium in the entire specimen. To address this issue we evaluate the SIF evolution

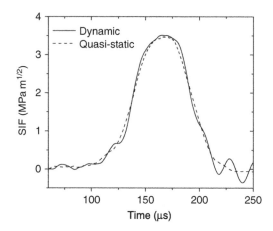

Figure 3.27 The evolution of SIF of the notched SCB specimen tested using a modified SHPB with both quasi-static analysis and dynamic analysis.

by dynamic finite element analysis, and compare the result with that from a quasi-static analysis (Fig. 3.27). The dynamic SIF exhibits no fluctuation at all in contrast to that shown in Figure 3.24. The evolutions of SIF from both static and dynamic methods match reasonably well.

From the above discussion, we have verified that with dynamic force balance in SHPB, the peak far-field load coincides with the fracture onset. The fracture toughness can thus be confidently deduced from the peak far-field load by virtue of quasi-static equations. For the case we examined, the dynamic fracture toughness is 3.47 MPa·m$^{1/2}$, with the loading rate of 79.7 GPa·m$^{1/2}$/s. It is also noted that when there is no pulse shaper, the failure time is at 76 μs. The corresponding dynamic stress intensity factor is 1.5 MPa·m$^{1/2}$ (Fig. 3.23). This value can not be used as the dynamic fracture toughness because it carries significant errors. First, the loading condition is not well defined due to the oscillation of the load. Secondly, the oscillation is due to the dispersion of stress waves in the bar system, and thus it only represents the accurate trend of the dynamic load but not the accurate force at individual measurement points. As a matter of fact, the static fracture toughness of this rock is about 1.5 MPa·m$^{1/2}$ (Nasseri and Mohanty, 2008). The dynamic fracture toughness should be much higher. Hence, we show again that the test without pulse-shaping is not reliable.

3.5.2.4 Fracture energy and propagation toughness

Figure 3.28 shows a typical loading history (P_2) and the corresponding CSOD history during a dynamic SCB fracture test. Because the dynamic force balance is achieved, the peak point of the loading (A) corresponds to the fracture initiation in the specimen, as in a quasi-static experiment. The temporal derivative of the CSOD history is the crack surface opening velocity (CSOV) history. The crack surface opening velocity (CSOV) increases with time and then approaches a terminal velocity of $v = 13.9$ m/s at the turning point B.

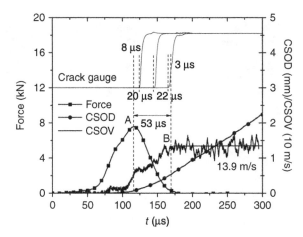

Figure 3.28 Typical loading history and CSOD history of the SCB specimen tested in SHPB. Inset: the crack gauge signals at three locations.

The two vertical lines passing through points A and B divide the whole deformation period into three stages I–III. We believe that in stage I the crack opens up elastically, in stage II the crack propagates dynamically, and in stage III the fracture separates the sample into two pieces and the two fragments rotate away from each other. The separation velocity of the two fragments (normal to the bar axis) is approximately the terminal velocity in CSOV (for small angle of rotation in stage III), and doubles the fragment velocity.

The crack propagation process lasts about $\Delta t_{AB} = 53\,\mu s$ as seen from CSOD and CSOV. Given the crack distance $L_s = R - a = 16\,mm$ for this test (Fig. 3.20), we estimate the average crack growth velocity v_f to be about 300 m/s. We also use crack gauges to estimate the fracture propagation velocity. Three cracks are mounted on the specimen (Fig. 3.20), separated by $\Delta l_1 = 5.36\,mm$ and $\Delta l_2 = 7.81\,mm$. The time separations between the arrivals of the fracture signals are $\Delta t_1 = 20\,\mu s$, and $\Delta t_2 = 22\,\mu s$, respectively. Thus the corresponding fracture velocities are $v_1 = 268\,m/s$, and $v_2 = 355\,m/s$. The fracture velocity appears to increase as the crack propagates during dynamic loading. The first gauge is cemented at about 2 mm away from the crack tip in order to avoid interfering the crack initiation. So there is 8 µs delay between the crack initiation and the breaking of the first crack gauge. The fracture velocity as measured with LGG is consistent with the crack gauge results. One advantage of the LGG is that it is a non-contact method.

A high speed camera (Photron Fastcam SA1) is used to monitor the fracture initiation and propagation process as well as the trajectories of the fragments. The high speed camera is placed perpendicular to the SHPB and specimen. Images are recorded at an inter frame interval of 8 µs; the sequence shown in Figure 3.29 represents only the frames of representative features. The first two images show the pre-fabricated notch and the crack opening can be barely seen. The opening of the SCB crack becomes visible at t > 40 µs. At 80 µs, the SCB specimen is split completely into two fragments.

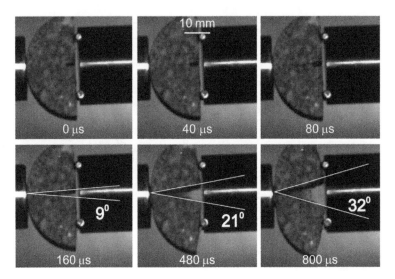

Figure 3.29 Selected high speed camera images showing the fracture and fragmentation of an SCB specimen.

The fragments then rotate about the contact point between the specimen and the incident bar. The rotation angle of the fragment is measured to be 9° at 160 μs, 21° at 480 μs, and 32° at 800 μs. This indicates that the angular velocity of the fragments is almost constant during the period (about 314 rad/s), and the motion of the fragments is rotational.

The high speed camera imaging indicates that the fragments rotate around the axis along the loading point. The LGG system measures CSOD and the fragment angular velocity can be deduced. The linear velocity of the two rotating fragments at the LGG point is approximately the terminal velocity in the CSOV curve (Fig. 3.28). The distance between the LGG and the rotating axis $\Delta l = 18$ mm, so the angular velocity $\omega = v/2/\Delta l = 313$ rad/s for the shot shown in Figure 3.29, in excellent agreement with the result obtained from high speed imaging.

We next use the energy conservation principle to calculate the propagation fracture energy and fracture toughness. A similar method was used by Zhang *et al.* (2000), who used a high-speed camera to estimate the fragment residual velocities. The elastic energy carried by a stress wave is (Song and Chen, 2006):

$$W = \int_0^t E\varepsilon^2 AC \, d\tau \tag{3.10}$$

The total energy absorbed by the specimen then is $\Delta W = W_i - W_r - W_t$, where i, r, and t denote incident, reflected and transmitted wave respectively. Part of the total energy absorbed is used to create new crack surfaces, called the total fracture energy (W_G); the other part remaining in the fragments as the residue kinetic energy (K). That is, $\Delta W = W_G + K$. For the rotating fragments, the moment of inertia is I, and the total

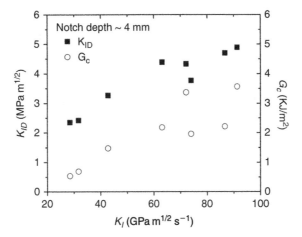

Figure 3.30 The effect of loading rate on the initiation fracture toughness and fracture energy.

rotational kinetic energy is $K = I\omega^2/2$, where the fragment angular velocity ω is estimated from the CSOD data. The average propagation fracture energy is $G_c = W_G/A_c$, where A_c is the area of the crack surfaces created. The average dynamic propagation fracture toughness is:

$$K_{IP} = \sqrt{G_c E/(1 - v^2)} \qquad\qquad (3.11)$$

where E and v are the Young's modulus and Poisson's ratio of the specimen respectively. Here we assume the plain-strain condition.

3.5.2.5   Results

Figure 3.30 shows the measured initiation and propagation fracture toughnesses at different loading rates; both of them increase linearly with increasing loading rates. The propagation fracture toughness also increases with the fracture velocity (Fig. 3.31). At the highest fracture velocity ($\sim$850 m/s), the fracture toughness value is 9.48 MPa m$^{1/2}$, about twice of those at slower fracture velocities near 300 m/s.

3.5.3   Cracked chevron-notched Brazilian disc (CCNBD) method

3.5.3.1   Methodology

Figure 3.32 shows the schematics of the spit Hopkinson pressure bar (SHPB) system and the laser gap gauge (LGG) system. The geometry of the CCNBD sample is shown in Figure 3.33.

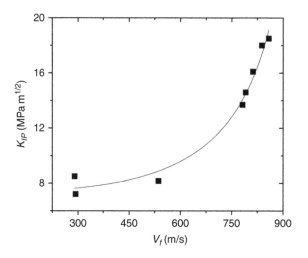

Figure 3.31 The variation of propagation fracture toughness with fracture velocity.

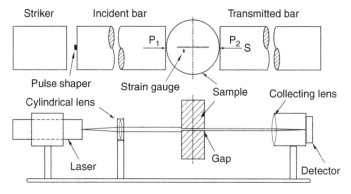

Figure 3.32 Schematics of the spit Hopkinson pressure bar (SHPB) system and the laser gap gauge (LGG) system.

Provided a quasi-static state of the specimen has been achieved during the SHPB test with pulse shaping, the initiation fracture toughness K_{IC} of CCNBD specimen is then determined by the ISRM suggested method (Fowell *et al.*, 1995):

$$K_{IC} = \frac{P_{\max}}{B\sqrt{R}} Y^*_{\min} \tag{3.12}$$

where $P_{\max}$ is the measured maximum load, B and R are the thickness and the radius of the disc respectively, $Y^*_{\min}$ is the minimum value of Y^*, and Y^* is the dimensionless

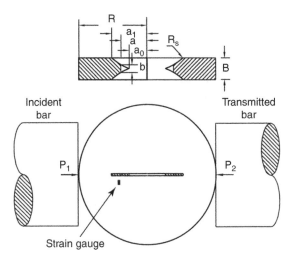

Figure 3.33 The CCNBD specimen in a SHPB system. R = radius of the disc, B = thickness of the disc, R_s = radius of the diamond saw for making notch, a = Length of crack, a_0 = initial half length of chevron notch, a_1 = final half length of chevron notch.

SIF and can be determined in advance by numerical calibrations according to Equation (3.12):

$$Y^* = K_I / \left(\frac{P}{B\sqrt{R}} \right) \qquad (3.13)$$

As a critical factor for determining fracture toughness, Y^*_{min} corresponds to the dimensionless SIF at the critical dimensionless crack length $\alpha_m (\alpha_m = a_m/R$, and a_m is the critical crack length), where the load is maximum.

For a given CCNBD sample configuration, Y^*_{min} can be found from ISRM suggested method (Fowell *et al.*, 1995). However, the corresponding critical dimensionless crack length α_m is not explicitly documented (Fowell, *et al.*, 1995). A commercial finite element analysis software ANSYS is used in this work to determine the critical dimensionless crack length α_m and the corresponding Y^*_{min}.

To achieve accurate SIF values, a sub-modeling technique is adopted to achieve a fine mesh zone around the crack front. A typical sub-modeling sequence is twofold in practice. A full-model, generally with a coarse mesh, is first analyzed. This is followed by analyzing the zone of interest sliced from the full model using a finely meshed sub-model. Sub-modeling is also known as the cut-boundary displacement method or the specified boundary displacement method. The boundary of the sub-model inherits the displacement obtained from the analysis of the full model (Manual, 1999).

Before we calculate the SIF of the CCNBD specimen, the analysis capabilities of the ANSYS sub-modeling technique on three dimensional crack problems are evaluated by several benchmark problems, involving the calculation of SIFs for a penny-shaped crack and an elliptic crack in an infinite domain under remote uniform traction. The

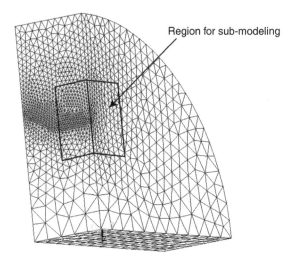

Figure 3.34 Mesh of one eighth of the CCNBD specimen as well as the cut-boundary of the sub-model.

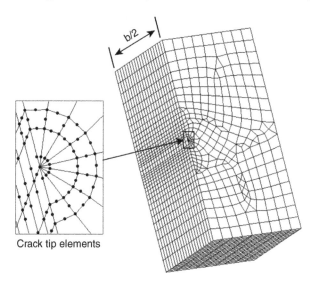

Figure 3.35 Mesh of the sub-model.

results are highly satisfactory, with the maximum error less than 0.4% compared to the theoretical results. We then conduct an elaborate analysis on the CCNBD specimen with the ANSYS sub-modeling technique. Due to symmetry, one eighth of the specimen is first modeled. Solid 92 elements (10-node tetrahedral structural solid) are used in the mesh. The total model is meshed with 34907 elements and 50427 nodes as shown in Figure 3.34.

We then cut a brick from the model enclosing the straight crack front (shown in Fig. 3.35) and analyze it as a sub-model. Solid 95 elements (20-node brick shaped element)

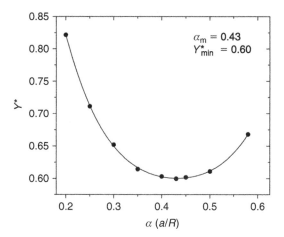

Figure 3.36 The calculated dimensionless SIFs vary with the dimensionless crack length α.

are used. This sub-model is meshed with 6258 elements and 25829 nodes as shown in Figure 3.35. Specifically, to simulate the stress singularity of $r^{-1/2}$ near the crack tip (r is the radius to the crack tip), quarter-nodal elements (Barsoum, 1977) are used to mesh the region adjacent to the crack front. For the CCNBD sample configurations used in this research, the calculated dimensionless SIFs vary with the dimensionless crack length α (Fig. 3.36). Y^*_{min} is found as 0.6 and the corresponding critical dimensionless crack length α_m is 0.43.

Following the same method described in section 5.2.4, we can calculate the fracture energy. The only difference is on the calculation of the residue kinetic energy in the two cracked fragments, K. The kinetic energy K for CCNBD test can be calculated with $K = mv^2/2$, where m is the mass of the specimen, v is the translation velocity of the fragment, which can be deduced from the CSOD history data using with our optical device.

3.5.3.2   Stable-unstable crack propagation transition

We employ pulse shaping technique for all our dynamic CCNBD tests. The dynamic forces on both loading ends of the sample are critically assessed. To compare the dynamic force histories of these two, the time zeros of the incident and reflection stress waves are shifted to the sample-incident bar interface and the time zero of the transmitted stress wave is shifted to the sample-transmitted bar interface invoking 1D stress wave theory. Hereafter, a typical dynamic CCNBD test is shown and discussed. Figure 3.37 compares the time-varying forces on both ends of the sample for this test. The dynamic forces on both sides of the samples are almost identical throughout the dynamic loading period. Obviously, the dynamic forces on both ends of the sample are balanced and the inertial effects are thus eliminated because there is no global force difference in the specimen to induce inertial force.

Figure 3.38 shows the measured CSOD by LGG as well as the strain gauge signal mounted on the sample, compared with the transmitted force (P_2) in the SHPB test.

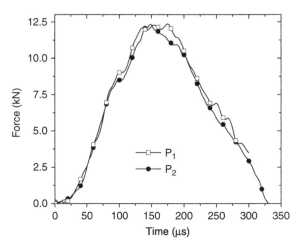

Figure 3.37 Dynamic force balance in a typical CCNBD-SHPB test with pulse shaping.

With dynamic force balance (Fig. 3.37), the transmitted force P_2 can be regarded as the loading to the sample, similar to the quasi-static case. The strain gauge signal of the sample surface is used to detect the fracture initiation and propagation. The fracture initiation from the notch tip will result in a decrease in the strain gauge signal, denoted as point C in Figure 3.38. This fracture initiation coincides with the turning point A in the sustaining load P_2. After this instant, to further drive the propagation of the crack, the load has to increase until the peak point B. At this instant, the crack reaches the critical crack length (with dimensionless crack length α_m) and the unloading starts due to transition of crack growth from stable to unstable. The peak B of P_2 occurs at time 149 µs, 4 µs after the critical crack length is reached as indicated on the strain gauge signal as point D. We believe that the peak of the loading corresponds to the moment the crack reaches the critical crack length. The delay in time between point B and D can be explained in this way. The load on the specimen increases before the propagating crack reaches critical crack length, when the release waves are emitted at the sound speed of the rock material. The distance between crack tip and the transmitted loading end is about 20 mm and it thus takes around 4 µs for the first release wave to reach the transmitted end of the specimen. It is noted also that the measured CSOD curve from the LGG system exhibits an obvious linear segment after point E is reached at 227 µs. The slope of this linear segment indicates constant departure velocity of the two fractured fragments. The point E thus designates the complete separation of the two fragments of the CCNBD specimen.

The dynamic fracture process of the CCNBD specimen in SHPB test can be divided into four stages, separated by three vertical lines through points A, B, and E (denoted by I–IV in Fig. 3.38). The elastic deformation of the CCNBD specimen dominates stage I. At the end of the stage I, the crack initiates from the notch tip, and propagates until the turning point B, when the propagating crack reaches the critical crack length a_m (stage II). We believe that point B designates the transition of stable to unstable crack propagation. During stage II, the crack propagates stably; while in stage III, the crack

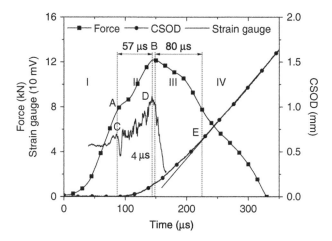

Figure 3.38 LGG measured CSOD and strain gage signal of the CCNBD sample surface, compared with the transmitted force in the SHPB test with pulse shaping.

propagates unstably. Finally, the sample is cracked completely into two half fragments in stage IV flying away from each other.

Both stable and unstable crack propagation velocities can be quantified. For the typical test, the stable crack propagation lasts around $\Delta t_s = 57\,\mu s$ in stage II and the distance of the crack propagation during this stage can be calculated by our finite element analysis: $L_s = a_m - a_0 = 5.1$ mm. The average velocity of the stable crack growth is then $V_s = L_s/\Delta t_s = 89$ m/s. The unstable crack propagation, shown in stage III, lasts around $\Delta t_{us} = 80\,\mu s$. The unstable crack growth distance $L_{us} = R - a_m = 11.4$ mm. The average unstable crack propagation velocity is thus determined as $V_{us} = L_{us}/\Delta t_{us} = 143$ m/s.

3.5.3.3 Results

Figure 3.39 illustrates the measured dynamic mode-I fracture initiation toughness and the average propagation toughness of LG with respect to the loading rates. The fracture loading rate is determined from the slope of the loading curve before fracture initiation. Within the range of loading rates from 30 to 70 GPa m$^{1/2}$ s^{-1}, both toughness values increase almost linearly with increasing loading rates.

Figure 3.40 shows the average stable-unstable fracture velocities with loading rates. The unstable fracture velocity is always larger than the stable fracture velocity for each test (two to three times).

3.5.4 Comparison of dynamic CCNBD results with dynamic SCB results

The measured fracture initiation toughness values from our dynamic CCNBD method are compared with those from dynamic SCB tests in Figure 3.41. The fracture initiation toughness is quite consistent with the measured results by SCB. We are thus confident

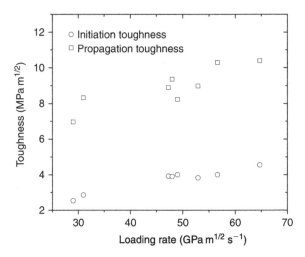

Figure 3.39 The effect of loading rates on the fracture initiation toughness and the average propagation toughness.

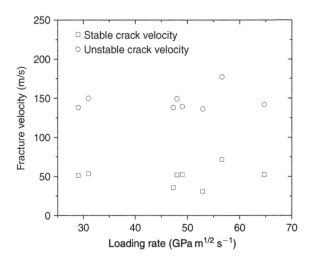

Figure 3.40 The effect of loading rates on the average stable and unstable crack velocities.

that the measured results from dynamic CCNBD tests are reliable. It is noted here that the determination of the loading rate for the CCNBD test is difficult: the loading rates before the transition and after the transition are different. We use the slope of the force after the transition the critical dimensionless SIF Y^* to calculate the loading rate. Cares should be thus taken to apply the dynamic CCNBD results.

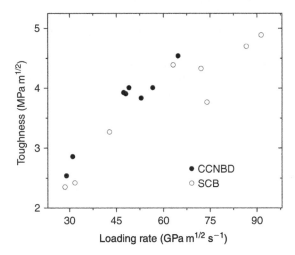

Figure 3.41 Comparison of the initiation toughness from dynamic CCNBD and dynamic SCB methods.

3.6 CONCLUSIONS

This chapter discussed the new advancements in testing techniques using the split Hopkinson pressure bar (SHPB) apparatus. These new techniques made the SHPB suitable for testing rocks and other brittle solids. Because of their small failure strain, the biggest challenge of applying SHPB for testing brittle solids has been to ensure the dynamic force balance during the dynamic loading. The pulse-shaping technique that involves using pulse shaper to slow down the rising of the loading pulse to achieve the dynamic force balance is thus extremely useful. The momentum-trap technique ensures valid post-mortem examination of the tested materials, and thus enables the establishments of the dynamic loading to the damage. Among various optical methods developed for SHPB, the laser gap gauge (LGG) is very useful for monitoring the opening of crack surfaces as demonstrated.

The traditional application of SHPB is the dynamic compression of materials. The choice of sample dimension and reduction of the friction between the sample and the bars are two critical issues in this application. These two problems are coupled in some sense. With the utilization of the new techniques such as pulse-shaping, it is shown in this work that with proper lubrication, the slenderness ratio of the sample can be chosen between 0.5 and 2, without affecting the accuracy of the result. The effect of friction on the measurement is demonstrated using lubricated, dry friction, and bonded bar/sample interfaces.

SHPB can also be used to measure the dynamic tensile strength of rocks using indirect tension methods. The dynamic Brazilian disc (BD) method is adopted in this work. It is demonstrated that the dynamic force balance is a prerequisite for quasi-static data analysis with the aid of numerical simulation and high speed photography. Given the dynamic force balance, the evolution of the stress at the center of the specimen synchronizes to the far-field load; the peak of the load can thus to be used to deduce the nominal tensile strength. In addition, the overshoot of the peak load first suggested for static BD method is worse in the dynamic case. It is necessary to use a strain gauge cemented on the end surface of the BD sample to detect the failure onset.

As a relatively new application, SHPB is applied in this work to measure the dynamic fracture parameters of rocks. Two methods are used to determine the dynamic rock fracture toughness: semi-circular bend (SCB) method and cracked chevron-notched Brazilian disc (CCNBD) method. The two methods give comparable dynamic fracture toughness values; the SCB sample features an unstable crack configuration and thus it is easier to determine the loading rate, while the CCNBD sample has a stable to unstable transition of the fracture propagation and it is difficult to detect the transition point and ambiguity exists on the meaning of the loading rate. Again the dynamic force balance is needed for valid testing condition for both methods. In light of the sample preparation and result interpretation, the dynamic SCB method is advantageous. However, CCNBD method is more applicable for testing fine-grained rocks.

In summary, the SHPB can be used to effectively determine the dynamic mechanical properties of rocks. As a universal prerequisite, the dynamic force balance is needed for all dynamic tests. It is then possible to extend the static tests to their dynamic counterparts. The dynamic compressive test, dynamic BD tension test, dynamic SCB fracture test, and dynamic CCNBD fracture test are used in this work as examples. Some of these testing methods can be potentially adopted by the ISRM as suggested methods.

REFERENCES

ASTM Standard E399-90: Standard test method for plane strain fracture toughness of metallic materials, *Annual book of ASTM Standards*, vol. 03.01. ASTM International, 2002.

Barsoum, R.S.: Triangular quarter-point elements as elastic and perfectly-plastic crack tip elements. *International Journal for Numerical Methods in Engineering* 11 (1977), pp.85–98.

Bell, J.F.: An experimental diffraction grating study of the quasi-static hypothesis of the split Hopkinson bar experiment. *Journal of the Mechanics and Physics of Solids* 14 (1966), pp.309–327.

Bertholf, L.D. and Karnes, C.H.: Two dimensional analysis of the SHPB system. *Journal of the Mechanics and Physics of Solids* 23 (1975), pp.1–19.

Bertram, A. and Kalthoff, J.F.: Crack propagation toughness of rock for the range of low to very high crack speeds. In: Buchholz, F. G., Richard, H. A., and Aliabadi, M. H., (Eds.), *Advances in Fracture and Damage Mechanics Key engineering materials*, vol. 251–252. Trans Tech Publications, Uetikon-Zurich, 2003, pp. 423–430.

Bieniawski, Z.T. and Bernede, M.J.: Suggested methods for determining the uniaxial compressive strength and deformability of rock Materials. *International Journal of Rock Mechanics and Mining Sciences* 16 (1979), pp.137–138.

Bieniawski, Z.T. and Hawkes, I.: Suggested methods for determining tensile strength of rock materials. *International Journal of Rock Mechanics and Mining Sciences* 15 (1978), pp.99–103.

Böhme, W. and Kalthoff, J.F.: The behavior of notched bend specimens in impact testing. *International Journal of Fracture* 20 (1982), pp.R139–R143.

Cai, M., Kaiser, P.K., Suorineni, F. and Su, K.: A study on the dynamic behavior of the Meuse/Haute-Marne argillite. *Physics and Chemistry of the Earth* 32 (2007), pp.907–916.

Chen, R., Xia, K., Dai, F., Lu, F. and Luo, S.N.: Determination of dynamic fracture parameters using a semi-circular bend technique in split Hopkinson pressure bar testing. *Engineering Fracture Mechanics* 76 (2009), pp.1268–1276.

Chen, W. and Ravichandran, G.: An experimental technique for imposing dynamic multiaxial-compression with mechanical confinement. *Experimental Mechanics* 36 (1996), pp.155–158.

Chen, W. and Ravichandran, G.: Failure mode transition in ceramics under dynamic multiaxial compression. *International Journal of Fracture* 101(2000), pp.141–159.

Christensen, R.J., Swanson, S.R. and Brown, W.S.: Split Hopkinson bar tests on rock under confining pressure. *Experimental Mechanics* 12 (1972), pp.508–541.

Dai, F., Xia, K.W., and Tang, L.Z.: Rate dependence of the flexural tensile strength of Laurentian granite. *International Journal of Rock Mechanics and Mining Sciences* 47 (2010), pp.469–475.

Davies, E.D. and Hunter, S.C.: The dynamic compression testing of solids by the method of the split Hopkinson pressure bar. *Journal of the Mechanics and Physics of Solids* 11 (1963), pp.155–179.

Dwivedi, R.D., Soni, A.K., Goel, R.K. and Dube, A.K.: Fracture toughness of rocks under sub-zero temperature conditions. *International Journal of Rock Mechanics and Mining Sciences* 37 (2000), pp.1267–1275.

Field, J.E., Walley, S.M., Proud, W.G., Goldrein, H.T. and Siviour, C.R.: Review of experimental techniques for high rate deformation and shock studies. *International Journal of Impact Engineering* 30 (2004), pp.725–775.

Fowell, R.J., Hudson, J.A., Xu, C. and Chen, J.F.: Suggested method for determining mode-I fracture toughness using cracked chevron-notched Brazilian disc (CCNBD) specimens. *International Journal of Rock Mechanics and Mining Sciences & Geomechanics Abstracts* 32 (1995), pp.57–64.

Frantz, C.E., Follansbee, P.S. and Wright, W.J.: New experimental techniques with the split Hopkinson pressure bar. *8th International Conference on High Energy Rate Fabrication*, San Antonio, 1984, pp. 17–21.

Frew, D.J., Forrestal, M.J. and Chen, W.: A split Hopkinson pressure bar technique to determine compressive stress-strain data for rock materials. *Experimental Mechanics* 41 (2001), pp.40–46.

Frew, D.J., Forrestal, M.J. and Chen, W.: Pulse shaping techniques for testing brittle materials with a split Hopkinson pressure bar. *Experimental Mechanics* 42 (2002), pp.93–106.

Gray, G.T.: Classic split-Hopkinson pressure bar testing, *ASM Handbook Vol 8, Mechanical Testing and Evaluation, vol. 8.* ASM Int, Materials Park, OH, 2000, pp. 462–476.

Hauser, F.E., Simmons, J.A., Dorn, J.E. and Zackay, V.F.: *Response of metals to high velocity deformation.* New York Interscience Pub, New York, 1960.

Hudson, J.A., Rummel, F. and Brown, E.T.: Controlled Failure of Rock Disks and Rings Loaded in Diametral Compression. *International Journal of Rock Mechanics and Mining Sciences* 9 (1972), pp.241–248.

Iqbal, M.J. and Mohanty, B.: Experimental calibration of ISRM suggested fracture toughness measurement techniques in selected brittle rocks. *Rock Mechanics and Rock Engineering* 40 (2007), pp.453–475.

Iqbal, N. and Mohanty, B.: Experimental calibration of stress intensity factors of the ISRM suggested cracked chevron-notched Brazilian disc specimen used for determination of mode-I fracture toughness. *International Journal of Rock Mechanics and Mining Sciences* 43 (2006), pp.1270–1276.

Jiang, F.C., Liu, R.T., Zhang, X.X., Vecchio, K.S. and Rohatgi, A.: Evaluation of dynamic fracture toughness K-Id by Hopkinson pressure bar loaded instrumented Charpy impact test. *Engineering Fracture Mechanics* 71 (2004), pp.279–287.

Jiang, F.C. and Vecchio, K.S.: Experimental investigation of dynamic effects in a two-bar/three-point bend fracture test. *Review of Scientific Instruments* 78 (2007), 063903.

Kolsky, H.: An investigation of the mechanical properties of materials at very high rates of loading. *Proceedings of the Royal Society A-Mathematical Physical and Engineering Sciences* B62 (1949), pp.676–700.

Kolsky, H.: *Stress waves in solids*. Clarendon Press, Oxford, 1953.

Li, X.B., Lok, T.S., Zhao, J. and Zhao, P.J.: Oscillation elimination in the Hopkinson bar apparatus and resultant complete dynamic stress-strain curves for rocks. *International Journal of Rock Mechanics and Mining Sciences* 37 (2000), pp. 1055–1060.

Lim, I.L., Johnston, I.W., Choi, S.K. and Boland, J.N.: Fracture testing of a soft rock with semicircular specimens under 3-point bending. 1. Mode-I. *International Journal of Rock Mechanics and Mining Sciences & Geomechanics Abstracts* 31 (1994), pp.185–197.

Malinowski, J.Z. and Klepaczko, J.R.: A unified analytic and numerical approach to specimen behavior in the split Hopkinson pressure bar. *International Journal of Mechanical Sciences* 28 (1986), pp.381–391.

Manual: *Advanced Analysis Techniques Guide*, ANSYS Inc., 1999.

Mellor, M. and Hawkes, I.: Measurement of Tensile Strength by Diametral Compression of Discs and Annuli. *Engineering Geology* 5 (1971), pp.173–225.

Meng, H. and Li, Q.M.: Correlation between the accuracy of a SHPB test and the stress uniformity based on numerical experiments. *International Journal of Impact Engineering* 28 (2003), pp.537–555.

Narayanasamy, R. and Pandey, K.S.: Phenomenon of barrelling in aluminium solid cylinders during cold upset-forming. *Journal of Materials Processing Technology* 70 (1997), pp.17–21.

Nasseri, M.H.B. and Mohanty, B.: Fracture toughness anisotropy in granitic rocks. *International Journal of Rock Mechanics and Mining Sciences* 45 (2008), pp.167–193.

Nasseri, M.H.B., Mohanty, B. and Robin, P.Y.F.: Characterization of microstructures and fracture toughness in five granitic rocks. *International Journal of Rock Mechanics and Mining Sciences* 42 (2005), pp.450–460.

Nemat-Nasser, S., Isaacs, J.B. and Starrett, J.E.: Hopkinson techniques for dynamic recovery experiments. *Proceedings of the Royal Society A-Mathematical Physical and Engineering Sciences* 435 (1991), pp.371–391.

Ouchterlony, F.: Suggested methods for determining the fracture toughness of rock. *International Journal of Rock Mechanics and Mining Sciences & Geomechanics Abstracts* 25 (1988), pp.71–96.

Owen, D.M., Zhuang, S., Rosakis, A.J. and Ravichandran, G.: Experimental determination of dynamic crack initiation and propagation fracture toughness in thin aluminum sheets. *International Journal of Fracture* 90 (1998), pp.153–174.

Ross, C.A., Tedesco, J.W. and Kuennen, S.T.: Effects of Strain-Rate on Concrete Strength. *Aci Materials Journal* 92 (1995), pp.37–47.

Ross, C.A., Thompson, P.Y. and Tedesco, J.W.: Split-Hopkinson Pressure-Bar Tests on Concrete and Mortar in Tension and Compression. *Aci Materials Journal* 86 (1989), pp.475–481.

Schey, J.A., Venner, T.R. and Takomana, S.L.: The Effect of Friction on Pressure in Upsetting at Low Diameter-to-Height Ratios. *Journal of Mechanical Working Technology* 6 (1982), pp.23–33.

Song, B. and Chen, W.: Loading and unloading split Hopkinson pressure bar pulse-shaping techniques for dynamic hysteretic loops. *Experimental Mechanics* 44 (2004), pp.622–627.

Song, B. and Chen, W.: Energy for specimen deformation in a split Hopkinson pressure bar experiment. *Experimental Mechanics* 46 (2006), pp.407–410.

Subhash, G., Ravichandran, G. and Gray, G.T.: Split-Hopkinson pressure bar testing of ceramics, *ASM Handbook Vol 8, Mechanical Testing and Evaluation*, vol. 8. ASM Int, Materials Park, OH, 2000, pp.1114–1134.

Tang, C.N. and Xu, X.H.: A new method for measuring dynamic fracture toughness of rock. *Engineering Fracture Mechanics* 35 (1990), pp.783–789.

Wang, Q.Z., Li, W. and Song, X.L.: A method for testing dynamic tensile strength and elastic modulus of rock materials using SHPB. *Pure and Applied Geophysics* 163 (2006), pp.1091–1100.

Weerasooriya, T., Moy, P., Casem, D., Cheng, M. and Chen, W.: A four-point bend technique to determine dynamic fracture toughness of ceramics. *Journal of the American Ceramic Society* 89 (2006), pp.990–995.

Xia, K., Chalivendra, V.B. and Rosakis, A.J.: Observing ideal "self-similar" crack growth in experiments. *Engineering Fracture Mechanics* 73 (2006), pp.2748–2755.

Xia, K., Nasseri, M.H.B., Mohanty, B., Lu, F., Chen, R. and Luo, S.N.: Effects of microstructures on dynamic compression of Barre granite. *International Journal of Rock Mechanics and Mining Sciences* 45 (2008), pp.879–887.

Zhang, Z.X., Kou, S.Q., Jiang, L.G. and Lindqvist, P.A.: Effects of loading rate on rock fracture: fracture characteristics and energy partitioning. *International Journal of Rock Mechanics and Mining Sciences* 37 (2000), pp.745–762.

Zhang, Z.X., Kou, S.Q., Yu, J., Yu, Y., Jiang, L.G. and Lindqvist, P.A.: Effects of loading rate on rock fracture. *International Journal of Rock Mechanics and Mining Sciences* 36 (1999), pp.597–611.

Zhao, J. and Li, H.B.: Experimental determination of dynamic tensile properties of a granite. *International Journal of Rock Mechanics and Mining Sciences* 37 (2000), pp.861–866.

Chapter 4

Modified Hopkinson bar technologies applied to the high strain rate rock tests

Ezio Cadoni and Carlo Albertini

4.1 INTRODUCTION

Rock excavations in tunnel construction or mining works are usually performed by means of the action of pressure waves generated from controlled explosions and by impact loading from excavation machines. As a consequence, the pressure wave amplitude and duration are important. Firstly, for economic reasons, we need to obtain the desired effects with the lowest costs. Secondly, for safety measures, risks from heavy impact or vibrations especially in urban underground excavations should be avoided. The optimised choice of the impact loading parameters of the excavation machines are important, and also for the same reasons, the proper choice and design of the excavation machines.

No matter which calculation method is employed, either analytical or numerical, the very basic data needed by the engineer for the above mentioned optimisations will be:

a) The precise values of the amplitude, duration and shape of the pressure pulse acting on the rock to simulate the type, weight and shape of the explosive charge or the impact loading parameters of the excavation machine.
b) The complete stress-strain curve until fracture of the rock material to be excavated. It should be precisely measured at the strain rate imposed by the acting pressure pulse by means of a special dynamic material testing apparatus, possibly taking into account also the effect on the dynamic stress-strain curves under the in situ stresses during real excavation works.

The theme of the present chapter is the presentation of some precision measurement methodologies of the acting pressure pulses and of the rock stress-strain curves described at the above points.

The impact pressure pulses generated by the charge explosion or by the specialised excavation machine, act on the rock/soil through stress wave propagation, which is the most important phenomenon to be taken into account for achieving a precise measurement of the parameters of the generated pressure pulses. The impact load generated by the dynamic material testing apparatus during the high strain rate test acts also on the rock specimen through stress wave propagation. As a consequence, the precise measurement of the stress-strain curve of the rock/soil specimens at high strain rate should be performed by means of testing devices. Those devices will have controlled stress wave propagation so that it can be precisely analysed by means of the one-dimensional elastic wave propagation theory.

The pioneering studies by Hopkinson, father (1872), Hopkinson, son (1914) in the early 20th century and by Davies (1948) in the mid 20th century, introduced a method similar to the bar technique adapted to modern instrumentation. It is scientifically the most recognized methodology for the precise measurement of pressure pulse parameters and dynamic material properties. It allows the recording of the pressure pulses under wave propagation control, avoiding the complications introduced by wave reflections and superposition. Also, it allows the record analysis by means of the well proofed one-dimensional elastic wave propagation theory (Davies, 1948; Kolsky, 1953). The Hopkinson bar method requires the generation of a stress wave pulse well controlled in amplitude and duration which, by means of an elastic bar system, is propagated without dispersion and uncontrolled reflections to load and deform until fracture of a specimen. Furthermore, in the Hopkinson bar method, the specimen should have a gauge length allowing a state of stress homogeneity to be reached along the gauge length at the early stages of the deformation. It means that in a short time with respect to the duration of the test, the state of stress homogeneity is obtained by means of the stress waves propagating forwards and backwards inside the specimen. The gauge length of the specimen is kept sufficiently short.

In the analysis of the Hopkinson bar, the propagation of the stress waves without dispersion and uncontrolled reflections, and the deformation of the specimens in a state of stress homogeneity, are the basic conditions to be satisfied for a correct implementation of the one-dimensional elastic stress wave propagation theory. Therefore accurate measurements are needed to determine the dynamic mechanical properties of materials. However, the basic principles of the Hopkinson bar method are normally not respected, especially in the impact rigs based on the use of drop weight or missiles. In fact they impinge directly on the specimen generating an impact load and use load cells as load transducers that are in contact with the specimen. The material properties under impact loading measured with drop weight impact rigs are of low accuracy. Therefore they are not reliable for the development and calibration of material constitutive laws, due to many noisy phenomena like resonant vibrations, rebounds, superposition of waves, and reflections (Birch, Jones and Jouri, 1988). The low accuracy of load and displacement measurements performed with drop weight impact rigs has been clearly shown by a benchmark exercise where the European Commission (EC) Joint Research Centre (JRC) and 14 European Laboratories performed the measurement of the load–displacement curves of equal structural components using respectively a large Hopkinson bar and drop weight/horizontal sledge impact rigs (Albertini, Hanefi and Wierzbicki, 1995; Albertini, Solomos and Labibes, 1998).

In this benchmark exercise, because of the earlier mentioned noisy phenomena which obliged the application of questionable filtering processes to the experimental records, the drop weight impact rigs were practically unable to give reliable measurements of the load-displacement curves. They show a large spread of the results especially in the first part of the load-displacement curves where unrealistic large load peaks were recorded (Albertini, Hanefi and Wierzbicki, 1995; Albertuni, Cadoni and Labibes, 1997; Albertini, Solomos and Labibes, 1998). These initial "false" load peaks present in the records of drop weight impact rigs will particularly affect the accuracy of the impact strength measurements of concrete, reinforced concrete, and geological

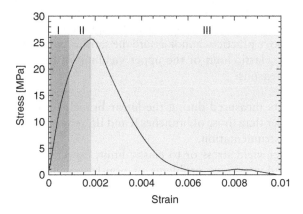

Figure 4.1 Dynamic stress versus strain curve of a rock specimen in tension (Cadoni, 2010).

material specimens. In fact, the resulting stress-strain curves will be strongly affected by the existence of the initial false load peak because its duration covers all the deformation phase (generally of low value) of these materials.

The large Hopkinson bar experimental records, taken during the benchmark exercise (Albertini, Hanefi and Wierzbicki, 1995; Albertini, Solomos and Labibes, 1998) were very clean. They do not require any filtering process before being analysed by means of the well proofed one-dimensional elastic wave propagation theory. No unrealistic first load peaks affected the records. The benchmark exercise demonstrated also the usefulness of the Hopkinson bar technique for the precise measurement of the amplitude, shape and duration of pressure pulses.

The Hopkinson bar technique has been modified during more than three decades in EC-JRC Ispra, Italy. It is intended to be applied successfully to a large range of materials extending from very ductile materials like metals to relatively fragile materials like concrete and ceramics.

The typical dynamic stress-strain curve of a rock specimen is shown in Figure 4.1, where it can be distinguished by three main branches:

i) a linearly increasing branch, which can be considered as elastic, characterised by the elastic modulus and by the elastic limit,
ii) a strain hardening branch, where stress increases non-linearly with strain up to the UTS and,
iii) a softening branch, where load decreases non-linearly with displacement, corresponding to the fracture propagation through the specimen cross-section.

The three above branches of the dynamic stress-strain curves are measured by means of a unique dynamic tensile test, extended up to fracture of the specimen with the Hopkinson bar at high strain rate. The Hopkinson bar governs the problem of stress wave propagation connected with the high strain rate testing in a way (exposed in more detail later) to lead back the dynamic test to a quasi-static test by means of numerous stress wave reflections inside the specimen, creating the need for homogeneous stress distribution along the specimen gauge length. This practice assures good measurement accuracy for the non-linear deformation phases II and III because they are characterised by displacements and loads of the same order of magnitude.

However, the above practice cannot assure the same accuracy for the branch I (the elastic one up to the elastic limit or the upper yielding) of the dynamic stress-strain curve for two main reasons:

- the displacements measured during the linear branch I are up to two orders of magnitude smaller than those of branches II and III, requiring specific displacement measurement instrumentation,
- the time to upper yield stress or to elastic limit, especially in high loading rate, is very short and may give problems of non-homogeneous stress distribution along the specimen gauge length which must be taken into account when defining material properties.

In the following sections, after a description of the principles of the JRC Modified Hopkinson Bar (MHB) and of its performance in testing plain concrete, some application proposals of the JRC-MHB for the dynamic mechanical characterisation of rocks, and of the acting pressure pulses in the case of excavation works, will be presented.

4.2 PRINCIPLES AND FUNCTIONING OF THE JRC-MHB WITH QUASI-STATICALLY PRE-STRESSED LOADING BAR

An innovative version of the modified Hopkinson bar (MHB) has been developed during the last three decades at EC-JRC. The JRC-MHB is capable of performing impact precision tests in tension, compression, bending and shear using the same supporting structures and the same measuring instrumentation.

The classical Hopkinson bar normally works only in compression and consists principally of, as shown in Figure 4.2, a projectile in order to generate a rectangular impact loading pulse by impinging on an input bar which transmits the load to a specimen inserted between the input and output bars (Davies, 1948; Kolsky, 1953). The main modification of the classical Hopkinson bar introduced at JRC (Albertini and Montagnani, 1974, 1977; Albertini and Labibes, 1997) consists in the substitution of the projectile, normally used to generate the impact loading pulse, with a statically

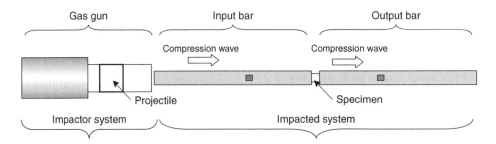

Figure 4.2 Traditional SHPB for rock specimen in compression (Cadoni, 2010).

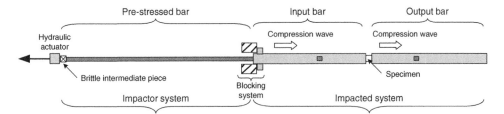

Figure 4.3 Modified SHPB for rock specimen in compression (Albertini and Labibes, 1997).

elastic pre-stressed bar which is the physical continuation of the input bar as shown in Figure 4.3.

The method of functioning the JRC modified Hopkinson bar in compression (Fig. 4.3) consists of the following phases:

Elastic energy is stored in the pre-stressed loading bar by statically tensioning the length of this bar between a blocking ring and a brittle intermediate piece connected to the hydraulic actuator.

A rectangular stress wave pulse is generated by suddenly breaking the brittle intermediate piece and propagates through the input bar, the specimen and the output bar, provoking a state of compressive stress in the specimen because the particles move from left to right.

Records are taken by the strain-gauge stations glued on the input and output bars of the elastic deformation ε_i provoked by the incident pulse propagating in the input bar, the elastic deformation ε_r provoked in the input bar by the part of the incident pulse reflected at the interface of input bar-specimen and the elastic deformation ε_t provoked in the output bar by the pulse transmitted through the specimen.

By applying the elastic one-dimensional stress wave propagation theory at the Hopkinson bar system, the forces F_1 and F_2 and the displacements D_1 and D_2 acting on the two faces of the specimen in contact with the input and output bars can be calculated, following the relationships below, using the recorded deformations ε_i, ε_r and ε_t of the elastic input and output bars:

$$F_1 = EA \left(\varepsilon_i + \varepsilon_r \right) \tag{4.1}$$

$$F_2 = EA \, \varepsilon_t \tag{4.2}$$

$$D_1 = C_0 \int \left(\varepsilon_i - \varepsilon_r \right) dt \tag{4.3}$$

$$D_2 = C_0 \int \varepsilon_t \, dt \tag{4.4}$$

Having realised the condition of the specimen deformation in a homogeneous stress state, the average stress σ, strain ε and strain rate $\dot{\varepsilon}$ in the specimen material can be determined with the following relationships:

$$\sigma = \frac{F_1 + F_2}{2A_0} = \frac{1}{2} E \frac{A}{A_0} \left(\varepsilon_i + \varepsilon_r + \varepsilon_t \right) \cong E \frac{A}{A_0} \varepsilon_t \tag{4.5}$$

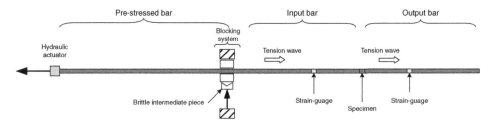

Figure 4.4 Modified SHPB for rock specimen in tension (Albertini and Montagnani, 1974, 1977).

$$\varepsilon = \frac{D_1 - D_2}{L} = \frac{C_0}{L}\int_0^T (\varepsilon_i - \varepsilon_r - \varepsilon_t)dt \cong \frac{2C_0}{L}\int_0^T \varepsilon_t\, dt \qquad (4.6)$$

$$\dot{\varepsilon} = \frac{C_0}{L}(\varepsilon_i - \varepsilon_r - \varepsilon_t) \cong \frac{2C_0}{L}\varepsilon_r \qquad (4.7)$$

where: L = gauge length of the specimen, A_0 = cross-sectional area of the specimen, T = test time, E = elastic modulus of the bar, A = cross-sectional area of the bar and C_0 = the elastic wave speed in the bars.

The approximations on the right hand side of the above equations (4.5), (4.6), (4.7) are based on the approximate equivalence of the forces on both faces of the specimen, so that:

$$F_1 \cong F_2 \qquad (4.8)$$

or

$$\varepsilon_i + \varepsilon_r \cong \varepsilon_t \qquad (4.9)$$

This approximation is valid for short specimen gauge length as required for having a homogeneous stress distribution in the specimen gauge length. The validity of this approximation can be checked experimentally by directly summing the indicated signal records.

In Figure 4.4, the scheme of the MHB (Albertini and Montagnani, 1974, 1977) has been shown as suitable for tensile tests. The tension device works in a similar way as the compression device described above. In this case, the tensile stress state of the specimen is obtained by placing the brittle intermediate piece in a device blocking the point of the bar which determines the separation between the pre-stressed bar and the input bar. In the physical reality, the pre-stressed bar and the input bar are a unique continuous bar which for a certain part of its length works as a pre-stressed bar. The remaining part works as an input bar (the functional subdivision is determined by the position of the blocking ring in the compression version of Figure 4.3 and by the position of the brittle intermediate piece in the tension version of Figure 4.4. By pulling the pre-stressed bar with the hydraulic actuator (Fig. 4.4) until the desired amplitude of the stored stress state is reached, and by successively fracturing the brittle intermediate piece, a stress wave pulse is generated propagating in the system that consists of the input bar, the specimen and the output bar. In this case, the input bar particle displacement is from right to left inducing a tensile stress state in the specimen.

The main difference of the JRC tension and compression MHB with respect to the classical bar lies mainly in the fact that the generation of the loading stress pulse is performed by means of a pre-stressed bar which is the physical continuation of the input bar instead of the launching of a projectile. This different way of generation of the loading pulse has allowed the generation of very long loading pulses by simply increasing the length of the pre-stressed bar. Loading pulses of 40 millisecond duration have been obtained by 100 m length of the pre-stressed bar (Albertini, Boone and Montagnani, 1985). If the projectile technique would have been used, it would have been necessary to launch a 100 m long projectile, and it would have been a very difficult and expensive solution. The generation of long duration loading pulses has been requested for testing very ductile materials and structural components.

The impact testing with the classical Hopkinson bar of low elongation materials like concrete and rocks requires a perfect plane impinging of the projectile on the input bar in order to have a clean record of the short duration mechanical response of the specimen. This condition is practically impossible to be realised with the projectile launching in engineering test conditions. However, in the JRC MHB, the loading pulse passes smoothly and in plane from the pre-stressed bar to the input bar because the two bars are the physical continuation of each other. This process assures the rising and the first part of the generated pulse remaining unspoiled from the oscillations generated by the imperfect projectile impinging. It is necessary to have precise measurements of the dynamic mechanical properties of the rock specimens.

4.3 TENSILE AND COMPRESSIVE IMPACT TESTS OF PLAIN CONCRETE WITH THE JRC-MHB

The two basic schemes of the JRC-MHB for tests in compression and tension, as shown in Figures 4.3 and 4.4, and described in Section 2, have been successfully applied to impact testing of plain concrete material of two cubic specimens with side length of 6 cm and 20 cm, respectively. The impact tensile tests of plain concrete cubes of 60 mm side are shown in Figure 4.5. The input and output aluminium bars having square cross-section of 60 mm side were constructed for better matching the mechanical impedances. A too large mismatch of the impedances would provoke a complete reflection of the incident wave at the input bar - specimen interface without any loading on the specimen.

Furthermore, the input and output aluminium bars were constructively subdivided by EDM (Electro-Discharging-Machine) into two bundles of input and output Hopkinson bars. The aim of this subdivision in two bundles was the measurement of the distribution of stress and strain over the cross-section of the specimen during both the ascending branch of the stress-strain curve and the softening branch characterized by the fracture initiation in a limited area of the specimen cross-section and then the propagation through the whole specimen cross section (Albertini, Cadoni and Labibes, 1997; 1999). The pre-stressed loading bar for practical reasons has been constructed

(a)   (b)

Figure 4.5 JRC-MHB for concrete specimen in tension (Albertini, Cadoni and Labibes, 1997).

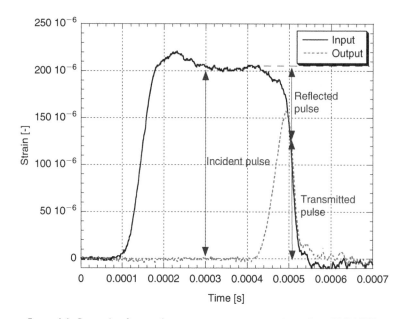

Figure 4.6 Records of a tensile test on a concrete specimen by a JRC-MHB.

in steel with a dimension assuring the mechanical impedance continuity between the steel pre-stressed bar and the aluminium input bar so that the generated stress wave pulse enters the input bar without any reflection.

The functioning of the device in Figure 4.5 is the same as that in Figure 4.4. The pre-stressed loading bar is quasi-statically pulled by the hydraulic actuator until the sudden fracture of the brittle intermediate piece generates stress wave pulses propagating along the input bar to the bar-specimen interface where some of them are reflected and some are transmitted through the specimen (deforming it until fracture) successively propagating along the output bar. The records of an impact test obtained from one bar

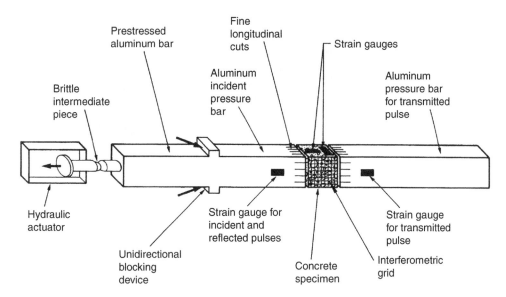

Figure 4.7 JRC-MHB for dynamic compression testing of concrete or rocks.

of the bundle of the device in Figure 4.5 are shown in Figure 4.6. The incident, reflected and transmitted pulses are clearly resolved. The application of the relationships (4.5) to (4.7) to the records of all the bars of the bundle allows the precise determination of the stress-strain curves of the plain concrete until fracture (Cadoni *et al.*, 1997, 2000; Albertini, Cadoni and Labibes, 1997, 1999).

The scheme in Figure 4.7 shows the Aluminium Hopkinson bar setup realised for testing the 6 cm cubes under compressive impact loads. The way of functioning is the same as described above for the general scheme in Figure 4.3.

A large MHB of high loading capacity (5 MN) is available at the EC-JRC Ispra, known by the name LDTF (Large Dynamic Testing Facility). It was originally designed for testing large steel specimens which need high dynamic load and large displacements for impact testing to failure. After a proper adaptation, it allows the tensile impact testing of large concrete specimens (cubes of 20 cm side) to study the effects of large aggregate size (similar to those used in real constructions) and of specimen size. The adaptation of the LDTF to impact testing of large plain concrete specimens has been realised by inserting a long (100 m) pre-stressed steel cable of LDTF, an input and an output aluminium Hopkinson bar of square cross sections of 20 cm side as shown in the scheme of Figure 4.8 (Cadoni *et al.*, 1997, 2000, 2001a, 2009). The input and output bars were subdivided each one in a bundle of 25 bars individually instrumented. This large device functions as follows:

–   the long steel cable has been pre-stressed;
–   an explosive bolt, inserted between the steel cable and the input aluminium bar, has been exploded in order to generate a stress wave pulse which shall propagate

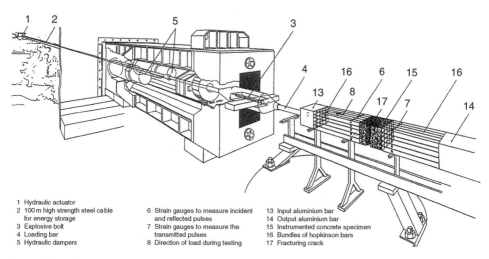

1 Hydraulic actuator
2 100 m high strength steel cable
  for energy storage
3 Explosive bolt
4 Loading bar
5 Hydraulic dampers

6 Strain gauges to measure incident
  and reflected pulses
7 Strain gauges to measure the
  transmitted pulses
8 Direction of load during testing

13 Input aluminium bar
14 Output aluminium bar
15 Instrumented concrete specimen
16 Bundles of hopkinson bars
17 Fracturing crack

Figure 4.8 JRC-Hopkinson Bar Bundle set-up for impact testing of large plain concrete specimen in tension.

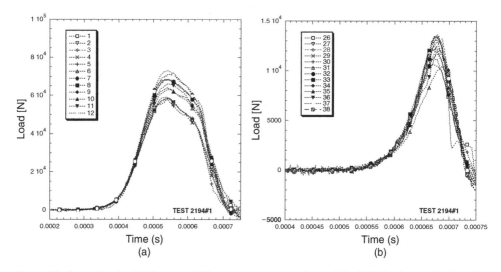

Figure 4.9 Some signals of (a) input and (b) output elementary bars of the JRC-Hopkinson Bar Bundle.

in the aluminium Hopkinson bar set-up and shall deform the concrete specimen until fracture;

– each one of the 25 Hopkinson input and output bars recorded the incident, reflected and transmitted pulses correlated to the portion of specimen cross section in front of the bar; Figure 4.9 (Cadoni *et al.*, 1997) shows the records of the pulses of one test;

– the application of the relationships (4.5) to (4.7) to the clear records like those of Figure 4.9 allow study of the stress-strain distribution over the cross section

9.97	10.75	10.35	9.36	8.30		6.13	6.61	6.85	6.68	5.79
8.22	10.25	10.50	9.78	9.17		5.00	6.11	6.31	6.26	5.86
9.05	9.85	9.46	9.87	9.23		4.60	5.52	5.44	5.80	5.72
8.04	8.20	8.99	9.40	7.87		4.19	5.02	5.01	5.34	4.51
4.19	7.33	9.49	9.64	9.80		4.24	4.61	4.99	5.13	4.95

TEST 2174#1 $\sigma_{av} = 9.09$ MPa;   TEST 2174#6 $\sigma_{av} = 5.47$ MPa;
L = 200 mm $\sigma_0 = 9.06$ MPa   L = 200 mm $\sigma_0 = 5.48$ MPa

Figure 4.10 Distribution of the maximum stress over the specimen cross-section measured by each bar of the JRC-Hopkinson Bar Bundle (Cadoni et al., 1997).

of the specimen both during the ascending branch and the descending branch of the stress-strain curve (Albertini, Cadoni and Labibes, 1999; Cadoni *et al.*, 2000). For example, Figure 4.10 shows the distribution of the stress over the specimen cross-section in correspondence with the maximum of the stress-strain curve. Some results concerning the effects of high strain rate on the mechanical properties of small and large specimens can be found in the work of Cadoni *et al.* (2000, 2001a, 2009). Also, the fracture initiation and propagation through the plain concrete specimen have been recorded with precision using the bundle Hopkinson bar as shown in Figure 4.11 (Cadoni *et al.*, 2000, 2001a, 2009).

–   The investigations conducted with the large aluminium Hopkinson bar installed in the LDTF were also extended to the study of the influence of humidity on the dynamic mechanical properties of plain concrete (Cadoni *et al.*, 2001b).

It is evident that the high loading capacity of the LDTF was needed for the impact testing of large concrete specimens of 20 cm side but that the 100 m length of the pre-stressed cable was oversized for this type of test because of the very small displacement needed to fracture plain concrete specimens. In fact, a pre-stressed bar of a couple of meters long would have been sufficient to bring to fracture the large plain concrete specimens.

In fact, more recently, an MHB which is more adequate for testing large concrete specimens has been constructed at JRC following the scheme sketched in Figure 4.12, where the pre-stressed bar, the input bar and the output bar are 2 m long. The total length of the equipment is about 8 m. This equipment has been used for testing large concrete specimens in compression (Cadoni *et al.*, 2009).

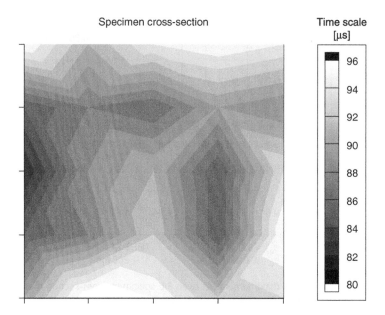

Specimen cross-section

Time scale
[µs]

Figure 4.11   Crack propagation of the specimen with $d_{max} = 25\,mm$ (strain rate = $10\,s^{-1}$).

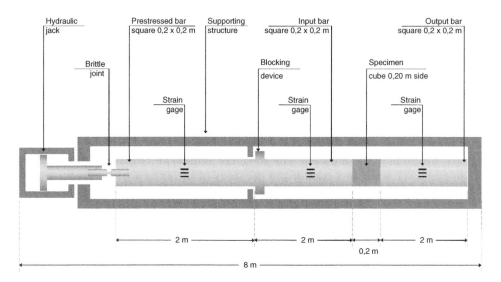

Figure 4.12   Scheme of the JRC-MHB for compressive impact test of large concrete specimens.

4.3.1   Special tensile tests with the **MHB** for high resolution measurement of elastic limit and elastic modulus of rocks

The tests at high strain rate are performed by means of the MHB. In the following section, we will concentrate on the definition of the optimised test conditions for

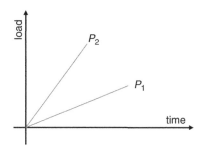

Figure 4.13 Representation of two loading rates $\dot{P}_1$ and $\dot{P}_2$.

the measurement of the mechanical properties of rocks at high strain rate using the MHB, in particular the measurement of the elastic modulus and the elastic limit, whose accurate measurement poses the major difficulties.

Whatever precise deformation mechanism is involved, the elastic limit of the linear branch of the stress-strain curve of a rock material corresponds to the transition from predominantly linear elastic to predominantly non-linear plastic deformation mode. The overall elastic strain rate just before the elastic limit and the nominal plastic strain rate just after the elastic limit should be nearly equal. However at the elastic limit, while the elastic strain will and must be homogeneously distributed over the specimen gauge length, the initial non-linear plastic strain will be confined to one or more restricted regions within the specimen gauge length. It takes some time to spread over the whole gauge length, with the consequence that a definition of an overall plastic strain rate in proximity of the elastic limit is impossible. In fact, in the regions of localised non-linear plastic strain of the specimen gauge length, the plastic strain rate will be very high while outside it will be very low, arriving to a stable value only after some time.

Therefore we can define the strain rate as follows:

$$\dot{\varepsilon}_E = \frac{1}{E}\frac{d\sigma}{dt} \tag{4.10}$$

obtained on the basis of the Hooke's law valid in the linear elastic field.

Considering that during the elastic straining phase the specimen cross-section (A) remains, with good approximations, constant, then Equation (4.10) can be written as:

$$\dot{\varepsilon}_E = \frac{1}{E}\frac{d\frac{P}{A}}{dt} = \frac{1}{EA}\frac{dP}{dt} \tag{4.11}$$

From Equation (4.11) we see that in order to obtain a constant strain rate in the linear elastic deformation phase, it is necessary to apply a loading pulse $\dot{P}(t)$ characterised by a constant loading rate. This means that the load P applied to the specimen will linearly increase with the time t as shown in Figure 4.13.

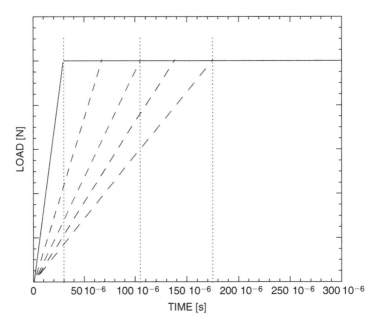

Figure 4.14 MHB loading pulse with different linear rise times.

An average value of the elastic strain rate over the time lap needed to reach the elastic limit can be defined as:

$$\dot{\varepsilon}_E = \frac{Y_D}{E \cdot t_y} \tag{4.12}$$

where Y_D = elastic limit, E = elastic Young's modulus, t_y = time to reach the elastic limit measured on the record of the specimen response.

As shown by Equations (4.10) or (4.11), for the accurate determination of the elastic modulus and the elastic limit of rock specimens at high constant elastic strain rates, it is necessary to apply linearly increasing loading pulses to the specimen, as shown in Figure 4.13.

The actual version of the MHB is designed for the generation of a loading pulse having the shape represented by the full line pulse in Figure 4.14. The load linearly increases in a time of about 30 μs to an amplitude A and then remains constant for a time T corresponding to the travel time of the elastic wave along the double length of the pre-tensioned bar. In order to change the value of the rise time of the linearly increasing loading pulse so as to perform tests at different elastic strain rates, we have to introduce in the MHB design an element to generate a load linearly increasing, with rise times larger than the actual ones as represented by the dashed lines in Figure 4.14. This element, called "pulse shaper", allowing the increase of the rise time, is sketched in two different versions in Figures 4.15a and 4.15b. Each version of the pulse shaper is intended to be inserted along the incident bar of the MHB.

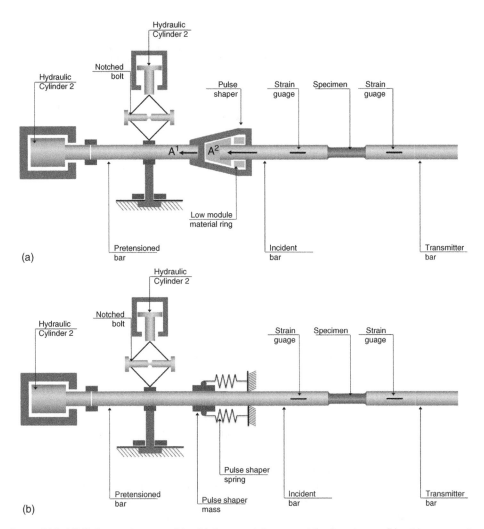

Figure 4.15 MHB for tension test: (a) with low modulus material pulse shaper; (b) with mass-spring pulse shaper.

The pulse shaper represented in Figure 4.15a is realised by interruption of the continuity of the incident bar at points $A^1 - A^2$ and by insertion between these points of the connection pieces re-establishing the continuity of the incident bar. The insertion of the connection pieces of the pulse shaper obliges the incident elastic plane stress wave propagating from the pre-tensioned bar to follow a path through the pulse shaper, indicated by the arrows in Figure 4.15a. The wave passes through a material ring of Young's modulus lower than the incident bar and then continues the propagation along the incident bar towards the specimen. The passing of the loading pulse through the ring of lower Young's modulus causes an increase in the rise time of the incident pulse arriving at the specimen, with respect to the loading pulse propagating without

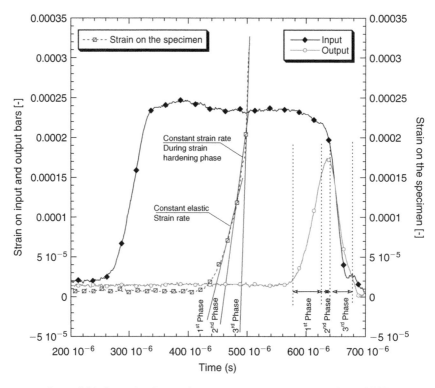

Figure 4.16 Records of a tensile test on a concrete specimen by an MHB.

insertion of the pulse shaper. By increasing the thickness of the ring with lower Young's modulus, pulses with increased rise time are obtained. It is noticed that, with the insertion of the pulse shaper with complete exclusion of the ring with low Young's modulus, it is possible to obtain pulses with extremely short rise time in order of few microseconds (≤ 10) allowing study of the elastic limit in case of extremely hard impact.

The second version of the pulse shaper shown in Figure 4.15b consists of a mass-spring system; the mass is bonded by friction to the incident bar and is connected to a spring resisting the movement of the incident bar. The inertia of the mass and the resistance of the spring will cause a linear increase of the rise time of the pulse. By changing the value of the mass-spring system, different loading rates can be realised corresponding to different elastic strain rates.

As shown in Figure 4.16, for a high strain rate test with the MHB, we observe that the incident pulse generated by the MHB is characterised by a linear increasing load during the rise time. It therefore meets the condition of constant loading rate desired for a constant elastic strain rate during the elastic deformation of the specimen. The constancy of the elastic strain rate is shown in Figure 4.16 by the linearity of the measured strain with time.

The implementation of the pulse shaper of Figure 4.15a or 4.15b allows changes of the slope of the rising branch of the incident pulse, to achieve different constant elastic strain rates for dynamic tests.

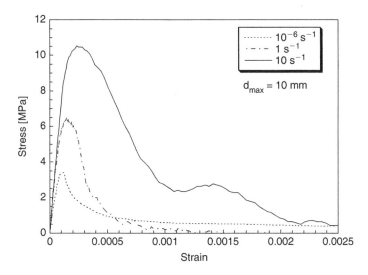

Figure 4.17 Stress versus strain curves of plain concrete in tension at different strain rates.

From Figure 4.17 and at the second phase of the transmitted loading pulse in Figure 4.16, we observe that the trends of these curve branches tend to flatten. Therefore in order to study these branches of the stress-strain curves at constant strain rate, a flattening to a constant value of the incident pulse is also needed.

For the incident pulse shown in Figure 4.16, after the linearly increasing load it follows a flat plateau. It satisfies the condition for a constant strain rate during the strain hardening phase of the dynamic stress-strain curve of Figure 4.17. The constancy of the strain rate during the strain hardening phase of the stress-strain curve is shown in Figure 4.16 by the linear trend of the measured strain with time, on the specimen during the strain hardening phase.

4.3.2 Important recommendations for analysis of test records of high strain rate tests performed with the MHB

It is important to note that, in the case of the Hopkinson bar tests of concrete and rocks, it is recommended that the complete analysis be applied with Equations (4.5), (4.6) and (4.7). We have in fact observed that the application of the simplified analysis procedure to the MHB test records, with the approximated form of Equations (4.5), (4.6), (4.7), do not give accurate values of strain rate and description of the strain rate trend during the different phases of the test.

To explain this important point, we have to look attentively at the records of Figure 4.16, where we distinguish the classic records of the incident, the reflected and the transmitted pulses of a Hopkinson bar test, together with the record of the strain versus time. The incident pulse has the correct shape in order to obtain a constant strain rate both during the linear elastic response and during the non-linear strain hardening response of the specimen to the applied incident pulse. That means a linear increasing load versus time (constant loading rate) followed by a constant load versus time.

In the record of the transmitted pulse, we can distinguish:

- a first phase where the load increases linearly with time,
- a second phase where the load increases non-linearly with time,
- a third phase where the load decreases linearly with time

On the record of the measured strain versus time, we observe:

- a first phase of time duration equal to the first phase of the transmitted pulse, where the strain increases linearly with time indicating a deformation at a constant strain rate,
- a second phase of time duration equal to the second phase of the transmitted pulse where the strain increases also linearly with time but with an increase of slope in respect to the first phase,
- a third phase where the strain increases linearly with time but with an increase of slope in respect to the second phase.

Coupling the records of the transmitted pulse and the strain, by applying the time shift and Equation (4.5), we can then affirm that:

- The coupling of the linear first phases of the transmitted pulse and of the strain gauge on the specimen generates the elastic phase of the stress-strain curve, which is then measured at the constant elastic strain rate given by the slope of the first phase of the strain gauge on the specimen.
- The coupling of the non-linear second phase of the transmitted pulse with the linear second phase of the strain gauge on the specimen generates the non-linear strain hardening phase of the stress-strain curve, which is also measured at the constant strain rate corresponding to the slope of the second phase of the strain gauge on the specimen.
- The linear third phase of the transmitted pulse coupled with the linear third phase of the strain gauge on the specimen generates the fracture propagation phase of the stress-strain curve, which seems to take place at a constant speed.

In conclusion we can observe that:

- The coupling of the records of the measured strain on the specimen and the resistance of the specimen measured on the transmitted bar generate an accurate stress-strain curve of the plain concrete specimen.
- The constant slope of the first and second phases of the record of the measured strain on the specimen gauge length demonstrates that the shape of the applied incident pulse is correct for performing a dynamic test at a constant elastic strain rate during the elastic phase and at a different strain rate during the strain hardening phase.
- On the contrary, when we apply the approximated version of Equation (4.7) for the simplified analysis procedure to the first and second phases of the record of the reflected pulse, we find the elastic strain rate and strain rate during the strain hardening phase always increase with time.

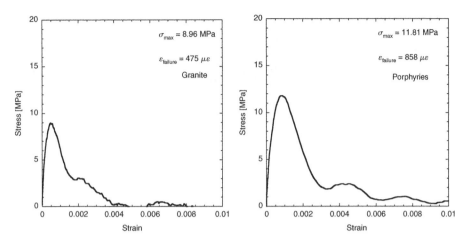

Figure 4.18 Stress versus strain curve for granite in tension obtained by means of JRC-MHB (Cadoni *et al.*, 2009b).

These results can also be directly deduced from the record of the reflected pulse in Figure 4.16. The amplitude of the reflected pulse, which is proportional to the strain rate, increases during the whole test time.

Therefore, taking the strain record as a reference, we can affirm that, for concrete specimens of about 50 mm length, it is necessary to analyse the MHB test records using the complete version of Equations (4.5), (4.6) and (4.7). It is also necessary to check the analysis with the strain measurements by strain gauges or high resolution optical instrumentation on the specimen gauge length.

4.4   LABORATORY MEASUREMENTS OF ROCKS UNDER STATIC MULTIAXIAL COMPRESSION

As mentioned earlier, the most precise measurements of load and displacement in high strain rate testing are those performed by a proper application of the Hopkinson bar testing methodology. This statement is particularly true for low ductility materials like concrete and rocks where the dynamic test takes place during the rising time and the first few microseconds of the loading pulse duration. The efficiency of the JRC-MHB in measuring the stress-strain curves of low ductility materials such as rocks and concrete has been shown in Section 3.

Some successful tensile impact tests were performed at JRC using the MHB for cubes of 60 mm side of rocks like porphyries, granite and gabbro. The results for porphyries and granite are presented in Figure 4.18 and by Cadoni, Solomos and Albertini (2009) showing that this type of test can be performed using the MHB installed in the DynaMat laboratory at SUPSI in Lugano. Recently more detailed results have been obtained with the JRC-MHB on yellow tuff (Asprone *et al.*, 2009), on orthogneiss (Cadoni, 2010) and on different marbles (Cadoni *et al.*, 2010).

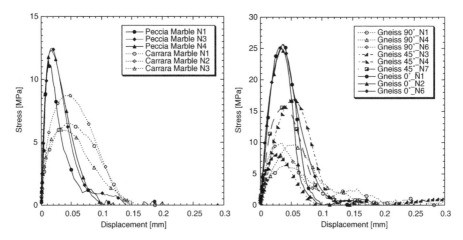

Figure 4.19 Stress versus displacement curves of Marbles and Orthogneiss in tension obtained by means of JRC-MHB (Cadoni *et al.*, 2010).

The conceptual schemes of a high load MHB that should allow the impact testing in compression or tension are presented:

–   of a cubic rock specimen of 100 mm side,
–   with the superposition of a biaxial or triaxial compression to the rock specimen before the impact test in order to simulate the natural confinement of in situ stresses.

The high load compression biaxial MHB, termed as ROCK-MHB is shown in Figure 4.20. It has the following features:

–   An MHB, having pre-stressed input and output bars of equal length of 2 m and equal square cross-sectional area of 100 mm side, generating a rectangular loading pulse of 3 MN amplitude and 800 μs duration which propagates through the input bar – rock specimen – output bar deforming the rock specimen at high strain rate to failure. The total length of the ROCK- MHB is about 8 m.
–   The addition of two bars, each one activated by a hydraulic actuator, orthogonally at the MHB axis, having the same cross-sectional area and the same length of the MHB input and output bars. It allows achievement of a static biaxial stress state (confining stresses) before the specimen undergoes the compressive impact test. In order to realise the biaxial static preloading, the output bar along the MHB axis is also provided with a hydraulic actuator.
–   The extension of the ROCK-MHB to a triaxial test is shown in the scheme in Figure 4.21, where two further output-confinement bars are placed along the second direction at 90° to the MHB axis.

The independent control of the confining loads along the three principal axes allow description of different loading paths in the principal stress space.

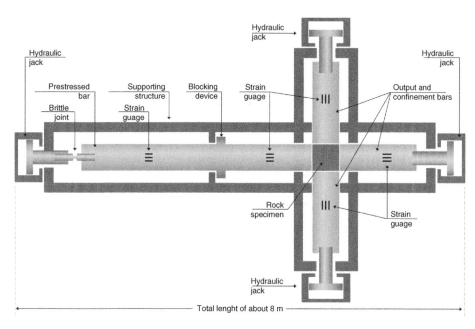

Figure 4.20 Scheme of the rock-MHB for compressive biaxial impact test of confined rock specimens.

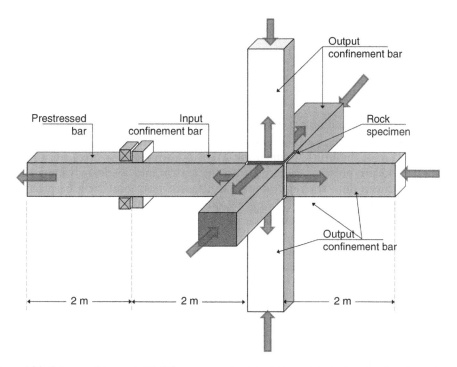

Figure 4.21 Scheme of the rock-MHB for compressive triaxial impact test of confined rock specimens.

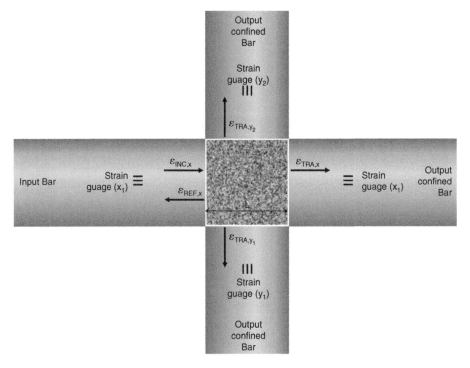

Figure 4.22 Stress wave paths in a biaxial rock-MHB impact test.

A test with the equipment shown in Figures 4.20 and 4.21 is performed in the following operational sequence:

- a quasi-static biaxial or triaxial stress state is introduced in the rock specimen by the hydraulic actuators of the output-confinement bars,
- the rupture of the brittle joint of the MHB pre-stressed bar gives rise to a rectangular square pulse propagating into the system and loading dynamically the rock specimen until fracture (in case of a tensile test, joints will be placed in a device blocking the point of the bar which determines the separation between the pre-stressed bar and the input bar, as shown in Fig. 4.4),
- the strain gauges on the input and output-confinement bars record the incident ε_i, reflected ε_r and transmitted ε_t pulses which allow the reconstruction of the equivalent stress-strain curves in the analysis shown below.

The scheme in Figure 4.22 shows the stress wave propagation in the case of a biaxial ROCK-MHB, where ε_{ix}, ε_{rx}, ε_{tx} are respectively the incident, reflected and transmitted pulses along the ROCK-MHB axis and ε_{ty-1}, ε_{ty-2} are the transmitted pulses along the output bars perpendicular to the ROCK-MHB axis.

After the rupture of the brittle joint, the stress wave propagates along the input and output bars placed along the X and Y axes. The one-dimensional elastic wave propagation theory applied to the configuration and Equations (4.5), (4.6) and (4.7) can be rewritten as follows:

$$\sigma_x = \frac{1}{2}E\frac{A}{A_0}(\varepsilon_{ix} + \varepsilon_{rx} + \varepsilon_{tx}) \tag{4.13}$$

$$\varepsilon_x = \frac{C_0}{L}\int_0^T (\varepsilon_{ix} - \varepsilon_{rx} - \varepsilon_{tx})dt \tag{4.14}$$

$$\dot{\varepsilon}_x = \frac{C_0}{L}(\varepsilon_{ix} - \varepsilon_{rx} - \varepsilon_{tx}) \tag{4.15}$$

$$\sigma_y = \frac{1}{2}E\frac{A}{A_0}(\varepsilon_{ty-1} + \varepsilon_{ty-2}) \tag{4.16}$$

$$\varepsilon_y = \frac{C_0}{L}\int_0^T (\varepsilon_{ty-1} - \varepsilon_{ty-2})dt \tag{4.17}$$

$$\dot{\varepsilon}_y = \frac{C_0}{L}(\varepsilon_{ty-1} - \varepsilon_{ty-2}) \tag{4.18}$$

where: E, A, A_0, L denote the elastic modulus and the cross-sectional area of the input and output bars elastic modulus, the cross-sectional area and the gauge length of the specimen, respectively.

The stresses σ_x and σ_y, the strains ε_x and ε_y, the strain rates $\dot{\varepsilon}_x$ and $\dot{\varepsilon}_y$ can then be combined following the equations of the chosen yielding-fracture criterion in order to obtain the dynamic equivalent stress-strain curves.

The illustrated analysis can be extended to the case of the triaxial ROCK-MHB of Figure 4.21. With a proper configuration of the rock specimen geometry it is possible to perform shear (Albertini et al., 1991) and bending (Chatani, Hojo and Tachiya, 1991) impact tests using the ROCK-MHB configuration sketched in Figures 4.20 and 4.21, using similar methods of analysis.

The ROCK-MHB can also be employed to impose repetitions of impact loading cycles simulating those generated by the rock excavation machines (e.g. hydraulic hammer). In this case, the pre-stressed bar of the ROCK-MHB might be configured as shown in Figure 4.23, where the hydraulic actuator and the brittle joint placed at the end of the pre-stressed bar shown in Figure 4.20 are substituted by a rotating cam. With the rotating cam, a store repetition at the desired frequency of half-cycle load in the pre-stressed bar can be realised and therefore a repetition at the desired frequency of half-cycle compressive loading pulses acting on the specimen can be generated. The ROCK-MHB can also have a similar configuration for the generation of tension half-cycles. The repetition of half-cycle loading pulses generated as shown in Figure 4.24 acts on the rock specimen in the same way as those generated by the rock excavation machine. In correspondence with each half cycle of the incident loading pulse, the related reflected and transmitted pulses are recorded before the arrival of the next half cycle of loading.

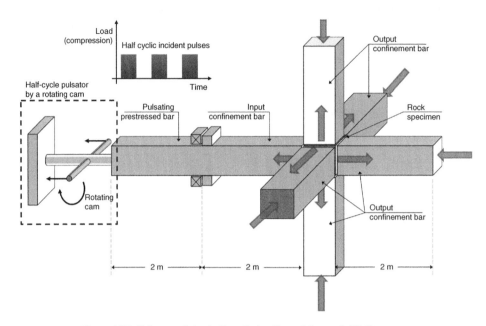

Figure 4.23 Scheme of the half-cyclic loading of the rock-MHB system.

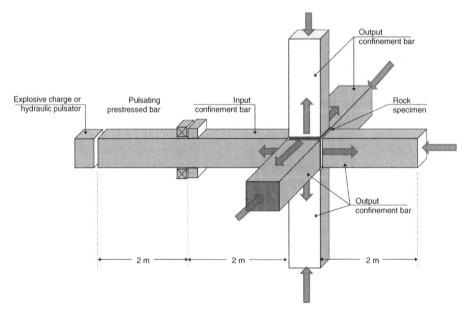

Figure 4.24 Setup for the laboratory recording with the rock-MHB of the pressure wave generated by an explosive charge or by the pulse of an excavation machine.

Therefore it is possible to measure in detail the mechanical response of the rock specimen for each half-cycle loading similar to that of the excavation machine, with the application of Equations (4.9) to (4.18). The shape of the half-cycle loading can be varied by changing the shape of the rotating cam. A repetition of complete impulsive loading cycles might also be generated by the introduction of the rotating cam and a change of cross-sectional area of the input bar as described by Albertini *et al.* (1988).

4.5   CONCLUSIONS

Appropriate experimental methods, based on the physics of stress wave propagation in the materials, are necessary in order to measure dynamic mechanical properties of rocks subjected to pressure pulses generated by explosive charges or by excavation machines. It is scientifically recognised that the most accurate methodology for the measurement of the dynamic mechanical parameters in presence of stress wave propagation is the Hopkinson bar.

The Modified Hopkinson Bar (MHB) with Pulse Shaper allows the generation of loading pulses with constant loading rate during the rise time followed by a constant amplitude of the pulse. The constant loading rate during the rise time is needed for the accurate determination of the elastic modulus and of the elastic limit of the rock specimen performed at constant elastic strain rate. The successive constant amplitude of the pulse is needed for the accurate determination during the strain hardening phase of the dynamic stress-strain curve.

The Modified Hopkinson Bar for rock tests (ROCK-MHB) allows the precise determination of the equivalent stress-strain curve at high strain rate, under single or cyclic pulses for rock specimens subjected to biaxial or triaxial quasi-static loads. It reproduces the in situ stress conditions existing in the excavation field. The ROCK-MHB also allows the precise calibration of the pressure pulses generated by explosive charges or by excavation machines, and provides the direct demonstration of the effect of such pulses on rocks.

REFERENCES

Albertini, C. and Labibes, K.: *Split Hopkinson Bar Compression Testing Apparatus*. European Patent EP 0849583, US patent No 6,109,093, 1997.

Albertini, C. and Montagnani, M.: *Dynamic Material Properties of Several Steels for Fast Breeder Reactors Safety Analysis, EUR 5787EN*, 1977.

Albertini, C. and Montagnani, M.: Testing techniques based on the Split Hopkinson bar. *Inst. of Physics, Conf. Ser. No 21*, London, 1974, pp.22–32.

Albertini, C., Boone, P.M. and Montagnani, M.: Development of the Hopkinson bar for testing large specimen in tension. *J. Phys. France 46 Colloq.C5* (Dymat85), 1985, pp.499–504.

Albertini, C., Cadoni, E. and Labibes, K.: Impact fracture process and mechanical properties of plain concrete by means of an Hopkinson bar bundle. *J. Phys. 4th France 7, colloque C3*, 1997, pp.915–920.

Albertini, C., Cadoni, E. and Labibes, K.: Precision measurements of vehicle crashworthiness by means of a large Hopkinson bar. *Journal de Physique*, IV, Vol.III, No.7, (1997), pp.79–84.

Albertini, C., Cadoni, E. and Labibes, K.: Study of the mechanical properties of plain concrete under dynamic loading. *Experimental Mechanics*, 25 (1999), pp.509–514.

Albertini, C., Hanefi, E.H. and Wierzbicki, T.: *Calibration of impact rigs for dynamic crash testing, EUR report 16347EN*, 1995.

Albertini, C., Labibes, K., Montagnani, M., Pizzinato, E.V., Solomos, G. and Viaccoz, B.: Biaxial direct tensile tests in a large range of strain rates - Results on a ferritic nuclear steel. *Journal de Physique IV* 10(PR9) (2000), pp.161–166.

Albertini, C., Montagnani, M., Eleiche, A.M. and Abdel Kader, S.: Shear cyclic testing through tensile – compressive fatigue loading. *8th Int. Conf. On Strength of Metals and Alloys*, University of Tampere, Finland, 1988, pp.653–658.

Albertini, C., Montagnani, M., Pizzinato, E.V. and Rodis, A.: Comparison of mechanical properties in tension and shear at high strain rate for AISI316 and ARMCO iron. *Proc. of Int. Conf. On Mechanical Behaviour of Materials*, The Society of Material Science of Japan, Kyoto, 1991, 351–356.

Albertini, C., Solomos, G. and Labibes, K.: Calibration of impact rigs. Crashworthiness testing of thin sheet metal boxes. *16th Int. Conf. on the Enhanced Safety of Vehicles*, Windsor, Canada, 1998, 763–768.

Asprone, D., Cadoni, E., Prota, A. and Manfredi, G.: Investigation on dynamic behavior of a Mediterranean natural stone under tensile loading. *Intern. J. of Rock Mechanics and Mining* 46 (2009), pp.514–520.

Birch, R.S., Jones, N. and Jouri, W.S.: Performance assessment of an impact rig. *Proceedings of the Institute of Mechanical Engineers*, vol.2, C4, (1988), pp.275-285.

Cadoni, E., Albertini, C., Dotta, M., Forni, D., Giorgetti, P. and Riganti, G.: Mechanical characterization of rocks at high strain-rate by the JRC modified Hopkinson bar: a tool in blast and impact assessment. In: J.A. Sanchidriàn (ed.): *Rock fragmentation by blasting*, Taylor & Francis, London, 2009, pp.35–41.

Cadoni, E., Antonietti, S., Dotta, M. and Forni, D.: Strain rate behaviour of three rocks in tension. *Proceedings of the 7th International Symposium on Impact Engineering*, Warsaw (Poland), 2010. (on CD).

Cadoni, E., Labibes, K., Albertini, C. and Solomos, G.: *Mechanical Response in Tension of Plain Concrete in a Large Range of Strain Rate*. Joint Research Centre Note I 97.194, 1997.

Cadoni, E., Labibes, K., Albertini, C., Berra, M. and Giangrasso, M.: Strain rate effect on the tensile behaviour of concrete at different relative humidity levels. *Materials and Structures* 34(2001a), pp. 21–26.

Cadoni, E., Labibes, K., Berra, M., Giangrasso, M. and Albertini, C.: High strain rate tensile behaviour of concrete. *Magazine of Concrete Research* 52(5) (2000), pp. 365–370.

Cadoni, E., Labibes, K., Berra, M., Giangrasso, M. and Albertini, C.: Influence of the aggregate size on the strain-rate tensile behaviour of concrete, *ACI Materials Journal* 98(3) (2001b), pp.220–223.

Cadoni, E., Solomos, G. and Albertini, C.: Mechanical characterization of concrete in tension and compression at high strain-rate using a modified Hopkinson bar. *Magazine of Concrete Research*, 61(3) (2009), pp.221–230.

Cadoni, E.: Dynamic characterization of orthogneiss rock subjected to intermediate and high strain rates in tension. *Rock Mechanics and Rock Engineering*, 43(6) (2010), pp.667–676.

Chatani, A., Hojo, A., and Tachiya, H.: The strength of hard steels at high strain rates. *Proc. of Int. Conf. on Mechanical Behaviour of Materials* The Society of Material Science of Japan, vol.1, Pergamon Press, 1991, pp.301–306.

Davies, R.M.: A critical study of the Hopkinson pressure bar. *Phil. Trans. Roy. Soc.* London Ser.A240, 1948, pp.375–457.

Hopkinson, B.: A method of measuring the pressure produced in the detonation of high explosives or by the impact of bullets. *Phil. Trans. R. Soc.* London A213, 1914, pp.437–456.

Hopkinson, J.: On the rupture of an iron wire by a blow. *Proc. Manch. Liter. Philos. Soc.*, 1872, pp.40–45.

Kolsky, H: *Stress waves in solids*. Clarendon press, Oxford, 1953.

Chapter 5

Wave shaping by special shaped striker in SHPB tests

Xibing Li, Zilong Zhou, Deshun Liu, Yang Zou and Tubing Yin

5.1 INTRODUCTION

The split Hopkinson pressure bar has been a very popular and promising experimental technique for the study of dynamic behaviours of metallic materials for its easy operation and relatively accurate results, with three basic assumptions (Davies and Hunter, 1963; Bazle, Sergey and John, 2004): (a) waves propagating in the bars can be described by the one-dimensional wave theory, (b) stress in the specimen is uniform, and (c) specimen inertia effect and friction between specimen and bars can be negligible.

Due to the advantages of SHPB in dynamic tests, it was gradually extended to the studies of brittle materials including rock, concrete and ceramic (e.g., Ravichandran and Subhash, 1994; Li and Gu, 1994; Tedesco and Ross, 1998). However, because of the brittle and heterogeneous characteristics of rock-like materials, the technique was affected by the following problems:

a) Difficulty in achieving stress uniformity and equilibrium in specimen. For rock-like geological materials, the grain size is usually large. Therefore, the specimen should be sufficiently large to represent the true mechanical properties. It is greatly suggested that the specimen size should at least be 5 times the maximum grain size. Accordingly, larger diameter bars of SHPB are needed to test rock and concrete specimens. Hence, the wave dispersion, inertia effect and stress non-uniformity introduced by large specimens and large diameter bars appear more prominent. It becomes difficult to fulfil the basic assumptions of SHPB technique.
b) Premature failure of specimen before stress equilibrium. Traditional SHPB with rectangular incident wave can provide useful experimental results for metals, whose compressive flow stresses happen at strains larger than a few percent. By contrast, brittle materials such as rocks, ceramics and concrete normally fail at strains less than 0.5 percent. With the steep front of a rectangular wave, a specimen of rock-like materials always fails before its stress equilibrium. Hence, the test results may not be reliable.
c) High oscillation of incident wave. Due to dispersion of the traditional rectangular wave in a large bar, the acquired incident wave, reflected wave and transmitted wave are usually oscillatory. These lead to jumpy stress-strain curves for rock-like materials, which actually contain the loading and unloading histories of the specimen.

d) Difficulty in ensuring the specimen deformation at constant strain rate. Rock-like brittle materials are usually rate sensitive. Only when the specimen deforms at constant strain rate during SHPB test can the obtained stress-strain results be regarded as obeying the constitutive relation correspondingly. However, the traditional SHPB with cylindrical striker does not always keep specimen deformation at constant strain rate. Therefore, the traditional SHPB needed to be improved.

Among the above 4 problems, problem (a) determines the applicability of SHPB for rock-like materials; problem (b) relates to the reliability of test results; problems (c) and (d) affect the accuracy of test results.

Focusing on these problems, efforts have been made to meet the requirements of big specimens, and large SHPB devices were constructed worldwide with bar diameters of 50 mm to 100 mm. The wave dispersion, stress equilibrium and inertia problems were studied accordingly (e.g. Gong, Malvern and Jenkins, 1990; Wu and Gorham, 1997; Meng and Li, 2003; Zhao and Gary, 1996; Zhao, 2003; Yang and Shim, 2005; Forrestal, Wright and Chen, 2007). In order to overcome the premature failure of specimen before its stress equilibrium, the incident wave has been modified by various techniques. One technique was the pulse shaper method (Frantz, Follansbee and Wright, 1984). Many researchers have investigated the response of pulse shapers with different materials and used them in tests (Follansbee and Frantz, 1983; Frew, Forrestal and Chen, 2002). The other technique uses a special shaped striker (Liu and Li, 1998). A series of studies have been done on theories and laboratory experiments (Liu, Peng and Li, 1998; Li, Lok and Zhao, 2000, 2005; Lok *et al.*, 2002; Li, Zhou and Lok, 2008; Li and Zhou, 2009). For the constant strain rate deformation of specimen in SHPB tests, it only attracted high attention recently after the strain rate sensitivity of rock-like material was recognized. By now, the shaped striker method and pulse shape method have gained good applications in SHPB testing (Li and Gu, 1994; Li, Lok and Zhao, 2000, 2005; Li, Zhou and Lok, 2008; Frew, Forrestal and Chen, 2002).

5.2 ADVANTAGE OF HALF-SINE WAVE FOR LARGE DIAMETER SHPB TESTS

5.2.1 Stress equilibrium during specimen deformation

The wave velocity in rock material is between 3000–6000 m/s. It results in a relatively long travel time of the wave in the specimen. At the same time, stress equilibrium in specimens in SHPB tests is usually reached after several reflections of wave in the specimen. This implies that, if a steep rising incident wave like a rectangular wave is applied to the specimen, the stress at the incident end of the specimen (the end near the input bar) will increase abruptly and be high enough to fail the material, while the other end of specimen (transmitted end) may have no stress disturbance yet. In this situation, the stress state of the specimen between the two ends is apparently non-uniform, especially for a brittle material with failure strain less than 0.5 percent. For a half-sine wave with slow rising slope, the responses of specimens differ greatly. Upon arrival of the wave front, the incident end has slightly higher stress than the other end. With wave reflection in the specimen, the stress at the transmitted end

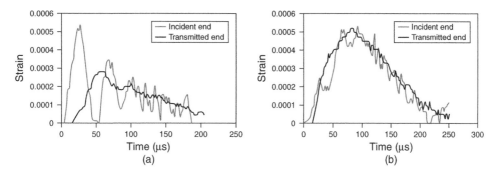

Figure 5.1 Stress at both ends of specimen with different incident wave: (a) rectangular incident wave; (b) half-sine incident wave.

increases gradually and accumulates gradually. After several reflections of wave in the specimen, the stress at both ends of the specimen reaches equilibrium with an average value still less than the failure stress of the specimen. Figure 5.1 gives typical stress histories at both ends of the specimen with rectangular and approximate half-sine wave from a shaped striker. It can be seen that the case with half-sine wave gives better stress equilibrium during specimen deformation, while the stress in specimen with a rectangular incident wave shows great deviation at the two ends during specimen deformation and failure, which violates the assumption of SHPB technique.

5.2.2   Less dispersion and better uniform stress at bar section

Different waves propagate in solids at different velocities, which depend on the solid material and the wave type. If the solid media has special geometries such as a cylinder or plate, dispersion happens. In SHPB tests, the incident bar, transmitting bar and absorbing bar are all cylindrical rods with dispersion effect. Researches indicated that waves in reality usually consist of a number of harmonic components with different frequencies, which can be decomposed with Fourier Transformation tools (Lifshitz and Leber, 1994; Gong, Malvern and Jenkins, 1990; Anderson, 2006). The harmonic wave components travel along a circular bar with different velocities as:

$$c = c_0 \left[1 - \mu^2 \pi^2 \left(\frac{a}{\lambda} \right)^2 \right] \tag{5.1}$$

where c_0 equals to $(E/\rho)^{1/2}$, E and ρ are the modulus and density of material respectively, μ is Poisson's ratio, a is the radius of the cylindrical rod, and λ is the wavelength; c/c_0 describes the wave dispersion along the bar.

Taking a rectangular incident wave as an example, it can be decomposed into a series of harmonic wave components as in Figure 5.2 (Li and Gu, 1994). From Equation (5.1), the component waves with different frequencies travel with different velocities individually. Finally, the original rectangular wave would be stretched and distorted, i.e. the wave oscillation appears (Fig. 5.3). More severely, the oscillation

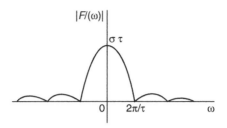

Figure 5.2 Frequency components of a rectangular wave.

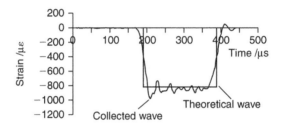

Figure 5.3 Oscillation of rectangular wave due to dispersion.

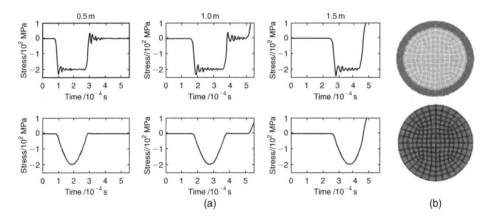

Figure 5.4 Dispersion effect of different incident waves in SHPB bars: (a) Signals along bar with different incident waves; (b) Stress distribution at bar section with different waves.

of the incident wave triggers oscillation in the reflected wave and transmitted waves correspondingly. Then the test results of SHPB calculated with the incident, reflected and transmitted waves will be deviated from true behaviour.

On the other hand, a sinusoidal wave has simple frequency, which travels at one determined velocity. Figure 5.4(a) gives the simulation results of signals travelling along an elastic rod with rectangular and half-sine incident waves. It can be seen that there is no dispersion for the half-sine wave travelling through a long bar.

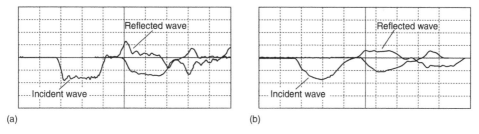

Figure 5.5 Signals obtained with two different strikers: (a) Signal obtained with a traditional cylindrical striker; (b) Signals obtained with a purpose-built striker.

Besides, wave dispersion from non-sinusoidal waves can lead to stress non-uniformity at bar section when the bar diameter is large. From the above, wave components of a non-sinusoidal wave travel with different velocities. In fact, wave components transfer not only particle motion, but also stress and energy. Simulation with rectangular and half-sine waves along a thick bar showed that the stress distribution at the middle sections of the bar are different, as shown in Figure 5.4(b). The half-sine wave gives uniform stress distribution and shows its advantage once again.

5.2.3 Constant strain rate deformation of specimens

Recent researches show that constant strain rate deformation of specimens can be achieved with a half-sine wave, which is vital for tests of rate-sensitive materials (Zhou, Li and Ye, 2010).

Figure 5.5 compares examples of signals produced by a traditional cylindrical striker and the new purpose-built striker.

It can be seen that the reflected wave sourced by the purpose-built striker has long and smooth segments, whereas the reflected wave sourced by the cylindrical striker has an apparent high-frequency overprint that makes it much more rugged in appearance. According to the principle of the SHPB device, the reflected waveform indicates the strain rate of the specimen directly. Therefore, the purpose-built striker is better than the traditional cylindrical striker for testing brittle materials.

5.3 GENERATING HALF-SINE WAVES BY SPECIAL SHAPED STRIKERS

5.3.1 Impact by striker and the generated stress wave

Practices like percussive drilling, piling and forging indicate that the impact-generated stress waves have intrinsic relations to the strikers, punch head or impactors.

In order to obtain the general relation between striker profile and stress wave, a non-prismatic striker is considered to impact a long cylindrical rod. As shown in Figure 5.6, the striker is divided into micro-segments, and then the characteristic line method is used to get the generated wave.

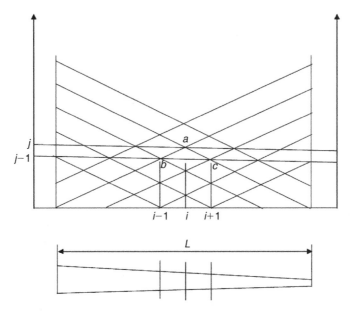

Figure 5.6 Stress analysis of non-prismatic striker with characteristic line method.

For any micro-segment in the striker, the following relation exists,

$$\left.\begin{aligned}
A\frac{\partial\sigma}{\partial x} + \sigma\frac{dA}{dx} + A\rho\frac{\partial v}{\partial t} &= 0 \\
\frac{\partial\varepsilon}{\partial t} + \frac{\partial v}{\partial x} &= 0 \\
\sigma &= E\varepsilon \\
E &= \rho c^2
\end{aligned}\right\} \quad (5.2)$$

Then,

$$\left.\begin{aligned}
\frac{\partial v}{\partial t} + \rho\frac{d\sigma}{dx} + \rho\frac{\sigma d\ln A}{dx} &= 0 \\
\frac{\partial v}{\partial x} + \frac{1}{\rho c^2}\frac{\partial\sigma}{\partial t} &= 0
\end{aligned}\right\} \quad (5.3)$$

Equation (5.3) gives the general wave equation for a continuous striker with arbitrary cross sections. The equation has the solving conditions as:

$$\left.\begin{aligned}
\sigma(x = 0, t) &= 0 \\
\sigma(t = 0, x) &= 0 \\
v(t = 0, x \le L) &= v_0 \\
v(t = 0, x > L) &= 0 \\
A = A(x), \quad 0 &\le x \le L \\
A = A_0, \quad x &> L
\end{aligned}\right\} \quad (5.4)$$

According to the characteristic line theory:

$$\left.\begin{array}{l} dx \mp cdt = 0 \\ d\sigma + \sigma d \ln A \pm \rho c dv = 0 \end{array}\right\} \tag{5.5}$$

For points a, b and c in Figure 5.6, their relations can be expressed as:

$$\left.\begin{array}{l} \sigma_{ij} - \sigma_{i-1,j-1} + \dfrac{\sigma_{ij} + \sigma_{i-1,j-1}}{2}(\ln A_i - \ln A_{i-1} + \rho c(v_{ij} - v_{i-1,j-1})) = 0 \\ \sigma_{ij} - \sigma_{i+1,j-1} + \dfrac{\sigma_{ij} + \sigma_{i+1,j-1}}{2}(\ln A_i - \ln A_{i+1} - \rho c(v_{ij} - v_{i+1,j-1})) = 0 \end{array}\right\} \tag{5.6}$$

The solving conditions can be determined as:

$$\left.\begin{array}{l} v_{i0} = v_0, \quad i \leq n_0 \\ v_{i0} = 0, \quad i > n_0 \\ \sigma_{i0} = 0, \quad i \geq 0 \\ \sigma_{0j} = 0 \\ v_{0j} = \sigma_{1,j-1}\left(1 + \dfrac{1}{2}\ln(A_0/A_1)\right)/(\rho c) + v_{1,j-1} \end{array}\right\} \tag{5.7}$$

With Equations (5.6) and (5.7), the stress wave generated by impact with an arbitrary striker can be obtained. This method can be called the forward design method.

5.3.2 Inverse design of striker for a specific wave

When a specific wave form is given beforehand, it is usually difficult to find the striker profile directly. This is a typical inverse problem.

Here a striker-rod system is established to investigate this inverse problem, as shown in Figure 5.7, where a non-prismatic striker with length of L impacts a long cylindrical rod of radius r_0. A spring, with stiffness k, is assumed to lie between the striker and the rod to simulate the impact contact between them.

The motion of the impact system is governed by the one-dimensional wave equation:

$$\left.\begin{array}{l} \rho A(x)\dfrac{\partial v}{\partial t} + \dfrac{\partial F}{\partial x} = 0 \\ \rho c^2 A(x)\dfrac{\partial v}{\partial x} + \dfrac{\partial F}{\partial t} = 0 \end{array}\right\} \tag{5.8}$$

The boundary and initial conditions are:

$$\left.\begin{array}{l} F(x = 0, t) = 0 \\ \dfrac{dF(x = L, t)}{dt} = k[v(x = L, t) - P(t)/(\rho c A_0)] \\ P(t) = F(x = L, t) \\ v(0 \leq x \leq L, t = 0) = v_0 \\ F(0 \leq x, t = 0) = 0, \quad A(x) = \pi r^2(x) \end{array}\right\} \tag{5.9}$$

where v, F are the velocity and the normal force of the studied section; ρ, c are the density and p-wave velocity of the striker and bar; k is the stiffness of the virtual spring; $A(x)$, $r(x)$ are the cross-sectional area and radius which are functions of the coordinate x; L is the length of the striker; A_0, r_0 are the cross-sectional area and radius of the rod; v_0 is the impact velocity of the striker; $P(t)$ is the force wave generated by impact between the striker and rod, propagating in the rod.

For simplicity, the above equations can be rewritten in non-dimensional forms as follows.

$$x^* = x/L, \quad t^* = ct/L, \quad v^* = v/v_0, \quad F^* = F/(\rho c A_0 v_0), \quad k^* = kL/\rho c^2 A_0,$$
$$P^* = P/(\rho c A_0 v_0), \quad r^* = r/r_0$$

Thus, Equations (5.8) and (5.9) can be transformed into:

$$\left. \begin{array}{l} r^{*2} \dfrac{\partial v^*}{\partial t^*} + \dfrac{\partial F^*}{\partial x^*} = 0 \\[2mm] r^{*2} \dfrac{\partial v^*}{\partial x^*} + \dfrac{\partial F^*}{\partial t^*} = 0 \\[2mm] F^*(x^* = 0, t^*) = 0, \quad \dfrac{dF^*(x^* = 1, t^*)}{dt} = k\,(v^* - P^*) \\[2mm] F^*(x^* = 1, t^*) = P^* \\[2mm] v(0 \le x^* \le 1, t^* = 0) = v_0, \quad F^*(0 \le x^*, t^* = 0) = 0 \\[2mm] r^* = r^*(x^*), \quad 0 \le x^* \le 1 \end{array} \right\} \tag{5.10}$$

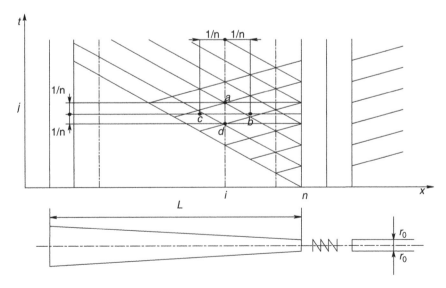

Figure 5.7 Striker-rod system and characteristic line sketch.

According to the characteristic line theory, the following differential equation can be established:

$$\left.\begin{array}{l} dt^* \mp dx^* = 0 \\ r^{*2}dv^* \pm dF^* = 0 \end{array}\right\} \tag{5.11}$$

Let a striker occupy $x \in [0, L]$, and consist of N uniform segments with constant cross-sectional area and equal transit time $\Delta t^* = 1/n$ as shown in Figure 5.7. The characteristic line grid can be established. For points a, b and c in Figure 5.7, they have the following relations:

$$F_{i,j}^* - F_{i-1,j-1}^* + r_i^{*2}(v_{i,j}^* - v_{i-1,j-1}^*) = 0, \quad i = 1, 2, \ldots, n \tag{5.12a}$$

$$F_{i,j}^* - F_{i+1,j-1}^* + r_{i+1}^{*}{}^2(v_{i,j}^* - v_{i+1,j-1}^*) = 0, \quad i = 0, 2, \ldots, n-1 \tag{5.12b}$$

Accordingly, the boundary and initial conditions can be specialized as:

$$\left.\begin{array}{l} F_{0j}^* = 0 \\ F_{nj}^* - F_{n,j-1}^* = k^*(v_{nj}^* - F_{nj}^*)/n \\ P^*(j/n) = P_j^* = F_{nj}^* \\ v_{i0}^* = 1, F_{i0}^* = 0, \quad i = 0, \ldots, n \\ r_i^* = r^*(i/n), \quad i = 0, \ldots, n \end{array}\right\} \tag{5.13}$$

With Equations (5.12) and (5.13), the radii of the striker can be determined. Therefore, the profile of the striker is obtained.

5.4 SHPB TESTS WITH SPECIAL SHAPED STRIKER

5.4.1 Configuration of SHPB system with special shaped striker

A standard SHPB (Fig. 5.8) consists of a striker bar, an input bar, an output bar, an absorption bar, a buffer, a gas gun and a data acquisition unit. Supporting foundation is needed to keep the bar system stable and coaxial. The specimen is sandwiched between the input and output bars. After ejecting from the gas gun, the striker bar impacts the free end of the input bar, thus generating a compressive longitudinal wave that propagates along the bar towards the specimen. Once the wave reaches the bar/specimen interface, a part of it is reflected, whereas another part goes through the specimen and transmits into the output bar and absorption bar.

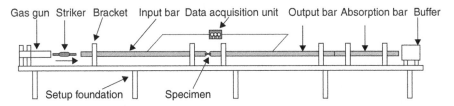

Figure 5.8 General view of SHPB with special shaped striker.

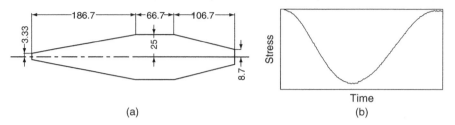

Figure 5.9 Typical special shaped striker: (a) Parameters of the striker; (b) Stress wave from the left striker.

The input, output and absorption bars are cylindrical with diameter D of 50 mm and length of 2000 mm, 2000 mm, 1000 mm respectively. It is suggested the specimen be cylindrical with diameter of 50 mm and length of 25 mm. The bars should be machined from steel with minimum yield stress of 1000 MPa. The supporting foundation shall be safely bolted to the floor with brackets and have a fine adjustment function to ensure the bar system is exactly coaxial. The striker bar should be longer than $5D$. The traditional cylindrical striker for metal tests should be avoided for the premature failure of specimens under a rectangular stress wave. A striker with the configuration shown in Figure 5.9(a) generating the approximate half-sine stress wave of Figure 5.9(b) is suggested.

Brackets on the setup foundation should be carefully designed and installed so that they can be adjusted up and down precisely, and be individually translated back and forth along bars. The quantity of brackets depends on the bar diameter and stiffness. Precision bar alignment is required for both stress uniformity and 1D wave propagation within the pressure bars and specimen during test. Lack of free movement of the bars will lead to forced clamping and result in a wrong signal and high background noise.

A pair of strain gauges should be glued diametrically at the middle points of the input bar and of the output bar to obtain the incident wave, reflected wave and transmitted wave. Strain gauge with length shorter than 2 mm is suggested for capturing details of stress waves. A strain meter with sampling frequency larger than 1 MHz should be used.

5.4.2 Test procedures on SHPB with special shaped striker

5.4.2.1 Specimen preparation

Specimens should be cored from the same rock block with no visible geological weakness. Specimens should be intact, petrographically uniform and representative of the rock. Ultra-sonic velocity should be measured to group the specimens of similar velocities.

A grinding machine or fine sandpaper should be used to ensure that ends of specimen are smooth and parallel. The ends of the specimen shall be flat to 0.02 mm and shall not depart from perpendicularity to the axis of the specimen by more than 0.001 radian or 0.025 mm in 25 mm. The side surface of the specimen shall be smooth and free of abrupt irregularities, and straight to within 0.02 mm over the full length of

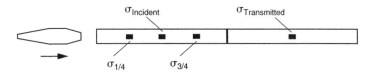

Figure 5.10 Sketch of system calibration.

the specimen. The diameter D should be measured to the nearest 0.02 mm by averaging two diameters obtained at right angles to each other at about the mid-height of the specimen. The thickness of the specimen should also be determined to the nearest 0.02 mm by averaging two thickness measurements at right angles to each other.

If tests cannot be finished for one time, specimens shall be stored in the same environment as the test for no longer than 30 days. When the specimens are used next time, their ultra-sonic velocity and density should be measured again. The number of specimens tested under a specified condition should be more than five.

5.4.2.2   System preparation

a) Gas gun

Gas source should be checked every time before test to ensure there is enough gas. Gas pressure should be high enough to drive the striker movement and generate the planned incident wave. The relations between the striker position, gas pressure and incident stress can be determined after the construction of apparatus, and verified every year.

b) Strain meter

After being turned on, the strain meter shall warm up for more than 10 minutes. If the gauge bridges cannot reach balance automatically, manual reset is needed to get balance before tests. Trigger channel, trigger level, gain level and sampling rate of the signal recorder should be properly chosen so as to capture and show the test signals correctly.

5.4.2.3   System calibration

System calibration is necessary to ensure all parts of the apparatus work synergistically and precisely. In calibration tests, the strain gauges can be deployed as shown in Figure 5.10. With strain gauges mounted on the middle of the input bar and the output bar, the stress (strain) histories σ_{Incident}, $\sigma_{\text{Reflected}}$ and $\sigma_{\text{Transmitted}}$ are measured for calculation. σ_{Incident} and $\sigma_{\text{Reflected}}$ are measured with the same strain gauge. At the same time, in order to identify the wave attenuation and distortion for misalignment impact, two additional strain gauges are suggested, glued at the position of 1/4 and 3/4 of the input bar. The obtained stresses are denoted as $\sigma_{1/4}$, $\sigma_{3/4}$.

Then the calibration of the SHPB system with the special shaped striker can be divided into 4 steps.

i)   System adjustment. Tune the brackets under the SHPB bars to keep the striker, input bar and output bar in the same line axially as much as possible; make sure that the strain gauges are well glued and the strain meter can get the signals properly.

ii) Wave distortion identification. After system adjustment, the striker is fired to impact the input bar which contacts the output bar without specimen, then $\sigma_{1/4}$, $\sigma_{3/4}$, σ_{Incident} and $\sigma_{\text{Transmitted}}$ are obtained. In order to distinguish these stresses from those in tests with specimens, they are denoted as $\sigma_{1/4}^0$, $\sigma_{3/4}^0$, $\sigma_{\text{Incident}}^0$ and $\sigma_{\text{Transmitted}}^0$ respectively.

By using the special shaped striker, the stress waves on the input bar are expected to have half-sine waveforms. $\sigma_{1/4}^0$, $\sigma_{3/4}^0$ and $\sigma_{\text{Incident}}^0$ should be very similar when there is no damping.

iii) Measurement calibration. The measurement correction mainly deals with the attenuation of waves during propagation. The measurement calibration factor K_1 can be defined as:

$$K_1 = \max\left(|\sigma_{1/4}^0|\right)/\max\left(|\sigma_{3/4}^0|\right) \tag{5.14}$$

Then the incident and reflected waves in normal tests with specimens can be corrected respectively as:

$$\sigma_{\text{Incident}} = \sigma_{\text{Incident}}/K_1 \tag{5.15}$$

$$\sigma_{\text{Reflected}} = K_1 \cdot \sigma_{\text{Reflected}} \tag{5.16}$$

iv) Transmission correction. Transmission error mainly comes from the stress loss caused by the small gap between the input bar and the output bar. Besides, the wave attenuation for the traveling distance between specimen and the strain gauge at the middle of the output bar also contributes to the experimental error. By considering the gap effect and the attenuation effect, the transmission calibration factor can be defined as:

$$K_2 = \max\left(|\sigma_{\text{Incident}}^0|\right)/K_1^2 \max\left(|\sigma_{\text{Transmitted}}^0|\right) \tag{5.17}$$

Then the measured transmitted wave in tests with specimens should be corrected as:

$$\sigma_{\text{Transmitted}} = K_2 \cdot \sigma_{\text{Transmitted}} \tag{5.18}$$

5.4.2.4 Testing

a) Specimen placement

The specimen should be clamped between the input and output bar coaxially. Before placing, both ends of specimen should be lubricated with grease. When placing, the specimen should be rotated more than 360 degrees with tight squeezing from the input/output bar to achieve uniform lubrication. For specimens with large density, a supporter of soft material is needed to avoid dropping of the well-lubricated specimen.

b) Alignment of bars

The absorption bar is gently moved to the output bar until their surfaces contact totally. There should be no initial stress or visible gaps between contacted surfaces of the input bar, the output bar and the specimen.

c) Gas gun ready

Push the striker back into the gas gun to the right position, and adjust the gas pressure to preset level.

d) Strain meter ready

Press READY button to keep the strain meter waiting for trigger.

e) Gas gun action

Turn around the on-off switch on the gas gun to release the pressured gas. This pushes the striker out of the gun barrel to impact the input bar.

f) Data saving and transmission

After the strain meter captures the signals, test data should be saved and transmitted for post-processing.

5.4.3 Data processing

a) Wave extraction

With captured signals, the incident, reflected and transmitted waves can be extracted, as shown in Figure 5.11. The data from the strain meter are usually voltage signals and should be converted into stress or strain signals.

b) Wave shifting and value calibration

The obtained signals represent the stress/strain histories in the positions of strain gauges other than the specimen. So the extracted waves should be shifted to the bar/specimen interfaces for calculation. The incident wave should be shifted forward for $L_e/2C_e$, where L_e, C_e are the length and wave velocity of the input/output bar respectively.

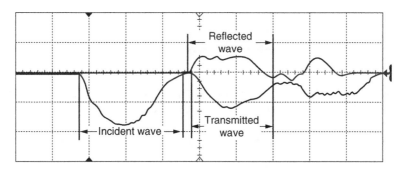

Figure 5.11 Incident, reflected and transmitted waves in captured signals.

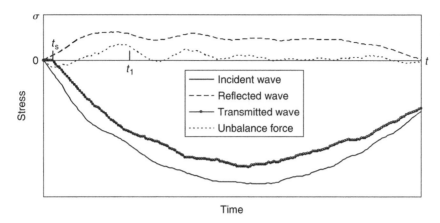

Figure 5.12 The incident, reflected, transmitted waves and the unbalance force in specimen.

Accordingly, the reflected and transmitted waves should be shifted backward for $L_e/2C_e$. Their values should be calibrated accordingly. Figure 5.12 shows a typical series of incident wave $\sigma_I(t)$, reflected wave $\sigma_R(t)$ and transmitted wave $\sigma_T(t)$, where they share the same time origin at which the wave propagates into the specimen from the input bar; t_s is the time needed for wave to go through the specimen.

c) Obtaining deformation variables of specimen

With the extracted incident, reflected and transmitted waves, the stress, strain and strain rate of the specimen can be derived according to SHPB principles:

$$\sigma(t) = \frac{A_e}{2A_s}[\sigma_I(t) + \sigma_R(t) + \sigma_T(t)] \tag{5.19}$$

$$\varepsilon(t) = \frac{1}{\rho_e C_e L_s} \int_0^t [\sigma_I(t) - \sigma_R(t) - \sigma_T(t)]dt \tag{5.20}$$

$$\dot{\varepsilon}(t) = \frac{1}{\rho_e C_e L_s}[\sigma_I(t) - \sigma_R(t) - \sigma_T(t)] \tag{5.21}$$

where A_e, ρ_e, C_e, E_e are the cross sectional area, density, wave velocity and Young's modulus of elastic bars, A_s, L_s are the cross-sectional area and length of the specimen.

5.5 DYNAMIC CHARACTERISTICS OF ROCK OBTAINED FROM SHPB WITH SPECIAL SHAPED STRIKER

5.5.1 Strain rate effect of rock under dynamic loading

Lots of dynamic tests were carried out on SHPB with the special shaped striker. Figure 5.13 shows a typical set of stress-stain curves of specimens under different

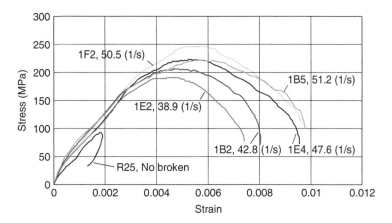

Figure 5.13 Typical stress-strain curves of specimen under dynamic loadings.

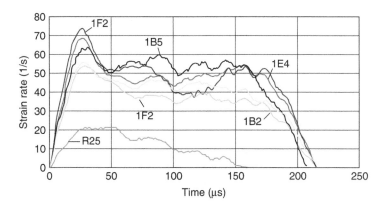

Figure 5.14 Strain rate histories of specimen in Figure 5.13.

strain rates. It can be observed from these curves that initially they follow the same path, and that the higher peak stresses apparently correspond to the higher strain rates. Figure 5.14 gives their strain rate histories respectively. For specimens that are totally fragmented, the strain rate increases to the maximum at a very short period of about 25 μs. Thereafter, the response remains constant over a relatively long period.

Figure 5.13 shows that although some results are scattered, there is a tendency that the strength increases with increasing strain rate. Regression analysis has further shown that the dynamic compression strength can be expressed as:

$$\sigma_f = K \left(\frac{d\varepsilon}{dt} \right)^{1/3} \tag{5.22}$$

where K is a constant and $d\varepsilon/dt$ is the strain rate.

5.5.2 Size effect of rock under dynamic loading

Rock material is usually heterogeneous with inner defects and different mineral grains, so its strength always shows a size effect. The size effect of static strength has been widely studied, while there are limits on dynamic size effect. With the special shaped striker on SHPB with bar diameter of 22 mm, 36 mm and 75 mm, size effects of granite, siltstone and limestone have been investigated. Figure 5.15 shows 4 different types of special shaped strikers used for SHPB tests with different bar diameters.

Figure 5.16 presents the size effect of granite under different strain rates. It can be seen that the relationship between dynamic strength and strain rate of rock can be described with Equation (5.22) when the same specimen dimension is used. However, when the specimen size is changed, the parameters would change accordingly. Under the same strain rate, the specimen with smaller size would have higher dynamic strength.

5.5.3 Dynamic strength of rock under coupling static and dynamic loads

Mining, geothermal exploitation and nuclear waste disposal lead us to deal with rock at great depth. Rock at great depth can endure high ground in situ stress, tectonic

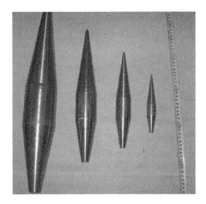

Figure 5.15 Strikers for dynamic size effect researches.

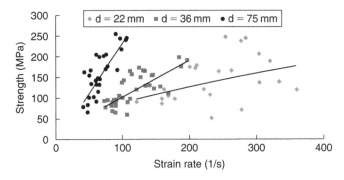

Figure 5.16 Size effect of granite under different strain rates.

stress, and dynamic loading of explosion and impact. Among these loads, some are static loads, while others are dynamic loadings. Therefore, rock behaviours under coupled static and dynamic loads have become of interest. The existing SHPB device was improved to conduct experiments on rock under coupled static and dynamic loads simultaneously (Li, Zhou and Lok, 2008).

Stress-strain curves of rock subjected to the same dynamic loading but different axial pre- stresses is shown in Figure 5.17.

The strength of rock under coupling loads changes when the axial static pressure is set to 80% of the specimen's static strength. The trend of this change is

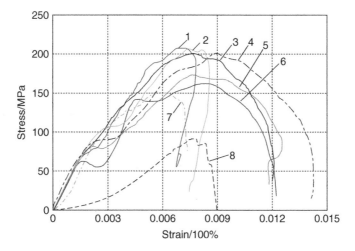

Figure 5.17 Stress-strain curves of rock subjected to same dynamic loading but different static pre-stresses.

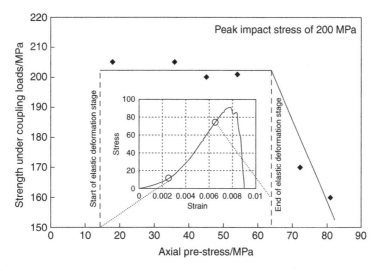

Figure 5.18 Dynamic strength of rock subjected to same dynamic loading but different static pre-stresses.

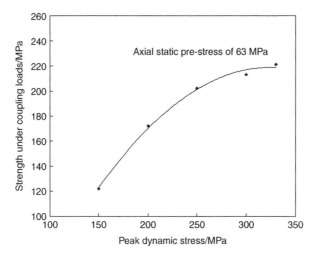

Figure 5.19 Strength of rock subjected to same axial pre-stress and different dynamic loadings.

evident in Figure 5.19. It can be seen that with coupling loads, the strength of rock increases with increasing dynamic peak stress. This phenomenon reinforces the original understanding that the material strength increases with increasing strain-rate.

5.6 CONCLUSIONS

Traditional SHPB devices with cylindrical strikers should be cautiously used for tests of brittle materials. Purpose-built special shaped strikers can be an excellent alternative. These typical strikers can generate an approximately half-sine wave, which is beneficial in overcoming premature failure of the specimen and realizing better stress uniformity in the specimen. The new method also appears to have the abilities of reducing signal oscillation and keeping specimen deformation at constant strain rate. The new system has been used to investigate the rate effect, size effect and coupled-load effect of rock materials, and good results have been obtained.

Acknowledgment: financial support from the National Natural Science Foundation of China (50934006, 50904079) and National Basic Research Program of China (2010CB732004) are greatly acknowledged.

REFERENCES

Anderson, S.P.: Higher-order rod approximations for the propagation of longitudinal stress waves in elastic bars. *Journal of sound and vibration* 290(1–2) (2006), pp.290–308.

Bazle, A.G., Sergey, L.L. and John, W.G.: Hopkinson bar experimental technique: A critical review. *Appl Mech Rev* 57(4) (2004), pp.223–250.

Davies, E. and Hunter, S.C.: The dynamic compression testing of solids by the method of the split Hopkinson pressure bar. *Journal of the Mechanics and Physics of Solids* 11 (1963), pp.155–179.

Follansbee, P.S. and Frantz C.: Wave propagation in the split Hopkinson pressure bar. *ASME J Eng Mater Technol* 105 (1983), pp.61–66.

Forrestal, M.J., Wright, T.W. and Chen, W.: The effect of radial inertia on brittle samples during the split Hopkinson pressure bar test. *International Journal of Impact Engineering* 34(3) (2007), pp.405–411.

Frantz, C.E., Follansbee, P.S. and Wright, W.J.: New experimental techniques with the split Hopkinson pressure bar. In: Berman, I. and Schroeder, J.W. (eds.), *Proc 8th Int Conf High Energy Rate Fabrication.* San Antonio, 1984, pp.17–21.

Frew D.J., Forrestal, M.J. and Chen W.: Pulse shaping techniques for testing brittle materials with a split Hopkinson pressure bar. *Experimental Mechanics* 42(1) (2002), pp.93–106.

Gong, J.C., Malvern, L.E. and Jenkins D.A.: Dispersion investigation in the split Hopkinson pressure bar. *ASME J Eng Mater Technol* 112 (1990), pp.309–314.

Li, X.B. and Gu, D.S.: *Rock Impact Dynamics.* Central South Univ Technol Press, Changsha, 1994.

Li, X.B. and Zhou Z.L.: Large diameter SHPB tests with special shaped striker. *ISRM News Journal* 12 (2009), pp.76–79.

Li, X.B., Lok, T.S. and Zhao J.: Dynamic characteristics of granite subjected to intermediate loading rate. *Rock Mech Rock Eng* 38(1) (2005), pp.21–39.

Li, X.B., Lok, T.S. and Zhao J.: Oscillation elimination in the Hopkinson bar apparatus and resultant complete dynamic stress-strain curves for rocks. *Int J Rock Mech Min Sci* 37 (2000), pp.1055–1060.

Li, X.B., Zhou, Z.L. and Lok T.S.: Innovative testing technique of rock subjected to coupled static and dynamic loads. *Int J Rock Mech Min Sci* 45(5) (2008), pp.739–748.

Lifshitz, J.M. and Leber, H.: Data processing in the split Hopkinson pressure bar tests. *International Journal of Impacting Engineering* 15(6) (1994), pp.723–733.

Liu, D.S., Peng, Y.D. and Li, X.B.: Inverse design and experimental study of impact piston. *Chinese Journal of Mechanical Engineering* 34(4) (1998), pp.78–84.

Lok, T.S., Li, X.B., Liu, D. and Zhao, P.J.: Testing and response of large diameter brittle materials subjected to high strain rate. *ASCE J Mater Civil Eng* 14(3) (2002), pp.262–271.

Meng, H. and Li, Q.M.: Correlation between the accuracy of a SHPB test and the stress uniformity based on numerical experiments. *International Journal of Impact Engineering* 28(5) (2003), pp.537–555.

Ravichandran, G. and Subhash, G.: Critical appraisal of limiting strain rates for compression testing of ceramics in a split Hopkinson pressure bar. *J America Ceram Soc* 77(1) (1994), pp.263–267.

Tedesco, J.W. and Ross, C.A.: Strain-rate dependent constitutive equations for concrete. *J Pressure Vessel Technol Trans ASME* 120(4) (1998), pp.398–405.

Wu, X.J. and Gorham, D.A.: Stress equilibrium in the split Hopkinson pressure bar test. *J Phys IV, France* C3 (1997), pp.91–96.

Yang, L.M. and Shim, V.P.: An analysis of stress uniformity in split Hopkinson bar test specimens. *Int J Impact Engg* 31 (2005), pp.129–150.

Zhao, H. and Gary, G.: On the use of SHPB techniques to determine the dynamic behavior of materials in the range of small strains. *Int J Solids Struct* 33(23) (1996), pp.3363–3375.

Zhao, H.: Material behaviour characterisation using SHPB techniques, tests and simulations. *Computers & Structures* 81(12) (2003), pp.1301–1310.

Zhou, Z.L., Li, X.B. and Ye, Z.: Obtaining constitutive relationship for rate-dependent rock in SHPB tests, *Rock Mechanics and Rock Engineering* 43(6) (2010), pp.697–706.

Chapter 6

Laboratory compressive and tensile testing of rock dynamic properties

Haibo Li, Junru Li and Jian Zhao

6.1 INTRODUCTION

Dynamic loads are usually associated with high amplitude and short duration. A proper understanding of the effect of strain/loading rate on the mechanical properties of rock is important in the analysis of rock behavior or the design of rock structures subjected to dynamic loads. For example, in rock blasting, a blast stress wave is generated and propagates through the rock mass; the rock mass and rock structure are subjected to blast shock loads at different strain/loading rates. Mechanical responses of a rock mass and a rock structure to various strain/loading rates are different (Serdengecti and Boozer, 1961; Price and Knill, 1966; Stowe and Ainsworth, 1968; Birkimer, 1971; Mellor and Hawkes, 1971; Shockey *et al.*, 1973; Vutukuri, Lama and Saluja, 1974; Janach, 1976).

Experiments have been conducted to study the effects of strain/loading rate on rock properties under dynamic loading. For example, Donath and Fruth (1971) conducted dynamic triaxial compression tests on a marble at confining pressure of 100 and 200 MPa, and strain rate from 10^{-7} to 10^{-3} s^{-1}. They reported that at confining pressure of 100 MPa, the strength increased by 30% when the strain rate increased by 5 orders of magnitude, while under the confining pressure of 200 MPa, the strength increased by 40%, for the same increment of strain rate. Similarly, Logan and Handin (1972) conducted dynamic triaxial compression tests on the Westerly granite at confining pressures up to 700 MPa. They found that the strength increases proportionally with increasing strain rate and the rising rate increases with increasing confining pressure. The same results are also reported by Masuda, Mizutani and Yamada (1987) on granite. The loading rate (or strain rate) has effects on rock fracturing mechanisms and mechanical properties including strengths and deformation modules. Extensive experimental results have indicated that rock strength increases with loading rate (e.g. Logan and Handin, 1972; Sangha and Dhir, 1975; Grady and Kipp, 1980; Lankford, 1981; Masuda, Mizutani and Yamada, 1987; Wu and Gao, 1987; Lajtai, Scott and Carter, 1991; Olsson, 1991; Ju and Wu, 1993; Yang and Li, 1994; Wu and Liu, 1996; Zhao and Zhao, 1998; Li, Zhao and Li, 1999; Li *et al.*, 2000; Li *et al.*, 2004; Zhao *et al.*, 1999a; Zhao *et al.*, 1999b; Cho, Ogata and Kaneko, 2003). The dynamic strength criteria would be different from the static ones. Attempts have been made to provide a dynamic version of Mohr-Coulomb and Hoek-Brown criteria (e.g. Zhao and Li, 2000), by including a loading rate depended term in these criteria.

Through dynamic Brazilian tests on dolerite and limestone, Price and Knill (1966) suggested that strength of both rocks generally increases with the increasing loading rate. For dolerite, the tensile strength at highest loading rate is 17% greater than that at the lowest loading rate. For limestone, the tensile strength at highest loading rate is 44% greater than that at the lowest loading rate. Wu and Liu (1996) studied the changes of tensile strength, Young's modulus and the failure strain for Longmen limestone at the loading rates from 10^{-3} MPa/s to 10^4 MPa/s by Brazilian test. It is reported that the tensile strength, Young's modulus and the failure strain for the rock clearly increase with loading rate. Based on the SHPB test for quartz monzonite rock and strain energy analysis for the failure of rock, Birkimer (1971) pointed out that the dynamic tensile strength of the rock increases with the cube root of the strain rates when strain rates range from 10^1 and $10^4\,\mathrm{s}^{-1}$. The similar results are also suggested by Grady and Kipp (1980).

In this chapter, dynamic compressive and tensile tests of rock material at moderate strain/loading rates are conducted. The apparatus, loading device, the sample preparation as well as the analysis of experimental results are presented.

6.2   DYNAMIC COMPRESSION TESTS FOR ROCK MATERIAL

A series of uniaxial and triaxial dynamic compression tests on the Bukit Timah granite of Singapore are conducted. The rock samples are mostly tested at 4 strain/loading rates and 7 confining pressures.

6.2.1   Test equipment

Figure 6.1 shows the layout of the system for dynamic compression tests. It consists of three parts: compressive loading frame, axial dynamic loading system and data acquisition system.

6.2.1.1   Compressive loading frame

The loading frame has a stiffness of 2.92 GN/m. It is capable of conducting uniaxial and triaxial dynamic compression testing of rock specimens up to 300 mm in height. During uniaxially compressive tests, a chamber is adopted to confine sample fragments.

6.2.1.2   Axial dynamic loading system

The axial dynamic loading system is driven hydraulically, as illustrated in Figure 6.2. The system mainly consists of two gas cylinders (A and B), a connecting piston, a release valve, a regulating valve, an oil cylinder and a loading piston. During the testing, pressures in the gas cylinders A and B are initially increased to a desired value, so as to produce a stress imposing on the rock sample at about three times its static strength. This is to ensure that the load generated by the loading system is sufficient to make the rock sample fail. When the release valve is opened, gas in the cylinder B escapes, and the pressure quickly reduces to zero. The connecting piston is then pushed down by the pressure in the cylinder A. The movement of the connecting piston creates a pressure in the oil cylinder and in turn, pushes the loading piston, thus applying a

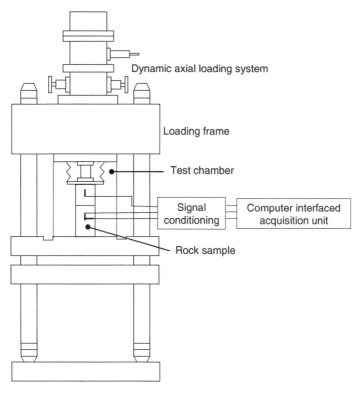

Figure 6.1 Schematic layout of a dynamic compressive machine and data acquisition system.

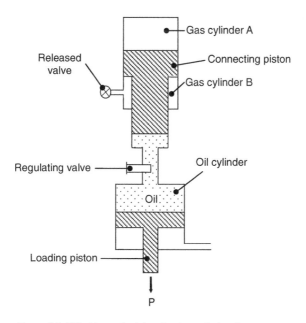

Figure 6.2 Working principle of a dynamic loading system.

dynamic load to the rock sample. The regulating valve controls the oil flow between the upper and lower parts of the oil cylinder, hence the piston is loaded. Four classes of the regulating valve are available to produce dynamic loads ranging from 10^0, 10^1, 10^3 and 10^5 MPa/s. The maximum axial dynamic load is 2,200 kN and the travel distance for the loading piston is 25 mm. The minimum rising time for an axial load of 2,200 kN is 12 ms.

6.2.1.3   Data acquisition system

The data acquisition system consists of a signal conditioning unit and an acquisition unit interfaced with a computer. The signal conditioning unit is used to convert the resistance signals of strain gauges to voltage signals. It has six connecting channels and is capable of recording strains of up to 100,000 uε. A computer is interfaced with an acquisition unit to acquire and store data. It contains six channels. The sampling data points and sampling time interval are selected according to the loading rate. The maximum number of sampling data points is 16,384 and the minimum sampling interval is 10 ms. The system can obtain more than 100 data points even at the highest loading rate (10^5 MPa/s) and more than 1,000 data points at a lower loading rate. Axial dynamic load, axial and circumferential strains are measured in the tests. The axial load is measured by a load cell consisting of a strain-gauged high strength steel block. Axial and circumferential strains are measured by two axial and two circumferential strain gauges which are glued on the rock samples at the mid-height and opposite sides diametrically.

6.2.2   Sample set-up and test technique

6.2.2.1   Sample preparation and test set-up

Cylindrical rock samples of 30 mm diameter and 60 mm length were cut from 52 mm diameter granite cores. The axial direction of each sample was the same as that of the rock cores. The ends of each sample were ground to be flat and parallel to each other. The deviation in the diameter and the undulation of ends were less than 0.2 mm. The vertical deviation was less than 0.001 radian. The samples were left air-dried over several days. Specific gravity and sonic velocity of the samples were measured before gluing the strain gauges. The sample stack, consisting of the sample, the load cell and the spacers, was then placed in the compression machine and connected to the data acquisition system before the commencement of testing.

6.2.2.2   Test procedure

The testing system was firstly calibrated before testing. Trial tests were conducted to obtain optimum parameters of the gain, the number of sampling points and the sampling interval of the acquisition system at different loading rates. The main test procedures are described below:

1)   The data acquisition system was connected to the sample stack,
2)   The loading piston was brought in contact with the sample stack,
3)   The regulating valve of the dynamic loading system was selected for the desired loading rate,

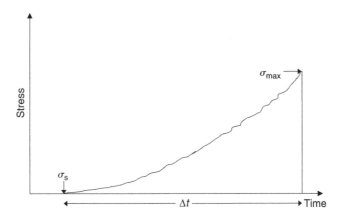

Figure 6.3 Schematic illustration of loading rate determination.

4) Pressures in the gas cylinders (A and B) were increased to the desired value,
5) The release valve was opened to apply dynamic loading to rock samples,
6) Stress and strain histories were recorded.

6.2.2.3   *Determination of loading rate*

Loading rate is determined from the stress history curve. The average loading rate is determined by

$$\bar{\dot{\sigma}} = \frac{\sigma_{max} - \sigma_s}{\Delta t} \tag{6.1}$$

where $\bar{\dot{\sigma}}$ denotes the average loading rate, σ_{max} is the maximum (failure) stress, σ_s is the initial stress, and Δt is the loading time. Figure 6.3 illustrates the definition of all parameters.

6.2.3   **Experimental results**

6.2.3.1   *Uniaxial compression*

A total of 12 dynamic uniaxial compression tests were performed at four different loading rates, i.e., 10^0, 10^1, 10^3 and 10^5 MPa/s. Figure 6.4 shows the typical stress and strain histories of the granite tested. The strength, Young's modulus and Poisson's ratio were obtained from the stress-strain curves. Test results are summarized in Figures 6.5 to 6.7, where Figure 6.5 shows the relationship between the uniaxial compressive strength and the loading rate. It can be seen from the figure that the uniaxial compressive strength increases with increasing loading rate. An average increase of 15% was observed when the loading rate was increased from 10^0 to 10^5 MPa/s. The increase of the uniaxial compressive strength with increasing loading rate has been observed by many other researchers. Figures 6.6 and 6.7 plot the values of Young's modulus and Poisson's ratio with the change of the loading rate. The results are scattered and

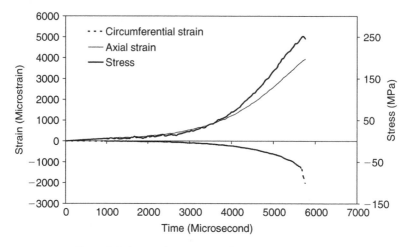

Figure 6.4 A typical stress-strain history measurement.

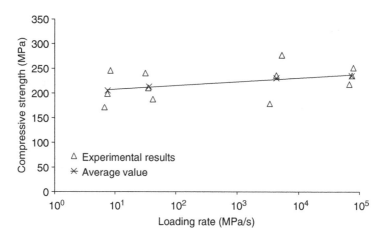

Figure 6.5 Change of uniaxial compressive strength with loading rate.

yet seem to indicate that the Young's modulus decreases slightly and the Poisson's ratio increases slightly with increasing loading rate. It was noted that the relationship between the uniaxial compressive strength and the loading rate could be described by the formula below:

$$\sigma_{dc} = A \log(\dot{\sigma}_{dc}/\dot{\sigma}_{sc}) + \sigma_{sc} \tag{6.2}$$

As shown in Figure 6.8, where σ_{dc} is the dynamic uniaxial compressive strength, $\dot{\sigma}_{dc}$ is the dynamic loading rate, $\dot{\sigma}_{sc}$ is the quasi-static loading rate (e.g. 5×10^{-5} MPa/s), σ_{sc} is the uniaxial compressive strength at the quasi-static loading rate, and A is a constant.

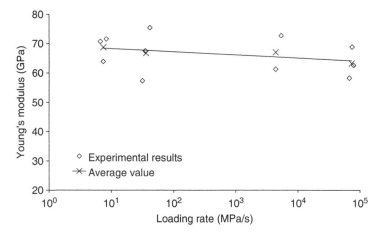

Figure 6.6 Change of the Young's modulus with loading rate.

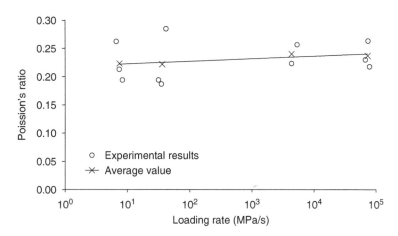

Figure 6.7 Change of the Poisson's ratio with loading rate.

For the granite tested, σ_{sc} is at 170 MPa obtained at 5×10^{-5} MPa/s. Regression analysis shown in Figure 6.8 indicates that A is 11.9.

However, it should be noted that the changes in the Young's modulus and Poisson's ratio of the granite with the loading rate are very small. Therefore, it can be considered that the Young's modulus and the Poisson's ratio are unaffected by the loading rate in the test range.

6.2.3.2 Triaxial compression

Figure 6.9 plots the typical stress-strain curves at different strain rates and confining pressures obtained from the tests. At a constant confining pressure, the triaxial compressive strength conclusively increases with increasing strain rate, as shown in

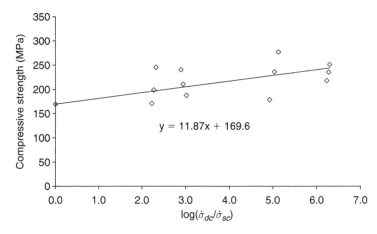

Figure 6.8 Relationship between the dynamic strength, static compressive strength and the normalized loading rate.

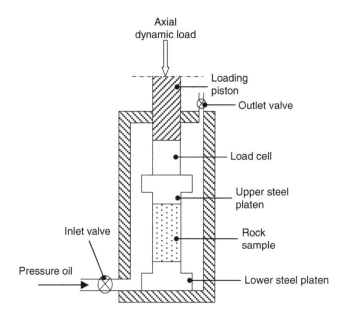

Figure 6.9 Schematic illustration of triaxial confining pressure cell system.

Figure 6.10. In addition, it is shown that the rising rates of compressive strength decrease with the increasing of confining pressures, that is, at the confining pressure of 20 MPa, the rising rate of strength reaches 50% when the strain rate increases by 5 orders of magnitude, while at the confining pressure of 170 MPa, the rising rate is about 5% for the same increment of strain rate. These results generally agree with the experimental observations by Yang and Li (1994), Sangha and Dhir (1975), and Ju and

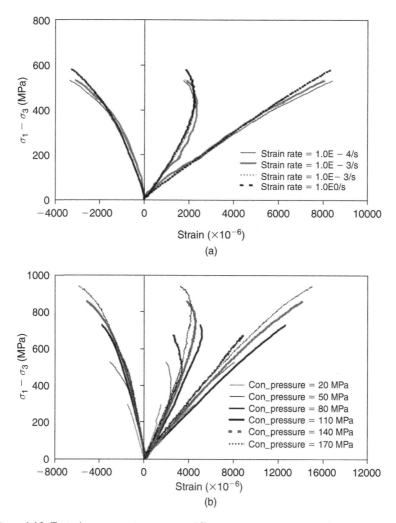

Figure 6.10 Typical stress-strain curves at different strain rates and confining pressures.

Wu (1993). At a constant strain rate, the compressive strengths clearly increase with increasing confining pressures, as shown in Figure 6.11. The change of compressive strength with confining pressures under dynamic loading condition seems follow the similar trend of static loading condition, i.e. the strength envelope complies with the Hoek-Brown criterion. It also appears that the tendencies of the strength changing with confining pressure are almost identical for different strain rates. Figures 6.12 and 6.13 present the change of the Young's modulus with strain rate and confining pressures. It is noted that the results are scattered. The Young's modulus shows no conclusive trend with increasing strain rate and it therefore can be suggested that the Young's modulus is not affected by strain rate. The Young's modulus appears to increase with increasing confining pressure, but the increment is very small.

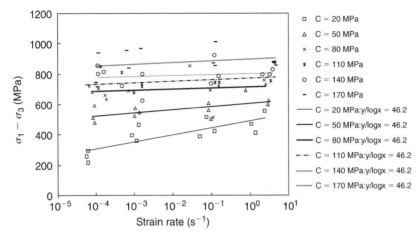

Figure 6.11 Variation of the axial deviator strength with the strain rate at different confining pressures.

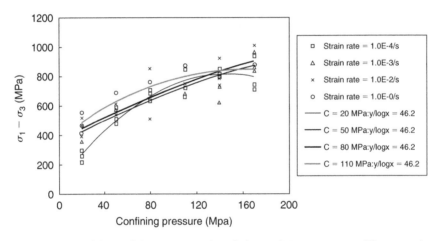

Figure 6.12 Variation of the axial deviator strength with the confining pressure at different strain rates.

Similarly, the results of the Poisson's ratio at different strain rates and confining pressures are scattered, as plotted in Figures 6.14 and 6.15. The Poisson's ratio seems to increase slightly with increasing strain rate and confining pressures. Nevertheless, the changes are not significant.

6.3   DYNAMIC TENSION TESTS FOR ROCK MATERIAL

A series of indirect tensile tests using the Brazilian method and the 3-point flexural method were conducted to determine the indirect tensile strength and Young's modulus of the Bukit Timah granite in Singapore. A total of twenty three Brazilian tests at 4

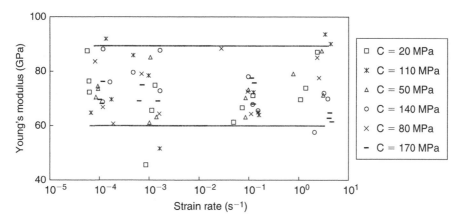

Figure 6.13 Variation of the Young's modulus with the strain rate at different confining pressures. (C denotes confining pressure)

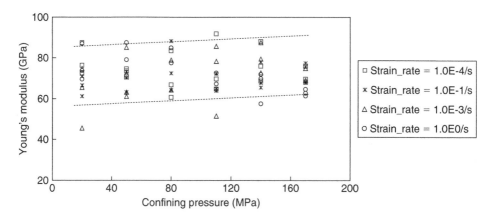

Figure 6.14 Variation of the Young' modulus with the confining pressure at different strain rates.

different loading rates, i.e. 10^{-1}, 10^0, 10^2, 10^3 MPa/s, and a total of twenty 3-point flexural tests at 4 loading rates, i.e. 10^0, 10^2, 10^3, 10^4 MPa/s were carried out.

6.3.1 Dynamic Brazilian test system and procedures

The Brazilian tensile test is schematically illustrated in Figure 6.17. The Brazilian tensile strength is determined by

$$\sigma_{td} = 2P_{max}/Dd \tag{6.3}$$

where σ_{td} is the Brazilian tensile strength, P_{max} is the maximum load, D and d are the diameter and thickness of the specimen, respectively.

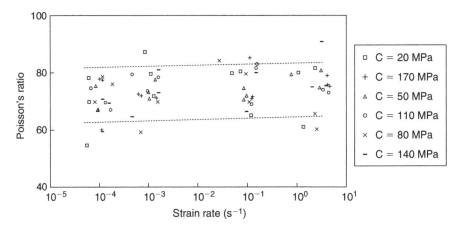

Figure 6.15 Variation of the Poisson's ratio with the strain rate at different confining pressures. (C denotes confining pressure).

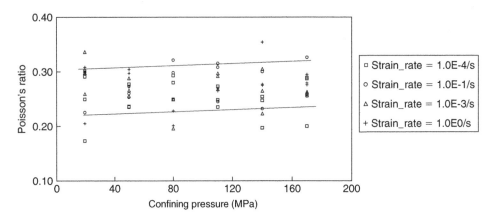

Figure 6.16 Variation of the Poisson's ratio with the confining pressure at different strain rates.

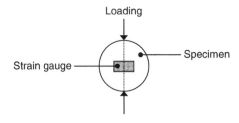

Figure 6.17 Schematic illustration of the Brazilian tensile test.

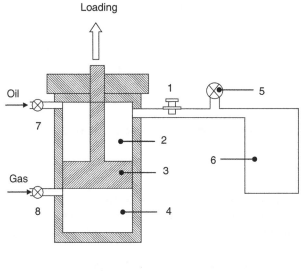

Figure 6.18 Principle of the dynamic loading machine.

1	Regulating valve	2	Oil chamber
3	Loading Piston	4	Gas chamber
5	Release valve	6	Oil reservoir
7	Oil valve	8	Gas valve

A set of Brazilian tests were carried out by using a 100 KN dynamic loading machine, as shown in Figure 6.18. The dynamic loading system was air- and oil-hydraulically driven. During the test, the pressure in the gas chamber was initially set to the desired value which can cause the specimen failure. The pressure of the oil chamber was also set to balance the gas chamber pressure. Subsequently, the release valve was opened, the oil in the oil chamber quickly flowed to the oil reservoir and the loading piston moved upward to impose a dynamic load on the specimen. The regulating valve controlled the oil flowing velocity and the loading time by the piston. As for the loading system used in the present study, the minimum rising time is 8 ms for the maximum axial loading of 100 KN, and the maximum piston travel distance is 25 mm.

The load was monitored by a load cell placed above the specimen. The tensile strain was measured by two electricity resistance strain gauges which were perpendicular to the loading axis, mounted at the mid-point of both sides of the specimen. The data acquisition system, which is the same as that described in the dynamic compression test,

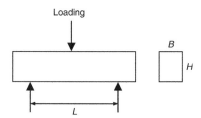

Figure 6.19 Schematic illustration of the 3-point flexural tensile test.

recorded the load and strain histories at the desired frequency. The average loading rate is calculated by the maximum load divided by the loading time. The stress and strength of the specimen can be calculated from Equation (6.3). The tensile Young's modulus is determined from the average slopes of the (more or less) straight-line portion in the stress-strain curves.

From the rock cores with diameter 50 mm, the specimen used in the Brazilian tests was cut into a disc with 50 mm diameter and 20 mm thickness. The two flat sides of the specimen were ground to parallel and their undulation was less than 0.03 mm.

The testing system was calibrated before tests. Trial tests were carried out to obtain the desired load and loading rate, and the optimum parameters of gain and sampling frequency at different loading rates.

6.3.2 Dynamic 3-point flexural test system and procedures

Figure 6.19 shows the schematic illustration for the 3-point flexural test. The 3-point flexural tensile strength of the specimen can be expressed as:

$$\sigma_{td} = 1.5 P_{\max} L / B H^2 \tag{6.4}$$

where σ_{td} is the 3-point flexural tensile strength, $P_{\max}$ is the maximum load, L is the span of the two supports, B and H are the width and height of the specimen, respectively.

The 3-point flexural tensile tests were also carried out on the same dynamic loading machine used for the Brazilian tests. In the tests, the same load cell used for the Brazilian tests was adopted to record the load. One strain gauge was pasted in the middle of the bottom side of the specimen to monitor the tensile strain. The test procedures are basically the same as those of the aforementioned Brazilian tests.

The average loading rate and Young's modulus of the 3-point flexural tensile test are calculated similarly to the methods in the Brazilian tests. The stress and strength of the specimen is also obtained by Equation (6.3).

The granite specimens used in the 3-point flexural tests were cut to cubes of size $140 \times 30 \times 15$ mm (length $\times$ height $\times$ width). The undulation for the surfaces of the specimens was less than 0.03 mm.

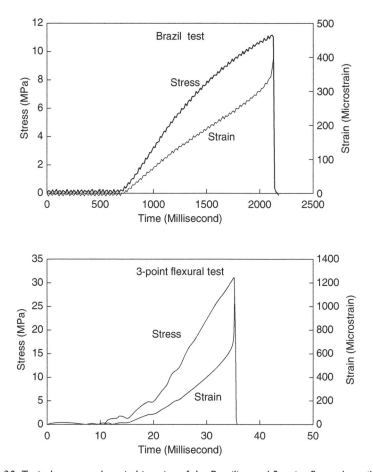

Figure 6.20 Typical stress and strain histories of the Brazilian and 3-point flexural tensile tests.

6.3.3 Experimental results

Figure 6.20 shows the typical stress and strain histories for the Brazilian and 3-point flexural dynamic tensile tests. The stress-strain curves, strength, Young's modulus and loading rate are interpreted from these histories.

Figure 6.21 shows the Brazilian and 3-point flexural tensile strengths at different loading rates. The results indicate that the tensile strengths obtained by both methods increase with increasing loading rate. It is also observed from the figure that the tensile strength determined by the 3-point flexural method is about 2.5 times of that determined by the Brazilian method when the loading rates are the same. In addition, although the tensile strengths obtained by the two methods are different, the trend of the dynamic tensile strength with the loading rate is almost the same, that is, when the loading rate increases by one order of magnitude, the tensile strengths of the granite increase by approximately 10%.

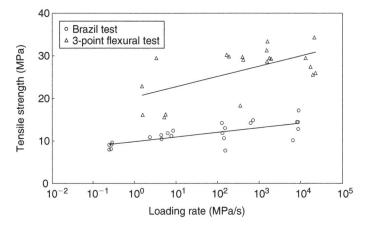

Figure 6.21 The Brazilian and 3-point flexural tensile strength at the different loading rates.

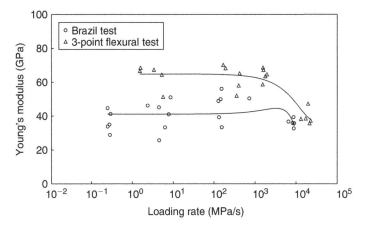

Figure 6.22 The Brazilian and the 3-point flexural tensile Young's modulus at the different loading rates.

The tensile Young's modulus at different loading rates is plotted in Figure 6.22. It can be seen that the Young's modulus obtained in the Brazilian tests tends to increase with increasing loading rate when the loading rate ranges from 10^{-1} to 10^2 MPa/s. When the loading rate is greater than 10^3 MPa/s, the Young's modulus begins to drop. The same tendency can also be observed from the 3-point flexural tests. It is found that the Young's modulus obtained by the Brazilian method is generally lower than that obtained by the 3-point flexural method. In addition, it should be noted that, except for the loading rates of 10^3 MPa/s for the Brazilian tests and 10^4 MPa/s for the 3-point flexural tests, the Young's modulus of the granite changes slightly with the loading rate, and the results are scattered.

6.4 SUMMARY

Dynamic compressive tests for the Bukit Timah granite were carried out at 4 moderate loading/strain rates and 6 confining pressures. It is concluded that the compressive strength generally increases with increasing loading/strain rate and confining pressure. The rate of increment of the compressive strength with strain rate is lower under a higher confining pressure. The strength envelopes at different strain rates, however, appear similar.

The results for the Young's modulus and the Poisson's ratio at different loading/strain rates and confining pressures are scattered. The Young's modulus increases slightly with increasing confining pressure and appears to be unaffected by loading/strain rate, while the Poisson's ratio increases with increasing loading/strain rate and confining pressure. Therefore, further tests are needed to overcome the scattering of the results to obtain conclusive indications on the change of the Young's modulus and the Poisson's ratio.

The dynamic tensile strengths of the Bukit Timah granite in Singapore were also determined by the Brazilian and 3-point flexural methods at different loading rates. The results show that the tensile strengths of the granite obtained by both methods increase with increasing loading rate and the trends for the change of the tensile strength with the loading rates are very close.

REFERENCES

Birkimer, D.L.: A possible fracture criterion for the dynamic tensile strength of rock. In: Clark G.B. (ed.): *Proc 12th U.S Symp Rock Mech*. Baltimore, Maryland, 1971, pp.573–590.

Cho, S.H., Ogata, Y. and Kaneko, K.: Strain-rate dependency of the dynamic tensile strength of rock. *Int J Rock Mech Min Sci*, 40(5) (2003), pp.763–777.

Donath, F.A. and Fruth, L.S.: Dependence of strain rate effects on deformation mechanism and rock type. *J Geol Soc* 79 (1971), pp.343–371.

Grady, D.E. and Kipp, M.E.: Continuum modeling of explosive fracture in oil shale. *Int J Rock Mech Min Sci* 17 (1980), pp.147–157.

Janach, W.: The role of bulking in brittle failure of rock under rapid compression. *Int J Rock Mech Min Sci* 13 (1976), pp.177–186.

Ju, Q.H. and Wu, M.B.: Experimental studies of dynamic properties of rocks under triaxial compression. *Chinese J Geotech Engng* 15 (1993), pp.73–80 (in Chinese).

Lajtai, E.Z., Scott Duncan, E.J. and Carter, B.J.: The effect of strain rate on rock strength. *Rock Mech Rock Eng* 24 (1991), pp.99–109.

Lankford, J.: The role of tensile microfracture in the strain rate dependence of the compressive strength of fine-grained limestone analogy with strong ceramics. *Int J Rock Mech Min Sci* 18 (1981), pp.173–175.

Li H.B., Zhao J., Li T.J. and Yuan J.X: Analytical simulation of the dynamic compressive strength of a granite using the sliding crack model. *International Journal for Numerical and Analytical Methods in Geomechanics* 25(2001), pp.853–869.

Li, H.B., Zhao, J. and Li, T.J.: Triaxial compression tests of a granite at different strain rates and confining pressures. *Int J Rock Mech Min Sci* 36(8) (1999), pp.1057–1063.

Li, H.B., Zhao, J., Li, J.R., Liu, Y.Q. and Zhou, Q.C.: Experimental studies on the strength for different rock under dynamic compression. *Int J Rock Mech Min Sci* 41(3) (2004), p.365.

Logan, J.M. and Handin, J.: Triaxial compression testing in intermediate strain rate. In: Clark, G.B. (ed.), *Dynamic Rock Mechanics*. Port City Press, USA, 1972, pp.167–194.

Masuda, K., Mizutani, H. and Yamada, I.: Experimental study of strain-rate dependence and pressure dependence of failure properties of granite. *J Phys Earth* 35 (1987), pp. 37–66.

Mellor, M. and Hawkes, I.: Measurement of tensile strength by diametrical compression of discs and annuli. *Eng Geol*, 5 (1971), pp.173–225.

Olsson, W.A.: The compressive strength of Tuff as a function of strain rate from 10^{-6} to 10^3/sec. *Int J Rock Mech Min Sci* 8(1) (1991), pp.115–118.

Price, D.G. and Knill, J.L.: A study of the tensile strength of isotropic rocks. *Proc 1st Congr Int Soci Rock Mech*, Lisbon, 1966, pp.439–442.

Sangha, CM and Dhir, RK. Strength and deformation of rock subject to multiaxial compressive stresses. *Int J Rock Mech Min Sci*, 1975, 12: 277–282.

Serdengecti, S. and Boozer, G.D.: The effects of strain rate and temperature on the behaviour of rocks subjected to triaxial compression. In: Hartman H.L. (ed.): *Proc 4th Symp Rock Mech.*, Penn State Univ, 1961, pp.83–97.

Shockey, D.A., Petersen, C.F., Curran, D.R. and Rosenberg, J.T.: Failure of rock under high rate tensile loads. In: Hardy, H.R. Jr and Stefanko, R. (eds.): *Proc 14th Symp Rock Mech*. Penn State Univ, Pennsylvania, 1973, pp.709–738.

Stowe, R.L. and Ainsworth, D.L.: Effect of rate loading on strength and Young's modulus of elasticity of rock. In: Cray K.E. (ed.): *Proc 10th U.S Symp Rock Mech*. Austin, Texas, 1968, pp.3–34.

Vutukuri, V.S., Lama, R.D. and Saluja, S.S.: Handbook on mechanical properties of rock. *Trans Tech Publications* 1 (1974), pp.93–95.

Wu, M.B. and Gao, J.G.: Experimental study on dynamic properties of the Yangquan coal. *Chinese J Coal* 3(1987), pp.31–37 (in Chinese).

Wu, M.B. and Liu, Y.H.: Experimental study on dynamic properties of the Longmen limestone. *Chinese J Rock Mech Engng* 15 (1996), pp.422–427 (in Chinese).

Yang, C.H. and Li, T.J.: The strain rate-dependent mechanical properties of marble and its constitutive relation. *Int Conf Computational Methods in Structural and Geotechnical Engineering*, Hongkong, 1994, pp.1350–1354.

Zhao, J. and Li, H.B.: Experimental determination of dynamic tensile properties of a Granite. *Int J Rock Mech Min Sci* 37 (2000), pp.861–866.

Zhao, J., Li, H.B., Wu, M.B. and Li, T.J.: Dynamic uniaxial compression tests on granite. *Int J Rock Mech Min Sci* 36(2) (1999a) pp.273–277.

Zhao, J., Zhou, Y.X., Hefny, A.M., Cai, J.G., Chen, S.G., Li, H.B., Liu, J.F., Jain, M., Foo, S.T., and Seah, C.C.: Rock dynamics research related to cavern development for ammunition storage. *Tunnelling and Underground Space Technology* 14(4) (1999b), pp.513–526.

Zhao, Y.H. and Zhao, J.: Compressive strength of rock materials at different strain rate. In: Yu, M.H. and Fan, S.C. (eds.), *Strength theory: Application, developments and prospect for the 21st Century*. Beijing, New York: Science Press, 1998, pp.497–505.

Chapter 7

Penetration and perforation of rock targets by hard projectiles

Chong Chiang Seah, Tore Børvik, Svein Remseth and Tso-Chien Pan

7.1 INTRODUCTION

Today, underground facilities in rocks are used extensively for military applications and civil defence (Zhao *et al.*, 1999). Because of their depth and hardened status, many of these strategic hard and deeply buried targets could only be put at risk by earth penetrating weapons (EPW). With the development of precision guidance systems, the scenario of an EPW hitting its target has become highly probable, and some weapons are even capable of penetrating deeper when assisted by a pilot-hole. This has posed great challenges to defence engineers and scientists working in both the field of weaponry and protective technology. Baty, Lundgren and Patterson (2003) recently reported on penetration tests conducted on concrete and in situ weathered granite with predrilled holes to understand the terra-dynamics of pilot-hole assisted penetration. Antoun, Lomov and Glenn (2003) have also undertaken a computational study to investigate the penetration efficiency of a sequence of penetrating bombs into granitic hard rock. However, the amount of non-classified literature on projectile penetration in rock materials remains small and limited.

Studies concerned with the penetration of projectiles into rock targets usually focus on the depth of penetration, penetration deceleration history or stresses on the nose (Young, 1969, 1997; Forrestal, Longcope and Norwood, 1981a, 1981b; Longcope and Forrestal, 1983; Frew, Forrestal and Hanchak, 2000). While several works have been published on the penetration and perforation of concrete targets (Hanchak *et al.*, 1992; Børvik *et al.*, 2002; Børvik, Gjørv and Langseth, 2006; Sjøl and Teland, 2003), it is not found in any open source literature that such ballistic experiments have been performed on rock targets of finite thickness, where the rear surface of the target exerts considerable influence on the deformation process during all (or nearly all) of the projectile motion. Hence, this chapter aims to extend the present knowledge of projectile penetration in rock materials to include the process of perforation. Sets of ballistic penetration experiments were conducted to study the penetration and perforation of $0.6 \text{ m} \times 0.6 \text{ m} \times 0.1 \text{ m}$ thick granite targets by 0.2 kg, 20 mm diameter hard projectiles. The spherical cavity-expansion model developed by Forrestal and Tzou (1997) for predicting stresses on the projectile nose and the final penetration depths of projectiles penetrating into concrete targets was extended to predict the perforation of granite target plates by hard projectiles with a three-stage analytical model.

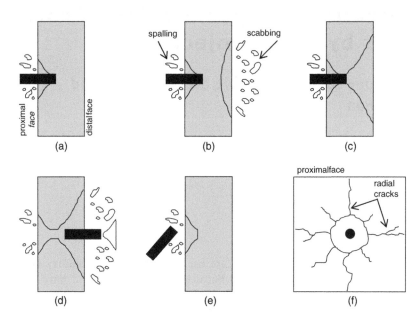

Figure 7.1 Projectile impact phenomena on rock-like targets: (a) penetration; (b) spalling and scabbing; (c) cone plugging; (d) perforation; (e) ricochet and rebound; (f) formation of radial cracks.

7.2 TERMINOLOGY

A large amount of work has been published in the field of impact and penetration mechanics, which also results in conflicting use of several technical terms. Therefore it is important that some of the terminology used in this field, which is essential to the subsequent discussions, be appropriately defined. Figure 7.1 shows the possible failure mechanisms in a rock target of finite thickness when impacted by a hard projectile.

- Penetration: Tunneling of a projectile into the target (the length of the tunnel measured from the proximal face is called the penetration depth *P*).
- Spalling: Ejection of target material from the proximal face of the target.
- Scabbing: Ejection of fragments from the distal face of the target due to reflected tensile waves.
- Cone plugging: Formation of a conical plug ahead of the projectile that is commonly observed at the distal face of a brittle target with intermediate thickness.
- Perforation: Complete passage of the projectile through the target.
- Ricochet or rebound: Deflection of the projectile from the target without being stopped.
- Radial cracking: Global cracks radiating from the impact point and appearing on either the proximal or distal face of the concrete slab or both, if the cracks develop through the target thickness.

The above-mentioned terms are defined as the local impact effects. In order to evaluate the projectile and target performance, limit velocities are often used, and we will consider and define the two following limit velocities:

- Ballistic limit (V_{bl}): The average of two striking velocities, one of which is the highest velocity giving a partial penetration and the other of which is the lowest velocity giving perforation.
- Protection limit (V_{pl}): The average of two striking velocities, one of which is the highest velocity such that daylight cannot be seen through the plate at the point of impact and the other of which is the lowest velocity such that the projectile creates a through hole but does not perforate the target.

7.3   EXISTING METHODS OF ANALYSIS AND PREDICTION

Generally, there are three methods of analysing and predicting the penetration of projectiles into rock targets:

 i) the empirical methods that are based upon experimental results obtained from impact and penetration tests,
 ii) the analytical methods, such as those using the cavity expansion theories, and
iii) numerical modelling.

In some instances, mixed methods are used, such as in the coupling of the PRONTO 3D finite-element code for the projectile, with spherical cavity expansion theory for the limestone target (Warren, 2002).

7.3.1   Empirical methods

In 1960, Sandia National Laboratories (SNL) began its terradynamics program with the objective of developing the technology to permit the design of a nuclear earth penetrating weapon (EPW). Based on the extensive experimental database, Young (1969) published a set of penetration equations which is still widely used today. The latest modifications of these equations and the associated technique for predicting the depth of penetration P by kinetic energy projectiles were summarised by Young (1997). It should be cautioned that the equations may not be applicable for projectile mass m less than 5 kg.

For striking velocity, $V_s < 61$ m/s:

$$P = 0.0008\, KSN(m/A)^{0.7} \ln(1 + 2.15\, V_s^2 \times 10^{-4}) \tag{7.1a}$$

For striking velocity, $V_s \geq 61$ m/s:

$$P = 0.000018\, KSN(m/A)^{0.7}(V_s^2 - 30.5) \tag{7.1b}$$

with $K = 0.46\, m^{0.15}$ if $m < 182$ kg, else $K = 1.0$. A is the average cross sectional area of the projectile body in m^2. $S = 2.7(\sigma_c Q)^{-0.3}$, where σ_c is the unconfined compressive strength of the intact rock sample in MPa. Q is a number that ranges from 0.1 to 1.0, and describes the rock quality as given in Table 7.1 below.

The selection of a value for Q is based largely on engineering judgment. If two or more of the terms are used to describe the rock quality, it is necessary to condense the aggregate descriptions into a single value of Q.

For Eq. (7.1a) and (7.1b), the nose performance coefficients N for ogive- and conical-nose projectiles are given by $N = 0.18(L_n/d) + 0.56$ and $N = 0.26(L_n/d) + 0.56$, respectively. L_n is the nose length and d is the diameter of the projectile.

Other than SNL, the following formulations for rock penetration were also proposed by the US Army Waterways Experiment Station (WES) (Bernard and Creighton, 1979).

$$P = \frac{m}{A} \cdot \frac{N}{\rho} \left[\frac{V_s}{3} \frac{\rho^{1/2}}{\sigma_{cm}^{1/2}} - \frac{4}{9} \ln\left(1 + \frac{3}{4} V_s \frac{\rho^{1/2}}{\sigma_{cm}^{1/2}}\right) \right] \tag{7.2a}$$

with

$$N = 0.863 \left[\frac{4(\psi)^2}{4\psi - 1} \right]^{1.4} \quad \text{for ogive-nose projectiles} \tag{7.2b}$$

$$N = 0.805(\sin \eta_c)^{-1/2} \quad \text{for conical-nose projectiles} \tag{7.2c}$$

$$\sigma_{cm} = \sigma_c (RQD/100)^{0.2} \tag{7.2d}$$

where N is the nose performance coefficient, ψ is the calibre-radius-head defined as the ratio of radius of curvature of the tangent ogive to the diameter of the projectile, η_c is the cone half-angle, and RQD the Rock Quality Designation. The units used are

Table 7.1 Descriptive Terms for Rock Quality Q (Young, 1997).

Descriptive Terms	Rock Quality Q
Massive	0.9
Interbedded	0.6
Joint Spacing < 0.5 m	0.3
Joint Spacing > 0.5 m	0.7
Fractured, blocky, or fissured	0.4
Highly fractured or jointed	0.2
Slightly weathered	0.7
Moderately weathered	0.4
Highly weathered	0.2
Frost shattered	0.2
Rock quality, very good/excellent	0.9
Rock quality, good	0.7
Rock quality, fair	0.5
Rock quality, poor	0.3
Rock quality, very poor	0.1

in m, kg, m/s, kg/m^3 and N/m^2. It should be noted that the WES formulation may not be applicable for $P < 3d$ or $RQD < 20$.

In addition to the two empirical formulae mentioned above, Kar (1978) also proposed the following set of equations for the depth of penetration in rock, igneous material and clay. The units used are in inch, lb, ft/s, and psi.

$$G = \frac{123.36}{\sqrt{\sigma_c}} \left(\frac{m}{d^{2.31}}\right) \left(\frac{E_s}{E_s'}\right)^{1.25} N \left(\frac{V_s}{1000}\right)^{1.25} \tag{7.3a}$$

with

$$G = (P/2d)^2 \quad \text{for } P/d \leq 2 \tag{7.3b}$$

$$G = [(P/d) - 1] \quad \text{for } P/d > 2 \tag{7.3c}$$

$$N = 0.72 + [(\psi)^{2.72}/1000] \tag{7.3d}$$

$E_s' = 30 \times 10^6$ psi, and E_s is the Young's Modulus of the projectile material.

7.3.2 Analytical methods

The most widely adopted analytical methods today are those that are based on the cavity expansion theory (CET). Other than the CET, there are others that are based on the differential area force law (DAFL) approach, which was adopted and modified by the US Army Waterways Experiment Station (WES) to provide a 2-D theory for the analysis of oblique impacts. The DAFL theory also forms the basis for the WES PENCO2D code which is still in use today. However, the number of publications on the CET far exceeds those of the DAFL. Since the early 1980s, numerous developments have been made in using CET to model projectile penetration in rock targets, most noticeably those by Forrestal and his co-workers at SNL. These developments on the CET will be described in this section.

The notion of analysing the penetration of an object in a semi-infinite medium by simulating it as a cavity expanding in that medium was first presented by Bishop, Hill and Mott (1945) more than 60 years ago. However, it was not until the 1970s that CET models such as Norwood (1974), Yarrington (1977), Yew and Stirbis (1978) and Davie (1979) were developed for penetration of projectiles into geological targets, particularly soil. Later, Forrestal, Longcope and Norwood (1981a) published a model to predict the forces on conical-nose projectiles for normal impact into dry rock targets based on a cylindrical cavity expansion approximation to model the target response. A closed-form solution to this model was subsequently given by Longcope and Forrestal (1981). Since then, several works (e.g., Forrestal, Longcope and Norwood, 1981b; Forrestal and Grady, 1982; Longcope and Forrestal, 1983; Forrestal 1986), on the use of cylindrical cavity expansion to study penetration of projectiles into rock targets were published, and reasonably good agreement was shown between the predictions and measured data.

More recently, Frew, Forrestal and Hanchak (2000) had used the model proposed by Forrestal *et al.* (1994) for concrete penetration to study the depth of penetration of ogive-nose steel rod projectiles into limestone targets. The equations for determining the depth of penetration P is given by

$$P = \frac{m}{2\pi a^2 \rho N} \ln \left(1 + \frac{N\rho V_1^2}{R}\right) + 4a \quad \text{for } P > 4a \tag{7.4a}$$

$$N = \frac{8\psi - 1}{24\psi^2}, \quad V_1^2 = \frac{mV_s^2 - 4\pi a^3 R}{m + 4\pi a^3 N\rho} \tag{7.4b}$$

$$R = \frac{N\rho V_s^2}{\left(1 + \frac{4\pi a^3 N\rho}{m}\right) \exp\left[\frac{2\pi a^2 (P - 4a) N\rho}{m}\right] - 1} \tag{7.4c}$$

in which the ogive-nose projectile is described by mass m, shank radius a, calibre-radius-head ψ, and striking velocity V_s. The target is described by density ρ and target strength constant R, which can be obtained from Eq. (7.4a) by measuring the striking velocity V_s and penetration depth P of a penetration experiment on the rock target. The accuracy of the prediction increases with the number of experiments being carried out.

7.3.3 Numerical modelling

While empirical and analytical approaches may provide a reasonably accurate and convenient engineering solution for a problem involving projectile impacting and penetrating a rock target, they are limited by their range of validity and assumptions made. Their applications are further impeded by the heterogeneous nature of rock materials, their complex constitutive behaviour, and the presence of micro-cracks and discontinuities. Hence, if a more comprehensive and complete solution, which includes simulating the penetration process and structural analysis of both the target and the projectile is required, numerical modelling has to be used. Huezé (1990) has presented an overview of the main numerical techniques and summarised the various computer programs developed in the 1970s and 1980s that had been used for penetration analyses in geological materials, with emphasis on rocks. Some of the computer programs such as ABAQUS, DYNA2D and DYNA3D and PRONTO2D have since been updated and widely used today in impact and penetration studies.

   While several recent works on the development of constitutive models can be found, and on the numerical simulations of penetration and perforation of concrete (e.g., Reidal et al., 1999; Gebbeken and Ruppert, 2000; Polanco-Loria, 2001; Fan, Zhou and Tan, 2003; Warren, Fossum and Frew, 2004; Warren, Hanchak and Poormon, 2004; Teng et al., 2005; Tham, 2005; Huang et al., 2006; Unosson and Nilsson, 2006; Rabczuk and Eibl, 2006; Leppänen, 2006), little has been done for rock materials. Recently, Warren (2002) developed a combined numerical and analytical technique by using the PRONTO 3D finite-element code to model the projectile and an analytical forcing function based on the dynamic expansion of a spherical cavity to represent the target. The results from the penetration simulations are compared with the corresponding results of the ballistic penetration experiments performed by Frew, Forrestal and Hanchak (2000) with limestone targets, and shown to be in good agreement. The combined techniques were later modified by Warren, Fossum and Frew (2004) and Warren, Hanchak and Poormon (2004) with a free surface effect model and used to simulate the penetration of limestone targets by ogive-nose steel projectiles at oblique angles. The results from the simulations are in reasonably good agreement with those from the experiments.

7.4 PENETRATION AND PERFORATION OF GRANITE TARGET PLATES

From the methods of analysis and prediction presented above, it is observed that previous studies concerning the penetration of projectiles into rock targets focused mainly on the penetration process. With the intent to extend the present knowledge of projectile penetration in rock to include the process of perforation, research was carried out to study the physical phenomena and failures taking place in a rock target of finite thickness when impacted, penetrated and perforated by a hard projectile (Seah, 2006). The research methodology involved an integrated use of experimental work (material tests and ballistic tests), and analytical modelling. Properties of the rock material obtained from material tests are used as inputs for the analytical model that is developed to make predictions for the ballistic tests.

7.4.1 Ballistic tests

7.4.1.1 Compressed gas gun facility

The compressed gas gun facility seen in Figure 7.2, and described by Børvik (2000) was used to carry out the ballistic tests. The main components of the facility consist

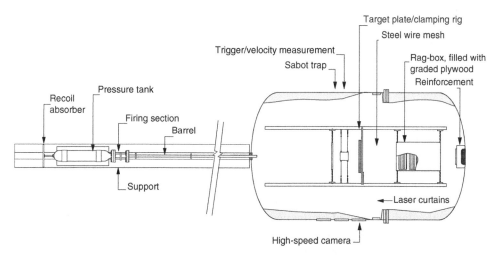

Figure 7.2 Sketch of compressed gas gun used in the experiments (Børvik, 2000).

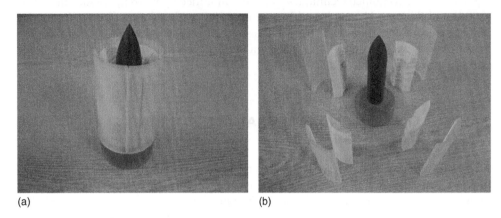

(a)   (b)

Figure 7.3 (a) The projectile package and (b) the design of the sabot.

of a 20 MPa pressure tank, a specially-designed firing section for compressed gas, a 10 m-long smooth barrel of bore diameter 50 mm, and a closed 16 m^3 impact chamber.

The projectile package comprises a 0.2 kg 20 mm-diameter projectile encased in an eight-piece serrated sabot with an obturator as shown in Figure 7.3. When the projectile package leaves the muzzle in the impact chamber, the sabot immediately separates from the projectile. A sabot trap, which is located at 1.5 m from the muzzle, catches the sabot parts. The projectile is then allowed to travel freely through the initial velocity measurement station before it finally impacts the target after about 2 m of free flight.

The initial velocity is measured by a photocell system consisting of two identical light-curtains with LED-light sources on one side and detectors on the other of the path of the projectile. An interruption will be caused in the light transmission between the sources and detectors when the projectile passes through the LED-light curtain, and the signals are recorded by an oscilloscope and a nanosecond counter. Knowing the distance between the two light curtains and the time that the projectile takes to pass through them, the striking velocity can be determined.

Previously, Børvik *et al.* (2002) had performed ballistic experiments on concrete targets with the compressed gas gun facility. Due to debris that triggered the optical devices ahead of the projectile when it perforated the concrete target, it was found to be difficult to obtain reliable residual velocity measurements from the tests. Similar problems were also encountered with the trial tests that were carried out on three granite targets when planning for the experimental programme. However, it is important to determine whether the projectile did perforate the granite target. A steel wire mesh was employed to cover the section between the rear side of the target and the rag-box to capture the projectile when it perforated the target as shown in Figure 7.1. The wire mesh also helped in the collection of the major debris ejecting from the distal face for studying the damage caused by the penetration and perforation process. During each test, a foam panel was placed at 0.6 m from the distal face of the granite target in the rag-box to provide some quantitative information on the energy in the debris by observing how far they travelled and the extent of damage on the foam.

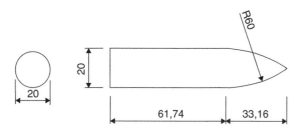

Figure 7.4 Dimensions of the ogive-nose projectile.

Table 7.2 Basic Properties of the Iddefjord Granite.

Young's Modulus E	Compressive Strength σ_c	Tensile Strength σ_t	Poisson's Ratio ν	Density ρ	Residual Shear Strength C_0	Friction Angle θ
54 GPa	163 MPa	7.1 MPa	0.27	2626 kg/m³	40.4 MPa	52.60

7.4.1.2 Projectiles and Targets

Ogive-nose projectiles were used in the ballistic tests. The projectiles were manufactured from Arne tool steel with a nominal mass and diameter of 197 g and 20 mm, respectively. The dimensions of the ogive-nose projectile are given in Figure 7.4. The ogive-nose projectile has a calibre-radius-head (CRH) of 3. After machining, the projectiles were oil-hardened to a nominal Rockwell C value of about 51–53, with yield strength of 1850 to 1900 MPa. They were then spray-painted bright red and equipped with fiducial marks for easier tracking by the high-speed cameras.

The granite target plates used in the experimental programme were quarried and cut from the Iddefjord granite. Material tests were performed on the Iddefjord granite and the properties are given in Table 7.2. All the granite targets are 0.6 m × 0.6 m × 0.1 m in dimension and weigh about 100 kg.

During each test, the granite target is fitted into a specially designed steel frame with screws and rigid steel plates and mounted onto a holding bracket which is fixed onto the supporting frame of the impact chamber as shown in Figure 7.5.

7.4.2 Analytical Modelling

The analytical model consists of a set of closed-form equations describing the three stages of the penetration and perforation process: cratering, tunnelling and plugging as shown in Figure 7.6. Generally, only catering and plugging are observed for thinner targets.

Post test observations of concrete and rock targets show that the cavity after penetration is a conical region with length about two projectile shank diameters $4a$ followed by a circular cylinder with diameter nearly equal to the projectile shank diameter $2a$

Figure 7.5 Experimental set-up for the granite targets during the tests.

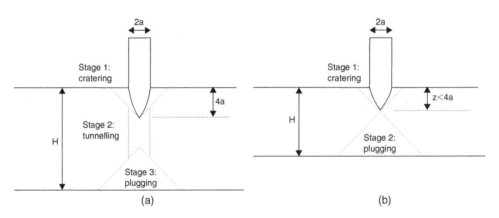

Figure 7.6 Penetration and perforation model (a) with and (b) without tunneling.

(Forrestal *et al.*, 1994). The depth of penetration z is measured from the target surface and P is the final penetration depth.

7.4.2.1 Projectile Force

The spherical cavity-expansion penetration model presented by Forrestal and Tzou (1997) for an incompressible material with elastic-crack-plastic response approximates the normal stress on the projectile nose by

$$\sigma_r(V_z, \phi) = \tau_0 \left[A + B V_z \cos \phi + C (V_z \cos \phi)^2 \right] \tag{7.5}$$

where τ_0, A, B and C are material parameters. From material tests, they are found to be 49.2 MPa, 44.7, 0.0663 and 0.0012, respectively for the Iddefjord granite. In

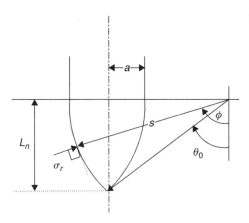

Figure 7.7 Normal stress on an ogive-nose projectile.

addition, it is assumed that there is no tangential stress on the nose from interface frictional resistance. The resulting axial force on the projectile nose at penetration depth z is then given as

$$F_z = 2\pi s^2 \int_{\theta_0}^{\pi/2} \left\{ \left[\sin\phi - \left(\frac{s-a}{s} \right) \right] \cos\phi \right\} \sigma_r(V_z, \phi) d\phi \tag{7.6}$$

where $\theta_0 = \sin^{-1}[(s-a)/s]$, V_z is the velocity of the projectile at penetration depth z and s, a and ϕ are as defined in Figure 7.7.

From observations of deceleration data from penetration tests carried out on antelope tuff (Longcope and Forrestal, 1983), the force on the projectile nose can be taken as

$$F_z = C_c z \quad \text{for } 0 < z \leq 4a \tag{7.7a}$$

$$F_z = 2\pi s^2 \tau_0 (f_1 + f_2 V_z + f_3 V_z^2) \quad \text{for } 4a < z \leq P \tag{7.7b}$$

where C_c, f_1, f_2 and f_3 are functions to be determined from the integration of Eq. (7.6), and they are given as

$$C_c = \frac{2\pi s \tau_0}{(\cos\theta_0 - \cos\phi)}(A' + B'V_z + C'V_z^2) \quad \text{for } 0 < z \leq L_n \tag{7.8a}$$

$$C_c = \frac{2\pi s^2 \tau_0}{z}(A' + B'V_z + C'V_z^2) \quad \text{for } L_n < z \leq 4a \tag{7.8b}$$

$$f_1 = A\left[\frac{1}{4}(1 - \cos 2\phi) - \frac{s-a}{s}\sin\phi + \frac{1}{2}\left(\frac{s-a}{s} \right)^2 \right] \quad \text{for } z < L_n \tag{7.9a}$$

$$f_1 = \frac{A}{2}\left(\frac{a}{s}\right)^2 \quad \text{for } z \geq L_n \tag{7.9b}$$

$$f_2 = B\left\{\frac{1}{3}(\cos^3\theta_0 - \cos^3\phi) - \frac{s-a}{2s}\left[\left(\frac{\sin 2\phi}{s} + \phi\right) - \left(\frac{\sin 2\theta_0}{2} + \theta_0\right)\right]\right\}$$
$$\text{for } z < L_n \tag{7.10a}$$

$$f_2 = B\left[\frac{1}{3}\cos^3\theta_0 - \frac{s-a}{2s}\left(\frac{\pi}{2} - \frac{\sin 2\theta_0}{2} - \theta_0\right)\right] \quad \text{for } z \geq L_n \tag{7.10b}$$

$$f_3 = C\left\{\frac{1}{4}(\cos^4\theta_0 - \cos^4\phi) - \frac{s-a}{s}\left[(\sin\phi - \sin\theta_0) - \frac{1}{3}(\sin^3\phi - \sin^3\theta_0)\right]\right\}$$
$$\text{for } z < L_n \tag{7.11a}$$

$$f_3 = C\left[\frac{1}{4}\cos^4\theta_0 - \frac{s-a}{s}\left(\frac{2}{3} - \sin\theta_0 + \frac{1}{3}\sin^3\theta_0\right)\right] \quad \text{for } z \geq L_n \tag{7.11b}$$

It should also be noted that V_z is given as a function of penetration depth z in Eq. (7.7b).

7.4.2.2 Plug resistance

It is well documented that the tensile strengths of rocks are much lower than their compressive strengths. When the projectile force reaches a critical value equal to the plug resistance, plugging occurs as shown in Figure 7.8.

The separation of a conical plug from the surrounding material is governed by the local principal stress and the tensile strength of the material. The simplest and most well-known criterion of failure in rocks is the Mohr-Coulomb criterion. From the Mohr-Coulomb failure envelope with a tensile cut-off at σ_t, the critical angle of the tensile crack is given as $\alpha = 45° + \theta/2$ and the plug resistance can then be derived by multiplying the vertical component of the tensile strength with the surface area of the truncated cone plug to give

$$F_p = \pi\sigma_t\left[H^*\tan\left(45° + \frac{\theta}{2}\right)\right]^2 \tag{7.12}$$

where $H^* = H - z$. For a given target thickness H the ballistic limit velocity V_{bl} is defined as the minimum impact velocity V_s for which the projectile will penetrate and perforate the target on the rear side of the target.

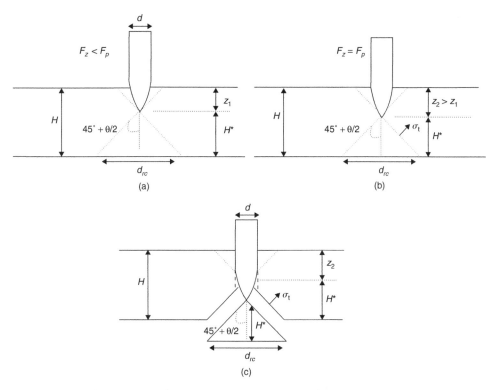

Figure 7.8 Illustration of a plugging process: (a) $F_z < F_p$; (b) $F_z = F_p$; and (c) plugging.

7.4.3 Limit velocities

7.4.3.1 Ballistic limit

By equating $F_z = F_p$, the penetration depth z when plugging will occur can be obtained. Then, the corresponding velocity V_z can be found, and this will become the residual velocity V_r. The solution procedure may seem rather tedious but can easily be done graphically by generating a series of projectile force versus penetration depth plots for different striking velocities with Equations (7.7a) and (7.7b), and superimposing the plug resistance versus penetration depth plot from Equation (7.12) onto them as illustrated in Figure 7.9. For each striking velocity V_s, there will be a residual velocity V_r. Finally, to determine the ballistic limit V_{bl}, the data obtained from the analytical model are curved fitted to a limit-velocity curve proposed by Recht and Ipson (1963) in Figure 7.10.

With the penetration depth z when plugging will occur already known, the height of the conical plug which is given by $H^* = H - z$ can then be calculated, together with the diameter of the rear crater d_{rc} which is given by

$$d_{rc} = 2H^* \tan\left(45° + \frac{\theta}{2}\right) \tag{7.13}$$

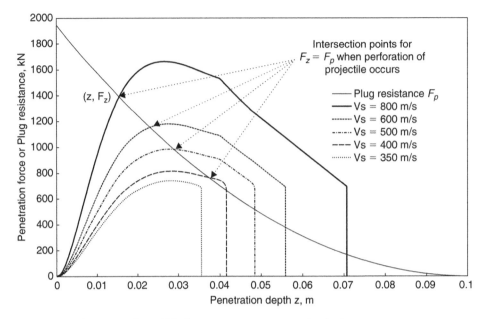

Figure 7.9   Force versus penetration depth plot.

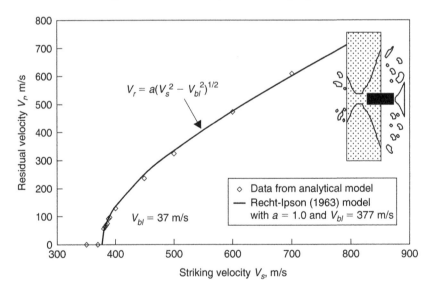

Figure 7.10   Determining of the ballistic limit V_{bl} with the limit-velocity curve.

7.4.3.2   Protection limit

If the striking velocity is less than the ballistic limit, the projectile is not expected to penetrate through to the other side of the target. At the instance when the projectile is suddenly brought to a stop, the stress at the projectile nose-target interface is doubled

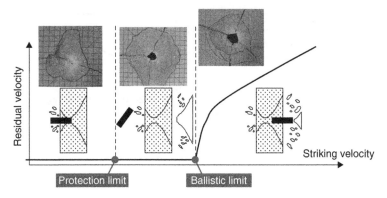

Figure 7.11 Illustration of the Ballistic Limit and Protection Limit of a rock target that is struck by a hard projectile.

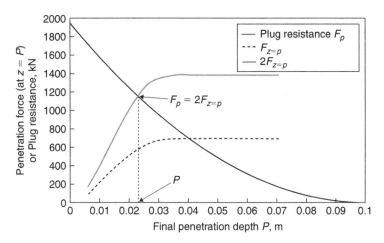

Figure 7.12 Determining the final penetration depth when a projectile suddenly stops in the rock target.

while the net displacement and particle velocity is zero. This net stress, if sufficiently large, will result in scabbing and create a through hole in the target. The minimum striking velocity at which this scenario can occur is the protection limit V_{pl} as illustrated in Figure 7.11. A graphical solution procedure can also be adopted to determine V_{pl}.

From the series of projectile force versus penetration depth plots shown in Figure 7.9, one can find the final penetration depth P and the corresponding projectile force $F_{z=p}$, for each striking velocity. Plots of final penetration depth versus striking velocity, and projectile force versus final penetration depth, can then be obtained. By equating $2 F_{z=p} = F_p$ and solving for the final penetration depth in Figure 7.12, we can finally determine the protection limit from Figure 7.13.

Similar to the case of the ballistic limit, the height of the conical plug and the diameter of the rear crater d_{rc} can be calculated, after the final penetration depth P when plugging occurs has been determined.

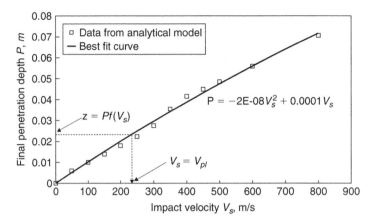

Figure 7.13 Plot of final penetration depth versus striking velocity.

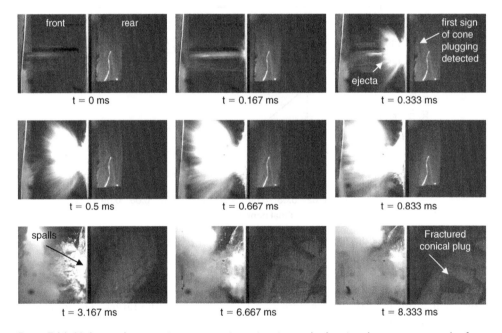

Figure 7.14 High-speed camera images at various time intervals showing the responses at the front and rear side of the granite target plate when impacted by ogive-nose projectile at $V_s = 279$ m/s.

7.5   RESULTS AND DISCUSSIONS

For every experiment, high-speed digital images as presented in Figure 7.14 were taken, and the striking velocities were measured. The debris collected in the steel-wire cage and the damage on the granite target plates were also documented as shown in Figure 7.15. Figure 7.16 shows the typical damage and failures on the granite targets that

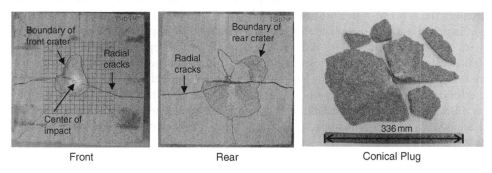

Front   Rear   Conical Plug

Figure 7.15 Documentation of the damage on the granite target plate and the debris collected.

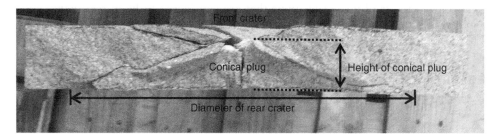

Figure 7.16 Typical damage and failures on the granite target plates that were observed in the ballistic tests.

Table 7.3 Comparison between predictions from analytical model and results from ballistic tests.

	Ballistic Limit			Protection Limit		
	V_{bl} (m/s)	H^* (mm)	d_{rc} (mm)	V_{pl} (m/s)	H^* (mm)	d_{rc} (mm)
Prediction	377	60	355	242	77	455
Experiment	406	55	297	288	65	312
% Diff	−7.1	9	19.5	−16.0	18.5	45.8

were observed in the ballistic tests. The ballistic limit and protection limit, and the dimensions of the conical plug associated with each of these limit velocities are summarised in Table 7.3 together with the predictions from the analytical model. Given the complexity of the penetration and perforation problems and the simplicity of the three-stage analytical model which was used, the predictions are considered to be in reasonably good agreement with the experimental results.

7.6 CONCLUDING REMARKS

Previous studies on projectile penetration in rock usually focussed on penetration depth, deceleration history or stresses on the nose of the projectile. This paper has extended the current knowledge of the penetration of projectiles into rock targets to include the process of perforation. The Iddefjord granite has been selected as the target material to be studied. The research work consists of experimental work (material tests and component tests), and analytical modelling. Properties of the rock material obtained from material tests are used as inputs for the analytical model that is developed to make predictions for the ballistic tests. Sets of ballistic penetration and perforation experiments were performed with $0.6\,m \times 0.6\,m \times 0.1\,m$ thick Iddefjord granite targets and $0.2\,kg$, $20\,mm$ diameter Arne tool steel projectiles. Ogive-nose projectiles with calibre-radius-head (CRH) of 3 were considered. In analytical modelling, a three-stage analytical model was proposed to make predictions for the experiments. Despite the complexity of the penetration and perforation problem, the three-stage analytical model is capable of taking into account the main physical phenomena and failures that take place when a rock target is impacted, penetrated and perforated by a hard projectile, and give predictions that are considered to be in reasonably good agreement with the experimental results.

REFERENCES

Antoun, T.H., Lomov, I.N. and Glenn, L.A.: Simulation of the penetration of a sequence of bombs into granitic rock. *International Journal of Impact Engineering*, 18(2) (1995), pp.141–230.

Baty, R.S., Lundgren, R.G. and Patterson, W.J.: On pilot-hole assisted penetration. *Proceedings of the 11th International Symposium on Interaction of the Effects of Munitions with Structures*, Mannheim, Germany (in CD Rom) (2003).

Bernard, R.S. and Creighton, D.C.: *Projectile penetration into soil and rock: analysis for non-normal impact*. Technical Report SL-79-15, US Army Engineer Waterways Experiment Station, Vicksburg (1979).

Bishop, R.F., Hill, R. and Mott, N.F.: The theory of indentation and hardness tests. *Proceedings of the Physical Society*, 57(3) (1945), pp.147–155.

Børvik, T.: *Ballistic penetration and perforation of steel plates*. Doctoral Thesis, Department of Structural Engineering, Norwegian University of Science and Technology, Trondheim (2000).

Børvik, T., Langseth, M., Hopperstad, O.S. and Polanco-Loria, M.A.: Ballistic perforation resistance of high performance concrete slabs with different unconfined compressive strengths. *Proceedings of the First International Conference on High Performance Structures and Composites*, Seville, Spain (2002).

Børvik, T., Gjørv, O.E. and Langseth, M.: Ballistic perforation resistance of high-strength concrete slabs by conical-nose steel projectiles. *Concrete International*, (2006).

Davie, N.T.: *A method for describing nearly normal penetration of an ogive nosed penetrator into terrestrial targets*. SAND 79-0721, Sandia National Laboratories, Albuquerque, NM (1979).

Fan, S.C., Zhou, X.Q. and Tan, G.E.B.: Numerical simulation of oblique perforation of concrete slab by projectile. *Proceedings of the Norway-Singapore Workshop on Protection*, Oslo, Norway (in CD Rom) (2003).

Forrestal, M.J., Longcope, D.B. and Norwood, F.R.: A model to estimate forces on conical penetrators into dry porous rock. *ASME Journal of Applied Mechanics* 48 (1981a), pp.25–29.

Forrestal, M.J., Norwood, F.R. and Longcope, D.B.: Penetration into targets described by locked hydrostats and shear strength. *International Journal of Solids and Structures* 17 (1981b), pp.915–924.

Forrestal, M.J. and Grady, D.E.: Penetration experiments for normal impact into geological targets. *International Journal of Solids and Structures*, 18(3) (1982), pp.229–234.

Forrestal, M.J.: Penetration into dry porous rock. *International Journal of Solids and Structures*, 22(12) (1986), pp.1485–1500.

Forrestal, M.J., Altman, B.S., Cargile, J.D. and Hanchak, S.J.: An empirical equation for penetration depth of ogive-nose projectiles into concrete targets. *International Journal of Impact Engineering* 15(4) (1994), pp.395–405.

Forrestal, M.J. and Tzou, D.Y.: A spherical cavity-expansion penetration model for concrete targets. *International Journal of Impact Engineering*, 34(21–32) (1997), pp.4127–4146.

Frew, D.J., Forrestal, M.J. and Hanchak, S.J.: Penetration experiments with limestone targets and ogive-nose steel projectiles. *ASME Journal of Applied Mechanics* 67 (2000), pp.841–845.

Gebbeken, N. and Ruppert, M.: A new material model for concrete in high-dynamic hydrocode simulations. *Archive of Applied Mechanics* 70 (2000), pp.463–478.

Hanchak, S.J., Forrestal, M.J., Young, E.R. and Ehrgott, J.Q.: Perforation of concrete slabs with 48 MPa and 140 MPa unconfined compressive strengths. *International Journal of Impact Engineering* 12(1) (1992), pp.1–7.

Heuzé, F.E.: An overview of projectile penetration into geological materials, with emphasis on rocks. *International Journal of Rock Mechanics and Mining Sciences & Geomechanics Abstracts* 27(1) (1990), pp.1–14.

Huang, F.L., Wu, H.J., Jin, Q.K. and Zhang, Q.M.: A numerical simulation on the perforation of reinforced concrete targets. *International Journal of Impact Engineering* 32(1–4) (2006), pp.173–187.

Kar, A.K.: Local effects of tornado generated missiles. *ASCE Journal of Structural Division*, 104(ST5) (1978), pp.809–816.

Leppänen, J.: Concrete subjected to projectile and fragment impacts: modelling of crack softening and strain rate dependency in tension. *International Journal of Impact Engineering*, 32(11) (2006), pp.1828–1841.

Longcope, D.B. and Forrestal, M.J.: Closed-form approximations for forces on conical penetrators into dry porous rock. *ASME Journal of Applied Mechanics* 48 (1981), pp.971–972.

Longcope, D.B. and Forrestal, M.J.: Penetration of targets described by a Mohr-Coulomb failure criterion with a tension cutoff. *ASME Journal of Applied Mechanics* 50 (1983), pp.327–333.

Norwood, F.R.: *Cylindrical cavity expansion in a locking soil*. SLA-74-0201, Sandia National Laboratories, Albuquerque, NM, (1974).

Polanco-Loria, M., 2001. Constitutive models for concrete materials under impact loading conditions. STF24 F00315, Structural Impact Laboratory(SIMLab), NTNU, Trondheim, Norway (confidential).

Polanco-Loria, M., Hopperstad, O.S., Borvik, T. and Berstad, T., 2006. Numerical predictions of ballistic limits for concrete slabs using a modified version of the HJC concrete model. Submitted for possible journal publication.

Rabczuk, T. and Eibl, J.: Modelling dynamic failure of concrete with meshfree methods. *International Journal of Impact Engineering* 32(11) (2006), pp.1878–1897.

Recht, R.F. and Ipson, T.W.: Ballistic Perforation Dynamics. *ASME Journal of Applied Mechanics* 30 (1963), pp.384–390.

Reidal, W.: *Beton unter dynamischen Lasten Meso- und makromechanische Modelle und ihre Paramter*. Doctoral thesis, Fakultät für Bauingenieur- und Vermessungswesen, Universität der Bundeswehr München (in German), (2000).

Seah, C.C.: *Penetration and Perforation of Granite Targets by Hard Projectiles*. Doctoral Thesis, Norwegian University of Science and Technology, Trondheim, (2006).

Sjøl, H. and Teland, J.A.: *Perforation of concrete targets*. FFI/RAPPORT-2001/05786, Norwegian Defence Research Establishment, Norway, (2003).

Teng, T.L., Chu, Y.A., Chang, F.A. and Shen, B.C.: Penetration resistance of reinforced concrete containment structures. *Annals of Nuclear Energy* 32(3) (2005), pp.281–298.

Tham, C.Y.: Numerical and empirical approach in predicting the penetration of a concrete target by an ogive-nosed projectile. *Finite Elements in Analysis and Design* 42(14–15) (2006), pp.1258–1268.

Unosson, M. and Nilsson, L.: Projectile penetration and perforation of high performance concrete: experimental results and macroscopic modelling. *International Journal of Impact Engineering* 32(7) (2006), pp.1068–1085.

Warren, T.L.: Simulations of the penetration of limestone targets by ogive-nose 4340 steel projectiles. *International Journal of Impact Engineering* 27 (2002), pp.475–496.

Warren, T.L., Fossum, A.F. and Frew, D.J.: Penetration into low-strength (23MPa) concrete: target characterisation and simulations. *International Journal of Impact Engineering* 30 (2004), pp.477–503.

Warren, T.L., Hanchak, S.J. and Poormon, K.L.: Penetration of limestone targets by ogive-nosed VAR 4340 steel projectiles at oblique angles: experiments and simulations. *International Journal of Impact Engineering* 30 (2004), pp.1307–1331.

Yarrington, P.: *A one-dimensional approximate technique for earth penetration calculations.* SAND 77-1126, Sandia National Laboratories, Albuquerque, NM, (1977).

Yew, C.H. and Stirbis, P.P.: Penetration of projectile into terrestrial target. *Journal of Engineering Mechanics Division ASCE*, EM2 (1978), pp.273–286.

Young, C.W.: Depth predictions for earth penetrating projectiles. *Journal of Soil Mechanics and Foundations* 95(SM3) (1969), pp.803–817.

Young, C.W.: *Penetration equations.* SAND97-2426, Sandia National Laboratories, Albuquerque, NM, (1997).

Zhao, J., Zhou, Y.X., Hefny, A.M., Cai, J.G., Chen, S.G., Li, H.B., Liu, J.F., Jain, M., Foo, S.T. and Seah, C.C.: Rock dynamics research related to cavern development for ammunition storage. *Tunnelling and Underground Space Technology* 14(4) (1999), pp.513–526.

Chapter 8

Incubation time based fracture mechanics and optimization of energy input in the fracture process of rocks

Yuri Petrov, Vladimir Bratov, Grigory Volkov and Evgeny Dolmatov

8.1 INTRODUCTION

The possibility of optimising the amount of energy required to fracture materials is of great interest in connection with many applications. Energy inputs for fracture induced by short impulse loadings are of major importance in such areas as percussive, explosive, hydraulic, electro-impulse and other means of mining, drilling, pounding etc. In these cases energy input usually accounts for the largest part of the process cost (e.g. Royal Dutch Petroleum Company Annual Report, 2003). Taking into consideration the fact that the efficiency of the above-mentioned processes rarely exceeds a few percent, the importance of energy inputs optimization becomes evident.

This chapter summarizes some results connected with the application of the incubation time approach to problems of dynamic fracture of rock materials. Incubation time based fracture criteria for intact media and media with cracks are discussed and a possibility of optimizing energy input for fracture is studied. It is shown that the minimal energy needed in order to initialize fracture in cracked rock media does strongly depend on amplitude and duration of an impact causing this rupture. Existence of optimal energy saving shapes for a single impact or a sequence of periodic impacts is demonstrated.

The purpose of the first section is to find and explore the amount of energy sufficient to initiate the propagation of a mode I loaded central crack in a plate subjected to plane strain deformation. Two ways to apply the dynamic load to the body are studied. In the first case the load is applied at infinity. The study involves the analysis of interaction of the wave package approaching from infinity with an existing central crack in a plane. The existing crack is oriented parallel to the front of the wave package. In the second case the load is applied at the crack faces. Tractions are normal to the crack faces. Following the superposition principle these two loading cases should produce identical stress-strain fields in the vicinity of the crack tip. It will be shown later that the amount of total energy applied to the body needed to initiate crack growth depends on the load application manner in different ways for the two cases under investigation.

In the second part of the chapter, one of the very first attempts to incorporate incubation time based fracture criterion into FEM (Finite Element Method) code is presented. Utilizing developed techniques, the conditions of SMART1 satellite impacting the moon surface are simulated. Received dimensions of a crater formed on the moon due to contact with SMART1 are the same as observed in reality.

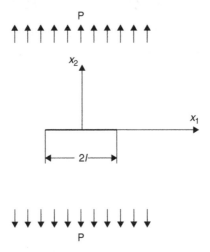

Figure 8.1 Experiment scheme. Central crack in an infinite plane is loaded by a wave approaching from infinity. Wave front is parallel to the crack plane.

In the final part of the chapter, as a result of using the incubation time approach we succeeded in giving an explanation for the experimentally observed effect of optimization of the energy spent on the fracture of materials by pulsed dynamic attacks. By example of the simplest problem about contact collision, we estimated the energy necessary for the fracture-pulse generation in the rock medium. It was found that this energy substantially depends on the attack duration and has the characteristic lowest value.

8.2   MODELING INTERACTION OF THE WAVE COMING FROM INFINITY WITH THE CRACK

Consider an infinite plane with a central crack (Fig. 8.1). The load is given by the wave, falling on the crack. Displacements of the plane are described by:

$$\rho u_{i,tt} = (\lambda + \mu)u_{j,ji} + \mu u_{i,jj},$$ (8.1)

where "," refers to the partial derivative with respect to time and spatial coordinates; ρ is the mass density, and the indices i and j assume the values 1 and 2. Displacements are given by u_i in the directions x_i respectively. T stands for time, λ and μ are Lame constants. Stresses and strains are coupled by Hooke's law:

$$\sigma_{ij} = \lambda\delta_{ij}u_{k,k} + \mu(u_{i,j} + u_{j,i}),$$ (8.2)

where σ_{ij} represents stresses in direction ij, and δ_{ij} is the Kronecker delta assuming value of 1 for $i=j$ and 0 otherwise. Boundary conditions are:

$$\sigma_{22}|_{|x_1|<l,x_2=0} = \sigma_{21}|_{|x_1|<l,x_2=0} = 0.$$ (8.3)

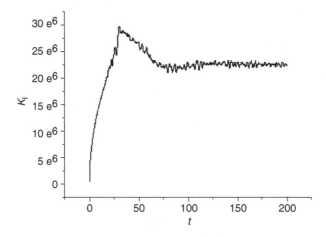

Figure 8.2 Typical stress intensity factor (Pa $\sqrt{m}$) time (μs) dependence in FE solution.

The impact is delivered to the crack by the falling wave:

$$\sigma_{22}|_{t<0} = P\left(H\left(t + \frac{x_2}{c_1}\right) + H\left(t - \frac{x_2}{c_1}\right) - H\left(t + \frac{x_2}{c_1} - T\right) - H\left(t - \frac{x_2}{c_1} - T\right)\right) \quad (8.4)$$

where c_1 is the longitudinal wave speed, H is the Heaviside step function and T is the impact duration. P represents the pressure pulse amplitude and has a dimension of Pa. The described problem is solved using the finite element method.

The process is analyzed utilizing the finite element method. The ABAQUS (see ABAQUS User Manual) finite element package was used to solve the problem. The task was formulated for a quarter sample using the symmetry of the problem about the x and y axes. Plane strain conditions were supposed. An area adjacent to the crack tip was meshed with triangular isoparametric quarter-point elements available in the ABAQUS package. Thus, the mesh in the vicinity of the crack tip may assume a square root singularity in stress/strain fields. The total of about 3×10^6 elements were used to model the cracked sample. The crack surface was represented by 50 nodes along the crack's half-length. Explicit time integration was utilized to solve the dynamical problem in question.

Computations were performed for granite ($E = 96.5$ GPa, $\rho = 2810$ kg/m³, $\upsilon = 0.29$, where E is the elasticity modulus and υ is the Poisson's ratio). The results of the investigation will qualitatively hold for a large variety of quasi-brittle materials.

In the conditions of plane strain, the interaction of the wave approaching from infinity with a central crack was investigated.

Firstly infinite impulse durations were supposed, i.e. $T = \infty$. Time dependence of the stress intensity factor K_I was studied. K_I used in a further analysis was calculated from the J-integral that is available as a direct output from ABAQUS solution. Computations were performed for different amplitudes of the loading pulse applied. Typical dependence of K_I on time is presented in Figure 8.2.

Apparently K_I is rapidly approaching the static level. Thus the time to approach the steady-state situation in the vicinity of a crack tip can be estimated as 5–10 times more than the time required by the wave to travel along the crack's half-length.

Fracture criterion fulfillment was checked for different load amplitudes and durations. Dependence of time-to-fracture T^* on the amplitude of the load applied was investigated. Time-to-fracture is the time from the beginning of interaction between the wave package and the crack to fracture initiation. Incubation time criterion of fracture (Petrov and Morozov, 1994; Morozov and Petrov, 2000) was adopted. A similar approach to be used in the case of short cracks is given by Petrov and Taraban (1997).

8.2.1 Incubation time fracture criterion

Incubation time based fracture theory proposed by Petrov and Utkin (1989), Petrov and Morozov (1994) and Morozov and Petrov (2000) is successfully used to describe fracture initiation in dynamic conditions (Petrov and Morozov, 1994; Morozov and Petrov, 2000; Bratov et al., 2004; Bratov and Petrov, 2007). Criterion for fracture at a point x at time t reads:

$$\frac{1}{\tau}\frac{1}{d}\int\limits_{x-d}^{x}\int\limits_{t-\tau}^{t}\sigma(x,t)dxdt \geq \sigma_c, \tag{8.5}$$

where τ is the fracture process incubation time (or microstructural time) – parameter characterizing response of a studied material on applied dynamical loads (i.e. τ is constant for a given material and does not depend on the problem geometry, the way a load is applied, the shape of a load pulse or its amplitude); d has the meaning of characteristic size of a fracture process zone and is constant for the given material and chosen spatial scale; σ is the stress at a point, changing with time and σ_c is its critical value (ultimate stress or critical tensile stress evaluated in quasistatic conditions); x^* and t^* are local coordinate and time.

Assuming

$$d = \frac{2}{\pi}\frac{K_{IC}^2}{\sigma_c^2}, \tag{8.6}$$

where K_{IC} is a critical stress intensity factor for mode I loading (mode I fracture toughness), measured in quasistatic experimental conditions, it can be shown that within the frames of linear fracture mechanics for the case of fracture initiation in a tip of an existing crack, loaded by mode I, criterion (8.5) is equivalent to:

$$\frac{1}{\tau}\int\limits_{t-\tau}^{t} K_I(t^*)dt^* \geq K_{IC}. \tag{8.7}$$

Condition (8.6) arises from a requirement that criterion (8.5) is equivalent to Irwin's criterion ($K_I \geq K_{IC}$) in quasistatic conditions ($t \to \infty$). This means that a certain size typical of fractured material appears. This size is believed to be associated with the size of a failure cell on the current spatial scale – all rupture sized significantly less than d cannot be called fracture on the current scale level.

Thus, by introducing τ and d, the time-spatial domain is discretized. Once the material and scale one is working with are chosen, τ gives time, such that energy accumulated during this time can be released by rupture of the cell that accumulated it. The d assigns dimensions for such a cell. Introduction of time and spatial domain discretization is very important. To our belief, a correct description of high loading rate effects is not possible if this time-spatial discreteness is not accounted for in some way. The advantage of the incubation time approach is that one can stay within the frames of continuum linear elasticity, utilizing all the consequent advantages and accounting for discreteness of the problem only in the critical fracture condition.

As was shown in earlier publications (e.g., Petrov, 1991; Petrov, Morozov and Smimov, 2003; Petrov and Sitnikova, 2005), criterion (8.5) can be successfully used to predict fracture initiation in brittle solids. For slow loading rates and, hence, times to fracture that are much bigger than τ, condition (8.7) for crack initiation gives the same predictions as the Irwin's criterion (Irwin, 1957) of the critical stress intensity factor. For high loading rates and times to fracture comparable with τ all the variety of effects experimentally observed in dynamical experiments (e.g. Smith, 1975; Ravi-Chandar and Knauss, 1984; Shockey et al., 1986; Kalthoff, 1986; Dally and Barker, 1988) can be received using condition (8.7) both qualitatively and quantitatively (Petrov, 2004). Application of condition (8.7) to description of real experiments or usage of (8.7) as a critical fracture condition in finite element numerical analysis gives a possibility for better understanding of the nature of fracture dynamics (e.g. Bratov et al., 2004) and even prediction of new effects typical for dynamical processes (e.g. Bratov and Petrov, 2007).

Another known approach to dynamic fracture, originating from works of Freund (Freund and Clifton, 1974) and later developed by Freund (Freund, 1990) and Rosakis (e.g. Owen et al., 1998) is based on an assumption that fracture toughness can be directly and unequivocally coupled with loading rate or stress intensity factor rate. Sometimes, for specific experimental conditions with stress intensity factor (or just stress) monotonously growing with time, such a dependency can be observed in reality. But, generally speaking, the majority of known experimental results for short pulse fracture demonstrate the inapplicability of this approach. In numerous experiments (Shokey et al., 1986; Zlatin and Pugachev, 1975; Berezkin et al., 2000), it is observed that fracture can initiate at a moment when the stress intensity factor (or stress, if concerning fracture of intact material, for example, in dynamic cleavage experiments) is decreasing, and hence is having a negative rate. Obviously these phenomena are impossible to describe presuming unequivocal dependency of fracture toughness (or critical stress) on stress intensity factor rate (or stress rate).

All this provides grounds to state that an incubation time based approach to fracture has the most potential of all currently known approaches in dynamic fracture.

Using criterion (8.7), the dependence of time-to-fracture on the amplitude of the load pulse applied was studied. Values of $K_{IC} = 2.4\,MPa\sqrt{m}$ and $\tau = 72\,\mu s$, typical for the granite under investigation, were used. Integration of the temporary dependence of the stress intensity factor was done numerically. In Figure 8.3, the x-axis represents the time from the beginning of interaction of the wave coming from infinity with the crack to the fracture initiation. The y-axis represents the corresponding amplitude of

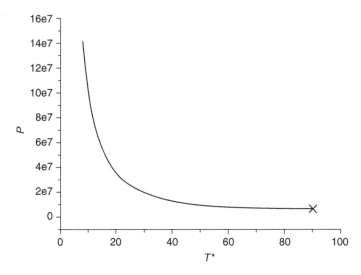

Figure 8.3 Curve limiting the pulses leading to crack propagation. Time-to-fracture (μs) vs. applied pressure amplitude (Pa).

the load applied at infinity. The point on Figure 8.3 marked with a cross corresponds to the maximum possible time-to-fracture for a given problem. As follows, for the investigated granite and studied experimental conditions, fracture is only possible for times less than 92 μs.

At the same time the critical (threshold) amplitude of the applied load was found. This amplitude corresponds to the maximum time-to-fracture possible. Loads with amplitudes less than the critical one do not increase the length of the crack.

8.2.2 Dependence of the energy inputs for fracture on the load amplitude and duration

At this point we examine the specific momentum transferred to the plane under investigation by a loading device. In our case:

$$P(t) = P(H(t) - H(t - T)),$$ (8.8)

therefore the specific (per unit of length) momentum of the impact will be:

$$R = PT.$$ (8.9)

The filled area on Figure 8.4 corresponds to a set of momentum values causing fracture. For the values outside of this area, crack propagation does not occur. The minimum value for the momentum incrementing the crack length (44.7 kg m/s) is reached at impulse with duration of 72 μs while the amplitude of the load exceeds the minimal amplitude by more than 10%.

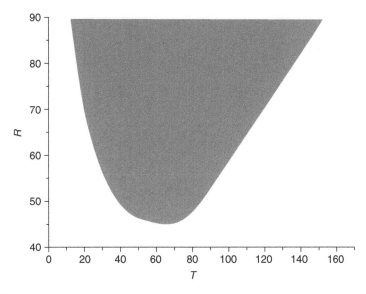

Figure 8.4 Filled area corresponds to a set of possible pulses leading to crack initiation. At $T = 72\,\mu s$ momentum R (kg m/s) needed to advance the crack is minimized.

Now we come to examination of the energy transmitted to the sample by a virtual loading device in the process of impact. The shape of the load applied is given by Equation (8.8). A specific (per unit of length) energy transmitted to the stripe can be calculated using the solution for the uniformly distributed load acting on a half plane. This problem can be easily solved utilising the D'Lambet method. The solution for a specific energy transmitted to the half plane appears to be:

$$\varepsilon_{spec} = \frac{1}{c\,\rho} \int\limits_0^T P^2(t)dt. \tag{8.10}$$

The c here is the same as c_1 and gives the longitudinal wave speed. This result can be used for the problem under investigation, as interaction of the loading device and the sample is finished before the waves reflected from the crack come back. Substitution of Equation (8.8) into Equation (8.10) gives $\varepsilon_{spec} = P^2 T/c\,\rho$.

Analogously to Figure 8.4, we plot a limiting curve for a set of energies that, being transmitted to the sample, cause the crack propagation (Fig. 8.5).

The minimum energy able to increment the crack length (172×10^6 J) is reached at load pulses with duration of $78\,\mu s$. As it is evident from Figure 8.5, the minimum energy required to propagate the crack by impacts with durations differing greatly from the optimal ones, significantly exceeds the minimum possible value. Thus the minimum energy initiating the crack for the load with duration of $92\,\mu s$ (at this impact duration crack propagation is possible with the impact of threshold amplitude), will

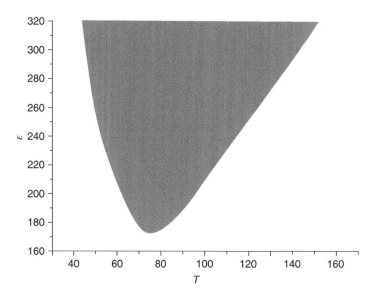

Figure 8.5 Filled area corresponds to a set of possible pulses leading to crack initiation. At $T = 72\,\mu s$ the energy ε (J/m^2) needed to advance the crack is minimized.

exceed the minimum energy possible by 10%, and at duration of $40\,\mu s$ it will be more than two times bigger.

8.3   THE CASE OF A LOAD APPLIED AT THE CRACK FACES

Now we consider a problem similar to the previous one, but with the load applied not at infinity but on the crack faces. The problem is solved numerically and in the same manner as the one for the load applied at infinity. Obviously, according to the superposition principle, the solution will coincide with the one for the stripe stretched by a load applied at infinity. Thus all the consequences of the previous solution are applicable, except for estimations of energy. Specific momentum transmitted to the sample will be the same as the one in the previous problem.

It is not possible to estimate energy transmitted to the sample analytically for the situation when the load is applied at the crack faces. However the finite element solution can be used in this case to estimate this energy. Figure 8.6 represents time dependence of full, kinetic and potential energies of deformation contained in a loaded sample for a particular pressure amplitude.

Firstly the kinetic energy increases linearly along with the potential one, in the same manner as happens in the case with the loaded half-plane. However at the moment of time equal to the time sufficient for a wave to travel along the crack length, the kinetic energy starts to transform into potential energy of deformation. Some part of the energy is returned to the loading device.

The limiting curve for the set of energies increasing the crack length is presented on Figure 8.7a. As can be noticed in the case of the load applied at the crack faces,

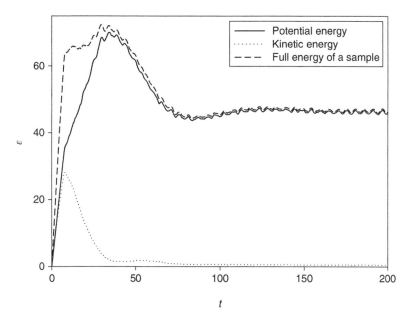

Figure 8.6 Transmitted energy (J) time (μs) dependence.

the energy input to increase the crack length has no marked minimum. The minimum energy needed to produce fracture in this case decreases with the growth of impulse duration. When the duration is equal to the maximum possible time-to-fracture, the energy reaches the minimum value.

Figure 8.7b enlarges the area adjacent to the point where the minimal energy is firstly reached on Figure 8.7a. As seen from Figure 8.7b for the pulse durations close to the maximal possible time-to-fracture (92 μs), the minimum energy input needed for crack propagation is not much different from the minimum value firstly achieved at 92 μs.

8.3.1 Optimization of the load parameters to minimize energy cost for the crack growth

With the majority of non-explosive methods used to fracture materials (drilling, grinding etc.), it is possible to control amplitude and frequency of impacts from the side of a rupture machine. The performed modeling shows that at a certain load duration (at impact fracture of large volumes of material, impulse duration is connected to the frequency of the machine impacts) energy input for crack propagation has a marked minimum.

Analogously to Figure 8.5, it is possible to plot the limiting curve for the set of energy values leading to propagation of a crack in the sample at different load amplitudes. This is done on Figure 8.8. Thus, it is possible to establish ranges of amplitudes and frequencies of load, at which energy costs for fracture of the material are minimized. These ranges are dependent on parameters of fractured material, predominant length of existing material cracks and the way the load is applied.

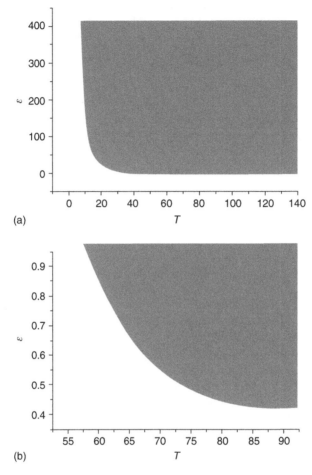

Figure 8.7 Energy minimization. Possible energy (J) quantities transmitted to a sample by a loading
device depending on load duration (μs). Figure 8.7b enlarges part of 8.7a.

Dependence of the optimal load parameters on the crack length was also studied. The results received are presented in Figures 8.9a and 8.9b. As observed from
Figure 8.9a, the duration of the load, that minimizes energy, and momentum inputs
are linearly or quasi-linearly dependent on the existing crack length. With the dis-
appearing crack length, the duration of the load minimizing momentum needed for
crack propagation approaches zero. At the same time, the duration optimal for the
energy inputs most probably tends to the microstructural time of the fracture process. The maximum possible time-to-fracture also tends to the microstructural time of
fracture.

Thus, considering intact media as the extreme case of media with cracks when the
crack length goes to zero, we find that the maximum possible time-to-fracture is the
same as the microstructural time of the fracture process. Durations of the loads being
optimal for the energy inputs for the fracture of intact media are also equal to the

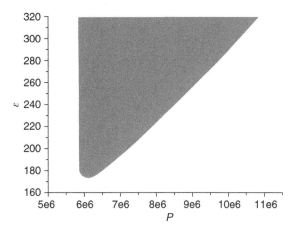

Figure 8.8 Finding optimal pulse amplitude. Possible energy (J/m^2) values for different pressure amplitudes P (Pa).

microstructure time of the fracture process. Amplitudes of loads, that minimize energy and momentum sufficient to increase the crack length, are presented in Figure 8.9b.

As expected, the amplitude of the threshold impulse is inversely dependent on $\sqrt{l}$, where l is the crack length. Dependence of amplitude, minimizing energy inputs, from the crack length is close to $1/\sqrt{l}$. The amplitude, minimizing the momentum, is back proportional to the crack length. When the crack length is close to zero, the amplitude of the load that minimizes the energy cost of the crack propagation, is close to threshold amplitude. However, the amplitude, minimizing the energy input, deviates from the threshold amplitude more and more with the growing crack length (Fig. 8.10).

8.3.2 Application to the problem of impact crater formation

In this section an attempt to incorporate incubation time approach into finite element (FE) code, and to simulate conditions of the satellite SMART1 lunar impact conducted by ESA in the year 2006 (ESA, 2006a, 2006b) is presented. The aim of the simulation is to compare dimensions of the crater created due to SMART1 contact with the moon surface to the results received using the FE method utilizing the ITFC as the critical rupture condition.

The traditional way to create a new surface in FE formulation is associated with splitting of existing nodes. Using this approach is reasonable in most cases, though this normally requires remeshing and remapping, which are rather time consuming procedures. For the studied problem the situation is different. To guarantee correct integration in condition (8.5), one should use small (as compared to τ) time steps. Thus the solution can result in a long series of tiny substeps. Solution (convergence) on every substep is achieved comparably fast – FE solver is almost not iterating. It was found that in this case it is more effective to use multiple nodes in the same location from the beginning, rather than split the node in question. Each element that the full model is constructed of, is not sharing nodes with other elements.

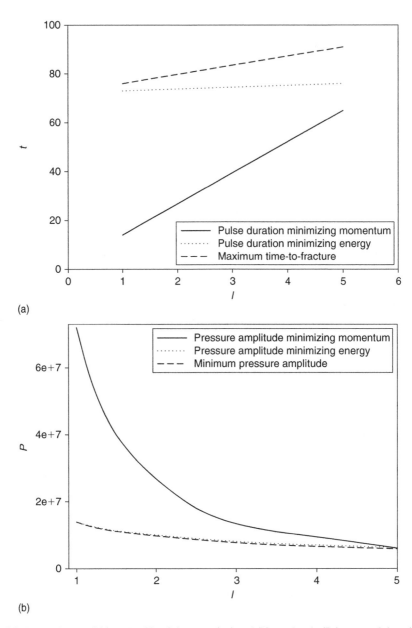

Figure 8.9 Dependence of (a) optimal load duration (μs) and (b) amplitude (Pa), on crack length (mm).

The 2-D problem with rotational symmetry is solved. Quadratic 4-node elements are used. The dimension of every element is exactly d times d (where d is given by (6)). Obviously, 4 nodes have the same location for inner points of a body and 2 nodes have the same location for the points belonging to the boundary. These nodes originally have their DOF's coupled. This results in exactly the same FE solution before

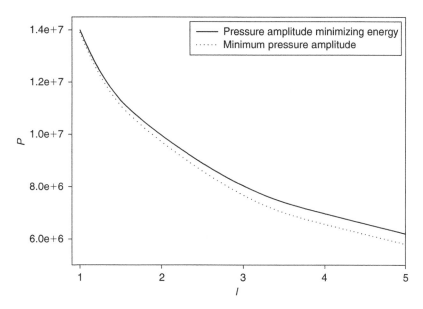

Figure 8.10 Dependence of optimal load amplitude (Pa) on crack length (mm).

the fracture condition is implemented in a respective point as if the elements had shared nodes. When the fracture condition is fulfilled, restriction on node DOF's is removed – a new surface is created. This is done automatically by FE code after every substep.

Figure 8.11 gives a schematic representation of internal points of a body. Originally all 4 nodes sharing the same location have all of their DOF's coupled. Condition (8.5) for this point can be written:

$$\frac{1}{\tau} \int_{t-\tau}^{t} \sigma_{ii}(t')dt' \le \sigma_c, \tag{8.11}$$

where i assumes values 1 and 2. Repeating indices do not dictate summation in this case. Spatial integration is removed, because the stress in the respective direction calculated by FE program is already a mean value over size d (since d is the element size being used). If condition (8.11) is fulfilled for σ_{11} and σ_{22} then displacements of nodes 1, 2, 3 and 4 on Figure 8.11 become uncoupled. If condition (8.11) is fulfilled for σ_{11}, two new couple sets consisting of nodes 1, 2 and 3, 4 are created. If condition (8.11) is fulfilled for σ_{22}, new couple sets are created for nodes 1, 3 and 2, 4. For later times, condition (8.11) in the applicable direction is traced for newly created couple sets separately. Contact between separated fragments is not modeled.

The problem is solved for half-space. The half-space representing the moon had the following material properties: $\sigma_c = 10.5$ MPa, $K_{IC} = 2.94\,MPa\sqrt{m}$, $\tau = 80\,\mu s$, $E = 60$ GPa, $\rho = 2850\,kg/m^3$, $\nu = 0.25$ typical for earth basalt. This results in $d = 5$ cm. Half-space is impacted by a cylinder with diameter of 1 meter and height of 1 meter.

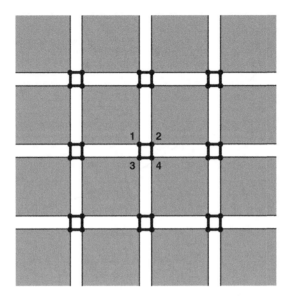

Figure 8.11 Model consisting of elements without shared nodes.

Density for the cylinder is chosen so that its mass is the same as that of the SMART1 satellite. We suppose the material of the cylinder is linear elastic and has no possibility of fracture. Elastic properties are: $E = 200$ GPa, $\upsilon = 0.32$, typical for steel. The SMART1 satellite had a form close to cubic with sides of 1 meter and had a mass of 366 kg. SMART1 impacted the moon surface at a speed of approximately 2000 m/s. In FE formulation the cylinder was given an initial speed of 2000 m/s prior its contact with the half-space boundary. Figure 8.12 gives an overview of the FE model. The size of the sample representing the half-space is chosen from the condition that the waves reflected from the sample boundaries are not returning to the region where the crater is formed in the process of the simulation. A total of 17,328 nodes and 17,252 elements were used in FE model. The time step was chosen to be equal to the time needed for the fastest wave to pass the distance equal to d.

ANSYS finite element package (ANSYS User's Guide (ANSYS, 2006)) was used to solve the stated problem. Separate ANSYS APDL subroutine was controlling the implementation of the fracture condition (Equation 8.11) in every point of the sample. The same subroutine was responsible for creation of a new surface once the rupture criterion is executed somewhere in the sample.

Figure 8.13 shows the sample state after the simulation is finished. Damage localized at the down part of the sample is due to the finite dimensions of a sample and represents cleavage fracture that occurred after compressive waves have reflected from the lower boundary. In Figure 8.14 locations of nodes where the fracture occurred are marked. This gives a possibility to assess the dimensions of the crater that is formed after the SMART1 impact. The damaged zone is found to be about 10 meters in diameter and about 3 meters deep. The zone where the material is fully fragmented (crater formed) can be assessed as being 7–10 meters in diameter and 3 meters deep. This

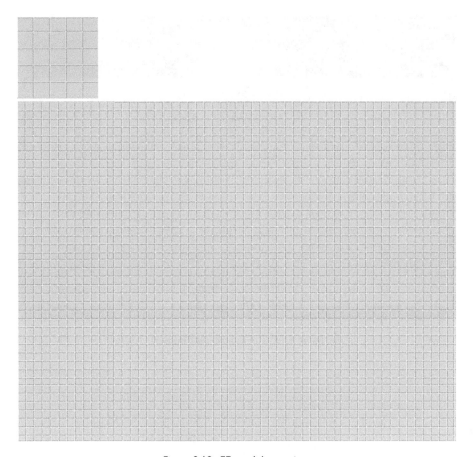

Figure 8.12 FE model overview.

result coincides with ESA estimations of dimensions of the crater formed due to the SMART1 impact (ESA, 2006a, 2006b).

8.3.3 Minimization of fracture energy in the case of contact interactions

In analyzing dynamic strength of materials, one is facing a contradiction between available experimental results and classical quasi-static approaches in fracture. Numerous experiments demonstrate that under high-rate dynamic loads, materials are able to endure loads significantly exceeding fracture loads in static (quasi-static) conditions. At the same time, in some of the experiments fracture in dynamic conditions is initiated at a moment when local stresses at a rupture point are significantly less as compared to stresses leading to fracture initiation in static (quasi-static) conditions. These obvious contradictions led to attempts to "correct" and "generalize" classical fracture criteria in order to make it applicable in the case of high-rate loads. This led to the appearance of a concept of "dynamic strength", depending not only on the material properties,

Figure 8.13 The sample after impact.

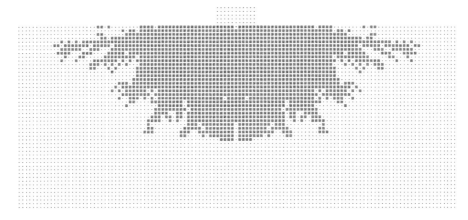

Figure 8.14 Locations of ruptured nodes.

but also on the loading rate and even the time-shape of the load pulse (Protasov, 2002; Latyshev, 2007). Practical utilization of this approach is rather complicated and often impossible, as there is no possibility of evaluating dynamic strength for all variety of loading rates and load shapes.

Most researchers dealing with problems of dynamic fracture are using fracture criteria based on extrapolation of quasi-static fracture criteria to dynamic conditions. Though they normally account for inertia and temporal characteristics of the load applied, temporal characteristics of the fracture process are usually not taken into consideration. Utilizing this kind of approach it is impossible to predict a critical situation, leading to fracture, applicable to both dynamic (high-rate loads) and quasi-static cases. In this section, an incubation time fracture criterion is used in order to predict fracture in the case of contact interactions. Employing this approach one does not need to worry about the time scale of the problem – the criterion gives correct predictions in a wide range of loading rates from static problems to the extremely dynamic ones. For the present analysis, we need to consider a wide range of loading rates and load durations. In this regard the incubation time fracture criterion provides a unique possibility to achieve correct estimations of conditions leading to fracture for the complex problem of spudding rocks.

In the simplest case, the incubation time fracture criterion is:

$$\frac{1}{\tau} \int_{t-\tau}^{t} \sigma(s)ds \leq \sigma_c, \tag{8.12}$$

where σ_c is the tensile strength of the material, evaluated for quasi-static conditions, and τ is the incubation time of the fracture process. Suppose that the shape of a loading pulse can be approximated by a smooth function:

$$\omega(t) = \begin{cases} \exp\left(\dfrac{1}{1-\left(\dfrac{2t}{t_0}-1\right)^{-2}}\right), & \left|t-\dfrac{t_0}{2}\right| \leq \dfrac{t_0}{2} \\ \\ 0, & \left|t-\dfrac{t_0}{2}\right| > \dfrac{t_0}{2} \end{cases},$$

where t_0 is for the load duration. Then the load is given by

$$\sigma(t) = \sigma_{max} \cdot \omega(t), \tag{8.13}$$

where σ_{max} is the load amplitude. Substituting Equation (8.13) into fracture criterion (8.12), one can obtain the critical (threshold) amplitude σ_{max} leading to fracture and corresponding to equality in Equation (8.12):

$$\sigma^* = \frac{\sigma_c \cdot \tau}{\max\limits_{t \in [0;t_0]} \int_{t-\tau}^{t} \omega(s)ds}$$

As an option for the way the energy is delivered to the fracture zone, consider a problem of impact interaction. Petrov, Morozov and Smirnov (2003) analyzed a problem for a spherical particle having radius R and velocity V impacting an elastic half-space using the classical Hertz contact scheme. The maximal stresses appearing in the half-space and the duration of interaction between the particle and the half-space were calculated.

According to the Hertz hypothesis, the contact force P arising between the particle and the half space can be presented as:

$$P(t) = kh^{\frac{3}{2}},$$

$$k = \frac{4}{3}\sqrt{R}\frac{E}{(1-v^2)}, \tag{8.14}$$

where h is a particle penetration and v is the Poisson's ratio of the elastic media. The maximal penetration h_0 can be found as:

$$h_0 = \left(\frac{5mV^2}{4k}\right)^{\frac{2}{5}}, \tag{8.15}$$

where m is the mass of a particle. The impact duration can be presented as

$$t_0 = \frac{2h_0}{V}\int_0^1 \frac{d\gamma}{1-\gamma^{\frac{5}{2}}} = 2,94\frac{h_0}{V}. \tag{8.16}$$

The dependence of time on penetration $h(t)$ can be approximated by:

$$h(t) = h_0\sin\left(\frac{\pi \cdot t}{t_0}\right). \tag{8.17}$$

Time-dependent maximum tensile stress generated in the impacted media can be estimated by

$$\sigma(V,R,t) = \frac{1-2v}{2}\cdot\frac{P(t)}{\pi a^2(t)}, \tag{8.18}$$

where the radius of the contact area $a(t)$ is given by:

$$a(t) = \left(3P(t)(1-v^2)\frac{R}{4E}\right)^{\frac{1}{3}}. \tag{8.19}$$

Knowing the duration and amplitude of the applied load, the mass and velocity of the impacting particle can be found from Equations (8.14)–(8.19):

$$R = \frac{t_0}{2,94}\left(\frac{6}{5}\frac{\sigma_{\max}}{\rho(1-2v)}\right)^{\frac{1}{2}},$$

$$V = \left(\frac{5}{4}\frac{\rho\pi(1-v^2)}{E}\right)^2\left(\frac{6}{5}\frac{\sigma_{\max}}{\rho(1-2v)}\right)^{\frac{5}{2}},$$

where ρ is a parameter of load intensity, having a dimension of mass density, and $\sigma_{\max}$ is the maximum stress (i.e. load amplitude). Evaluating the initial kinetic energy of the

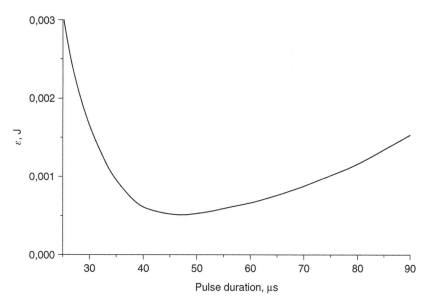

Figure 8.15 Energy (J) necessary to fracture versus load pulse duration (μs) for gabbro-diabase.

spherical particle, one can estimate the energy required in order to create fracture in the impacted media:

$$\varepsilon = \alpha \cdot \frac{t_0^3 \sigma_{max}^{\frac{13}{2}}}{\rho^{\frac{3}{2}} E^4},$$

where $\alpha = \frac{2}{3} \frac{\pi^5}{(2,94)^3} \left(\frac{5(1-\nu^2)}{4} \right)^4 \left(\frac{6}{5(1-2\nu)} \right)^{\frac{13}{2}}$ is a dimensionless coefficient. This energy, corresponding to the value $\rho = 2400\,\text{kg/m}^3$, is plotted versus impact duration in Figure 8.15. The properties of the material are taken to be equal to the properties of gabbro-diabase (Petrov *et al.*, 2004) ($E = 6.2 \cdot 10^9\,\text{N/m}^2$, $\sigma_c = 44.04 \cdot 10^6\,\text{N/m}^2$, $\nu = 0.26$ and $\tau = 440\,\mu\text{s}$).

8.4   CONCLUSIONS

The results received indicate a possibility of optimizing the energy consumption of different fracture-connected industrial processes (e.g., pounding, drilling). It is shown that the energy cost of crack propagation strongly depends on the amplitude and duration of the load applied. For example, in the studied problem when the duration of the load differs from the optimal one by 10%, the energy cost of initiating the crack is exceeding the minimum value by more than 10%. The obtained dependencies of the optimal characteristics of a load pulse on the existing crack length can help in predicting energy-saving parameters for the fracture processes by investigating the predominant crack size in a fractured material. Knowing the fracture incubation time for the particular material we can select the most energetically favorable mode of treatment. In particular, adjusting the duration of impacts, we can optimize the operation of rupture devices of the impact type. Similarly, it is possible to choose the vibration modes for decreasing the energy losses during processing of various materials. Thus, it was demonstrated that the incubation time approach is providing a possibility to predict the strength of rocks in a wide range of loading rates as well as to optimize the energy input needed to create rupture of rock media.

REFERENCES

ABAQUS USER MANUAL, Version 6.4, Hibbit, Karlson & Sorensen, Inc.

Bratov, V. and Petrov, Y.: Application of incubation time approach to simulate dynamic crack propagation. *Int J Fract* 146 (2007a), pp.53–60.

Bratov, V. and Petrov, Y.: Optimizing energy input for fracture by analysis of the energy required to initiate dynamic mode I crack growth. *International Journal of Solids and Structures* 44 (2007b), pp.2371–2380.

Bratov, V., Gruzdkov, A., Krivosheev, S. and Petrov, Y.: Energy balance in the crack growth initiation under pulsed-load conditions, *Doklady Physics* 49(5) (2004), pp.338–341.

Dally, J.W. and Barker, D.B.: Dynamic measurements of initiation toughness at high loading rates. *Experimental Mechanics* 28 (1988), pp.298–303.

ESA, 2006a: Impact landing ends SMART-1 mission to the Moon. Available online. http://www.esa.int/esaCP/SEM7A76LARE_index_0.html Accessed 10 Feb 2010.

ESA, 2006b: Intense final hours for SMART-1, Available online. http://www.esa.int/esaCP/SEMV386LARE_index_0.html Accessed 10 Feb 2010.

Freund, L.B. and Clifton, R.: On the uniqueness of plane elastodynamic solutions for running cracks. *Journal of Elasticity* 4 (1974), pp.293–299.

Freund, L.B.: *Dynamic Fracture Mechanics*. 1st Cambridge Press, Cambridge, 1990.

Irwin, G.: Analysis of stresses and strains near the end of a crack traversing a plate. *Journal of Applied Mechanics* 24 (1957), pp.361–364.

Latyshev, O.G.: *The Fracture of Rocks*. Moscow, 2007. (in Russian)

Morozov, N. and Petrov, Y.: *Dynamics of Fracture*. Springer-Verlag, Berlin-Heidelberg-New York, 2000.

Owen, D.M., Zhuang, S., Rosakis, A.J., and Ravichandran, G.: Experimental determination of dynamic crack initiation and propagation fracture toughness in aluminum sheets. *International Journal of Fracture* 90 (1998), pp.153–174.

Petrov Y.V., Smirnov V.I., Krivosheev S.I., Atroshenko S.A., Fedorovsky G.D. and, Utkin A.A.: Impact loading of rocks. In: *Proceedings of International Conference on Shock Waves in Condensed Matter* St.-Petersburg, 2004, pp.17–19.

Petrov, Y. and Morozov, N. (1994): On the modeling of fracture of brittle solids. *ASME J Appl Mech* 61 (1994), pp.710–712.

Petrov, Y. and Taraban, V.: On double–criterion models of the fracture of brittle materials. *St.-Petersburg University Mechanics Bulletin* 2 (1997), pp.78–81.

Petrov, Y. and Sitnikova, E.: Temperature dependence of spall strength and the effect of anomalous melting temperatures in shock-wave loading. *Technical Physics* 50 (2005), pp.1034–1037.

Petrov, Y.V.: Incubation time criterion and the pulsed strength of continua: fracture, cavitation, and electrical breakdown. *Doklady Physics* 49 (2004), pp.246–249.

Petrov, Y.V., Morozov N.F. and Smirnov, V.I.: Structural macromechanics approach in dynamics of fracture. *Fatigue Fract Engng Mater Struct* 26 (2003), pp.363–372.

Petrov, Y.V. and Utkin, A.A.: On the rate dependences of dynamic fracture toughness. *Soviet Material Science* 25(2) (1989), pp.153–156.

Petrov, Y.V.: On "Quantum" nature of dynamic fracture of brittle solids. *Dokl Akad Nauk USSR* 321(1) (1991), pp.66–68 [*Sov. Phys. Dokl.* 36, 802 (1991)].

Protasov, Y.I.: *The Fracture of Rocks*. Moscow, 2002. (in Russian).

Ravi-Chandar, K and Knauss, W.G.: An experimental investigation into dynamic fracture. *International Journal of Fracture* 25 (1984), pp.247–262.

Royal Dutch Petroleum Company *Annual Report* 2003, (2003).

Shockey, D.A. *et al.*: Short pulse fracture mechanics. *Journal of Engineering Fracture Mechanics* 23 (1986), pp.311–319.

Smith, G.C.: An *Experimental Investigation of the Fracture of a Brittle Material*. Ph.D. Thesis, California Institute of Technology, 1975.

Zlatin, N.A. and Pugachev, G.S.: Temoral dependency of metal strength. *Solid State Physics* 17 (1975), pp.2599–2602.

Chapter 9

Discontinuous approaches of wave propagation across rock joints

Xiaobao Zhao, Jianbo Zhu, Jungang Cai and Jian Zhao

9.1 INTRODUCTION

Rock masses often consist of multiple, near-parallel, planar joints and on most occasions, such a set (or sets) of parallel joints control the physical behaviour of rock masses. When a wave propagates through jointed rock masses, it is greatly attenuated (and slowed) due to the presence of joints. For rock engineering, the damage criteria of rock structures are generally regulated according to the threshold values of wave amplitudes, such as peak particle displacement, peak particle velocity and peak particle acceleration. Therefore, the prediction of wave attenuation across jointed rock masses is very important in assessing stability of, and damage to, rock structures under dynamic loads.

Wave propagation in layered media containing welded interfaces has been extensively studied, where stresses and displacements are continuous across the interfaces (Ewing, Jardetzky and Press, 1957; Brekhovskikh, 1980). However, joints have displacements (opening, closure and slip) under normal and shear stresses. Therefore, joints cannot be treated as welded interfaces, but should be referred to as displacement discontinuous boundaries (also termed non-welded interfaces, incompletely-welded interfaces or slip interfaces by some researchers).

Usually, the effects of joints on wave propagation have been modelled by equivalent medium methods (White, 1983; Schoenberg and Muir, 1989; Schoenberg and Sayers, 1995), which treat problems from the viewpoint of entirety. The effects of joints are lumped into effective elastic moduli of the equivalent medium. This assumption inherently results in two limitations. One limitation is the loss of discreteness of wave attenuation at individual joints, and the other limitation is the loss of the intrinsic frequency-dependent property at the joints. The frequency dependent property is attributed to two mechanisms. One mechanism is that the joints have an intrinsic frequency dependent property caused by the displacement discontinuity, and the other mechanism is aroused by multiple reflections between the joints.

As an alternative to the equivalent medium methods, the displacement discontinuity methods (Schoenberg, 1980) treat each joint as a non-welded interface of zero thickness. Thus, the stresses across the interface are continuous, but displacements across the interface are discontinuous. The discontinuity in displacement is equal to the average applied stress divided by the joint stiffness. When the joint stiffness approaches infinity, the interface becomes a welded boundary. When the joint stiffness

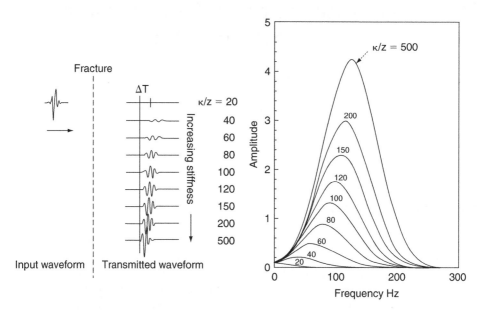

Figure 9.1 Illustration of the effects of a single joint on the transmitted waveform and corresponding amplitude spectra (after Myer, Pyrak-Nolte and Cook, 1990).

approaches zero, the interface becomes two free boundaries. Therefore, a non-welded interface can be thought as a generalized interface.

At the microscopic scale, a single joint appears as a planar collection of void spaces and asperities (asperities refer to contacts between two fracture surfaces). A comparison between wave scattering methods and displacement discontinuity methods was conducted by analyzing wave transmission across a periodic array of collinear micro-cracks (Angel and Achenbach, 1985a,b; Achenbach and Zhang, 1990). They found that solutions obtained by displacement discontinuity methods coincide with the far field solutions obtained by wave scattering methods, provided that the crack size and spacing are small relative to the incident wavelength. The agreement between two kinds of theories improves when crack size and spacing decrease. Myer (2000) carried out an ultrasonic test on wave reflection and transmission at a joint with different asperity separations. His results verified the conclusions obtained by Angel and Achenbach (1985a, b).

The effects of a single joint on wave propagation have been widely studied with full consideration of different joint deformational behaviour (Miller, 1977, 1978; Schoenberg, 1980; Myer, Pyrak-Nolte and Cook, 1990; Pyrak-Nolte, Myer and Cook, 1990b; Pyrak-Nolte, 1996; Gu *et al.*, 1996; Zhao and Cai, 2001). Figure 9.1 shows a numerical example to illustrate the effects of a linear deformational joint in a homogeneous medium on a normally incident wave, where k is the joint stiffness, Z is the wave impedance and ΔT is the time delay. For an incident pulse shown at the extreme left, transmitted waves were calculated for a range of joint stiffness values. Shown on the right in the figure are the amplitude spectra of transmitted waves. For a high joint stiffness, the transmitted wave is essentially identical to the incident pulse. The case of infinite stiffness corresponds to the welded boundary condition. As the joint stiffness decreases, the transmitted wave is both slowed and attenuated. The attenuation is characterized by both decreasing amplitude and filtering of high frequency components of the pulse. Gu *et al.* (1996) performed a study of wave reflection, transmission and conversion of harmonic wave incidence upon a joint at arbitrary angles. They found that a head wave or an inhomogeneous P-interface wave appears, when an SV-wave (SV-wave is a kind of S-wave with particle movement in the vertical plane) is incident at or beyond the critical angle, which is determined by the Poisson's ratio of the rock material.

In comparison, the effects of multiple joints on wave propagation become complicated due to multiple reflections occurring between joints. Actually, the transmitted wave across parallel joints can be treated as a wave superposition of transmitted waves arriving at different times, which are caused by the multiple reflections. However, it is difficult to explicitly determine the wave superposition. A simplified method was proposed by ignoring the multiple reflections as a short-wavelength approximation (Hopkins, Myer and Cook, 1988; Pyrak-Nolte, Myer and Cook, 1990a; Myer *et al.*, 1995). Thus, the transmission coefficient across parallel joints is calculated as the product of transmission coefficients of individual joints. Laboratory experiments conducted by Hopkins, Myer and Cook (1988), Pyrak-Nolte, Myer and Cook (1990a), Myer *et al.* (1995) and Nakagawa, Nihei and Myer (2000) verified that the T^N-method is valid, when the firstly arriving wave is not contaminated by the multiple reflections. In addition, Hopkins, Myer and Cook (1988) and Myer *et al.* (1995) observed that $|T_N|$ is larger than $|T_1|^N$, when joint spacing is small relative to incident wavelength.

In the present study, normal transmission of P-waves across parallel joints with linear deformational behaviour is examined, where the fractures are assumed to be planar, dry, and of a large extent and small thickness relative to the incident wavelength. In theoretical formulation, the method of characteristics is used to develop a set of recurrence equations with respect to particle velocities and normal stress. These equations are then numerically solved.

9.2 METHOD OF CHARACTERISTICS FOR ONE-DIMENSIONAL P-WAVE PROPAGATION ACROSS JOINTED ROCK MASSES

The method of characteristics has been widely used to study one-dimensional wave propagation in a continuous medium (e.g., Ewing, Jardetzky and Press, 1957; Brekhovskikh, 1980; Kennett, 1983; Bedford and Drumheller, 1994). Based on the one-dimensional wave equation, relations between particle velocity and stress along right- and left-running characteristics can be built. In an ideally elastic medium, a one-dimensional P-wave equation is

$$\frac{\partial^2 u}{\partial t^2} = \alpha_p^2 \frac{\partial^2 u}{\partial x^2}, \tag{9.1}$$

where u is the displacement, α_p is the P-wave velocity, x is the distance, and t is the time. Alternatively, Equation (9.1) can be expressed by particle velocity and strain:

$$\frac{\partial v}{\partial t} = \alpha_p^2 \frac{\partial \varepsilon}{\partial x}, \tag{9.2}$$

where $v = \partial u/\partial t$ is the particle velocity, and $\varepsilon = \partial u/\partial x$ is the strain. Apparently, the variables v and ε are related by

$$\frac{\partial v}{\partial x} = \frac{\partial \varepsilon}{\partial t}. \tag{9.3}$$

Therefore, the derivative of $(v - \alpha_p \varepsilon)$ is

$$d(v - \alpha_p \varepsilon) = \frac{\partial(v - \alpha_p \varepsilon)}{\partial t} dt + \frac{\partial(v - \alpha_p \varepsilon)}{\partial x} dx = \left(\frac{\partial v}{\partial t} - \alpha_p \frac{\partial \varepsilon}{\partial t} \right) dt + \left(\frac{\partial v}{\partial x} - \alpha_p \frac{\partial \varepsilon}{\partial x} \right) dx \tag{9.4}$$

Equations (9.2) and (9.3) are substituted into Equation (9.4):

$$d(v - \alpha_p \varepsilon) = \left(\frac{\partial v}{\partial x} - \frac{1}{\alpha_p} \frac{\partial v}{\partial t} \right) (dx - \alpha_p dt). \tag{9.5}$$

Equation (9.5) indicates that $d(v - \alpha_p \varepsilon)$ equals to zero, if $dx/dt = \alpha_p$. The quantity of $(v - \alpha_p \varepsilon)$ is a constant along any straight line with slope of $1/\alpha_p$ (right-running characteristic) in the x-t plane. Similarly, the quantity of $(v + \alpha_p \varepsilon)$ is a constant along any straight line with slope of $-1/\alpha_p$ (left-running characteristic) in the x-t plane. An illustration of right- and left-running characteristics in the x-t plane is shown in Figure 9.2.

$(v - \alpha_p \varepsilon)$ and $(v + \alpha_p \varepsilon)$ are multiplied by P-wave impedance (Z_p):

$$Z_p(v - \alpha_p \varepsilon) = Z_p v - Z_p \alpha_p \varepsilon = Z_p v + \sigma = \text{ constant} \tag{9.6}$$

along a right-running characteristic in the x-t plane, and

$$Z_p(v + \alpha_p \varepsilon) = Z_p v + Z_p \alpha_p \varepsilon = Z_p v - \sigma = \text{constant} \tag{9.7}$$

along a left-running characteristic in the x-t plane, where σ is the normal stress, $Z_p = \rho \alpha_p$, $\alpha_p^2 = E/\rho$, Z_p is the P-wave impedance, ρ is the rock density, and E is the Young's modulus of rock material. It is notable that normal stress is defined to be positive for compressive stress, and negative for tensile stress. The definition is consistent with that commonly used in rock mechanics.

Compared with the effects of a single joint, the effects of multiple parallel joints on wave propagation become complicated due to the multiple reflections occurring between the joints. A general model of the method of characteristics is introduced to solve the problem. In the x-t plane, new variables, nondimensional distance (n) and nondimensional time (j), are imported and defined as

$$j = \frac{t}{\Delta t}, \tag{9.8}$$

$$n = \frac{x}{\alpha_p \Delta t}, \tag{9.9}$$

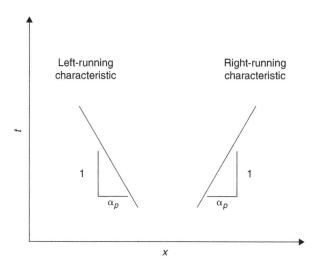

Figure 9.2 Right- and left-running characteristics in the *x-t* plane.

where Δt is the time interval. It is assumed that a finite number of interfaces are located at integral values of nondimensional distance in a half space with its left boundary at $n=0$, the first interface at $n=1$, the second interface at $n=2$ and the last interface at $n=l$ (l is an integer). The interfaces could be joints or welded interfaces, which can be treated as joints with infinite joint stiffness. Figure 9.3 shows conjunction points of right- and left-running characteristics at integral values of nondimensional distance and nondimensional time. Particle velocities and normal stresses are evaluated at these points. However, this does not mean that solutions can be obtained only at the interface positions. If the field between two adjacent interfaces is further divided into a number of uniform layers, solutions can be obtained at the boundaries of these layers, which are considered as joints with infinite joint stiffness.

The characteristic model shown in Figure 9.3 consists of two characteristics, and can be applied in the study of joints with different deformational models, e.g., joints with linear deformational behaviour (Cai and Zhao, 2000; Zhao, Zhao and Cai, 2006a), joints with nonlinear deformational behaviour described by the static BB model (Zhao and Cai, 2001; Zhao, Zhao and Cai, 2006b), and joints with Coulomb Slip behaviour (Zhao *et al.*, 2006). In order to simplify the model, it is assumed that joints and elastic media on both sides of the joints have identical properties.

Along the right-running characteristic *ab* and the left-running characteristic *ac* shown in Figure 9.3, two relations between particle velocities and normal stresses at points *a*, *b* and *c* are built:

$$Z_p v^-(n, j+1) + \sigma^-(n, j+1) = Z_p v^+(n-1, j) + \sigma^+(n-1, j), \tag{9.10}$$

$$Z_p v^+(n, j+1) - \sigma^+(n, j+1) = Z_p v^-(n+1, j) - \sigma^-(n+1, j), \tag{9.11}$$

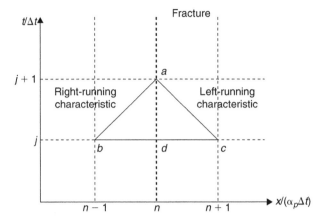

Figure 9.3 Conjunction points of right- and left-running characteristics at nondimensional distance *n* and nondimensional time *j* in the *n-j* plane.

where $v^-(n, j+1)$ and $v^+(n, j+1)$ are particle velocities at time $j+1$ before and after the joint at distance *n*. Similarly, $\sigma^-(n, j+1)$ and $\sigma^+(n, j+1)$ are normal stresses at time $j+1$ before and after the joint at distance *n*.

In the far field of dynamic responses induced by an explosion or vibration, the magnitude of the stress wave is too small to mobilize nonlinear deformation of the joints, so linear joint behaviour is adopted in the present study. The displacement discontinuous model is applied at point *a*:

$$\sigma^-(n, j+1) = \sigma^+(n, j+1) = \sigma(n, j+1), \tag{9.12}$$

$$u^-(n, j+1) - u^+(n, j+1) = \frac{\sigma(n, j+1)}{k_n}, \tag{9.13}$$

where k_n is the normal joint stiffness, $u^-(n, j+1)$ and $u^+(n, j+1)$ are displacements at time $j+1$ before and after the joint at distance *n*.

By considering Equation (9.12), the addition of Equations (9.10) and (9.11) is

$$Z_p v^-(n, j+1) + Z_p v^+(n, j+1) = Z_p v^+(n-1, j) + \sigma^+(n-1, j)$$
$$+ Z_p v^-(n+1, j) - \sigma^-(n+1, j). \tag{9.14}$$

The differentiation of Equation (9.13) with respect to *t* is

$$v^-(n, j+1) - v^+(n, j+1) = \frac{1}{k_n} \frac{\partial \sigma(n, j+1)}{\partial t}. \tag{9.15}$$

If Δt is small enough that the differential part in Equation (9.15) can be expressed as

$$v^-(n, j+1) - v^+(n, j+1) = \frac{1}{k_n} \frac{\partial \sigma(n, j+1)}{\partial t} = \frac{\sigma(n, j+1) - \sigma(n, j)}{k_n \Delta t}, \tag{9.16}$$

then Equation (9.16) is rewritten as

$$\sigma(n, j + 1) = \sigma(n, j) + k_n \Delta t[v^-(n, j + 1) - v^+(n, j + 1)]. \tag{9.17}$$

Substituting Equation (9.17) into Equation (9.10) gives

$$(k_n \Delta t + Z_p)v^-(n, j + 1) - k_n \Delta t v^+(n, j + 1) + \sigma(n, j) = Z_p v^+(n - 1, j) + \sigma(n - 1, j). \tag{9.18}$$

Equations (9.14) and (9.18) form a linear equation group with respect to particle velocities at point a before and after the joint. After the equation group is solved, expressions of particle velocities at point a are obtained:

$$v^-(n, j + 1) = (Z_p v^+(n - 1, j) + \sigma(n - 1, j) - \sigma(n, j) + \frac{k_n \Delta t}{Z_p}(Z_p v^-(n + 1, j)$$
$$+ Z_p v^+(n - 1, j) + \sigma(n - 1, j) - \sigma(n + 1, j)))/(2k_n \Delta t + Z_p), \tag{9.19}$$

$$v^+(n, j + 1) = ((Z_p v^+(n - 1, j) - \sigma(n + 1, j) + \sigma(n - 1, j)$$
$$+ Z_p v^-(n + 1, j))\frac{k_n \Delta t + Z_p}{Z_p} + \sigma(n, j) - \sigma(n - 1, j)$$
$$- Z_p v^+(n - 1, j))/(2k_n \Delta t + Z_p), \tag{9.20}$$

By substituting Equations (9.19) and (9.20) into Equation (9.17), the expression of normal stress is obtained:

$$\sigma(n, j + 1) = \sigma(n, j) + \frac{k_n \Delta t}{2k_n \Delta t + Z_p}(Z_p v^+(n - 1, j) + \sigma(n - 1, j)$$
$$- Z_p v^-(n + 1, j) + \sigma(n + 1, j) - 2\sigma(n, j)). \tag{9.21}$$

Equations (9.19), (9.20) and (9.21) show that the responses at point a are determined by those at points b, c and d. Meanwhile, it indicates that responses at time $j + 1$ can be calculated from those at time j. With input velocity of $v(0, j)$ and initial conditions of $v^+(n, 0)$, $v^-(n, 0)$ and $\sigma(n, 0)$, Equations (9.19), (9.20) and (9.21) are applied to determine particle velocities and stress at any point through an iterative computation, which can be implemented by a self-developed computer program.

The accuracy of differential calculation requires the time interval to be infinitely small. A smaller time interval can be achieved by further dividing the field between two adjacent interfaces into a number of uniform layers with sufficiently small ratio (γ) of layer thickness to incident wavelength. However, an extremely small γ may cost a lot of computation time. Therefore, it is necessary to determine a reasonable value of γ to achieve a balance between computation efficiency and accuracy. A parametric study on P-wave attenuation across a linear deformational joint ($|T_1|$) is performed in the following part to calibrate the computer program, and at the same time, to select a suitable value of γ.

Schoenberg (1980) and Pyrak-Nolte, Myer and Cook (1990) derived an analytical expression of transmission coefficient for normally incident harmonic P-wave across a single joint in an identical rock material:

$$T_1 = \frac{2(k_p/(Z_p\omega))}{-i + 2(k_p/(Z_p\omega))},\qquad(9.22)$$

where T_1 is the transmission coefficient across a single joint and ω is the angular frequency of the harmonic wave. Equation (9.22) shows that the transmission coefficient is dependent on the ratio (K_n) of normal joint stiffness to the product of P-wave impedance and angular frequency. The ratio (K_n) is named as normalized normal stiffness.

In the following calculations, it is assumed that rock density is $2650\,kg/m^3$ and P-wave velocity is $5830\,m/s$, as typical properties of the Bukit Timah granite of Singapore (Zhao, 1996). Normal joint stiffness is varied from 1 to 15 GPa/m to achieve different normalized normal stiffness ($K_n = k_n/(Z_p\omega)$). Because a harmonic wave only exists in the frequency domain and can be treated as a pulse with innumerous cycles, it is impossible to input a harmonic wave in the calculation. Hence, a one-cycle sinusoidal wave with unit amplitude is applied at the left boundary as the incident wave, where the frequency of the sinusoidal wave is 50 Hz. Figure 9.4 shows the incident waveform (in the left figure) and corresponding amplitude spectra (in the right figure). In order to obtain the analytical solution of the one-cycle sinusoidal wave across a single joint, the incident wave is firstly transformed into frequency domain by FFT (fast Fourier transform). In the frequency domain, the one-cycle sinusoidal wave is treated as the sum of a series of harmonic waves with dominant frequency of 42.5 Hz. Then, based on Equation (9.22), transmitted waves of all harmonic components across the joint are obtained. Finally, an inverse transform for these transmitted waves is conducted to get the transmitted wave of the one-cycle sinusoidal wave by IFFT (inverse fast Fourier transform). Although the frequency of the sinusoidal wave is closely related to the dominant frequency of the one-cycle sinusoidal wave, it is reasonable to use the dominant frequency to represent the frequency of the incident wave, instead of the frequency of the sinusoidal wave. Therefore, $f = 42.5$ Hz is used in the following calculations.

During the computation, incident, reflected and transmitted waves are obtained at two receiving points before and after the joint. The two points should be carefully chosen, so that the receiving waves have no superposition with each other, and are not contaminated by reflected waves from computation boundaries.

The comparisons between computed results for different γ (ratio of layer thickness to incident wavelength) and theoretical solution of $|T_1|$ (the magnitude of transmission coefficient across a single joint) as a function of normalized normal stiffness are shown in Figure 9.5. From the plots, it is found that the computed results agree well with the analytical solutions. The average error percentages (the error percentage is defined as a ratio of the difference of theoretical solution and computed result to the theoretical solution) for γ equal to 1/20, 1/50, 1/100 and 1/200 are 4.1, 2.1, 1.4 and 1.1%, respectively. By considering both computation efficiency and accuracy, $\gamma = 1/100$ is selected for the calculations in the next section.

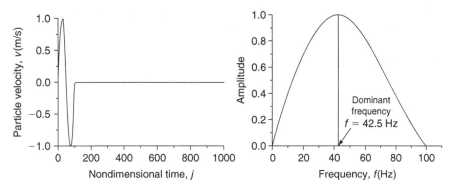

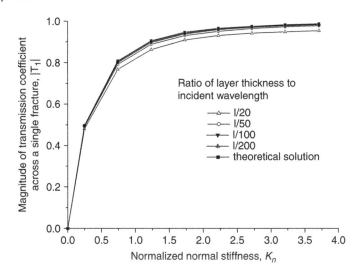

Figure 9.4 Illustration of incident wave (a one-cycle sinusoidal wave) and corresponding amplitude spectra.

Figure 9.5 Comparisons between computed results for different γ and theoretical solution of $|T_1|$ as a function of normalized normal stiffness.

Since particle velocity is the commonly used parameter in assessing stability and damage of rock structures under dynamic loads, the magnitudes of transmission and reflection coefficients in the present study are defined by particle velocities as ratios of amplitudes of transmitted and reflected waves to the amplitude of the incident wave, which is assumed to have unit amplitude.

9.3 PARAMETRIC STUDIES ON WAVE ATTENUATION ACROSS PARALLEL JOINTS

Because nondimensional variables are used in the model, nondimensional joint spacing (ξ), the ratio of joint spacing to incident wavelength, is imported to study the effects of joint spacing. During the calculation, ξ is assumed to be an integral multiple of γ and the value of ξ varies from 1/100 to 50/100. Therefore, the field between two adjacent joints could be further divided into a number of uniform layers with γ of 1/100 (e.g. when ξ is 1/100, there is only one layer between two joints, and when ξ is 50/100, there are 50 layers between the two joints).

Studies on wave attenuation across two parallel joints separated with different nondimensional joint spacing at different normalized normal stiffness ($K_n = 0.247$, 0.494, 0.988, 1.482 and 1.976) are conducted. In addition, energies of transmitted and reflected waves are examined. The transmitted and reflected energy rates are defined as

$$e_{tra} = \frac{E_{tra}}{E_{inc}} = \frac{\int_{t_{tra}^0}^{t_{tra}^1} Z_p (v_{tra})^2 dt}{\int_{t_{inc}^0}^{t_{inc}^1} Z_p (v_{inc})^2 dt} = \frac{\sum\limits_{j=t_{tra}^0}^{j=t_{tra}^1} Z_p (v_{tra})^2 \Delta t}{\sum\limits_{j=t_{inc}^0}^{j=t_{inc}^1} Z_p (v_{inc})^2 \Delta t}, \tag{9.23}$$

$$e_{ref} = \frac{E_{ref}}{E_{inc}} = \frac{\int_{t_{ref}^0}^{t_{ref}^1} Z_p (v_{ref})^2 dt}{\int_{t_{inc}^0}^{t_{inc}^1} Z_p (v_{inc})^2 dt} = \frac{\sum\limits_{j=t_{ref}^0}^{j=t_{ref}^1} Z_p (v_{ref})^2 \Delta t}{\sum\limits_{j=t_{inc}^0}^{j=t_{inc}^1} Z_p (v_{inc})^2 \Delta t}, \tag{9.24}$$

where e_{tra} and e_{ref} are transmitted and reflected energy rates, E_{tra}, E_{ref} and E_{inc} are energies of transmitted, reflected and incident waves, Z_p is the P-wave impedance, v_{tra}, v_{ref} and v_{inc} are particle velocities of transmitted, reflected and incident waves, t_{tra}^0, t_{ref}^0 and t_{inc}^0 are initial times of transmitted, reflected and incident waves, and t_{tra}^1, t_{ref}^1 and t_{inc}^1 are final times of transmitted, reflected and incident waves.

Calculation results for the magnitude of transmission coefficient across two parallel joints ($|T_2|$) as a function of ξ at different K_n are shown in Figure 9.6. From the results, it is found that:

1 $|T_2|$ increases with increasing K_n.
2 Two important indices of ξ, threshold value (ξ_{thr}) and critical value (ξ_{cri}), are identified. They divided the area of ξ into three parts: the individual joint area ($\xi \geq \xi_{thr}$), the transition area ($\xi_{thr} > \xi > \xi_{cri}$) and the small spacing area ($\xi \leq \xi_{cri}$).
3 In the individual joint area, $|T_2|$ remains constant with changing ξ, and it indicates that the wave superposition of transmitted waves arriving at different times has no effects on $|T_2|$.
4 When $\xi < \xi_{thr}$, the wave superposition has obvious effects on $|T_2|$, so the transition area and small spacing area are together named as superposition area. In the transition area, $|T_2|$ increases from the constant value to a maximum value with decreasing ξ. While in the small spacing area, $|T_2|$ decreases from the maximum value with decreasing ξ.

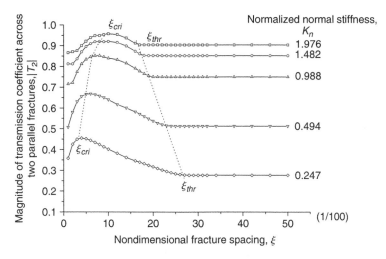

Figure 9.6 $|T_2|$ as a function of nondimensional joint spacing at different normalized normal stiffness.

5   When $\xi \to 0$, the two joints act together as an equivalent joint with effective normal joint stiffness of $k_n/2$.

6   When K_n is large, $|T_2| \approx |T_1|^2$ ($|T_1|$ for K_n equal to 0.247, 0.494, 0.988, 1.482 and 1.976 are 0.494, 0.696, 0.860, 0.921 and 0.949, respectively) in the individual joint area. When K_n is small, $|T_2| > |T_1|^2$ in the individual joint area. The phenomena will be explained in the following discussion section.

7   Values of ξ_{thr} and ξ_{cri} vary with K_n, as shown in Figure 9.6 by two dotted lines. Generally, ξ_{thr} decreases and ξ_{cri} increases with increasing K_n.

Correspondingly, the transmitted and reflected energy rates (e_{tra} and e_{ref}) are calculated as functions of ξ at different K_n, as shown in Figure 9.7. From Figure 9.7, the following observations can be noted:

1   e_{tra} and e_{ref} are functions of ξ and K_n. Generally, e_{tra} increases and e_{ref} decreases with increasing K_n.

2   The balance of energy rates (energy conservation) is always preserved (e.g. $e_{ref} + e_{tra} \approx 1$). The energy conservation coincides with the basic assumptions: (a) rock material is ideally elastic without material attenuation and, (b) joint has elastic deformational behaviour.

3   The increase (or decrease) of e_{tra} is not always consistent with that of $|T_2|$. This is because multiple reflections may lead to a transmitted wave with a low amplitude but long lasting time, which has large e_{tra}.

4   Actually, the sum of energy rates ($e_{ref} + e_{tra}$) are not exactly equal to 1 (e.g. $e_{ref} + e_{tra} = 0.976$, when K_n is 1.976 and ξ is 7/100). This is because the value of γ equal to 1/100 causes some numerical error.

If more parallel joints are further incorporated, the magnitude of transmission coefficient across multiple parallel joints ($|T_N|$) is calculated for different numbers of

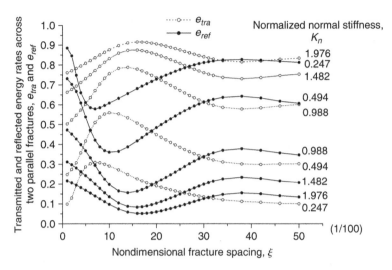

Figure 9.7 e_{tra} and e_{ref} as functions of nondimensional joint spacing at different normalized normal stiffness.

joints (N). Figure 9.8 shows $|T_N|$ as a function of ξ for $K_n = 1.482$ and $N = 2 \sim 10$. It is found that:

1 $|T_N|$ decreases with increasing N.
2 Two important indices of ξ, threshold value (ξ_{thr}) and critical value (ξ_{cri}), are identified. They divided the area of ξ into three parts: the individual joint area $(\xi \geq \xi_{thr})$, the transition area $(\xi_{thr} > \xi > \xi_{cri})$ and the small spacing area $(\xi \leq \xi_{cri})$.
3 In the individual joint area, $|T_N|$ remains constant with changing ξ, and indicates that wave superposition of transmitted waves arriving at different times has no effects on $|T_N|$.
4 When $\xi < \xi_{thr}$, the wave superposition has obvious effects on $|T_N|$, so the transition area and small spacing area are together named as the superposition area. In the transition area, $|T_N|$ increases from the constant value to a maximum value with decreasing ξ, while in the small spacing area, $|T_N|$ decreases from the maximum value with decreasing ξ.
5 When $\xi \to 0$, the joints act together as an equivalent joint with effective normal joint stiffness of k_n/N.
6 In the individual joint area, $|T_N|$ decreases substantially with increasing N. The dependence of $|T_N|$ on N becomes weak in the transition area, and becomes much weaker in the small spacing area, especially for the values of $|T_N|$ in circle A or circle B as shown in Figure 9.8. The phenomena in the small spacing area will be explained by the equivalent medium method in the following discussion section.
7 When N is small, $|T_N| \approx |T_1|^N$ ($|T_1|$ for K_n equal to 1.482 is 0.921) in the individual joint area. When N is large, $|T_N| > |T_1|^N$ in the individual joint area. The phenomena will be explained in the following discussion section.
8 Values of ξ_{thr} and ξ_{cri} vary with N, as shown in Figure 9.8 by two dotted lines. Generally, ξ_{thr} decreases and ξ_{cri} increases with decreasing N.

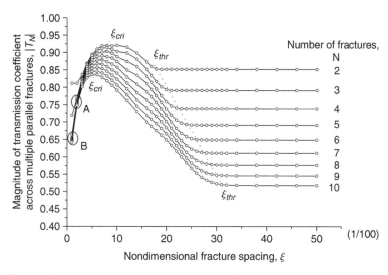

Figure 9.8 $|T_N|$ as a function of nondimensional joint spacing for different numbers of joints ($K_n = 1.482$).

9.4   EFFECTS OF SINGLE JOINT AND PARALLEL JOINTS ON WAVE TRANSMISSION

The transmitted wave across parallel joints can be treated as a wave superposition of transmitted waves arriving at different times, which are caused by multiple wave reflections. Therefore, waveforms of differently arriving transmitted waves are important, and should be studied in detail.

Figures 9.9 and 9.10 show the incident, reflected and transmitted waves upon a single joint at different normalized normal stiffness ($K_n = 0.988$ and 1.976). It can be seen that

1   The reflected and transmitted waves are very different from the incident wave. Compared with the incident wave, the transmitted wave has small amplitude and low frequency; while the reflected wave is a three-bulb pulse, which has small amplitude and high frequency.
2   The transmitted and reflected waves change with K_n. When K_n becomes small, the transmitted wave has small amplitude and low frequency; while the reflected wave has large amplitude and low frequency.

When the incident wave is transformed into frequency domain by FFT (fast Fourier transform), the one-cycle sinusoidal wave can be treated as the sum of a series of harmonic waves with dominant frequency of 42.5 Hz. The joint acts as a frequency filter as shown in Figure 9.1: it reduces the high frequency components of the incident wave. Meanwhile, the frequency of the reflected wave becomes higher than that of the incident wave. The phenomenon can also be observed as the broadening of transmitted waveforms or the shortening of reflected waveforms in Figures 9.9 and 9.10

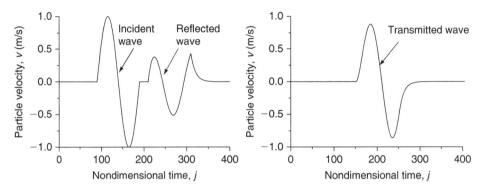

Figure 9.9 Incident, reflected and transmitted waves upon a single joint ($K_n = 0.988$).

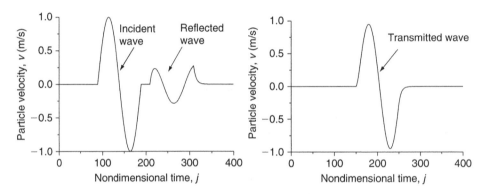

Figure 9.10 Incident, reflected and transmitted waves upon a single joint ($K_n = 1.976$).

(e.g. the incident waveform lasts for 100 nondimensional time units, but the transmitted waveforms or reflected waveforms in Figures 9.9 and 9.10 last more or less than 100 nondimensional time units). In addition, the decrement of transmitted wave amplitude is observed in Figures 9.9 and 9.10 ($|T_1| = 0.86$ for $K_n = 0.988$ and $|T_1| = 0.949$ for $K_n = 1.976$).

Figures 9.11 and 9.12 show the incident, reflected and transmitted waves upon two parallel joints at different normalized normal stiffness ($K_n = 0.988$ and 1.976). Large nondimensional joint spacing is adopted in the calculation, so that differently arriving waves can be shown separately in the figures. From the plots, it is found that:

1 Compared with the transmitted wave across a single joint, the firstly arriving transmitted wave across two parallel joints has small amplitude and low frequency.
2 The secondly arriving transmitted wave is a three-bulb pulse, and it has high frequency compared with the firstly arriving transmitted wave. In addition, the secondly arriving transmitted wave changes with K_n. When K_n is small, it has large amplitude and low frequency.

Figure 9.1 shows that the frequency of transmitted wave decreases with decreasing K_n. Figures 9.11 and 9.12 verify the conclusion and show the decrement of the

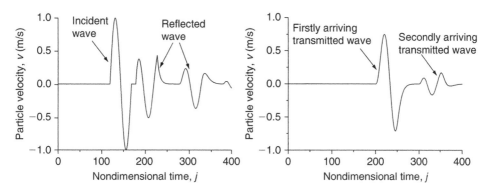

Figure 9.11 Incident, reflected and transmitted waves upon two parallel joints ($K_n = 0.988$).

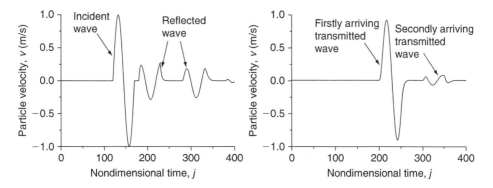

Figure 9.12 Incident, reflected and transmitted waves upon two parallel joints ($K_n = 1.976$).

frequency of the firstly arriving transmitted wave with decreasing K_n. Furthermore, it is understandable to conclude that the frequency of the firstly arriving transmitted wave decreases with increasing N. Therefore, when K_n is small or N is large, the frequency of the firstly arriving transmitted wave decreases significantly, and it causes $|T_N| > |T_1|^N$ in the individual joint area.

The effects of multiple reflections are applicable to explain the change of $|T_N|$ with joint spacing. If joints are sparsely placed relative to the incident wavelength, i.e. $\xi \geq \xi_{thr}$ (in the individual joint area), multiple reflections have no effects on the amplitude of the transmitted wave. This means that each of the joints contributes individually to wave attenuation like a singe joint. However, if $\xi < \xi_{thr}$ (in the superposition area), the effects of multiple reflections become significant due to the close joint spacing. Since the arriving-time difference between differently arriving transmitted waves depends on the joint spacing, the superposition of transmitted waves arriving at different times cause an increase in $|T_N|$ if $\xi_{thr} > \xi > \xi_{cri}$ (in the transition area) and a decrease in $|T_N|$ if $\xi \leq \xi_{cri}$ (in the small spacing area) with decreasing ξ.

Equivalent medium methods have been commonly applied to determine the overall properties of fractured rock masses in the long wavelength limit (Schoenberg, 1983;

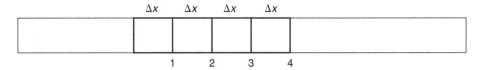

$$\Delta x \quad \Delta x \quad \Delta x \quad \Delta x$$

1 2 3 4

Figure 9.13 Illustration of effective length of an equivalent medium.

Pyrak-Nolte, Myer and Cook, 1990; Schoenberg and Sayers, 1995). When the incident wavelength is much larger than the fracture spacing (e.g. in the small spacing area, $\xi \leq \xi_{cri}$), an equivalent medium method is developed to explain wave phenomena in this area.

Generally, the effective length of an equivalent medium is defined as the sum of N fractures and N fracture spacings. Figure 9.13 illustrates the definition of effective length of an equivalent medium, which has four fractures numbered from 1 to 4. Therefore, the effective length of the equivalent medium includes four fractures, three fracture spacings ($3\Delta x$) between the fractures and an extended fracture spacing (Δx) on the left side of fracture 1. According to the definition, the effective length of an equivalent medium is $N\Delta x$, and it can be rewritten as $N\xi\lambda_p$. Since nondimensional variables are used, nondimensional effective length is defined as the ratio of effective length to incident wavelength, and is equal to $N\xi$. Subsequently, the effective Young's modulus of the equivalent medium (E_e) is defined as

$$E_e = \frac{\sigma}{\varepsilon_e} = \frac{\sigma}{\frac{Nd}{N\xi\lambda_p} + \varepsilon} = \frac{\sigma}{\frac{\frac{\sigma}{k_n}N + \frac{\sigma}{E_r}N\xi\lambda_p}{N\xi\lambda_p}} = \frac{1}{\frac{1}{k_n\xi\lambda_p} + \frac{1}{E}}, \tag{9.25}$$

where E_e is the effective Young's modulus, ε_e is the effective strain, d is the fracture closure, ε is the strain of rock material, N is the number of fractures, σ is the normal stress, k_n is the normal fracture stiffness, ξ is the nondimensional fracture spacing, λ_p is the P-wave wavelength in rock material, and E is the Young's modulus of rock material.

Additionally, the portion of $k_n\xi\lambda_p$ in Equation (9.25) can be rewritten as

$$k_n\xi\lambda_p = \frac{k_n\xi\alpha_p}{f} = \frac{Z_p 2\pi}{Z_p 2\pi} \cdot \frac{k_n\xi\alpha_p}{f} = Z_p\alpha_p\frac{k_n}{Z_p\omega}2\pi\xi = Z_p\alpha_p K_n 2\pi\xi = EK_n 2\pi\xi, \tag{9.26}$$

where α_p is the P-wave velocity in rock material, f is the P-wave frequency, $E = \rho\alpha_p^2 = Z_p\alpha_p$, and ρ is the rock density. Equations (9.25) and (9.26) show that E_e is a function of ξ and K_n.

In Figure 9.6, $|T_2|$ changes with K_n at a specific ξ value in the small spacing area, their nondimensional effective lengths are the same and equal to 2ξ. However, the effective Young's modulus is a function of K_n. According to the knowledge of one-dimensional wave propagation across a layered medium, $|T_N|$ increases with increasing effective Young's modulus. Therefore, $|T_2|$ increases with increasing K_n at a specific ξ value in the small spacing area. In Figure 9.8, $|T_N|$ changes with N at a specific ξ value in the small spacing area, their effective Young's moduli are the same, but nondimensional lengths are different and proportional to N. Similarly, according to the knowledge of one-dimensional wave propagation across a layered medium, $|T_N|$ decreases with increasing nondimensional effective length. The decreasing rate becomes small, when nondimensional effective length is large. After the nondimensional effective length exceeds a certain value, $|T_N|$ keeps constant with changing nondimensional effective length. Hence, $|T_N|$ decreases with increasing N, but the dependence of $|T_N|$ on N becomes much weaker in the small spacing area.

When $\xi \to 0$, $\frac{1}{K_n 2\pi\xi} >> 1$. Then, E_e is approximated as

$$E_e = \frac{1}{\frac{1}{EK_n 2\pi\xi} + \frac{1}{E_r}} = \frac{E}{\frac{1}{K_n 2\pi\xi} + 1} \approx EK_n 2\pi\xi. \tag{9.27}$$

Meanwhile, the equivalent medium can be treated as an equivalent fracture with effective normal fracture stiffness (k_{ne}):

$$k_{ne} = \lim_{\xi \to 0} \frac{E_e}{N\xi\lambda_p} = \lim_{\xi \to 0} \frac{EK_n 2\pi\xi}{N\xi\lambda_p} = E\frac{K_n}{N}\frac{2\pi}{\lambda_p} = \frac{K_n}{N}z_p\omega = \frac{k_n}{N}, \tag{9.28}$$

Equation (9.28) shows that when $\xi \to 0$, the equivalent medium including N fractures with normal fracture stiffness of k_n can be treated as an equivalent fracture with effective normal fracture stiffness of k_n/N.

9.5 OUTLOOKS

Besides the method of characteristics, other approaches, e.g. the scattering matrix method, the virtual wave source method, and the superposed analytical method, can also be used to study wave propagation across parallel joints combined with the displacement discontinuity model, where multiple wave reflections among joints are taken into account.

The scattering matrix method (SMM) (Fig. 9.14), which is also termed the propagation matrix method, is adopted to study wave propagation across rock joints (Aki and Richards, 2002; Perino, Barla and Orta, 2010; Zhao *et al.*, 2011). Combined with the displacement discontinuity model, the method of plane wave analysis and propagator matrix are applied to develop relations between the first layer (incident wave) and the nth layer with respect to potential amplitudes or displacements and stresses in matrix form. Then, with initial boundary conditions, potential amplitudes in any layer or displacements and stresses at any point can be obtained by solving corresponding matrices. In the case of a planar discontinuity like a joint, incident, reflected and transmitted plane waves have the same transverse wave-vector. The respective amplitudes are related by a 2×2 block matrix:

$$\begin{pmatrix} c_1^- \\ c_2^+ \end{pmatrix} = \begin{pmatrix} S_{11} & S_{12} \\ S_{21} & S_{22} \end{pmatrix} \begin{pmatrix} c_1^+ \\ c_2^- \end{pmatrix} \tag{9.29}$$

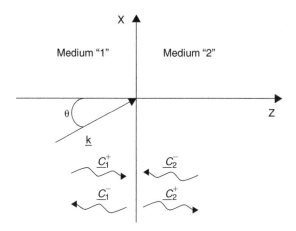

Figure 9.14 Wave propagation through an interface with the SMM.

where c_1^+ and c_2^- are the amplitudes of the waves incident on the discontinuity, whereas c_1^- and c_2^+ are the amplitudes of the scattered waves (reflected and transmitted), S_{ii} have the meaning of reflection coefficients at the two sides of the discontinuity, and S_{ij} have the meaning of transmission coefficients. Since elastic waves have three possible polarization states (P, SV, SH), the submatrices have a size of 3×3. In order to extend the scattering matrix method to the case of multiple parallel joints, one may compute the scattering matrix for each discontinuity. Then, by using a "chain rule" procedure, the global scattering matrix is defined. This is a combination of the components of the scattering matrix for each discontinuity and represents the effect on elastic wave propagation due to the N discontinuities. The global scattering matrix contains the global reflection and transmission coefficients of a set of parallel discontinuities, where multiple wave reflections among the joints are taken into account.

Combined with the equivalent medium model, the concept of virtual wave source (VWS) is introduced to study normally incident wave propagation across one joint set, where multiple wave reflections among the joints were considered (Li, Ma and Zhao, 2010). The VWS produced one reflected wave each time an incident wave arrives at the joint. The transmitted wave was derived by the effective moduli of the equivalent model which includes a rock joint and rock material with length equal to joint spacing. The virtual wave source method (VWSM) can also be extended to study wave propagation across joints in combination with the displacement discontinuity model (Zhu et al., 2011). As shown in Figure 9.15, in order to obtain the theoretical solution of the transient wave (v_I) transmitting across one joint set, the incident transient wave is firstly transformed into frequency domain by FFT (fast Fourier transform). In frequency domain, the incident transient wave can be transformed as the sum of a series of harmonic waves:

$$v_I = \sum_{i=-\infty}^{\infty} v_{Ii} = \sum_{i=-\infty}^{\infty} A_i e^{i\omega_i t} \tag{9.30}$$

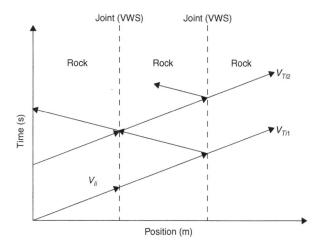

Figure 9.15 Scheme of jointed rock mass with VWSM.

where v_{Ii} is one harmonic wave, A_i and ω_i are the amplitude and angular frequency of v_{Ii}. VWSM is used to take into account multiple wave reflections for wave propagation across one joint set. However, the definition of VWS here is different from that of Li, Ma and Zhao (2010). VWS exists at the joint position and represents the mechanical properties of the joint. Each time an incident wave propagates across the joint, VWS produces 2 new waves for normally incident wave or 2 or 4 new waves for obliquely incident wave, which can be directly derived by using the reflection and transmission coefficients of a single joint. Thus, the transmitted harmonic wave across one joint set is the result of wave superposition of different transmitted waves created by VWS:

$$v_{Ti} = \sum_{j=1}^{\infty} v_{Tij} \qquad (9.31)$$

where v_{Ti} is the transmitted wave and for the incident harmonic wave v_{Ii}, v_{Tij} is the transmitted wave arriving at a different time. Finally, an inverse transform for these waves is conducted to get the transmitted waves of the incident transient wave by IFFT (inverse fast Fourier transform), which can transform one series of harmonic waves into one transient wave (v_T):

$$v_T = \sum_{i=-\infty}^{\infty} v_{Ti} \qquad (9.32)$$

A methodology termed the superposed analytical method (SAM) is also introduced to study wave propagation across multiple parallel joints, where multiple wave reflections among joints are considered (Zhu, 2011). Assuming but not limiting that the background rock media of the opposite sides of each joint are identical, the mechanical properties are the same for every joint, and joints are equally spaced, all the transmitted waves across 2^{n-1} joints form a geometric sequence, and all the reflected waves across 2^{n-1} joints except the first one also form a geometric sequence. Therefore, the reflection and transmission coefficients across 2^n joints, which are considered as basic solutions, can be expressed as a function of $R_{2^{n-1}}$ and $T_{2^{n-1}}$:

$$R_{2^n} = R_{2^{n-1}} + \frac{T_{2^{n-1}}{}^2 R_{2^{n-1}} e^{i4\pi\xi}}{1 - R_{2^{n-1}}{}^2 e^{i4\pi\xi}}, \tag{9.33}$$

$$T_{2^n} = \frac{T_{2^{n-1}}{}^2 e^{i2\pi\xi}}{1 - R_{2^{n-1}}{}^2 e^{i4\pi\xi}}, \tag{9.34}$$

where, ξ is the ratio of the joint spacing to the wavelength and is termed the nondimensional joint spacing. Therefore, if the reflection and transmission coefficients across a single joint are known, e.g. those obtained in Eqs. (9.22), the reflection and transmission coefficients across 2^n joints can be derived. However, it does not mean that the reflection and transmission coefficients can only be obtained for 2^n joints. The reflection and transmission coefficients across other numbers of joints can be derived through basic solutions. For example, R_3, T_3 can be obtained from R_1, T_1, R_2 and T_2:

$$R_3 = R_1 + \frac{T_1^2 R_2 e^{i4\pi\xi}}{1 - R_1 R_2 e^{i4\pi\xi}} = R_2 + \frac{T_2^2 R_1 e^{i4\pi\xi}}{1 - R_1 R_2 e^{i4\pi\xi}} \tag{9.35}$$

$$T_3 = \frac{T_1 T_2 e^{i2\pi\xi}}{1 - R_1 R_2 e^{i4\pi\xi}} \tag{9.36}$$

The SAM can also be extended to study obliquely incident waves, but the parameters in Equations (9.33)–(9.36) should be changed to matrix form. It should be noted that the superposed analytical method is different from communication theory (e.g., Treitel and Robinson, 1966; Luco and Aspel, 1983), because basic solutions are available and hence the computational speed will be much faster.

The above discontinuous approaches are limited to studying wave propagation across a single joint set. However, for practical jointed rock masses, there usually exist several sets of joints. When a stress wave is incident to the jointed rock masses, wave phenomena will become more complicated due to wave conversion, and multiple reflections and refractions between joints. Numerical computation and experimental testing should be adopted to study these complicated wave propagation problems.

REFERENCES

Aki, K. and Richards P.G.: *Quantitative Seismology*, University Science Books, California, 2002.

Achenbach, J.D. and Zhang, C.H.: Reflection and transmission of ultrasound by a region of damaged material. *Journal of Nondestructive Evaluation* (1990), 9(2/3), pp.71–89.

Angel, Y.C. and Achenbach, J.D.: Reflection and transmission of elastic waves by a periodic array of cracks. *Journal of Applied Mechanics* 52 (1985a), pp.33–40.

Angel, Y.C. and Achenbach, J.D.: Reflection and transmission of elastic waves by a periodic array of cracks: Oblique incidence. *Wave Motion* 7 (1985b), pp.375–382.

Bandis, S.C., Lumsden, A.C. and Barton, N.R.: Fundamentals of rock fracture deformation. *International Journal of Rock Mechanics and Mining Sciences & Geomechanical Abstracts* 20(6) (1983), pp.249–268.

Barton, N.R., Bandis, S.C. and Bakhtar, K.: Strength, deformation and conductivity coupling of rock joints. *International Journal of Rock Mechanics and Mining Sciences & Geomechanical Abstracts* 22(3) (1985), pp.121–140.

Bedford, A. and Drumheller, D.S.: *Introduction to Elastic Wave Propagation*. John Wiley and Sons, Chichester, 1994.

Brekhovskikh, L.M: *Waves in Layered Media*. Academic Press, New York, 1980.

Cai, J.G. and Zhao, J.: Effects of multiple parallel joints on apparent wave attenuation in rock masses. *International Journal of Rock Mechanics and Mining Sciences* 37(4) (2000), pp.661–682.

Ewing, W.M., Jardetzky, W.S. and Press, F.: *Elastic Waves in Layered Media*. New York: McGraw-Hill, 1957.

Gu, B., Suárez-Rivera, R., Nihei, K.T. and Myer, L.R.: Incidence of plane waves upon a joint. *Journal of Geophysical Research* 101(B11) (1996), pp.25337–25346.

Hopkins, D.L., Myer, L.R. and Cook, N.G.W.: Seismic wave attenuation across parallel joints as a function of joint stiffness and spacing. *EOS Transaction AGU* 69(44) (1988), pp.1427–1438.

Kennett, B.L.N.: *Seismic Wave Propagation in Stratified Media*. Cambridge University Press, London, 1983.

Li, J.C., Ma, G.W. and Zhao, J.: An equivalent viscoelastic model for rock mass with parallel joints. *Journal of Geophysical Research* 115 (2010), B03305, doi:10.1029/2008JB006241.

Luco, J.E. and Apsel R.J.: On the Green's functions for a layered half-space. Part I. *Bulletin of the Seismological Society of America* 73(1983), 909–929.

Miller, R.K.: An approximate method of analysis of the transmission of elastic waves through a frictional boundary. *Journal of Applied Mechanics* 44 (1977), pp.652–656.

Miller, R.K.: The effects of boundary friction on the propagation of elastic waves. *Bulletin of Seismological Society of America* 68(4) (1978), pp.987–998.

Myer, L.R.: Fractures as collections of cracks. *International Journal of Rock Mechanics and Mining Sciences* 37 (2000), pp.231–243.

Myer, L.R., Hopkins, D., Peterson, J.E. and Cook, N.G.W.: Seismic wave propagation across multiple joints. *Joints and Jointed Rock Masses* (1995), pp.105–110.

Myer, L.R., Pyrak-Nolte, L.J. and Cook, N.G.W.: Effects of single joint on seismic wave propagation. In: *Proceedings of ISRM Symposium on Rock Joints*, Loen, 1990, pp.467–473.

Nakagawa, S., Nihei, K.T. and Myer, L.R.: Stop-pass behaviour of acoustic waves in a 1D joint system. *Journal of Acoustic Society of America* 107(1) (2000), pp.40–50.

Perino, A., Barla, G. and Orta, R.: Wave propagation in discontinuous media. *Proceedings of European Rock Mechanics Symposium (EUROCK) 2010*, Lausanne, Switzerland, 2010, pp.285–288.

Pyrak-Nolte, L.J.: The seismic response of joints and the interrelations among joint properties. *International Journal of Rock Mechanics and Mining Sciences & Geomechanical Abstracts* 33(8) (1996), pp.787–802.

Pyrak-Nolte, L.J., Myer, L.R. and Cook, N.G.W.: Anisotropy in seismic velocities and amplitudes from multiple parallel joints. *Journal of Geophysical Research* 95(B7) (1990a), pp.11,345–11,358.

Pyrak-Nolte, L.J., Myer, L.R. and Cook, N.G.W.: Transmission of seismic waves across single natural joints. *Journal of Geophysical Research* 95(B6) (1990b), pp.8617–8638.

Schoenberg, M.: Elastic wave behaviour across linear slip interfaces. *Journal of Acoustic Society of America* 68(5) (1980), pp.1516–1521.

Schoenberg, M.: Reflection of elastic waves from periodically stratified media with interfacial slip. *Geophysical Prospecting* 31 (1983), pp.265–292.

Schoenberg, M. and Muir, F.: A calculus for finely layered anisotropic media. *Geophysics* 54(5) (1989), pp.581–589.

Schoenberg, M. and Sayers, C.M.: Seismic anisotropy of jointed rock. *Geophysics* 60(1) (1995), pp.204–211.

White, J.E.: *Underground Sound*. New York: Elsevier, 1983.

Treitel, S. and Robinson, E.A.: Seismic wave propagation in layered media in terms of communication theory. *Geophysics* 31(1) (1966), pp.17–32.

Zhao, J.: Construction and utilization of rock caverns in Singapore, part A: bedrock resource of the Bukit Timah granite. *Tunnelling and Underground Space Technology* 11(1) (1996), pp.65–72.

Zhao, J. and Cai, J.G.: Transmission of elastic P-waves across single joint with a nonlinear normal deformational behaviour. *Rock Mechanics and Rock Engineering* 34(1) (2001), pp.3–22.

Zhao, J., Zhao, X.B. and Cai, J.G.: A further study of P-wave attenuation across parallel fractures with linear deformational behaviour. *International Journal of Rock Mechanics and Mining Sciences* 43(5) (2006a), pp.776–788.

Zhao, X.B., Zhao, J. and Cai, J.G.: P-wave transmission across fractures with nonlinear deformational behaviour. *International Journal for Numerical and Analytical Methods in Geomechanics* 30(11) (2006b), pp.1097–1112.

Zhao, X.B., Zhao, J., Hefny, A.M. and Cai, J.G.: Normal transmission of S-wave across parallel fractures with coulomb slip behaviour. *Journal of Engineering Mechanics-ASCE* 132(6) (2006), pp.641–650.

Zhao, X.B., Zhu, J.B., Cai, J.G. and Zhao, J.: Study of wave attenuation across parallel fractures with propagator matrix method, *International Journal for Numerical and Analytical Methods in Geomechanics* (2011), DOI: 10.1002/nag.1050.

Zhu, J.B.: *Theoretical and numerical analyses of wave propagation in jointed rock masses*. PhD thesis, École Polytechnique fédérale de Lausanne (EPFL), Lausanne, Switzerland (2011).

Zhu, J.B., Zhao, X.B., Li, J.C., Zhao, G.F. and Zhao, J.: Normally incident wave propagation across one joint set with virtual wave source method. *Journal of Applied Geophysics* 73(3) (2011), pp.283–288.

Chapter 10

Equivalent Medium Model with Virtual Wave Source Method for wave propagation analysis in jointed rock masses

Jianchun Li, Guowei Ma and Jian Zhao

10.1 INTRODUCTION

Rock mass usually consists of multiple, parallel planar joints, known as joint sets, which govern the mechanical behavior of the rock mass. The dynamic behavior and wave propagation across jointed rock masses are of great interest in geophysics, mining and underground constructions. It is also significant to assess the stability and damage of rock structures under dynamic loads. Because of the discontinuity of the joints, the dynamic response of jointed rock masses is a complicated process. It is of significance to develop an efficient and explicit model to represent the dynamic property of the jointed rock mass.

Currently, the methods for analyzing the effect of joints on the properties of rock masses can be divided into two categories. One is the displacement discontinuity method (DDM) (Miller, 1977; Schoenberg, 1980) and the other is the effective moduli method (White, 1983; Schoenberg and Muir, 1989; Pyrak-Nolte, Myer and Cook, 1990; Cook, 1992). In the displacement discontinuity method, the stresses across the interface are continuous, while the displacements across the interface are discontinuous. Generally, the displacement discontinuity method treats joints, particularly the dominant sets, as discrete entities. It predicts well the effect of joints on the transmission of seismic waves (Pyrak-Nolte, 1988; Cook, 1992). Successful applications of this method have been reported for wave transmission across a single joint (Miller, 1977; Pyrak-Nolte, 1988; Pyrak-Nolte, Myer and Cook, 1990; Cook, 1992; Zhao and Cai, 2001; Li, Ma and Huang, 2010; Ma, Li and Zhao, 2011) and multi-parallel joints (Cai and Zhao, 2000; Zhao, Zhao and Cai, 2006a, 2006b; Zhao et al., 2006). In all these applications, the joints were considered to be linear or nonlinear elastic, and the rock between each two joints was intact and elastic. Based on the displacement discontinuity method, the derivation of wave propagation equations is straightforward and it is in a differential form, which may not display an explicit expression of the solutions.

The effective moduli methods predict the aggregate effects of many joints or joint systems within a representative elementary volume (REV), so as to make the analysis of continuum problems contractible. Using a static approach, Zhao, Zhao and Cai (2006a) deduced the effective normal joint stiffness in a rock mass with parallel joints and small joint spacing. Pyrak-Nolte, Myer and Cook (1990) derived the time delay between two joints by using DDM and obtained the effective velocity for a normal incident wave through a set of parallel joints. Cook (1992) showed that the effective moduli methods account for the effect of joints on wave velocities while ignoring their influence on wave dissipation. The demerits of the effective moduli methods (White, 1983; Schoenberg and Muir, 1989; Pyrak-Nolte, Myer and Cook, 1990; Zhao, Zhao and Cai, 2006a) are that they simplify the discontinuous rock mass to an elastic medium, which is effective only if the frequency-dependence and the discreteness of joints, or multiple reflections among the joints, are negligible.

Pyrak-Nolte, Myer and Cook (1990) indicated that the frequency dependence can be accounted for with an assumption of an equivalent viscoelastic medium. By conducting extensive laboratory tests on ultrasonic wave transmission across natural joints, Pyrak-Nolte (1988) and Pyrak-Nolte, Myer and Cook (1990) suggested that the natural rock joints may possess elastic as well as viscous coupling across the interface. The definition of a linear viscoelastic solid is that it is a material for which the stress and the strain components are related by linear differential equations which involve the stress, the strain, and their derivatives with respect to time (Kolsky, 1953). The wave propagation in linear viscoelastic solids has been investigated by Kolsky (1953) and Tsai and Kolsky (1968), in which the Voiget solid model, the Maxwell solid model and some more general solid models were proposed. By analysis, it is found that the auxiliary spring placed in series with the Voiget model is a more appropriate equivalent model for a rock mass with one joint set, which can display both the attenuation and the frequency dependence of the transmitted wave (Li, Ma and Zhao, 2010).

This chapter proposes a viscoelastic equivalent medium model (EMM) for wave propagation through rock masses with parallel joints. The new model is in general form to describe the normal and shear effective property of the rock mass. The EMM combines a linear viscoelastic solid model with the concept of virtual wave sources (VWS), in which the frequency dependence and the discreteness of joints in rock mass are taken into account. The parameters in the equivalent model are derived by analysis of longitudinal (P-) and shear (S-) wave propagation normally across REV by using DDM and EMM. To verify the proposed EMM, the results of the transmitted waves through a set of equally spaced parallel joints are compared with those from DMM. The effects of VWS are also demonstrated.

10.2   CONVENTIONAL EFFECTIVE ELASTIC MODULI METHODS

The effective elastic moduli methods are commonly used to determine the overall properties of the jointed rock mass. For a rock mass with a set of equally-spaced parallel joints, a representative elementary volume (REV) is one part of the rock mass. REV is composed of a joint and the rock between two adjacent joints. The length of the REV is defined as the joint spacing. Hence, the associated equivalent medium is defined as the continuum medium with the summary of the representative elementary volume.

There are two main conventional effective elastic moduli methods used to derive the elastic constants of the rock mass: one is static and the other is dynamic. For the statically effective elastic moduli methods, the rock mass with a set of parallel joints is replaced by a transversely isotropic medium model, which can be expressed by five elastic constants. These five elastic constants are described in terms of the properties of joints and rock material (Pyrak-Nolte, Myer and Cook, 1990).

Therefore, if an incident wave normally propagates across parallel and linear deformable rock joints, the propagation velocity of the wave is expressed as (Amaderi and Goodman, 1981; Pyrak-Nolte, Myer and Cook, 1990)

$$\rho V^2 = C_{3333} \tag{10.1}$$

where V is the magnitude of the wave velocity, ρ is the density of the rock material and C_{3333} is the elastic constant of the transversely isotropic medium model, which can be obtained from

$$\frac{1}{C_{3333}} = \frac{1}{E} + \frac{1}{SK_p} \tag{10.2}$$

where E is the Young's modulus of the intact rock, S is the joint spacing, and K_p is the normal specific stiffness of the joints.

In the dynamic effective elastic moduli method (Crampin, McGonigle and Bamford, 1980; Ciccotti and Mulargia, 2004), the effective modulus of the rock mass is calculated by

$$E_{eff} = \rho V_e^2 \tag{10.3}$$

where the subscript 'eff' denotes the effective property and 'e' denotes the value obtained by experimental or field measurement.

The effective modulus E_{eff} of the representative element is then extended to the entire rock mass. Therefore, the effective velocity of the stress wave propagating in the jointed rock mass is the same as given in Equation (10.3).

As mentioned above, the material constants can be obtained by using the static and dynamic effective elastic moduli method in Equations (10.2) and (10.3), respectively. However, as a result of the presence of the joint, the jointed rock mass is no longer purely elastic. Accordingly, the dynamic effective moduli are not the same as the static effective moduli. Therefore, it is not possible that the effective velocities obtained by these two methods coincide with each other. Several laboratory measurements showed that the static effective elastic modulus appeared to be lower than the dynamic effective modulus (Eissa and Kazi, 1988; Ciccotti and Mulargia, 2004).

10.3 EQUIVALENT VISCOELASTIC MEDIUM MODEL FOR ROCK MASS WITH PARALLEL JOINTS

Natural jointed rock mass is discontinuous and may include one or more joint sets. When an incident body wave propagates across the rock mass, the transmitted wave depends not only on the characteristics of the incident wave but also on the properties of the joints, such as the stiffness and spacing (Pyrak-Nolte, Myer and Cook, 1990; Cai and Zhao, 2000; Zhao, Zhao and Cai, 2006a, 2006b; Zhao et al., 2006).

Besides the Voiget and the Maxwell solid models, two extended linear viscoelastic solid models are also used for solid medium to describe the stress-strain relation (Kolsky, 1953). One is an auxiliary spring in parallel with the Maxwell model, and the other is an auxiliary spring placed in series with the Voiget model. By comparing the Maxwell model, the Voiget model and their extended forms, it is found that the auxiliary spring placed in series with the Voiget model is a more appropriate equivalent model for a rock mass with one joint set, which can display both the attenuation and the frequency dependence of the transmitted wave. This viscoelastic model was used by Li, Ma and Zhao (2010) to describe the stress-stain relation for a rock mass with one joint, when the incidence was only a P-wave normally across the joints. In order to consider the effect of the wave reflections between joints, the concept of virtual wave source (VWS) was introduced by Li, Ma and Zhao (2010). For a normally incident S-wave, the concept of viscoelastic model and VWS can also be adopted to describe the dynamic shear property of a jointed rock mass. In this chapter, the incident waves are assumed to be P- and S-wave normally across a set of parallel joints within a rock mass. The objective of the chapter is to introduce a general EMM for wave propagation through rock masses with a set of parallel joints.

10.3.1 Wave equations for linear viscoelastic medium

The equivalent mechanical model of the auxiliary spring placed in series with the Voiget model is shown in Figure 10.1, where Figure 10.1(a) is the viscoelastic model with normal property and Figure 10.1(b) is the viscoelastic model with shear property. E and G are respectively the Young's modulus and the shear modulus of the rock material, E_{vp} and η_{vp} are the normal stiffness and the viscosity contributed by the joint, respectively, E_{vs} and η_{vs} are the shear stiffness and the shear viscosity of the joint, respectively. For simplification, when k is p and s, E_{vk} is adopted as a unified symbol for E_{vp} and E_{vs}, respectively, so does η_{vk} for η_{vp} and η_{vs}.

If the Young's modulus E and the shear modulus G of the rock material are expressed as a unified symbol E_{ak} ($k = p$ for E and $k = s$ for G), the stress-strain relation for the model shown in Figure 10.1 can be expressed in a general form, that is:

$$(E_{ak} + E_{vk})\sigma_k + \eta_{vk}\frac{\partial \sigma_k}{\partial t} - \eta_{vk}E_{ak}\frac{\partial \varepsilon_k}{\partial t} - E_{vk}E_{ak}\varepsilon_k = 0, \quad (k = p, s) \qquad (10.4)$$

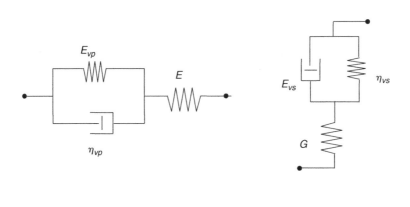

(a) With normal property   (b) With shear property

Figure 10.1 Equivalent mechanical model of an auxiliary spring in series with Voiget model.

where σ_k and ε_k are the normal stress and strain when $k=p$ and the shear stress and strain when $k=s$. Considering the longitudinal motion equation for one-dimensional problem, there is:

$$\rho\frac{\partial v_k}{\partial t} = \frac{\partial \sigma_k}{\partial x_k} \tag{10.5}$$

where ρ is the density of the medium, v_k is the particle velocity along normal or shear direction when $k=p$ or s, respectively, and t is time.

Differentiating Equation (10.4) with respect to x_k and substituting for $\partial \sigma_k / \partial x_k$ with Equation (10.5) yields

$$\rho\eta_{vk}\frac{\partial^2 v_k}{\partial t^2} + \rho(E_{ak} + E_{vk})\frac{\partial v_k}{\partial t} - \eta_{vk}E_{ak}\frac{\partial^2 v_k}{\partial x_k^2} - E_{vk}E_{ak}\int \frac{\partial^2 v_k}{\partial x_k^2}dt = 0 \tag{10.6}$$

Defining $\tau_k = \eta_{vk}/E_{vk}$ as the time of retardation of the Voiget element, when a trial solution has the form of

$$v_k = A_k \cdot \exp(\beta_k x_k)\exp[i(\omega_k t - \alpha_k x_k)] \tag{10.7}$$

where A_k ($k=p$, s) are the amplitudes of the incident wave, $\omega_k = 2\pi f_k$ and f_k is the frequency of the wave, $k=p$ and s are for P- and S-waves, respectively. It is found that Equation (10.6) can be solved if

$$\begin{cases} \alpha_k = \left\{ \frac{\rho\omega_k^2}{2E_{ck}E_{ak}}\left[\left(\frac{E_{ak}^2 + E_{ck}^2\omega_k^2\tau_k^2}{1 + \omega_k^2\tau_k^2}\right)^{1/2} + \frac{E_{ak} + E_{ck}\omega_k^2\tau_k^2}{1 + \omega_k^2\tau_k^2}\right]\right\}^{\frac{1}{2}} \\ \\ \beta_k = -\left\{ \frac{\rho\omega_k^2}{2E_{ck}E_{ak}}\left[\left(\frac{E_{ak}^2 + E_{ck}^2\omega_k^2\tau_k^2}{1 + \omega_k^2\tau_k^2}\right)^{1/2} - \frac{E_{ak} + E_{ck}\omega_k^2\tau_k^2}{1 + \omega_k^2\tau_k^2}\right]\right\}^{\frac{1}{2}} \end{cases} \tag{10.8}$$

where

$$\frac{1}{E_{ck}} = \frac{1}{E_{ak}} + \frac{1}{E_{vk}}. \tag{10.9}$$

The α_k gives the phase shift per unit length; and the minus sign of β_k indicates the wave attenuation. It is shown in Equations (10.7) and (10.8) that the wave propagation in a viscoelastic solid is frequency-dependent and its amplitude attenuates during the wave propagation process.

10.3.2 Virtual wave source (VWS)

Assume a rock mass contains one joint set, i.e. equally-spaced multiple parallel joints. When a longitudinal or shear incident wave normally reaches a joint, a transmitted wave and a reflected wave are created and propagate in two opposite directions to the neighboring joints as two new incident waves. Multiple transmitted and reflected waves are repeatedly created among the joints. Because of the discreteness of the joints, the newly created waves have different amplitudes and phase shifts to the initial incident waves. Across the joint set, the final transmitted wave is the superposition of two parts, one is from the direct transmission of the initial incident wave and the other part is from the multiple reflections among the joints. Although the frequency-dependence and wave attenuation have been shown in Equations (10.7) and (10.8), the effect of the discreteness of joints on wave propagation in the viscoelastic solid still cannot be reflected in the two equations.

In order to solve this problem, the concept of virtual wave source (VWS) is proposed in the equivalent viscoelastic medium model. The VWS exists at each joint surface and produces a new wave (in the opposite direction of the incident wave) at each time when an incident wave propagates across the VWS. The distance between two adjacent VWSs is equal to the joint spacing S. The equivalent length of the medium is defined as the product of joint number N and the joint spacing S, i.e. NS. Figure 10.2 shows a rock mass with three parallel joints and the corresponding equivalent medium with and without VWS, where the equivalent length is $3S$. The concept of VWS can be interpreted as that a reflected wave is created from the virtual wave source when either a positive wave or a negative wave arrives at the VWS.

Assume there is an incident harmonic wave

$$v_{Ik}(t, 0) = A_k \exp(-i\omega_k t) \tag{10.10}$$

from the left side a of the equivalent medium in Figure 10.2. According to Equation (10.7), along the direction of the incident wave the particle velocity at point b is

$$v_{ek}(t, S) = A_k \exp(\beta S) \exp[i(-\omega_k t + \alpha_k S)] \tag{10.11}$$

where the phase shift of $v_{ek}(t, S)$ and $v_{Ik}(t, 0)$ is $\alpha_k S$. According to the energy conservation of simple harmonic waves (Cook, 1992), the amplitude of the reflected wave at the interface b is $A_k\{1 - [\exp(\beta_k S)]^2\}^{1/2}$, if the interface b is a discontinuous boundary. From the Kramer-Kronig relation (a statement of causality), any change in the amplitude of a wave must be accompanied by a change in phase. Since the phase shift between the reflected and transmitted waves is $\pi/2$ (Pyrak-Nolte, Myer and Cook, 1990; Cook, 1992), the reflected wave at b can be expressed as

$$v'_{ek}(t, S) = A_k\sqrt{1 - [\exp(\beta_k S)]^2} \exp[i(-\omega_k t + \alpha_k S + \pi/2)] \tag{10.12}$$

where $v'_{ek}(t, S)$ is regarded as the wave produced from the VWS at b. Then, $v_{ek}(t, S)$ and the created wave $v'_{ek}(t, S)$ propagate along two opposite directions as newly incident waves to the adjacent interfaces c and a, where new waves are repeatedly created and propagate to their adjacent interfaces. The transmitted wave at the right side d of the equivalent medium is a wave superposition of $v_{ek}(t, 3S)$ arriving at different times, which is the summation of multiple waves created from the three VWSs

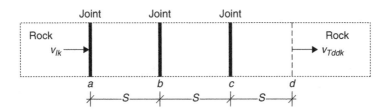

(a) Jointed rock mass

(b) Equivalent medium model without virtual wave source

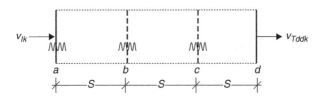

(c) Equivalent medium model with virtual wave source

Figure 10.2   Scheme of jointed rock mass and equivalent medium ($k = p$ for incident P-waves and $k = s$ for incident S-waves).

and the transmitted wave from the incident wave $v_{Ik}(t, 0)$ propagating across the viscoelastic medium.

10.4   DETERMINATION OF THE PARAMETERS

In the present study, the joint is assumed to be planar, large in extent and small in thickness compared to the wavelength, and the joint and the intact rock are linear elastic. For the equivalent medium model in Equations (10.10)–(10.12), E_{ak} is a known parameter, where E_{ap} is equal to the Young's modulus E and E_{as} is the shear modulus G of the intact rock; E_{vk} and η_{vk} need to be determined by comparing the transmitted wave through the equivalent medium with the existing solutions of discontinuous rock mass.

10.4.1 Single joint case

If an incident P- or S-wave at the boundary with the form of $v_{Ik} = A_k \exp(i\omega_k t)$ propagates in a rock mass with one joint, the transmitted wave after the joint was derived and written as (Pyrak-Nolte, Myer and Cook, 1990; Cook, 1992)

$$v_{Tk1} = \frac{2K_k/z_k}{-i\omega_k + 2K_k/z_k} A_k \exp[i(-\omega_k t + x_k \omega_k/C_k)] \tag{10.13}$$

or

$$v_{Tk1} = \frac{2K_k/z_k}{\sqrt{\omega_k^2 + (2K_k/z_k)^2}} A_k \exp[i(-\omega_k t + x_k \omega_k/C_k + \theta_k)] \tag{10.14}$$

where K_k is the joint normal or shear stiffness when $k = p$ or s; z_k is the wave impendence and $z_k = \rho C_k$; C_k is the wave velocity in the intact medium; x_k is the length of the rock mass along the wave propagation path; $\theta_k = \arctan[-\omega_k/(2K_k/z_k)]$. When $x = S$, the transmitted wave can be expressed as Equation (10.14) for a rock mass with one joint or Equation (10.11) for the corresponding equivalent viscoelastic medium. Thus, the amplitude and phase in Equation (10.11) should respectively be equal to those in Equation (10.14), i.e.

$$\begin{cases} \beta_k = \dfrac{1}{S} \ln\left[\dfrac{2K_k/z_k}{\sqrt{\omega_k^2 + (2K_k/z_k)^2}} \right] \\[3mm] \alpha_k = \dfrac{\omega_k}{C_k} + \dfrac{1}{S} \arctan\left(\dfrac{\omega_k}{2K_k/z_k} \right) \end{cases} \tag{10.15}$$

where α_k and β_k are shown in Equation (10.8).

10.4.2 Parameter determination from single joint analysis

Defining g_1 and g_2 as the functions of the two parameters, E_{vk} and η_{vk}, Equation (10.15) can be rewritten as

$$\left. \begin{aligned} g_1(E_{vk}, \eta_{vk}) &= \beta_k - \frac{1}{S} \ln\left[\frac{2K_k/z}{\sqrt{\omega^2 + (2K_k/z)^2}} \right] \\ g_2(E_{vk}, \eta_{vk}) &= \alpha_k - \frac{\omega_k}{C_k} - \frac{1}{S} \arctan\left(\frac{\omega_k}{2K_k/z} \right) \end{aligned} \right\} . \tag{10.16}$$

The Newton-like method (Kelley, 2003) is adopted to generate a sequence $\{(E_{vk})_n \quad (\eta_{vk})_n\}$ until the converged solutions are obtained:

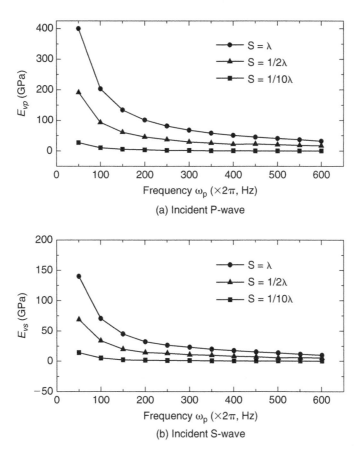

Figure 10.3 The relation between parameters E_{vk} and frequency ω_k $(k=p,s)$.

$$\begin{Bmatrix} (E_{vk})_{n+1} \\ (\eta_{vk})_{n+1} \end{Bmatrix} = \begin{Bmatrix} (E_{vk})_n \\ (\eta_{vk})_n \end{Bmatrix} - G_k^{-1} \begin{Bmatrix} g_2[(E_{vk})_n, (\eta_{vk})_n] \\ g_2[(E_{vk})_n, (\eta_{vk})_n] \end{Bmatrix} \tag{10.17}$$

where G_k is a matrix as follows:

$$G_k = \left\{ \begin{array}{cc} \dfrac{\partial g_1}{\partial(E_{vk})_n} & +\dfrac{\partial g_1}{\partial(\eta_{vk})_n} \\[3mm] \dfrac{\partial g_2}{\partial(E_{vk})_n} & \dfrac{\partial g_2}{\partial(\eta_{vk})_n} \end{array} \right\} \tag{10.18}$$

In the following calculations, it is assumed that the rock density ρ is 2650 kg/m³, the P-wave velocity C_p is 5830 m/s and the S-wave velocity C_s is 2950 m/s, the joint normal stiffness K_p is 3.5 GPa/m and joint shear stiffness K_s is 2.46 GPa/m. The parameters E_{vk} and η_{vk} are predicted with different incident wave frequency ω_k and joint spacing S. Figures 10.3(a) and 10.3(b) respectively show the relations between E_{vk} and

Table 10.1 Coefficients for curve fittings of $E_{vk} \sim \omega_k$, $\eta_{vk} \sim \omega_k$ in Figures 10.3 and 10.4.

	$S = 1/10\lambda_k$	$S = 1/2\lambda_k$	$S = \lambda_k$
(a) Incident P-wave			
B_{1p} (GPa)	92.7	454.8	829.00
B_{2p} (GPa)	20.5	124.2	196.80
B_{3p} (GPa)	−0.1	16.3	22.99
B_{4p} (MPa·s)	15.7	117.7	648.56
B_{5p} (MPa·s)	64.9	550.4	267.81
B_{6p} (MPa·s)	0.8	3.4	18.22
F_{1p} (Hz)	166.4	191.9	239.3
F_{2p} (Hz)	713.3	861.5	1.3×10^3
F_{3p} (Hz)	965.7	1.14×10^3	253
F_{4p} (Hz)	276.6	215.5	935.6
(b) Incident S-wave			
B_{1s} (GPa)	42.9	151.8	52.2
B_{2s} (GPa)	2.45	23.7	283.2
B_{3s} (GPa)	−0.184	−4.54	−10.4
B_{4s} (MPa·s)	4.5	55.7	9.4
B_{5s} (MPa·s)	32.1	127.6	124.3
B_{6s} (MPa·s)	0.3	2.9	3.45
F_{1s}	0.79	0.93	12.7
F_{2s}	7.39	13.06	0.98
F_{3s}	3.02	2.36	9.64
F_{4s}	0.82	0.46	2.1

ω_k, η_{vk} and ω_k, when S is $1/10\lambda_k$, $1/2\lambda_k$ and λ_k, where $k = p$ and s are for incident P- and S-waves, respectively. From the two figures, it can be seen that either E_{vk} or η_{vk} depends on the incident wave frequency ω_k and the joint spacing S. For a given S, E_{vk} and η_{vk} decrease with increasing ω. For a given ω, E_{vk} and η_{vk} increase with increasing S.

The relation between E_{vk} and S, η_{vk} and S can respectively be derived by the least square regression method as the two exponential forms,

$$E_{vk} = B_{1k} \exp(-\omega_k/F_{1k}) + B_{2k} \exp(-\omega_k/F_{2k}) + B_{3k} \tag{10.19}$$

and

$$\eta_{vk} = B_{4k} \exp(-\omega_k/F_{3k}) + B_{5k} \exp(-\omega_k/F_{4k}) + B_{6k} \tag{10.20}$$

where B_{ik} and F_{jk} ($i = 1 \sim 6$, $j = 1 \sim 4$, $k = p, s$) are the coefficients from the curve fitting, which are listed in Table 10.1.

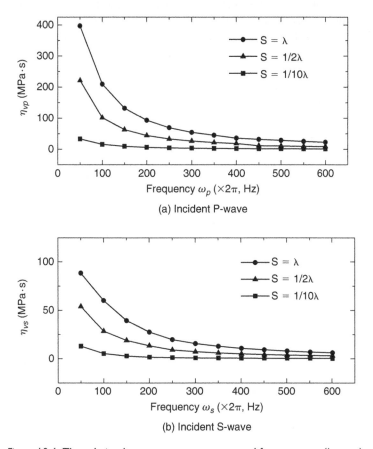

Figure 10.4 The relation between parameters η_{vk} and frequency ω_k ($k=p, s$).

10.5   VERIFICATIONS OF EMM WITH VIRTUAL WAVE SOURCE METHOD

10.5.1   Periodical function expression for an arbitrary incident wave

Using the Fourier and inverse Fourier transforms, any arbitrary incident wave can be expressed as the sum of periodical functions. Assume the incident P-wave applied at the left side a in Figure 10.2 is a half-cycle sinusoidal wave with the form of

$$v_{Ip}(t,0) = \begin{cases} I\sin(\omega_0 t) \\ 0 \end{cases}, \quad \text{when} \quad \begin{array}{l} 0 \leq t \leq \pi/\omega_0 \\ t > \pi/\omega_0 \end{array} \tag{10.21a}$$

and the incident S-wave applied at the left side a in Figure 10.2 is a full-cycle sinusoidal wave with the form of

$$v_{Is}(t,0) = \begin{cases} I\sin(\omega_0 t) \\ 0 \end{cases}, \quad \text{when} \quad \begin{array}{l} 0 \leq t \leq 2\pi/\omega_0 \\ t > 2\pi/\omega_0 \end{array} \tag{10.21b}$$

where I is the amplitude of the incident wave and equal to 1 m/s; ω_0 is the angular frequency of the incident wave. In order to obtain the periodical function expression of the incident waves, Equations (10.21a) and (10.21b) are firstly transformed in the frequency domain by the Fourier transform, i.e.

$$F[v_{Ip}(t,0)] = \frac{2}{\sqrt{2\pi}} \frac{\omega_0 I}{\omega_0^2 - \omega^2} \cos\frac{\pi\omega}{2\omega_0} e^{-i\frac{\pi\omega}{2\omega_0}}, \quad \text{for incident P-wave} \quad (10.22a)$$

$$F[v_{Is}(t,0)] = \frac{I}{\sqrt{2\pi}} \frac{\omega_0}{\omega^2 - \omega_0^2} (e^{-2\pi\frac{\omega}{\omega_0}i} - 1), \quad \text{for incident S-wave} \quad (10.22b)$$

The harmonic wave form with innumerable cycles of Equation (10.18) is derived by using the inverse Fourier transform of Equation (10.19) in the time domain.

$$v_{Ip}(t,0) = F^{-1}\{F[v_{Ip}(t,0)]\} = \frac{2}{\pi} \int_0^{+\infty} \frac{\omega_0 I}{\omega_0^2 - \omega^2} \cos\left(\frac{\pi\omega}{2\omega_0}\right) \cos\left(\omega t - \frac{\pi\omega}{2\omega_0}\right) d\omega,$$

$$\text{for incident P-wave} \quad (10.23a)$$

$$v_{Is}(t,0) = F^{-1}\{F[v_{Is}(t,0)]\} = \frac{I}{\pi} \int_0^{+\infty} \frac{2\omega_0}{\omega^2 - \omega_0^2} \sin\left(\frac{\pi\omega}{\omega_0}\right) \sin\left(\omega t - \frac{\pi\omega}{\omega_0}\right) d\omega,$$

$$\text{for incident S-wave} \quad (10.23b)$$

If the frequency interval $\Delta\omega$ is sufficiently small, Equation (10.20) can be rewritten as

$$v_{Ipa}(t,0) = \frac{2I}{\pi} \sum_{j=1}^{+\infty} \left\{ \frac{\omega_0}{\omega_0^2 - \omega_j^2} \cos\left(\frac{\pi\omega_j}{2\omega_0}\right) \cos\left(\omega_j t - \frac{\pi\omega_j}{2\omega_0}\right) \Delta\omega \right\},$$

$$\text{for incident P-wave} \quad (10.24a)$$

$$v_{Isa}(t,0) = \frac{2I}{\pi} \sum_{j=1}^{+\infty} \left\{ \frac{\omega_0}{\omega_j^2 - \omega_0^2} \sin\left(\frac{\pi\omega_j}{\omega_0}\right) \sin\left(\omega_j t - \frac{\pi\omega_j}{\omega_0}\right) \Delta\omega \right\},$$

$$\text{for incident S-wave} \quad (10.24b)$$

When the number of the frequency ω_j is sufficiently large, so that the main frequencies in the frequency domain of the incident wave are included, $v_{Ika}(t,0)$ is approximately equal to $v_{Ik}(t,0)$, i.e. $v_{Ika}(t,0) \cong v_{Ik}(t,0)$.

10.5.2 Result comparison and verification of the equivalent medium model

In order to verify the general EMM, the transmitted wave calculated by the proposed model is compared with that determined by DDM. Assuming the incident velocity wave at the boundary a in Figure 10.2 has the form of Equation (10.18) and $\omega_0 = 2\pi \times 100\,\text{Hz}$, the transmitted wave v_{Tddk} through the discontinuous rock mass with joint spacing S is obtained from the displacement discontinuity method. To analyze the wave propagation across the equivalent medium, the waves given by Equations (10.24a) and (10.24b) are chosen as the incident waves. Considering Equations (10.24a) and (10.24b) and the wave propagation equations (10.8) and (10.10)–(10.12) for the equivalent medium model, the transmitted wave at the interface b in Figure 10.2 (c) by the incident wave is

$$
v_{ebp} = \sum_{j=1}^{+\infty} \left[I\Delta\omega \frac{2}{\pi} \frac{\omega_0}{\omega_0^2 - \omega_j^2} \cos\left(\frac{\pi\omega_j}{2\omega_0}\right) \exp(\beta_p S) \cos\left(\omega_j t - \frac{\pi\omega_j}{2\omega_0} - \alpha_p S\right) \right],
$$

$$
\text{for incident P-wave} \qquad (10.25a)
$$

$$
v_{ebs} = \sum_{j=1}^{+\infty} \left[\frac{2I\Delta\omega}{\pi} \frac{\omega_0}{\omega_j^2 - \omega_0^2} \sin\left(\frac{\pi\omega_j}{\omega_0}\right) \exp(\beta_s S) \sin\left(\omega_j t - \frac{\pi\omega_j}{\omega_0} - \alpha_s S\right) \right],
$$

$$
\text{for incident S-wave} \qquad (10.25b)
$$

and the reflected wave at interface b is

$$
v'_{ebp} = \sum_{j=1}^{+\infty} \left\{ \frac{2I\Delta\omega}{\pi} \frac{\omega_0}{\omega_0^2 - \omega_j^2} \cos\left(\frac{\pi\omega_j}{2\omega_0}\right) \sqrt{1 - [\exp(\beta_p S)]^2} \right.
$$

$$
\left. \cos\left(\omega_j t - \frac{\pi\omega_j}{2\omega_0} - \alpha_p S - \frac{\pi}{2}\right) \right\}, \quad \text{for incident P-wave} \qquad (10.26a)
$$

$$
v'_{ebs} = \sum_{j=1}^{+\infty} \left\{ \frac{2I\Delta\omega}{\pi} \frac{\omega_0}{\omega_j^2 - \omega_0^2} \sin\left(\frac{\pi\omega_j}{\omega_0}\right) \sqrt{1 - [\exp(\beta_s S)]^2} \right.
$$

$$
\left. \sin\left(\omega_j t - \frac{\pi\omega_j}{\omega_0} - \alpha_s S - \frac{\pi}{2}\right) \right\}, \quad \text{for incident S-wave} \qquad (10.26b)
$$

The forward wave v_{ebk} and the backward wave v'_{ebk} then move toward the interfaces c and a respectively as new incident waves, where $k = p$ and s for incident P- and S-waves, respectively. The interfaces a, b and c then perform as VWSs respectively which generate a backward wave once an incident wave reaches the joint. The waves superpose with each other in the medium and the final transmitted wave v_{Tek} at the interface d is obtained.

The transmitted waves v_{Tddk} based on DDM and v_{Tek} based on EMM are compared in Figures 10.5 to 10.8, where Figures 10.5 and 10.6 are respectively for incident P- and S-waves across the rock mass with different joint number, and Figures 10.7 and

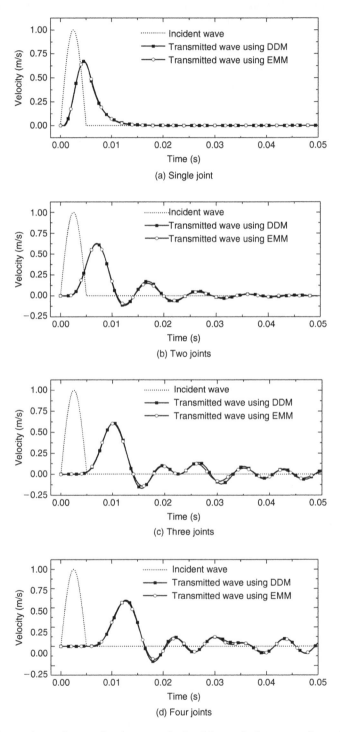

Figure 10.5 Comparison of transmitted waves obtained from displacement discontinuity method (DDM) and equivalent medium method (EMM) with different joint number ($S = 1/10\lambda_p$, and incident P-wave).

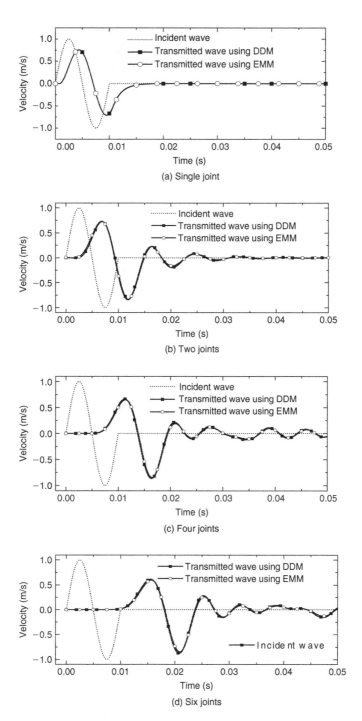

Figure 10.6 Comparison of transmitted waves obtained from displacement discontinuity method (DDM) and equivalent medium method (EMM) with different joint number N ($S = 1/10\lambda_s$, and incident S-wave).

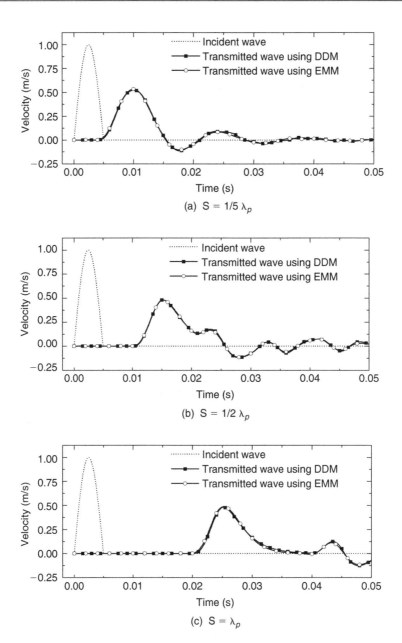

Figure 10.7 Comparison of transmitted waves obtained from displacement discontinuity method (DDM) and equivalent medium method (EMM) with different joint spacing S (Incident P-wave).

10.8 are for incident P- and S-waves across the rock mass with different joint spacing. In Figures 10.5 and 10.6, the joint spacing is one-tenth of the elastic wave length λ_k and $\lambda_k = 2\pi C_k/\omega_0$ ($k = p$ and s for incident P- and S-waves, respectively), while the joint number changes from one to six to reflect the medium length effect. In Figure 10.5, the

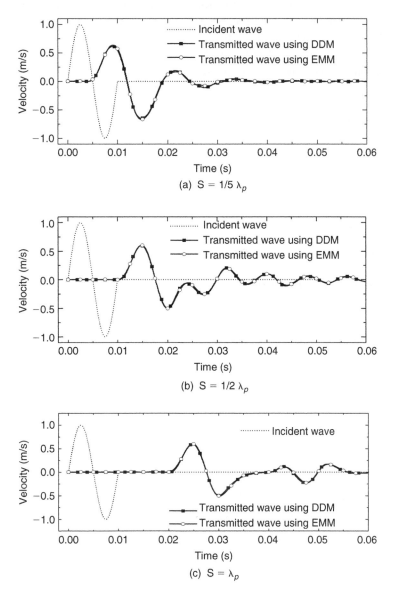

Figure 10.8 Comparison of transmitted waves obtained from displacement discontinuity method (DDM) and equivalent medium method (EMM) with different joint spacing S (Incident S-wave).

transmission coefficients, $\max(v_{Tep})/\max(v_{Ip})$, are about 0.67 for the one-joint case, 0.626 for the two-joint case, 0.602 for the three-joint case, and 0.59 for the four-joint case. In other words, the transmitted wave attenuates quickly when the rock mass has one and two joints while the attenuation of the transmitted wave becomes slow with increasing number of joints.

Figures 10.7 and 10.8 show the relationship between the transmitted waves and the joint spacing, or the joint frequency when two joints are considered. In Figure 10.7, the time delay of the transmitted P-wave across two joints with $S = \lambda_p$ is about 0.022s, which is equal to that of $2S \cdot \partial \alpha_p / \partial \omega_p |_{\omega_p = \omega_0}$ as given in Equation (10.7). The transmitted wave depends mainly on the original incident wave and the effect of the reflected waves between the two joints appears minor for a large joint spacing, e.g. when $S = \lambda_p$.

It is found from Figures 10.5 to 10.8 that the waveforms of v_{Tek} agree very well with those of v_{Tddk} for all the cases studied. The comparisons between v_{Tek} and v_{Tddk} verify that the EMM proposed in the present study is effective for studying wave propagation, and the proposed EMM can describe the longitudinal and shear wave propagation in a rock mass with one set of parallel joints without losing accuracy.

10.6 APPLICATIONS AND OUTLOOKS

10.6.1 Transmitted wave

In the following discussions, only the case of incident P-waves is considered. Define $v_{Tep,1}$ as the transmitted wave based on the equivalent viscoelastic medium without considering the effects of the VWSs; and define $v_{Tep,2}$ as the transmitted wave due to the reflections between the VWSs. From Equation (10.25a), the transmitted wave $v_{Tep,1}$ can be expressed as

$$v_{Tep,1} = \sum_{j=1}^{+\infty} \left[I \Delta \omega \frac{2}{\pi} \frac{\omega_0}{\omega_0^2 - \omega_j^2} \cos\left(\frac{\pi \omega_j}{2\omega_0}\right) \exp(N\beta_p S) \cos\left(\omega_j t - \frac{\pi \omega_j}{2\omega_0} - N\alpha_p S\right) \right]$$

(10.27)

where N is the number of VWS. When two VWS are in the equivalent medium and $S = 1/10\lambda_p$, the curves of $v_{Tep,1}$, $v_{Tep,2}$ and v_{Tep} which is the superposition of $v_{Tep,1}$ and $v_{Tep,2}$ are plotted in Figure 10.9. The v_{Tep} in Figure 10.9 is exactly the same as the transmitted wave shown in Figure 10.5(b). The $v_{Tep,1}$ in Figure 10.9 is the further attenuation with time shift of the transmitted wave shown in Figure 10.5(a), which is from an equivalent medium with only one VWS. It implies that $v_{Tep,1}$ is purely from the original incident wave.

When the VWS spacing S is λ_p and the length of an equivalent medium is $2\lambda_p$ or $4\lambda_p$, the transmitted waves across the equivalent medium with and without the VWS are derived as shown in Figure 10.10. For both medium lengths, the waveform of $v_{Tep,1}$ is very similar to the first part of v_{Tep}. Comparison of v_{Tep} in Figures 10.9 and 10.10 indicates that the VWS spacing influences the transmitted waveform, while the effect of the number of VWS on the transmitted waveform is minimal when the VWS spacing is larger than a critical value. Therefore, when the VWS spacing is very large, the transmitted wave can be directly calculated from Equation (10.27).

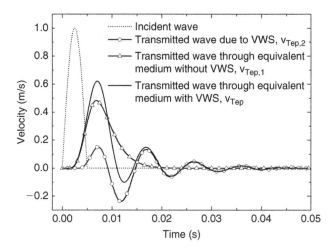

Figure 10.9 Effect of virtual wave source (VWS) on transmitted waveforms ($S = 1/10\lambda_p, N = 2$).

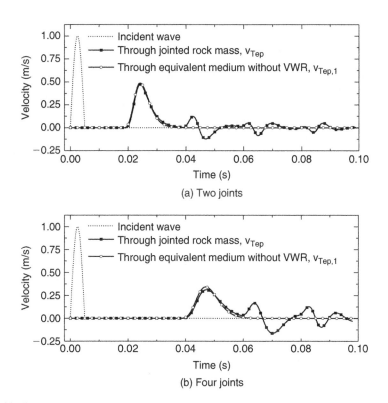

Figure 10.10 Transmitted waves by using displacement discontinuity method and equivalent medium model without virtual wave source ($f = 100\,\text{Hz}, S = \lambda_p$).

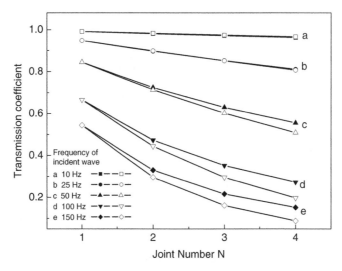

Figure 10.11 Transmission coefficient with different frequency of incident waves ($k_n = 3.5$ GPa/m and $S = 8\lambda_p$).

10.6.2 Transmission coefficient

In rock engineering, the damage criteria of rock structures are generally regulated according to the threshold values of wave amplitudes, such as the peak particle velocity. Therefore, the transmission coefficients for the waves propagating through the rock mass with multiple joints are important.

One simplified method to calculate the transmission coefficient is the T^N-method (Hopkins, Myer and Cook, 1988; Pyrak-Nolte, Myer and Cook, 1990; Myer *et al.*, 1995), which considers the transmission coefficient across parallel joints as the product of transmission coefficients of individual joints:

$$|T_N| = |T_1|^N \tag{10.28}$$

where T_N denotes the transmission coefficient after N joints and T_1 denotes the transmission coefficient after a single joint. Laboratory experiments verified that the T^N-method is valid when the first arriving wave is not contaminated by the multiple reflections, which can be satisfied only when the joint spacing is relatively larger than the incident wavelength.

When ω_0 is 100π for the incident wave in Equation (10.21a), and the joint stiffness k_n varies from 1.0, 3.5, 6.0, 8.5 to 11.0 GPa/m, the transmission coefficient is calculated and shown in Figure 10.11.

In Figure 10.11, the solid marks represent the transmission coefficients obtained from the EMM and the hollow marks represent the results from the T^N-method. It is also seen from the figure that the transmission coefficients obtained from the T^N-method are close to the transmission coefficient from the equivalent medium method.

10.6.3 Effective velocities

The effective velocity C_e for the incident wave v_{Ip} in an equivalent medium is a function of the medium length to the time difference between the two peak velocities of incident and transmitted waves, i.e.

$$C_e = \frac{NS}{t_T - t_I} \tag{10.29}$$

where t_I and t_T are the time spots for the peak velocities of v_{Ip} and v_{Tep}, respectively.

From the present study, the effective velocity of the equivalent medium is affected by the ratio of the wave length λ_p over the spacing S. For example, if the incident wave in the form of Equation (10.29) has the frequency $f = 100$ Hz or $\lambda_p = 58.3$ m and the spacing S is $1/10\lambda_p$, the effective velocity C_e can be calculated from Figure 10.5. Based on Equation (10.29), C_e is about 3320 m/s for the one-joint case, which approaches the effective velocity $C_{eff} = 3322$ m/s given by Pyrak-Nolte, Myer and Cook (1987):

$$C_{eff} = C_p \left[\frac{1 + \left(\dfrac{z_p \omega_0}{2K_p}\right)^2}{1 + \left(\dfrac{z_p \omega_0}{2K_p}\right)^2 + \dfrac{C_p}{S}\dfrac{z_p}{2K_p}} \right] \tag{10.30}$$

where the interaction between joints and multiple reflections are ignored. The C_e of the two-joint case is calculated from Figure 10.5(b) to be approximately 2780 m/s, about 2365 m/s for the three-joint case from Figure 10.5(c), and about 2180 m/s for the four-joint case from Figure 10.5(d). It is clear that the effective velocity is sensitive to the ratio of λ_p/S. The reason is that the peak values of the transmitted wave are affected by the reflected waves created by the VWSs, as shown in Figure 10.9. It is the advantage of the present equivalent medium model to accurately account for the discreteness of the joints.

If the joint spacing S is larger, e.g. equals to λ_p as shown in Figure 10.7(c), the effective velocity C_e is calculated about 5420 m/s which matches with the effective velocity $C_{eff} = 5421$ m/s given by Pyrak-Nolte, Myer and Cook (1987). Therefore, if the VWS spacing is sufficiently large, the influence of the multiple reflections among joints on the main transmitted wave is minimal, and the effective velocity C_e is the same as that given by Pyrak-Nolte, Myer and Cook (1987). On the other hand, if the VWS is not considered in the equivalent medium model, the effective velocity derived from the present study agrees very well with the previous results by Pyrak-Nolte, Myer and Cook (1987).

10.6.4 Outlooks

The EMM has obvious advantages in describing the dynamic property of jointed rock masses and for considering the stress wave attenuation and the viscous loss in the rock masses. By comparison with the results from the DDM method, the EMM method is able to give satisfactory results in terms of transmitted wave form, transmission coefficients and effective velocity. However, this model is only applied for one dimensional problems and linear joints. For more complex geological conditions with nonlinear joints, or two or more sets of parallel joints, the applicability of EMM is yet to be verified. Therefore, future work will focus on developing the EMM for a rock mass, in which the joints with linear or nonlinear properties are randomly distributed.

10.7 SUMMARY

An equivalent viscoelastic medium model is proposed in this chapter for determining the P- and S-wave transmission through a rock mass containing equally-spaced parallel joints. In the proposed equivalent medium model, the linear viscoelastic medium having shear and normal properties is combined with the concept of virtual wave sources.

Most existing analytical and numerical studies on wave propagation in jointed rock mass approximate the medium as a continuous elastic solid with the effective Young's modulus based on either a quasi-static deformation approach for a composite material or an effective seismic/acoustic velocity approach. These traditional equivalent models, as limited by the elastic assumption, are not able to describe the effect of the discreteness of the joints. Based on the present study, the equivalent viscoelastic medium model not only predicts accurately the transmission coefficients but also derives analytically the transmitted waveforms.

The proposed equivalent viscoelastic medium model also has obvious advantages over the displacement discontinuity method. When the effect of the VWSs is not prominent, the proposed equivalent model is much more efficient in analyzing wave propagations. When the VWSs are considered, the proposed model is still able to give analytical solutions of the wave propagations without losing efficiency and accuracy.

Although the present study involved only simplified cases with equally-spaced parallel joints, it demonstrated that the current viscoelastic equivalent medium model is able to produce results in wave propagation analysis as accurately as those from the displacement discontinuity models. The present study with the explicit form of solutions can serve as a benchmark for verification of relative numerical or analytical studies of stress wave propagation through rock mass. Further exploration is underway to extend the current model for more complicated joint forms and incident waves.

REFERENCES

Amaderi, B. and Goodman, R.E.: A 3-D constitutive relation for fractured rock masses. *In: Proceedings of the International symposium on Mechanical Behavior of Structured Media.* Balkema, 1981, pp.249–268.

Cai, J.G. and Zhao, J.: Effects of multiple parallel fractures on apparent wave attenuation in rock masses. *International Journal of Rock Mechanics and Mining Sciences* 37(4) (2000), pp.661–682.

Ciccotti, M. and Mulargia, F.: Differences between static and dynamic elastic moduli of a typical seismogenic rock. *Geophysical Journal International* 157 (2004), pp.474–477.

Cook, N.G.W.: Natural joint in rock: mechanical, hydraulic and seismic behaviour and properties under normal stress. *International Journal of Rock Mechanics and Mining Sciences and Geomechanics Abstracts* 29(3) (1992), pp.198–223.

Crampin, M., Mcgonigle, R. and Bamford, D.: Estimating crack parameters from observations of P-wave velocity anisotropy. *Geophysics* 45(3) (1980), pp.345–360.

Eissa, E.A. and Kazi, A.: Relation between static and dynamic Young's moduli for rocks. *International Journal of Rock Mechanics and Mining Science & Geomechanics Abstracts* 25 (1988), pp.479–482.

Hopkins, D.L., Myer, L.R. and Cook, N.G.W.: Seismic wave attenuation across parallel fractures as a function of fracture stiffness and spacing. *EOS Transaction AGU* 69(44) (1988), pp.1427–1438.

Kelley, C.T.: *Solving nonlinear equations with Newton's method*, SIAM, Philadelphia, 2003.

Kolsky, H.: *Stress waves in solids*, Clarendon Press, Oxford, 1953. (Dover reprint in press).

Li, J.C., Ma, G.W. and Huang, X.: Analysis of Wave Propagation through a Filled Rock Joint. *Rock Mechanics and Rock Engineering* 43(6) (2010), pp.789–798.

Li, J.C., Ma, G.W. and Zhao, J.: An Equivalent Viscoelastic Model for Rock Mass with Parallel Joints. *Journal of Geophysical Research* 115(B03305) (2010) (DOI: 10.1029/2008JB006241).

Ma, G.W., Li, J.C. and Zhao, J.: Three-phase medium model for filled rock joint and interaction with stress waves. *International Journal for Numerical and Analytical Methods in Geomechanics* 35(1) (2011), pp.97–110.

Miller, R.K.: An approximate method of analysis of the transmission of elastic waves through a frictional boundary. *Journal of Applied Mechanics-ASME* 44(4) (1977), pp.652–656.

Myer, L.R., Hopkins, D., Peterson, J.E. and Cook, N.G.W.: Seismic wave propagation across multiple fractures. *Fractured and Jointed Rock Masses* (1995), pp.105–110.

Pyrak-Nolte, L.J., Meyer, L.R. and Cook, N.G.W.: Seismic visibility of fractures. In: I.W. Farmer, J.J.K. Daemen, C.S. Desai, D.E. Glass and S.P. Neuman (eds.): *Rock Mechanics: Proc. U.S. Symp.*, Balkeman, Rotterdam, 1987, pp. 47–56.

Pyrak-Nolte, L.J., Myer, L.R. and Cook, N.G.W.: Anisotropy in seismic velocities and amplitudes from multiple parallel fractures. *Journal of Geophysical Research* 95(B7) (1990), pp.11345–11358.

Pyrak-Nolte, L.J.: *Seismic visibility of fractures*, Ph.D. thesis, Univ. of Calif., Berkeley, 1988.

Schoenberg, M. and Muir, F.: A calculus for finely layered anisotropic media. *Geophysics* 54(5) (1989), pp.581–589.

Schoenberg, M.: Elastic wave behaviour across linear slip interfaces. *The Journal of the Acoustical Society of America* 68(5) (1980), pp.1516–1521.

Tsai, Y.M. and Kolsky, H.: Surface wave propagation for linear viscoelastic solids. *Journal of the Mechanics and Physics of Solids* 16 (1968), pp.99–109.

White, J.E.. *Underground sound*, Elsevier, New York, 1983.

Zhao, J. and Cai, J.G.: Transmission of elastic P-waves across single fractures with a nonlinear normal deformational behaviour. *Rock Mechanics and Rock Engineering* 34(1) (2001), pp.3–22.

Zhao, J., Zhao, X.B. and Cai, J.G.: A further study of P-wave attenuation across parallel fractures with linear deformational behaviour. *International Journal of Rock Mechanics and Mining Sciences* 43(5) (2006a), pp.776–788.

Zhao, X.B., Zhao, J. and Cai, J.G.: P-wave transmission across fractures with nonlinear deformational behaviour, *International Journal for Numerical and Analytical Methods in Geomechanics* 30(11) (2006b), pp.1097–1112.

Zhao, X.B., Zhao, J., Hefny, A.M. and Cai, J.G.: Normal transmission of S-wave across parallel fractures with coulomb slip behavior. *Journal of Engineering Mechanics-ASCE* 132(6) (2006), pp.641–650.

Chapter 11

Polycrystalline model for heterogeneous rock based on smoothed particle hydrodynamics method

Guowei Ma, Xuejun Wang and Lei He

11.1 INTRODUCTION

Rock materials are multiphase heterogeneous composites, consisting of mineral grains with preexisting defects in the forms of voids and cracks in their microstructures as shown in Figure 11.1. With the advance of experimental techniques, close observations on the specimen's microscopic behaviour during rock failure processes become possible. For instance, Tapponnier and Brace (1976) investigated stress-induced microcrack developments within different mineral components in the Westerly granite. Wong (1982) further investigated the faulting mechanisms of different minerals in Westerly granite with different confining pressures and temperatures and concluded that the failure mechanism was related to both mineralogy and mineral grain orientation. Several numerical approaches have also been put forward to study the fracture processes of such heterogeneous materials, such as the lattice-based model (Blair and Cook, 1998), the RFPA (Realistic Failure Process Analysis) code (Tang *et al.*, 2000), the local degradation model based on the FLAC (Fast Lagrangian Analysis of Continua) code (Fang and Harrison, 2002), the synthetic rock mass model (SRM) based on the PFC (Particle Flow Code) (Potyondy and Cundall, 2004). Although they succeeded in one aspect or another, most of these models were based on finite element or discrete element methods that in some respects have certain limitations. For instance, the continuum-based finite element models do not work well to capture the failure process featured by distortion and large deformation, fracture propagation and fragmentation. The discrete element-based methods, on the contrary, are not well defined in processing

Figure 11.1 Typical Bukit Timah granite in Singapore.

the continuous deformations. Besides, few of these approaches can properly model the aggregates or grains in their microstructures and appropriately account for their effects in rock failure simulations.

Microscopically, the intact rock material is a multiphase composite consisting of various mineral grains or aggregates with different sizes. Traditionally, by using the SEM (scanning electronic microscope) or X-ray CT, one can capture the sample's internal microstructure information such as the shapes and spatial distributions of the components, etc. Numerical specimens can therefore be rebuilt by using those already acquired digital images (Chen, Yue and Tham, 2004). However, these techniques are still costly for practical applications. The numerical specimen generated by using these techniques can only reflect the characteristics on the given small piece of the real material, and their representations are not general.

If the statistical microstructure information for these multiphase materials, for instance their statistical components sizes and contents, can be acquired then the artificial microstructure of the specimen may be numerically constructed based on these data to resemble the real material without losing its general characteristics. The work done by Li *et al.* (2003) can be classified in this category, in which they modeled the microstructure of granite and analyzed its failure process by the RFPA code. The sizes and shapes of mineral grains in their model were chosen based on statistical analysis of tested specimens. The success of their work inspires the current study in further investigating polycrystalline rock failure by using a modified smoothed particle hydrodynamics (SPH) method.

The present study first introduces the SPH method and its governing equations in the continuum mechanics framework. A microstructure model is then put forward to construct artificial specimens for polycrystalline rocks. The microstructure model is applied to generate artificial granite specimens. An elasto-plastic damage model to reflect the strength behavior of rock materials is adopted. Numerical simulations on Brazilian splitting failure are performed to verify the developed code. Uniaxial compression tests are subsequently simulated for the constructed artificial specimens and results are discussed. The focus of the present study is not only on the macro-mechanical behavior of the specimens, but also on the characteristics of crack developments during the failure process and the final failure patterns.

11.2 SMOOTHING PARTICLE HYDRODYNAMICS (SPH) METHOD

The theoretical basis of the SPH method is the interpolation theory. By introducing an interpolation function (kernel function W) that gives the "kernel estimate" of the field variables at a point, the properties of each particle are evaluated by the integrals or the sums over the values of its neighboring particles. Considering a problem domain Ω that is discretized by a group of particles, assuming W has a compact supporting domain with a radius of kh, approximations of a function $f(x)$ and its differential form $\langle \nabla f(x) \rangle$ at point i can be expressed by the discretized particles as

$$\langle f(x) \rangle_{x=x_i} = \sum_{j=1}^{N} \frac{m_i}{\rho_j} f(x_j) W_{x_i}(x_i - x_j, h) \tag{11.1}$$

$$\langle \nabla f(x) \rangle_{x=x_i} = -\sum_{j=1}^{N} \frac{m_j}{\rho_j} (f(x_i) - f(x_j)) \nabla [W_{x_i}(x_i - x_j, h)] \tag{11.2}$$

where the summation is over all the particles (with a total number of N, including particle i) within the supporting domain of the given particle i; the label j denotes those influenced particles which are the neighboring particles of the particle i; h is called the smoothing length which defines the supporting domain of the particle; and $W(x - x', h)$ is the smoothing kernel function. The interpolation kernel named B-spine (Monaghan and Lattanzio, 1985) is adopted in this study. This kernel interpolates to the second order of the smoothing length $h(o(h^2))$ and it is always nonnegative.

The mass and momentum conservation equations of continuum mechanics give

$$\frac{d\rho}{dt} = -\rho \cdot \frac{\partial v_\alpha}{\partial x_\alpha}$$
$$\frac{dv_\alpha}{dt} = \frac{1}{\rho} \cdot \frac{\partial \sigma_{\alpha\beta}}{\partial x_\beta} \tag{11.3}$$

In the above equations, dependent variables are the density (ρ), the velocity (v_α), and the stress tensor $\sigma_{\alpha\beta}$ which is defined as $\sigma_{\alpha\beta} = p\delta_{\alpha\beta} + S_{\alpha\beta}$ in terms of the pressure $p = Tr(\sigma)/3$ and the traceless symmetric deviator stress $S_{\alpha\beta}$. The independent variables are the spatial coordinates (x_α) and the time (t). The total time derivative operator (d/dt) is determined in the moving Lagrangian framework. The Greek subscripts α and β are used to denote the coordinate directions. $\delta_{\alpha\beta} = 1$ if and only if $\alpha = \beta$, $\delta_{\alpha\beta} = 0$ otherwise. The summation is implemented over repeated Greek indices. Stresses are positive in tension.

Based on Equation (11.2) the density approximation given by Gray, Monaghan and Swift (2001) is adopted when the calculation involves heterogeneous materials as:

$$\frac{d\rho_i}{dt} = \rho_i \sum_{j=1}^{N} \frac{m_j}{\rho_j} [(v_\alpha)_i - (v_\alpha)_j] \frac{\partial W_{ij}}{\partial (x_\alpha)_i} \tag{11.4}$$

According to Libersky et al. (1993), the strain rate tensor $\dot{\varepsilon}_{\alpha\beta}$ can be expressed by the derivatives of the velocity in the SPH approximation as

$$(\dot{\varepsilon}_{\alpha\beta})_i = -\frac{1}{2} \sum_{j=1}^{N} \frac{m_j}{\rho_j} \left([(v_\alpha)_i - (v_\alpha)_j] \frac{\partial W_{ij}}{\partial (x_\beta)_i} + [(v_\beta)_i - (v_\beta)_j] \frac{\partial W_{ij}}{\partial (x_\alpha)_i} \right) \tag{11.5}$$

Once $\dot{\varepsilon}_{\alpha\beta}$ is determined, the stress rate $\dot{\sigma}_{\alpha\beta}$ tensor is derived through the adopted constitutive model. Consequently, the stress tensor $\sigma_{\alpha\beta}$ is calculated by an explicit time integration approach.

Similarly, the SPH formulation for momentum evolution is derived as

$$\frac{d(v_\alpha)_i}{dt} = \sum_{j=1}^{N} m_j \left(\frac{(\sigma_{\alpha\beta})_i}{\rho_i^2} + \frac{(\sigma_{\alpha\beta})_j}{\rho_j^2} + \Pi_{ij}\delta_{\alpha\beta} \right) \frac{\partial W_{ij}}{\partial (x_\beta)_i} \tag{11.6}$$

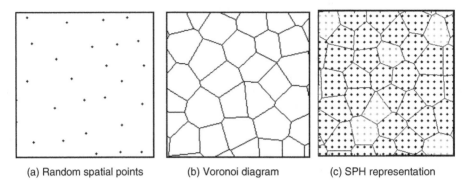

(a) Random spatial points   (b) Voronoi diagram   (c) SPH representation

Figure 11.2 2-D Voronoi diagram construction and physical domain's microstructure representation by SPH particles.

where Π is the artificial viscous pressure introduced by Gingold and Monaghan (1977) to smooth shocks and stabilize numerical solutions.

11.3 ARTIFICIAL MICROSTRUCTURE FOR MULTIPHASE MATERIALS

11.3.1 2D-domain discretization based on Voronoi diagram

A Voronoi diagram composed of many convex polygons is used to explicitly discretize a physical domain. These polygons are constructed by a set of spatial points $\{\vec{p}\}$ filled in the domain. Each cell V_i is associated with one point $\vec{p}_i$. Any point in V_i is almost closer to the point $\vec{p}_i$ than to any other point $\vec{p}_j$ in $\{\vec{p}\}$. Such a relationship can be described as

$$V_i(x) = \cap_{j \neq i}\{\vec{x}|s(\vec{x}_i, \vec{x}) \leq s(\vec{x}_j, \vec{x})\} \tag{11.7}$$

where, x_i are the coordinates of $\vec{p}_i$; $s(\vec{x}_i, \vec{x})$ is the Euclidean distance between $\vec{p}_i$ and any point in the interior of V_i. The Voronoi diagram is directly constructed from a set of points by using the software Matlab as illustrated in Figures 11.2(a) and 11.2(b). Obviously, the coordinates of these points are vital because they control the shapes as well as the area of generated polygons. In order to guarantee the areas of these arbitrarily generated polyhedrons within a prescribed range, the distance of any two spatial points in $\{\vec{p}\}$ must be checked to meet

$$L_{\min} \leq s(\vec{x}_i, \vec{x}_j) \leq L_{\max} \tag{11.8}$$

One simple method to generate these set of points is to insert randomly generated points sequentially until they eventually saturate the whole domain. When a new point is to be inserted, Equation (11.8) must be performed against all the accepted points. When the domain is nearly saturated, the probability of the acceptable point becomes low and more trial points may be rejected. Hence, the insertion of new points becomes difficult. When there is no satisfied point after a fairly long time, the Voronoi diagram can then be constructed based on all accepted points.

11.3.2 Microstructure representation by SPH particles

After the Voronoi diagram is constructed, the generated polygons are classified to represent different components, for example, different mineral grains according to their volume ratios in rock. Considering the spatial variations in the distributions of these components, the current work takes a stochastic approach.

A series of pseudo-random numbers is generated corresponding to these polygons one by one. According to the statistical contents of these minerals in the polycrystalline rock, these pseudo-random numbers are divided into several groups by their values. Each group corresponds to one type of mineral. Thus, each polygon is specified to be one type of artificial component according to its associated pseudo-random number value. Once the ratio of these generated artificial components meets the requirement, such a process stops. Otherwise, it needs repeating more times.

In order to use the SPH particles to represent the physical domain's microstructure, it must be determined to which polygon each particle belongs. If the particle's center falls inside the polygon, it is assigned to the component's properties represented by this polygon. By using the above method, the physical domain can be represented by different clusters of particles to resemble the specimen's microstructure. Figure 11.2(c) gives an illustration of such a process by using regularly packed particles.

11.3.3 2-D granite microstructure generation

Granitic rock is generally light gray and medium grained which can be discerned easily by the naked eye. Statistical data (Zhao, 1999) show that the Bukit Timah granite in Singapore (as shown in Fig. 11.1) is mainly composed of feldspar, quartz and biotite grains in a ratio of around 6:3:1 with densities of 2570, 2648 and 2800 kg/m^3, respectively. The grain size generally ranges from 3.0 mm to 5.0 mm.

Two rectangular artificial rock specimens are randomly constructed based on the above-described procedure to resemble the Bukit Timah Granite. These two specimens have the same size, with a width of 0.05 m and a height of 0.1 m. Each is discretized by 31,250 regularly distributed SPH particles with the same smoothing length of 0.4 mm.

Random points are first generated to fill in the domain with L_{Min} and L_{Max} of 3 mm and 5 mm, respectively, to meet the requirement of typical granite grain size. After the Voronoi polygons are ready, they are classified into different mineral types by the values of the pseudo-random numbers associated with them according to the prescribed ratio of the three kinds of minerals.

Figure 11.3 shows the two constructed artificial granite samples, named N1 and N2 respectively. In the figure, the black grains are biotite, quartzite grains are in light gray and the feldspar ones are in dark gray. The mineral contents of the two artificial granite samples are close to the expected ratio as presented in Table 11.1. Although their microstructures are different, these two artificial samples have the common features in mineral component contents and typical grain size.

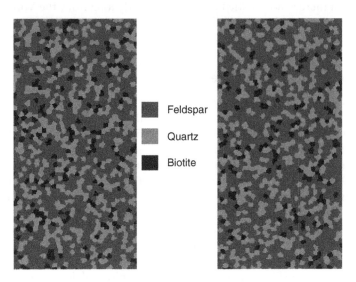

Figure 11.3 Two artificial granite specimens.

Table 11.1 Ratios of mineral grains in the two artificial granite specimens.

Specimen	Feldspar	Quartz	Biotite
N1	59.52%	29.94%	10.54%
N2	61.77%	29.38%	8.85%

11.4   ELASTO-PLASTIC DAMAGE MODEL

An elasto-plastic-damage model that can effectively represent the mechanical behavior of rock-like material failure is presented. Similar to the Johnson-Holmquist model (Holmquist, Templeton and Bishnoi, 2001) and the concrete damage model (Malvar *et al.*, 1997), the current model comprises two surfaces to represent the strengths of intact and fractured materials. It also includes a damage scalar that describes the evolution of the material from an intact state to a fractured state. The strength criterion is based on an extension form of the unified twin shear strength (UTSS) criterion (Yu and He, 1991; Yu *et al.*, 2002), which includes two hydrostatic pressure-dependent meridians representing the generalized tensile and compressive strength states, respectively.

11.4.1   Generalized unified twin shear strength criterion

The UTSS criterion (Yu and He, 1991; Yu *et al.*, 2002) was established based on the assumption that the strength behavior of material was governed by the two larger principal shear stresses and associated normal stresses. A weighting coefficient b in the range of 0 to 1 was introduced to reflect the effect of the second principal shear stress. The trajectories of the UTSS criterion on the deviatoric plane with different b

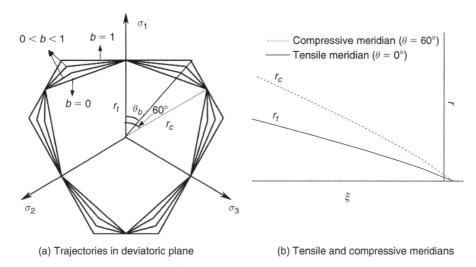

(a) Trajectories in deviatoric plane   (b) Tensile and compressive meridians

Figure 11.4 Unified twin shear strength criterion.

values are shown in Figure 11.4(a). From Figure 11.4(a), with the variation of the weighting coefficient b, many existing failure criteria, such as the Tresca criterion, the Mohr-Coulomb criterion, the Mises-Schleicher, and Drucker-Prager criteria, etc., can be approximated by the UTSS criterion. Besides, new yield criteria can be derived by assigning a suitable value of b from 0 to 1. The generalized tensile and compressive meridians corresponding to $\theta = 0°$ and 60°, respectively, are plotted in Figure 11.4(b), where $r = \sqrt{2J_2}$, $\xi = I_1/\sqrt{3}$.

According to the UTSS criterion, when $b = 1$, it gives the upper bound of the convex failure surface; when $b = 0$, it yields the lower bound. If the two meridians, i.e. the generalized tensile meridian r_t and the generalized compressive meridian r_c with respect to the hydrostatic pressure p are determined, the two bounds with $b = 0$ and 1 respectively on the deviatoric plane can be obtained by geometry analysis. Any failure criterion satisfying the convex requirement can be derived by a linear combination of the lower and upper bounds using the parameter b. According to Fan and Wang (2002), the UTSS criterion on the deviatoric plane is expressed as

$$r = \begin{cases} \dfrac{r_t r_c \sin 60°}{r_t \sin\theta + r_c \sin(60° - \theta)}(1 - b) + b\dfrac{r_t}{\cos\theta} & \text{when } 0° \leq \theta \leq \theta_b \\[4mm] \dfrac{r_t r_c \sin 60°}{r_t \sin\theta + r_c \sin(60° - \theta)}(1 - b) + b\dfrac{r_c}{\cos(60° - \theta)} & \text{when } \theta_b \leq \theta \leq 60° \end{cases}$$

(11.9)

In Equation (11.9), $\theta = \frac{1}{3}\arccos\left(\dfrac{3\sqrt{3}J_3}{2(\sqrt{J_2})^2}\right)$. The value of θ_b which corresponds to the corners of the outer shape in Figure 11.4(a) can be determined as

$$\theta_b = \arctan\left[\frac{1}{\sqrt{3}}\left(\frac{2r_c}{r_t} - 1\right)\right]$$

(11.10)

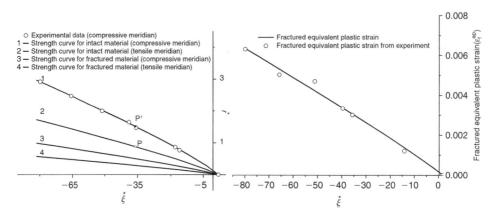

Figure 11.5 Regressed curves from experimental data.

The derived convex shape on the deviatoric plane has threefold symmetry. If its shape in the range of $0 \leq \theta \leq 60°$ is given, the full deviatoric plane is obtained. The value of b for a specific material is determined by curve-fitting of experimental results. Yu *et al.* (2002) suggested that for rock materials b took a value between 0.5 and 1.0. In the current model, b takes 0.6 unless it is explicitly stated otherwise.

The UTSS criterion with the above simplification of the deviatoric plane is adopted to represent rock strength behavior in the present study. The two meridians, i.e. the variation of r_t and r_c with respect to the hydrostatic pressure are determined by application of the uniaxial and triaxial compression test results of the considered granite.

11.4.2 Determination of the meridians and the damage model

The proposed elasto-plastic damage model includes two failure surfaces for intact material and fully fractured/damaged material respectively. The former one corresponds to initial failure of the material, while the other one gives the residual strength of the material. Each surface has two meridians with respect to $\theta = 0°$ and $\theta = 60°$ respectively. The determined meridians of the two surfaces are depicted in Figure 11.5(a). The strength of the partially damaged material falls in-between the two surfaces which is characterized by a damage scalar D.

The compressive meridian indicating initial failure in terms of normalized variables is expressed as

$$r_{ic}^* = A\left(1 - \frac{\xi^*}{\sqrt{3}}\right)^N. \tag{11.11}$$

The compressive meridian representing the residual strength of fully damaged material is given by

$$r_{fc}^* = B\left(-\frac{\xi^*}{\sqrt{3}}\right)^M \tag{11.12}$$

where, r_{ic}^* represents the strength of the intact material, while r_{fc}^* the residual strength; A, N, B, and M are the material constants; ξ^* is the normalized pressure and $\xi^* = \sqrt{3}p/p_T$; p is the pressure variable (positive in tension); and p_T is the ultimate volumetric tensile pressure, which performs as a cut-off tensile pressure in the model. Considering that brittle failure occurs at a low volumetric tensile pressure such as the uniaxial tensile condition, it gives $p_T = \sigma_t/3$, and σ_t is the uniaxial tensile strength.

The normalized strengths (r^*, r_{ic}^*, r_{fc}^*, r_D^*, r_f^*) with regard to the radius on the deviatoric plane have the general form of

$$r^* = \frac{r}{\sigma_c} = \frac{\sqrt{2J_2}}{\sigma_c} \tag{11.13}$$

where σ_c is the uniaxial compressive strength.

It should be mentioned that the compressive meridian with respect to $\theta = 60°$ corresponds to a stress state of $\sigma_3 \leq \sigma_2 = \sigma_1$ where σ_1, σ_2, and σ_3 are principal stresses and positive in tension. This stress state can be obtained with quasi-triaxial testing machines. The tensile meridian with $\theta = 0°$ and $\sigma_3 = \sigma_2 \leq \sigma_1$ requires higher confining pressure which is difficult to be achieved in a conventional triaxial compressive test. Therefore, the tensile meridian is derived by scaling the compressive meridian with a constant r_{tc}, which will be given later.

The normalize strength r_D^* for a damaged material is defined as

$$r_D^* = r_i^* - D(r_i^* - r_f^*) \tag{11.14}$$

where D is the scalar damage variable ($0 \leq D \leq 1.0$) and is defined as

$$D = \sum \frac{\Delta \bar{\varepsilon}_p}{\bar{\varepsilon}_p^f} \tag{11.15}$$

where $\Delta \bar{\varepsilon}_p$ is the effective plastic strain during a cycle of integration, and $\bar{\varepsilon}_p^f$ is the equivalent plastic strain to fracture under a constant pressure p. The expression of $\bar{\varepsilon}_p^f$ is

$$\bar{\varepsilon}_p^f = D_1 \left(1 - \frac{\xi^*}{\sqrt{3}}\right)^{D_2} \tag{11.16}$$

where D_1 and D_2 are two constants. The material cannot undertake any plastic strain at $p = p_T$. The incremental equivalent plastic shear strain is defined as

$$\Delta \bar{\varepsilon}_p = \int \sqrt{\frac{2}{3}(\varepsilon'_{\alpha\beta})_p : (\varepsilon'_{\alpha\beta})_p} \tag{11.17}$$

From the above introduction, it is seen that this model has the advantage to account for the material strength degradation from the intact state to the fully damaged state by employing a scalar damage variable induced by the effective plastic strain. The pressure-cutoff failure criterion is used to reflect the volumetric tensile

Table 11.2 Material parameters for Singapore granite.

Parameters	Symbol	Unit	Granite
Density	ρ	kg/m^3	2670
Young's modulus	E	GPa	75.20
Poisson's ratio	ν	–	0.2
Tensile strength of intact rock	σ_t	MPa	16.1
Uniaxial compressive strength of intact rock	σ_c	MPa	157.0
Hydrostatic tensile pressure limit	p_T	MPa	5.367
Normalized strength parameter for compressive meridian in the intact state	A	–	0.1334
Normalized strength parameter for compressive meridian in the fully fractured state	B	–	0.04446
Ratio of the tensile meridian radius to the compressive meridian radius	r_{tc}	–	0.58
Strength parameter for the intact material (exponent)	N	–	0.8536
Strength parameter for the fractured material (exponent)	M	–	0.8536
Parameter for damage model	D_1	–	1.748×10^{-4}
Parameter for damage model (exponent)	D_2	–	0.9326
Weighting coefficient in UTSS criterion	b	–	0.6

failure. Thus, it captures both tension-induced brittle cracks and compression-shear dominated crushing failure.

Parameters in the expression of the two meridians corresponding to the initial failure and in the damage model can be determined by curve fitting of the experimental results. For a typical granitic rock in Singapore (Zhao, 1999), the uniaxial compressive strength σ_c is 157 MPa, and the tensile strength σ_t is 16.1 MPa. The compressive meridians for the intact material and the fully damaged material are shown in Figure 11.5(a). For the tensile meridian curve of the intact material, we assume its shape is similar to the compressive one. By determining a ratio r_{tc} of the tensile radius r_t to the compressive r_c at the same pressure, the tensile meridian can be derived by scaling the compressive meridian with r_{tc}.

According to the UTSS criterion, the equal biaxial compression test result that falls onto the tensile meridian, which can be used to determine the ratio r_{tc}. The biaxial compression strength is assumed to be 1.15 times the uniaxial compressive strength. Its corresponding coordinates on the tensile meridian curve are denoted by ξ^* and r^* located at point $P(-36.37, 0.94)$ (Fig. 11.5(a)). The corresponding point P' on the compressive meridian at the same pressure is $(-36.37, 1.61)$. Therefore, $r_{tc} = 0.58$ is then obtained as the ratio of the value r_t at P to that of r_c at P'.

For the fractured material, it is assumed that the residual strength of the fractured rock is one-third of the intact strength as shown in Figure 11.5(a). The regressed representation of the equivalent plastic strain to fracture by Equation 11.16 is depicted in Figure 11.5(b).

The parameters used for the two-surface failure model are listed in Table 11.2. It should be mentioned that these approximated parameters in Table 11.2 are derived with limited experimental data. With more experimental data for the meridians, more accurate parameters can be obtained. More details about a multi-surface strength model can be found in the works by Holmquist, Templeton and Bishnoi (2001), Malvar *et al.* (1997) and Fan and Wang (2002).

11.5 NUMERICAL SIMULATIONS

11.5.1 Heterogeneity treatments in the artificial specimen

As mentioned earlier, since the heterogeneities in rock-like materials cause different micro-cracking activities and hence macro-mechanical responses in their failures, the heterogeneities should be properly presented and modeled in numerical analysis. Heterogeneity modeling by using the Weibull distribution law has been successfully employed in rock-like material failure simulations (Tang *et al.*, 2000; Fang and Harrison, 2002). The current approach is to model the heterogeneity due to a specimen's aggregates or grains. Heterogeneity in different minerals is modeled separately. Hence, the strength-related parameters in mineral grains with the same type are assigned to random values according to the Weibull's function as,

$$f(\omega) = \frac{\beta}{\mu} \left(\frac{\omega}{\mu}\right)^{\beta-1} e^{\left[-\left(\frac{\omega}{\mu}\right)^{\beta}\right]} \tag{11.18}$$

where μ is the scale parameter giving the characteristic value of distribution ω; β is the shape parameter describing the spatial concentration and dispersion degree of ω. With increasing β, the generated data are more concentrated. Hence, β is called the homogeneous index.

In the current model, the variations of the modulus, strength meridians and the fracture plastic strains in specimen's microstructure are considered to be the most significant factors to influence its macroscopic behaviors. Therefore, the modulus parameter E, parameters σ_c, σ_t and D_1 are selected to characterize the heterogeneity in different types of mineral grains.

Mineral characteristic properties must be determined in order to model the heterogeneity on grains they represent using the Weibull distribution law. Due to lack of reliable data, the mechanical properties of these three minerals in the artificial specimens are assumed to be proportional to its Mohs hardness scale value multiplied by a coefficient R. R depends on the artificial specimen's heterogeneity configurations. Its value can be determined by the uniaxial compressive strength σ_c (taken as 157 MPa) divided by the strength of a trial artificial specimen with $R = 1$.

The hardness of the feldspar, quartz and biotite are 6, 7 and 3, respectively. Correspondingly, the ratios of their characteristic modulus and strength-related parameters (E, σ_c, σ_t and D_1) to the granite macroscopic ones are $1.125R$, $1.3125R$ and $0.5625R$, respectively. Other parameters are the same as the granite macroscopic values given in Table 11.2.

Assume that the three normalized Weibull distributions for feldspar, quartz and biotite are ω_f, ω_q and ω_b, respectively. The model parameters in those mineral grains have the relationship with the corresponding granite macroscopic ones as listed in Table 11.2. Consequently, each SPH particle will be assigned to the same parameters of the mineral grain it belongs to.

Table 11.3 Parameters of mineral grains in artificial granite specimen.

Mineral components	Ratio of mineral's parameter value to that of the corresponding granite macroscopic one	
	$E, \sigma_c, \sigma_t, D_I$	*others*
Feldspar grains	$1.125\omega_f R$	1.0
Quartz grains	$1.3125\omega_q R$	1.0
Biotite grains	$0.5625\omega_b R$	1.0

11.5.2 Verification by simulating the Brazilian splitting test

To verify the model, numerical simulations of the Brazilian splitting test are performed for a circular specimen which is cut out from the central part of the specimen N1. Experimental studies found that the quartz grains have more pre-existing micro-cracks than those in the biotite and feldspar grains. Hence, the quartz grains are given a more heterogeneous Weibull distribution of the material properties. The homogenous index of the feldspar and biotite grains are 50 and that of the quartz is 5. Model parameters of the specimen's particles follow those shown in Table 11.3, where R takes 2 by a trial simulation. The specimen has a radius of 25 mm, containing 12,281 SPH particles with the same smoothing length of 0.4 mm. The geometry and loading conditions for these two specimens are illustrated in Figure 11.6(a). In the simulation, the specimen was sandwiched between two rigid walls acted upon by velocity boundaries symmetrically. Particles at the upper and lower boundaries were given a constant velocity of 0.0025 m/s. The average compressive force on the two boundaries was recorded and taken as the loading force.

The loading force versus the loading displacement curve is plotted in Figure 11.6(b). The profile of the simulated acoustic emission (AE) count is also plotted in the figure by recording the number of damaged particles. It is commonly believed that, during the specimen's deformation process, a variety of micro-activities including dislocation, twists and crack formation will cause such AE events as indicated by Cox and Meredith (1993). By associating a single AE event with the micro-crack formation, one can deduce the specimen's damage by the AE event record. When the particle fails, the strain energy carried by the particle is released accordingly. Therefore, each particle failure event can be regarded as an AE count.

The loading force increases almost linearly with the loading displacement. When reaching the peak value, it drops down abruptly. As can be observed, the damaged particles are predominately concentrated near the peak force region.

Figure 11.7 shows the process of crack initiation and propagation and the final failure pattern of specimen B1. The predicted cracks are represented by damaged particles in white. The crack initiates at 98% of the peak force. It is located on the boundaries between the quartz and biotite grains around the disc center, as indicated by the dashed circle. Unlike the homogeneous case where the crack occurs exactly at the disc center, crack initiation in a heterogeneous specimen largely depends on the local stress conditions affected by the interactions of mineral grains and their mechanical properties. Therefore, the first crack may take place where the stress condition is most severe as

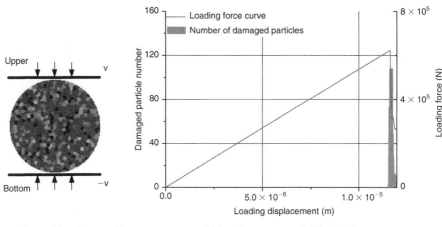

(a) Geometry and loading condition   (b) Loading curve and AE activities

Figure 11.6 Verification by simulating Brazilian splitting test, specimen B1.

discussed by Andreev (1995). With the increase in loading force, the central crack propagates rapidly towards both ends of the disc along the loading diameter until it splits
the whole disc. Again, due to the specimen's heterogeneity effect, it is very interesting
to observe that the crack deviates from its original path at 92% of the peak load in
the post-peak region. After the specimen is completely failed, those damaged particles
separate naturally from each other and form up a macro-crack as shown in the final
failure pattern. Hence, the white gaps between the fragments represent numerically
predicted fracture zones.

Figure 11.8 presents the profiles of the horizontal and vertical stress component
distributions along the loading diameter during the specimen's failure process and
compares these with the theoretical solutions.

As shown in Figure 11.7 and Figure 11.8(a), the crack may not exactly initiate from
the disc center due to the heterogeneity effects in the specimen's microstructure. Since
a crack creates new open boundaries, those particles adjacent to the open boundaries
will release their stresses. Hence, at the peak load, horizontal stresses of some particles
near the disc center unload to zero (Fig. 11.8(b)). In the post-peak region, more and
more particles along the loading diameter are affected by the developing boundaries,
as seen from Figures 11.8(c) and 8(d). Features in such a heterogeneous specimen are
different from those in a homogeneous one. The simulation results have shown that
the microstructure modeling method can well capture detailed failure process of the
granite specimen. It is also shown that the numerical results match very well with
theoretically predicted stress distributions along the loading diameter.

11.6 SIMULATIONS OF THE UNIAXIAL COMPRESSION TESTS

The validated polycrystalline model and the developed code are subsequently applied
in simulating rock specimen failure in unixial compression tests. An artificial specimen

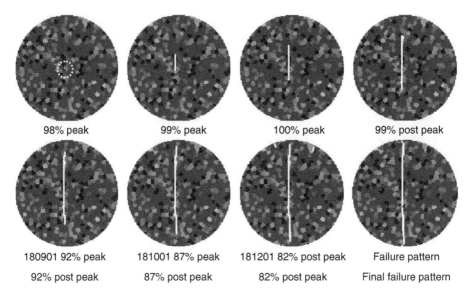

98% peak	99% peak	100% peak	99% post peak

180901 92% peak	181001 87% peak	181201 82% post peak	Failure pattern
92% post peak	87% post peak	82% post peak	Final failure pattern

Figure 11.7 Failure process of specimen B1.

N1-1 is generated. The homogenous index of both feldspar and biotite grains are 50 and that of the quartz remains 5. Model parameters of the particles follow those shown in Table 11.3, where R takes 2. Each specimen contains 31,250 SPH particles with a smoothing length of 0.4 mm. In the simulation, the specimen is sandwiched between two rigid walls acted upon by velocity boundaries without friction between the wall and specimen's ends as shown in Figure 11.9(a). Loading velocity is kept to a constant of 0.01 m/s. The average compressive stress in the specimen's ends is taken as the specimen's macroscopic compressive stress.

11.6.1 Predicted axial stress-strain curve and failure process

The axial stress-strain curve of specimen N1-1 as well as the record of the simulated AE count with respect to the axial strain is plotted in Figure 11.9(b). The predicted curve again shows brittle failure of the specimen. It almost keeps linear until the peak stress and then drops down abruptly. The AE activities become active at about 68% peak stress (marked as the first circle on the stress-strain curve). Most of these AE events appear around its peak stress.

The specimen's failure process is depicted in different frames in Figure 11.10 corresponding to those circle marks in Figure 11.9(b). It can be seen in Figure 11.10 that cracks first occur in the quartz and the boundaries between the quartz and feldspar grains, as indicated by the dashed white circles. With further load (point 'b'), those cracks propagate along the loading direction. Subsequently, they become the transgranular cracks. At the peak stress (point 'c'), new cracks are initiated at the quartz and the boundaries of quartzes and biotites. At the post-peak stage (point 'd'), those cracks on the right side of the specimen propagate rapidly and extend over several

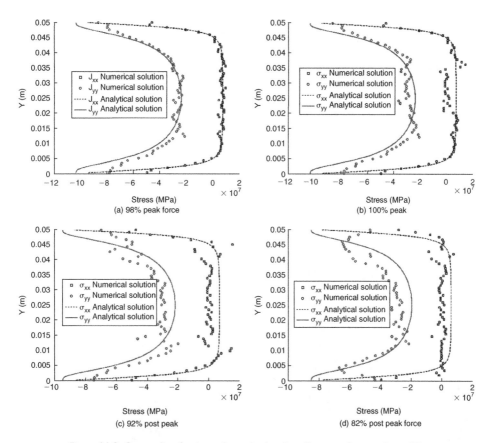

Figure 11.8 Stress distributions along the loading diameter for specimen B1.

grains. At 95% of post peak stress (point 'e'), many long cracks can be found in arrays and some form a cluster. Finally, several vertical cracks split the specimen. The specimen falls into several large pieces eventually. Spalling can also be observed at the specimen's edges. Further observation finds that, in the pre-peak stress stage, few cracks are in the biotite grains. Beyond the peak strength, the crack kinks in the biotite grains and leads to small shear faults in the specimen as observed in the final failure pattern. Because the quartz grains are more heterogeneous, cracks are first initiated among these grains. The phenomenon that cracks occur along grain boundaries is due to the stiffness mismatches between different mineral grains (Janach, 1977).

11.6.2 Parametric studies

The microstructures in the two generated artificial specimens N1 and N2 reflect two different spatial distributions of granite components. For a real granite specimen, they are also subjected to other conditions, such as weathering and pre-stress induced cracks, which cause their strength reduction. To investigate the above factors on granite failures, parametric studies are performed on another three specimens, namely N1-2,

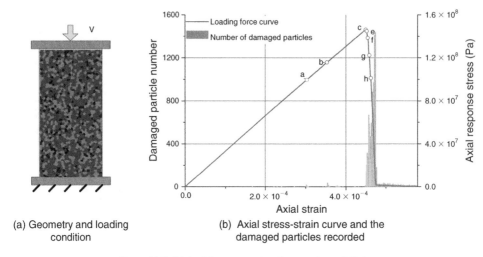

(a) Geometry and loading condition

(b) Axial stress-strain curve and the damaged particles recorded

Figure 11.9 Uniaxial compression for specimen N1-1.

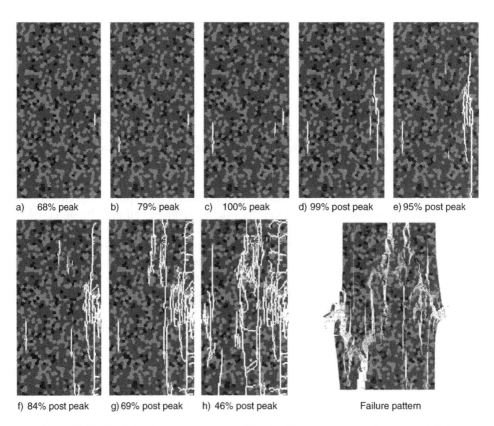

a) 68% peak   b) 79% peak   c) 100% peak   d) 99% post peak   e) 95% post peak

f) 84% post peak   g) 69% post peak   h) 46% post peak   Failure pattern

Figure 11.10 Predicted fracture process and the final fracture pattern of specimen N1-1.

Table 11.4 Heterogeneous configurations in mineral grains of specimens.

Specimen	Homogeneous index in mineral grains		
	Feldspar	Quartz	Biotite
N1-2	50	5	50
N2-1	10	3	10
N2-2	50	5	50

N2-1 and N2-2. Their configurations are listed in Table 11.4. Among them, N1-1 and N1-2 have the same microstructure as N1. Similarly, the microstructures in N2-1 and N2-2 are the same as that of N2. Obviously, N1-1 and N2-1 are more heterogeneous than N1-2 and N2-2.

Specimen N2-1 has a different microstructure compared to N1-1. The failure process of specimen N2-1 is plotted in Figure 11.11(a). At 96% of the peak stress, the crack appears at the boundaries of quartz and feldspar grains. With a further load at the peak stress level, another crack occurs within the quartzes near the specimen's middle height close to the left side. More cracks are initiated in the post-peak region. They propagate rapidly, predominately along the loading direction. It can be seen that a crack's propagation is more difficult in relatively hard and homogeneous feldspar grains than in other mineral grains. The first two cracks are almost halted when they enter the feldspar grains. The kinking and coalitions of these cracks can also be observed. Appearance of a parallel array with step-like propagation paths of these cracks strongly suggests that the failure has an axial splitting mode with some small shear faults. Although the specimen N1-1 and N2-1 have different microstructures, their final failure patterns are similar.

Specimen N1-2 has the same microstructure as N1-1. The failure process of N1-2 is plotted in Figure 11.11(b). The cracks appear firstly at quartz grains at 99% of its peak stress. When the compressive stress approaches the specimen's peak stress, the transgranular cracks among the quartz and biotite grains can be found. The formed major crack propagates rapidly parallel to the loading direction in the post-peak region. Crack nucleation is observed within quartz grains at 82% peak stress in the post-peak stage. Subsequently, many single isolated axial cracks are created. These cracks become a cluster around the well-developed major crack and eventually form a fault as shown in the final failure pattern.

Figure 11.11(c) presents the failure process of the specimen N2-2. As can be observed, the first evident crack appears at the quartz grains at the pre-peak stage around 99% of its peak stress. It propagates into different mineral grains rapidly along the loading direction after the peak stress. At 99% of its peak stress in the post-peak stage, one can observe that the crack deviated from its original path as indicated by the dashed circles. Such deviations might be resulted by stress concentrations and redistributions due to mismatched properties among these different mineral grains. This failure mode is also mainly due to the axial splitting of cracks. Meanwhile, some small shear faults are also formed.

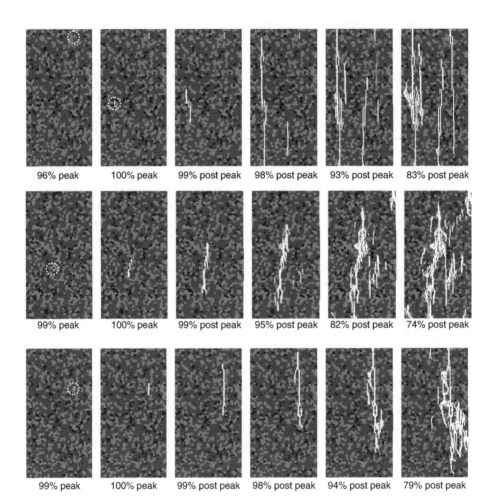

Figure 11.11 Predicted failure process of specimen: (a) N2-1 (top), (b) N1-2 (middle) and (c) N2-2 (bottom).

The above results show that the specimen's failure process is influenced by the spatial distributions of different grains in their microstructures as well as the specimen's heterogeneity. It is clear that, in N1-2 and N2-2, specimen's failure is due to axial crack and the consequent shear faults by the nucleation of an array of short cracks mainly in biotite grains. However, in N1-1 and N2-1, due to the existence of many grains of different strength-weakness, the specimens are split into many pieces by parallel vertical cracks. The cracks are more often developed in relatively heterogeneous quartz grains. In addition, they may take place at the boundaries between different grains due to stress concentrations induced by the mismatching stiffness. The transgranular cracks can be intensively observed during the post-peak stage. One can also observe that most biotite grains keep intact until the specimen fails, while some may kink and slide to form shear faults.

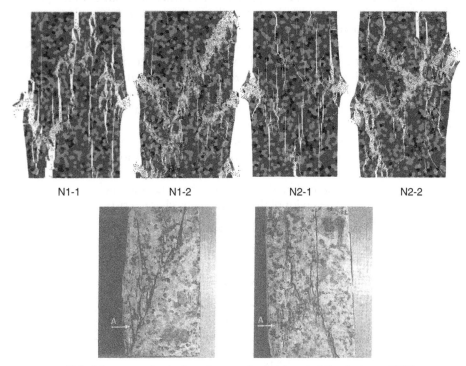

N1-1   N1-2   N2-1   N2-2

Uniaxial compression test results on rock experiments (after Andreev, 1995)

Figure 11.12 Predicted final failure patterns and comparison with experimental results.

The predicted failure process agrees well with observations by other researchers. Hallbauer, Wagner and Cook (1973) described how a fault is formed by stepwise joining of the growing fractures with the existing macro-cracks close to the specimen's failure. Janach (1977) also proposed a failure model to account for the granite specimen's failure process. He explained that the observed shear fault was caused by the formation of a diagonal array of tipped elements due to the stiffness mismatches among those different mineral components. The predicted final failure patterns of these specimens are plotted in Figure 11.12. Their differences reflect the variations in the rock specimens in terms of spatial distribution of microstructures as well as the heterogeneous strength distributions among their mineral components. As can be seen, these failure patterns agree well with those observed in the experiments (Andreev, 1995).

11.7 CONCLUSIONS

In this chapter, a polycrystalline model for simulation of multiphase material failure has been presented by taking into account the actual contents of different components and their characteristic sizes. A treatment of the heterogeneities in these components is also addressed by introducing a statistical method using the Weibull distribution law.

The polycrystalline model has been applied to construct artificial granite specimens and it has been verified by simulating a Brazilian splitting test. Uniaxial compression tests are further simulated by using the artificial specimens to investigate the effect of material microstructure and mineral component spatial distributions. Results show that both the microstructure and mineral component heterogeneities have an influence on the specimen's failure process and macroscopic mechanical response. The simulation results demonstrate that the developed microstructure modeling method can well represent those micro-cracking activities as well as macro-failure behavior and final failure pattern.

Although the generated microstructure of the granite specimens can only be regarded as an approximation of a real rock specimen, it shows potential applications in accurately simulating brittle failure of polycrystalline rock specimens, as far as detailed information is obtained on the specimen characteristics including major component contents, strength behavior of the components, characteristic sizes of the grains, etc. The present approach can be further extended to three-dimensional applications.

REFERENCES

Andreev, G.E.: *Brittle Failure of Rock Materials: Test Results and Constitutive Models.* Rotterdam; Brookfield, VT: A.A. Balkema, 1995.

Blair, S.C. and Cook, N.G.W.: Analysis of compressive fracture in rock using statistical techniques: Part I. A non-linear rule-based model. *International Journal of Rock Mechanics and Mining Sciences* 35 (1998), pp.837–848.

Chen, S., Yue, Z.Q. and Tham, L.G.: Digital image-based numerical modeling method for prediction of inhomogeneous rock failure. *International Journal of Rock Mechanics and Mining Sciences* 41 (2004), pp.939–957.

Cox, S.J.D. and Meredith, P.G.: Microcrack formation and material softening in rock measured by monitoring acoustic emissions. *International Journal of Rock Mechanics and Mining Science & Geomechanics Abstracts* 30 (1993), pp.11–24.

Fan, S.C. and Wang, F.: A new strength criterion for concrete. *ACI Structural Journal* 99 (2002), pp.317–326.

Fang, Z. and Harrison, J.P.: Development of a local degradation approach to the modelling of brittle fracture in heterogeneous rocks. *International Journal of Rock Mechanics and Mining Sciences* 39 (2002), pp.443–457.

Gingold, R.A. and Monaghan, J.J.: Smoothed particle hydrodynamics: theory and application to non-spherical stars. *Monthly Notices of the Royal Astronomical Society* 181 (1977), pp.375–389.

Gray, J.P., Monaghan, J.J. and Swift, R.P.: SPH elastic dynamics. *Computer Methods in Applied Mechanics and Engineering* 190 (2001), pp.6641–6662.

Hallbauer, D.K., Wagner, H. and Cook, N.G.W.: Some observations concerning the microscopic and mechanical behavior of quartzite specimens in stiff, triaxial compression tests. *International Journal of Rock Mechanics and Mining Science & Geomechanics Abstracts* 10 (1973), pp.713–726.

Holmquist, T.J., Templeton, D.W. and Bishnoi, K.D.: Constitutive modeling of aluminum nitride for large strain, high-strain rate, and high-pressure applications. *International Journal of Impact Engineering* 25 (2001), pp.211–231.

Janach, W.: Failure of granite under compression. *International Journal of Rock Mechanics and Mining Science & Geomechanics Abstracts* 14 (1977), pp.209–215.

Li, L., Lee, P.K.K., Tsui, Y., Tham, L.G. and Tang, C.A.: Failure process of granite. *International Journal of Geomechanics* 3 (2003), pp.84–98.

Libersky, L.D., Petschek, A.G., Carney, T.C., Hipp, J.R. and Allahdadi, F.A.: High Strain Lagrangian Hydrodynamics: A Three-Dimensional SPH Code for Dynamic Material Response. *Journal of Computational Physics* 109 (1993), pp.67–75.

Malvar, L.J., Crawford, J.E., Wesevich, J.W. and Simons, D.: A Plasticity concrete material model for DYNA3D. *International Journal of Impact Engineering* 19 (1997), pp.847–873.

Monaghan, J.J. and Lattanzio, J.C.: A refined particle method for astrophysical problems. *Astronomy and Astrophysics* 149 (1985), pp.135–143.

Potyondy, D.O. and Cundall, P.A.: A bonded-particle model for rock. *International Journal of Rock Mechanics and Mining Sciences* 41 (2004), pp.1329–1364.

Tang, C.A., Liu, H., Lee, P.K.K., Tsui, Y. and Tham, L.G.: Numerical studies of the influence of microstructure on rock failure in uniaxial compression – Part I: effect of heterogeneity. *International Journal of Rock Mechanics and Mining Sciences* 37 (2000), pp.555–569.

Tapponnier, P. and Brace, W.F.: Development of stress-induced micro cracks in Westerly Granite. *International Journal of Rock Mechanics and Mining Science & Geomechanics Abstracts* 13 (1976), pp.103–112.

Wong, T.F.: Micromechanics of faulting in westerly granite. *International Journal of Rock Mechanics and Mining Science & Geomechanics Abstracts* 19 (1982), pp.49–64.

Yu, M. and He, L.: A new model and theory on yield and failure of materials under the complex stress state. In: Jono, M. and Inoue, T. (eds): *Mechanical Behaviour of Material – VI*, Pergamon: Oxford, 1991. pp.841–846.

Yu, M.H., Zan, Y.W., Zhao, J. and Yoshimine, M.: A Unified Strength criterion for rock material. *International Journal of Rock Mechanics and Mining Sciences* 39 (2002), pp.975–989.

Zhao, J.: *Strength and Deformation Characteristics of the Bukit Timah Ggranite under Static and Dynamic Compression*. Technical Report, Nanyang Technological University, Singapore, 1999.

Chapter 12

Finite Element Method modeling of rock dynamic failure

Chun'an Tang and Yuefeng Yang

12.1 INTRODUCTION

Numerical methods adopted for rock dynamic analysis can generally be classified into the vibration method and the wave method. The essential difference between the two methods is whether to consider the wave propagation in the medium under study. The vibration method ignores the wave propagation and applies dynamic loading on the area under study. As the problem is simplified, the vibration method can easily be coupled with the finite element method. Since the constraint on time step is far less than that of the wave method, the vibration method has been widely applied in research and engineering projects. On the other hand, the wave method considers the effect of wave propagation. The wave takes some time to propagate to other locations. Its theory is rigorous and can reflect the response process of a medium under dynamic loading more accurately. Therefore, the wave method has a promising prospect and profound significance in researches.

In short, the wave method describes the wave propagation process in the medium. The solution for wave problems can be determined by two general types of methods. One is to solve the wave equations by integration or other mathematical methods. The advantage of this is that various types of waves can be studied separately. However, it requires advanced knowledge of mathematics as complex derivations are involved. In addition, analytic solutions may not be found for some problems or sometimes the results are complex functions. With the help of numerical methods, although the computational work is less, the result is only for the wave field at a specific point. If the spatial variation of the wave field needs to be investigated, point-by-point calculation is necessary, which leads to higher cost.

The other method is numerical simulation of the wave propagation process by using finite the difference method or finite element method. The semi-discrete method with separate numerical time-space treatment is applied to transform the wave equations into second order ordinary differential equations. Its advantage is that the numerical solution of the entire wave field in the time-space domain can be obtained and various characteristics of the wave field can be illustrated intuitively. However, the discrete method leads to errors in solutions for continuous media. One of the major problems is the "low-pass filtering" phenomenon (Liao and Liu, 1986; Liu and Liao, 1989, 1990). Each element has a critical frequency, and the wave components in the transient wave with frequency higher than the critical frequency would be filtered while only those with lower frequency can be passed. However, many high-frequency wave components have great influence on wave problems and generally they should not be discarded. In this case, a refined mesh is required to increase the critical frequency, which increases the computational work without adding computation difficulties.

The focus of this chapter is to investigate wave propagation in rock by using the finite element method. First, the finite element method is briefly introduced. Two major problems in wave propagation are then discussed, namely, artificial boundary and impact dynamic contact. The two-step method of visco-elastic boundary and dynamic contact force proposed by Liu and others of Tsinghua University, China, is employed (Liu and Wang, 1995a, 1995b). However, the theory and relevant researches on this method is limited to homogeneous materials and cannot be applied directly to heterogeneous media. In the two-dimensional rock failure process analysis (RFPA2D) program, the Weibull distribution and mesoscopic linearity are adopted to reflect the macroscopic non-linearity. RFPA2D has been applied to many studies on heterogeneous media, which have yielded fruitful achievements and been consistent with laboratory test results (Tang, 1997a, 1997b; Tang and Kaiser, 1998; Fu, 2000). On the basis of the two-step method and RFPA2D analysis, the study on wave propagation is extended to heterogeneous media. Finally, dynamic fracturing process analysis of rock material under the Brazilian tensile test condition is conducted using RFPA2D. The analysis aims to investigate the influence of applied stress wave amplitude on the fracturing process and failure induced in the rock material. Heterogeneity of the rock material is taken into account and its influence on stress wave propagation is first discussed. Then, the dynamic failure process analyses are extended to investigate the influence of waveforms in terms of stress wave amplitude on failure modes. These simulations reveal that the failure modes are affected strongly by the stress wave amplitude. A stress wave of higher amplitude generates fracturing earlier. Consequently, the higher wave amplitude leads to intense fracture initiation near the loading point, whereas lower amplitude leads to intense fracture initiation near the bottom where the wave is reflected. The observation obtained from the simulations provides a new insight on the fracturing and failure mechanism of rock under the dynamic Brazilian tensile test. The simulation also shows that RFPA has the potential to simulate and study material failure under dynamic loading conditions.

12.2   RFPA DYNAMIC MODELING APPROACH

12.2.1   Finite element solutions for elastic wave

For dynamic analysis by the wave method, the transient wave propagation is one of the main issues. The finite element method is usually adopted to solve the complex transient wave propagation problem, including the central difference method, Newmark method and Wilson-θ method. Depending on whether the stiffness matrix is decomposed to determine the solutions, it can be divided into the explicit integration scheme and the implicit integration scheme. The explicit integration scheme is represented by the central difference method, while the Newmark method and Wilson-θ method are often adopted in implicit integration schemes. The explicit scheme is usually used in combination with the lumped mass method. The inverse of the stiffness matrix is not required to determine the solutions and even assembling of the overall stiffness matrix is not needed. Under the premise of stability, it has high computation efficiency and precision. On the other hand, the inverse of the stiffness matrix has to be calculated in the implicit integration scheme. If the consistent mass matrix is adopted, the implicit integration would greatly increase the computation effort. However, with better stability of the implicit integration and by setting the integration parameters in the Newmark method, the systematic high-frequency reflection can be effectively filtered out in the high-frequency range (Fang, 1992; Fang and Chen, 1993). Moreover, with technological advances, many efficient iterative methods have been developed, among which the conjugate gradient (CG) method is a representative one. By applying different preprocessing techniques, various preconditioned conjugate gradient (PCG) methods have been proposed, among which the most important one is the symmetric successive over-relaxation preconditioned conjugate gradient (SSOR-PCG) method. Using this method, the computation precision can be assured and the number of iterations and computation work can be greatly reduced. Furthermore, with the introduction of vector computers and parallel computers, the previous "serial mode" (the second operation can only be started after the first one is completed) is changed, and a large amount of matrices can be computed simultaneously. Hence, the low efficiency of the implicit method due to the large amount of computation work is improved. Consequently, the implicit integration method, such as the Newmark method, is widely applied not only in earthquake analysis, but also in the impact problem. In this chapter, the Newmark method is employed to investigate wave propagation in rock under dynamic loading.

According to Hamilton's variational principle, after discretization in the space domain by using the finite element method, the dynamic equation can be expressed in the following form:

$$M\ddot{u} + C\dot{u} + Ku = Q \tag{12.1}$$

The following relationship of velocity and displacement is imported in the Newmark method:

$$\ddot{u}_{t+\Delta t} = \frac{1}{\beta \Delta t^2}(u_{t+\Delta t} - u_t) - \frac{1}{\beta \Delta t}\dot{u}_t - \left(\frac{1}{2\beta} - 1\right)\ddot{u}_t$$

$$\dot{u}_{t+\Delta t} = \frac{\gamma}{\beta \Delta t}(u_{t+\Delta t} - u_t) + \left(1 - \frac{\gamma}{\beta}\right)\dot{u}_t - \left(\frac{\gamma}{2\beta} - 1\right)\ddot{u}_{t+\Delta t}\Delta t \tag{12.2}$$

Substituting into the dynamic equation, we can have

$$\widehat{K}u_{t+\Delta t} = \widehat{Q}_{t+\Delta t}$$

$$\widehat{K} = K + \frac{1}{\beta \Delta t^2}M + \frac{1}{\beta \Delta t}C$$

$$\widehat{Q}_{t+\Delta t} = Q_{t+\Delta t} + M\left[\frac{1}{\beta \Delta t^2}u_t + \frac{1}{\beta \Delta t}\dot{u}_t + \left(\frac{1}{2\beta} - 1\right)\ddot{u}_t\right]$$

$$+ C\left[\frac{\gamma}{\beta \Delta t}u_t + \left(\frac{\gamma}{\beta} - 1\right)\dot{u}_t + \left(\frac{\gamma}{2\beta} - 1\right)\ddot{u}_t\right] \tag{12.3}$$

where u_t, $\dot{u}_t$, $\ddot{u}_t$ are the displacement, velocity and acceleration vector at time t, respectively; K, M and C are the stiffness matrix, mass matrix and damping matrix of the system, respectively; γ and β are the integral coefficients in the Newmark method. When $\gamma \geq \frac{1}{2}, \beta \geq \frac{1}{4}\left(\gamma + \frac{1}{2}\right)^2$, the Newmark method is unconditionally stable.

12.2.2   Brief description of the RFPA2D model

RFPA is a program which can numerically simulate the progressive failure process of rock. It brings the heterogeneity parameters of rock into the basic element and realizes description of the visco-elastic boundary element method by heterogeneity at meso-scale. The macroscopic nonlinearity of a material during the deformation process intuitively reflects the entire process from mesoscopic basic element damage to macroscopic failure. In order to manifest the mesoscopic heterogeneity of rock material, the rock material is considered to be composed of uniform quadrilateral elements and its properties are assumed to satisfy the Weibull distribution (Tang, 1997b; Chau et al., 2004):

$$\varphi(\alpha) = \frac{m}{\alpha_0} \cdot \left(\frac{\alpha}{\alpha_0}\right)^{m-1} \cdot e^{-\left(\frac{\alpha}{\alpha_0}\right)^m} \tag{12.4}$$

where $\varphi(\alpha)$ is the statistical distribution density of meso-elements which have a certain mechanical property α; α represents the mechanical property of meso-elements in the material (such as strength, elastic modulus, Poisson's ratio, bulk density etc.); α_0 is the average value of the property of the meso-elements; m defines the shape of the distribution function and its physical meaning reflects the homogeneity of rock material. m can be defined as the coefficient of homogeneity; the larger the value of m is, the more homogeneous the rock material is; the smaller m is, the more heterogeneous the rock material is. Equation (12.4) reflects the distribution of the components of a heterogeneous material.

In the RFPA2D program, the properties of elements in meso-scale satisfy the elasto-plastic or elasto-brittle constitutive model. When the element meets the damage criterion, damage occurs in the element. In the numerical analysis, the Mohr-Coulomb criterion and the maximum tensile stress criterion are adopted to judge whether damage or failure occurs in the element (Zhu et al., 2006). For damaged elements, the stiffness is reduced according to the elastic damage model. The next step of loading is then applied until the entire analysis process is completed.

12.2.3   Elastic damage constitutive law of meso-elements in the RFPA2D model

The meso-element is initially an elastomer and the mechanical property can be expressed by elastic modulus and Poisson's ratio. With increasing stress in the element, the stress or strain state of the element will satisfy the given damage threshold and element damage occurs. Two criteria are adopted in the RFPA program (Zhu et al., 2006): one is the maximum tensile strain (stress) criterion, i.e. when the maximum tensile strain (stress) of the meso-element reaches the given limit value, tensile damage occurs in the element; the other one is the Mohr-Coulomb criterion, i.e. when

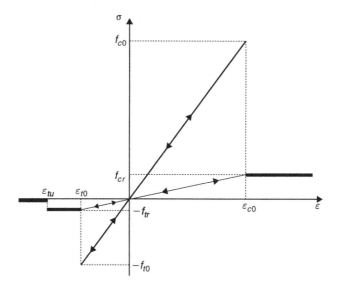

Figure 12.1 Elastic damage constitutive law of element under uniaxial stress state.

the stress state of the meso-element meets the Mohr-Coulomb criterion, shear damage occurs in the element. Meanwhile, the maximum tensile strain criterion has the priority, in other words, when the meso-element satisfies the maximum tensile strain criterion, there is no need to consider whether the element meets the Mohr-Coulomb criterion. Figure 12.1 shows the elastic damage constitutive law of an element under uniaxial compression and tension.

According to the elastic damage theory, the elastic modulus of an element gradually reduces with increasing damage. The damage equation is as follows:

$$E = (1 - D)E_0 \tag{12.5}$$

where E and E_0 are the elastic modulus after damage and the initial elastic modulus, respectively; D is the damage variable.

When the tensile stress in the element reaches the uniaxial tensile strength f_{t0}, i.e. $\sigma_3 \leq -f_{t0}$, the damage variable can be written in the following form:

$$D = \begin{cases} 0 & \varepsilon > \varepsilon_{t0} \\ 1 - \lambda\varepsilon_{t0}/\varepsilon & \varepsilon_{tu} < \varepsilon \leq \varepsilon_{t0} \\ 1 & \varepsilon \leq \varepsilon_{tu} \end{cases} \tag{12.6}$$

where λ is the coefficient of residual strength and can be determined by $f_{tr} = \lambda f_{t0}$, ε_{t0} is the tensile strain corresponding to the elastic deformation limit, ε_{tu} is the ultimate tensile strain, upon which the element loses its tensile capability. $\varepsilon_{t0} = \eta\varepsilon_{tu}$, where η is the ultimate tensile strain coefficient.

Under multiaxial stress state, the principal strain of the element may be higher than the ultimate tensile strain ε_{t0}. In this case, the equivalent principal strain $\bar{\varepsilon}$ shall be calculated:

$$\bar{\varepsilon} = -\sqrt{\langle -\varepsilon_1 \rangle^2 + \langle -\varepsilon_2 \rangle^2 + \langle -\varepsilon_3 \rangle^2} \tag{12.7}$$

where ε_1, ε_2 and ε_3 are the three principal strains. The $\langle\,\rangle$ is a calculation operator, and can be written as follows:

$$\langle x \rangle = \begin{cases} x & x \geq 0 \\ 0 & x < 0 \end{cases} \tag{12.8}$$

In this case, the damage variable can be expressed as:

$$D = \begin{cases} 0 & \bar{\varepsilon} > \varepsilon_{t0} \\ 1 - \lambda \varepsilon_{t0}/\varepsilon & \varepsilon_{tu} < \bar{\varepsilon} \leq \varepsilon_{t0} \\ 1 & \bar{\varepsilon} \leq \varepsilon_{tu} \end{cases} \tag{12.9}$$

In order to analyze the damage behavior of the element under compression and (or) shear, the Mohr-Coulomb criterion is selected as the second strength criterion:

$$\sigma_1 - \sigma_3 \frac{1 + \sin\phi}{1 - \sin\phi} \geq f_{c0} \tag{12.10}$$

where σ_1 is the major principal stress, σ_3 is the minor principal stress, ϕ is the friction angle, f_{c0} is the uniaxial compressive strength. The damage variable is then in the following form:

$$D = \begin{cases} 0 & \varepsilon < \varepsilon_{c0} \\ 1 - \lambda \varepsilon_{c0}/\varepsilon & \varepsilon \geq \varepsilon_{c0} \end{cases} \tag{12.11}$$

where ε_{c0} is the compressive strain when the elastic deformation limit is reached, λ is the coefficient of residual strength.

Under multiaxial stress state, the major principal strain can be calculated by the following equation:

$$\varepsilon_{c0} = \frac{1}{E_0} \left[f_{t0} + \frac{1 + \sin\phi}{1 - \sin\phi} \sigma_3 - v(\sigma_1 + \sigma_3) \right] \tag{12.12}$$

where v is the Poisson's ratio and other parameters are defined as above.

In the RFPA program, it is assumed that the development of shear damage is only related to the major principal strain ε_1. Extended further from the above derivations, the triaxial shear damage variable is:

$$D = \begin{cases} 0 & \varepsilon_1 < \varepsilon_{c0} \\ 1 - \lambda \varepsilon_{c0}/\varepsilon & \varepsilon_1 \geq \varepsilon_{c0} \end{cases} \tag{12.13}$$

12.3   TRANSIENT WAVE PROPAGATION IN INFINITE MEDIUM

For simulation of transient wave propogation in an infinite medium, it is common practice to take the near-field calculation area with finite size from the infinite medium and introduce an artificial boundary around the calculation area. The finite element method or other techniques can then be applied to discretize differential equations of motion and physical boundary conditions in the time and space domains. Consequently, the direct simulation of real wave propagation is realized. The commonly used artificial boundaries include the paraxial boundary, the transmitting boundary, the viscous boundary and the visco-elastic boundary. The paraxial boundary is an approximation from the standard wave equations and the travelling wave is diffracted by the physical wave velocity of the corresponding homogeneous medium. However, the paraxial boundary cannot be applied when multiple physical wave velocities exist on the boundary. The transmitting boundary is a kind of kinematic simulation. It can directly model the transmitting process for various travelling waves, including surface waves and body waves. Its precision can be guaranteed by multiple transmissions (Liao and Liu, 1989; Liao and Yang, 1994; Liao, 2002) and is the highest among the four boundaries. However, when multiple transmission equations are used, the effect of finite element discretization in the space domain will result in zero-frequency drift and high-frequency oscillatory instability. Therefore, a damping layer in the boundary and smoothing techniques are required to eliminate the high-frequency oscillatory instability. Application of the transmitting boundary requires good understanding of the theory and of programming, and is not suitable for secondary development on the existing large-scale softwares and programs. The viscous boundary, which has a clear concept, is easily implemented and widely used. However, it only considers absorption of scattered waves and may lead to low-frequency drift phenomenon. Hence, it is not able to simulate elastic restoration of a semi-infinite medium. The visco-elastic boundary excels the viscous boundary. It has faster computation speed and can easily be implemented (Liu and Lu, 1997, 1998; Liu, Gu and Du, 2006). The visco-elastic boundary can be embedded into the existing program or software conveniently (Liu, Du and Yan, 2007) and the accuracy can meet the engineering requirements.

On the basis of visco-elastic boundary, Liu, Gu and Du (2006) derived the consistent visco-elastic artificial boundary and its boundary element and proposed an

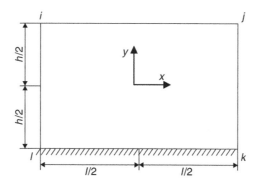

Figure 12.2 Rectangular finite element.

equivalent visco-elastic artificial boundary element. Hence, the ordinary finite element can be used to generate the visco-elastic boundary, which has the same calculation accuracy with the lumped visco-elastic artificial boundary.

12.3.1 Element with equivalent stiffness

Liu, Gu and Du (2006) verified that the thickness of the artificial boundary element has no significant effects on the simulation results. As the nodes l and k of an element are fixed, the explicit plane stiffness matrix can be obtained. Removing the stiffness matrix elements related to l and k, one can have the equivalent stiffness of the element corresponding to the nodes i and j. Equation (12.14) can be adopted for calculation and the conversion formulas for relevant parameters can be found in Equation (12.15):

$$[\tilde{K}] = \frac{l\tilde{\rho}}{6h} \begin{bmatrix} 2\tilde{c}_s & 0 & \tilde{c}_s & 0 \\ 0 & 2\tilde{c}_p & 0 & \tilde{c}_p \\ \tilde{c}_s & 0 & 2\tilde{c}_s & 0 \\ 0 & \tilde{c}_s & 0 & 2\tilde{c}_p \end{bmatrix} \tag{12.14}$$

$$\begin{cases} \tilde{\rho}\tilde{c}_s = hK_{BT} \\ \tilde{\rho}\tilde{c}_P = hK_{BN} \\ \tilde{G} = hK_{BT} = \alpha_T h \dfrac{G}{R} \\ \tilde{E} = \dfrac{(1+\tilde{\nu})(1-2\tilde{\nu})}{1-\tilde{\nu}} hK_{BN} = \alpha_N h \dfrac{G}{R} \dfrac{(1+\tilde{\nu})(1-2\tilde{\nu})}{1-\tilde{\nu}} \\ \alpha = \alpha_N/\alpha_T \\ \tilde{\nu} = \begin{cases} \dfrac{\alpha-2}{2(\alpha-1)}, & \alpha \geq 2 \\ 0, & \alpha < 2 \end{cases} \end{cases} \tag{12.15}$$

where $\tilde{\rho}$ is the mass density of the equivalent element; $\tilde{c}_s$ and $\tilde{c}_P$ are the S-wave and P-wave velocity of the medium composed of equivalent elements, respectively; $\tilde{G}$, $\tilde{E}$ and $\tilde{\nu}$ are the equivalent shear modulus, equivalent elastic modulus and equivalent Poisson's ratio of the equivalent visco-elastic boundary, respectively; h is the thickness of the equivalent element; R is the distance between the wave source and the artificial boundary; G is the shear modulus of the medium; K_{BT} and K_{BN} are the tangential and normal spring constants of the equivalent physical system, respectively; α_T and α_N are the tangential and normal coefficient of the visco-elastic artificial boundary, respectively. The range recommended by Liu, Gu and Du (2006) is [0.35, 0.65] and [0.8, 12], respectively.

12.3.2 Element with equivalent damping

The damping matrix of the equivalent boundary element which has the damping proportional to the stiffness is adopted. Let

$$[C]_B = [\tilde{\eta}][K]_B \tag{12.16}$$

$$[\tilde{\eta}] = \begin{bmatrix} \tilde{\eta}_{BT} & & & \\ & \tilde{\eta}_{BN} & & \\ & & \tilde{\eta}_{BT} & \\ & & & \tilde{\eta}_{BN} \end{bmatrix} \tag{12.17}$$

$$\begin{cases} \tilde{\eta}_{BT} = \dfrac{C_{BT}}{K_{BT}} = \dfrac{\rho c_s R}{\alpha_T G} \\[2mm] \tilde{\eta}_{BN} = \dfrac{C_{BN}}{K_{BN}} = \dfrac{\rho c_p R}{\alpha_N G} \end{cases} \tag{12.18}$$

where $\tilde{\eta}_{BT}$ and $\tilde{\eta}_{BN}$ are the proportional coefficient related to the tangential and normal stiffness, respectively; c_s and c_p are S-wave and P-wave velocity, respectively; G is the shear modulus of the medium; ρ is the mass density of the medium; α_T and α_N are the tangential and normal coefficient of the visco-elastic boundary.

According to the derivations by Liu, Gu and Du (2006), an element is extended from the visco-elastic boundary in the RFPA2D numerical model. The relevant element matrix can be calculated similarly to the 2D equivalent consistent visco-elastic boundary element (Equations (12.14)–(12.18)).

The Lamb's problem is an ideal model in geophysics firstly proposed by Lamb (1904). The medium in the Lamb's problem is a semi-infinite fully elastic homogeneous medium with free surface, which can be classified into four types according to the form of wave sources, namely: a point source acting on the medium surface, a line source acting on the medium surface, a point source acting in the interior of the medium and a line source acting in the interior of the medium. The earlier Lamb's problem only considered the vertical concentrated force, and later the other wave activiation forms including the horizontal concentrated force and the logitudinal wave source. This is called the generalized Lamb's problem. By solving different types of Lamb's problems, some fundamental characteristics of artificially-generated seismic wave fields have been identified.

For solving a wave propagation problem in an infinite medium, it is a common practice to take the near-field calculation area of finite size from the infinite medium and introduce a visco-elastic boundary around the calculation area. The purpose of the artificial boundary is to keep the simulation result in the near field consistent with that from infinite domain so that wave propagation in infinite domain can be modelled realistically. As the visco-elastic boundary is theoretically derived on the basis of cylindrical waves, the result has to be compared with that from infinite domain. In this study, the comparison method in Liu and Lu (1997, 1998) and Liu, Gu and Du (2006) are adopted. The result of infinite domain is obtained by the extended finite element method (i.e. to enlarge the calculation area so as to approach the exact solution). The extended solution is calculated by the large-scale software ANSYS and the reliability of comparison analysis can be ensured. For convenience, the extended solution is denoted by E.S. and the result from the visco-elastic boundary is denoted by V.S. with the postfixed character standing for the key observation point.

Figure 12.3 shows a typical numerical model, where the dashed line represents the visco-elastic boundary. The following parameters are adopted: elastic modulus $E = 2.5\,\text{MPa}$, Poisson's ratio is 0.25, the density is $1\,\text{kg/m}^3$, model size $4\,\text{m} \times 2\,\text{m}$, size of finite elements $\Delta x = \Delta y = 0.05\,\text{m}$ and $\Delta t = 0.01\,\text{s}$. The calculation time is $10\,\text{s}$. The

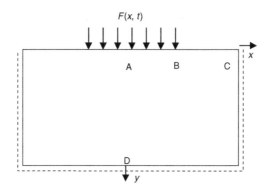

Figure 12.3 Numerical model.

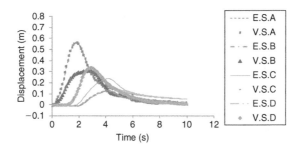

Figure 12.4 Displacement history of different observation points.

wave source is a distributed load $F(x,t) = S(x)T(t)$ acting on the free surface along y-axis. $S(x)$ and $T(t)$ are functions shown in Equation (12.19). It can be seen from Figure 12.4 that the simulation results for key points A, B and D are very close to the extended solutions, while the simulation result for point C is smaller than the extended solution. As the visco-elastic boundary is derived with assumption of cylindrical waves, the flat artificial boundary would lead to loss in accuracy. However, its accuracy is higher than that of the viscous boundary (Liu and Lu, 1997, 1998; Liu, Gu and Du, 2006).

$$S(x) = \begin{cases} 1 & |x| = 1 \\ 0 & |x| \neq 1 \end{cases}$$

$$T(t) = \begin{cases} t & 0 \leq t \leq 1 \\ 2 - t & 1 \leq t \leq 2 \\ 0 & \text{others} \end{cases} \tag{12.19}$$

12.3.3 Infinite domain problem considering heterogeneity of medium

The model used is similar to that for verification for homogeneous media, as shown in Figure 12.3. A finite element model with extended boundary is established and the

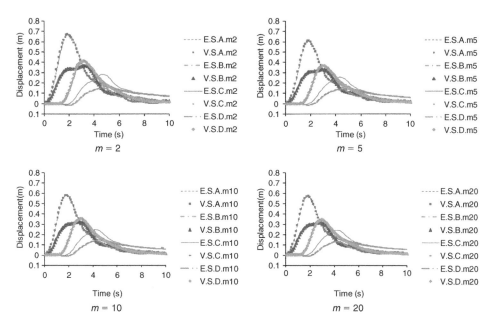

Figure 12.5 Comparison of simulated displacement along y-axis and the extended solution for different observation points in media of different degrees of heterogeneity.

elastic modulus of the medium satisfies the Weibull distribution. The computation model for the visco-elastic boundary is a section taken from the extended model. In order to show the generality of comparison, four cases with different Weibull shaper parameter m are considered, namely, $m = 2$, 5, 10 and 20, respectively.

Figure 12.5 compares the results from the visco-elastic boundary and the extended solutions. It can be seen that the accurary of the visco-elastic boundary is still reasonably high. Therefore, the visco-elastic boundary can be applied for heterogeneous material and its accuracy can meet the engineering requirements.

It is shown in Figure 12.6 that the shape parameter m affects the response of the model and is directly related to material heterogeneity. The smaller m is, the more heterogeneous the material is. When $m = 20$, the response of the model is close to that of homogeneous material. When smaller m is adopted, the peak displacement of the key point is larger. It can be seen from Figure 12.7 that the more homogeneous the material is, the smoother the wavefront is; the more heterogeneous the material is, the more irregular the wavefront is.

12.4 DYNAMIC CONTACT PROBLEM

In numerical dynamic analysis, solution of contact problems involves two major processes: one is iteration of the boundary condition or contact state; another one is calculation of contact force. In earlier days, the direct stiffness method in finite element analysis was used to solve the contact problem (Ohte, 1973), which requires

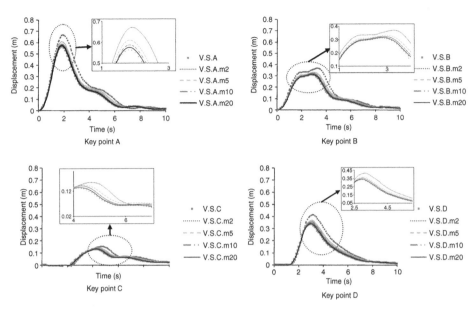

Figure 12.6 Displacement history along *y*-axis of different observation points in media with different degrees of heterogeneity.

repeated modifications and solutions of the overall stiffness matrix and involves a large amount of computation work. Later the hybrid method was proposed, in which the overall stiffness matrix is solved by applying unit force on the contact point and thus the flexibility matrix is obtained. In this way, less iterations are required, so the computation efficiency is improved (Ou and Gong, 1988). Among various contact element methods, the penalty function method and the Lagrange multiplier method are most widely used. The penalty function method is simple and suitable for explicit calculation. However, it affects the critical step time in the explicit calculation and the singularity of the coefficient matrix in the implicit calculation. Furthermore, selection of the proper penalty factor affects the reliability of calculation results (Chen, Pan and Duan, 2006). The Lagrange multiplier method is accurate in calculating the contact force. However, it is not compatible with the explicit equations and requires special numerical treatment. When dealing with contact problems, zero diagonal items occur in the asymmetrical coefficient matrix. Thus, it would require a huge amount of calculation, and the contact surface can only be determined by trial and error (Ou and Gong, 1988; Wen and Gao, 1994; Tworzydloa *et al.*, 1998; Chen, Pan and Duan, 2006). In short, for any iterative method which needs to evaluate the initial contact state and reassemble the stiffness matrix, even if the number of iterations can be reduced by some techniques, the assembly of the overall stiffness matrix and solution of control equations still results in an increase of workload. It is of great significance for engineering practices and scientific researches to find a dynamic contact method which is highly efficient, highly accurate and can easily be implemented. Liu and Wang (1995a, 1995b) proposed a two-step method for dynamic contact force. The contact

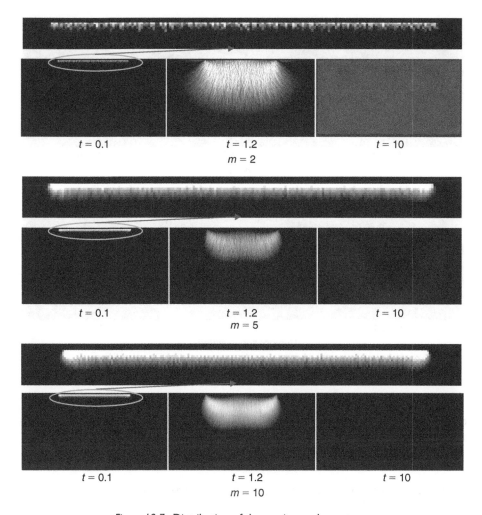

Figure 12.7 Distribution of the maximum shear stress.

force is only calculated for the contact point without reassembling the stiffness matrix. Hence, the computation workload is greatly reduced. Liu and Wang (1995a, 1995b) applied the two-step method in the dynamic response of contactable cracks and verified diffraction of incident P-wave on linear cracks. The method can describe the collision between contact surfaces and reflect the effect of static and dynamic friction between crack surfaces. Furthermore, Zhang, Chen and Tu (2004) applied the method to analysis of seismic resistance of an arch dam and effectively considered the dynamic interaction between blocks likely to slide in the dam abutment and the dam body. In our study, the dynamic contact method is imported into the collision contact problem and it is verified by comparison with ANSYS_DYNA modeling. In combination with the principle of reflection of macroscopic nonlinearity by mesoscopic linearity in

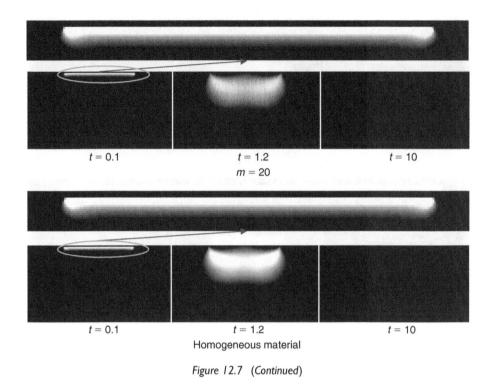

Figure 12.7 (Continued)

the RFPA program, a simple method for collision and contact between heterogeneous materials is proposed.

12.4.1 Relevant theories of dynamic contact model

Liu and Wang's studies (1995a, 1995b) were based on a mechanical model which contains cracks. Nodes on both sides of the crack form a node pair, and the slip displacement of the node cannot be too large. As the interaction time is short and the relative slip is small in many impact problems, the method can be widely applied. For cases with large slip displacement, the contact searching algorithm can be adopted to search the possible contact surfaces and pair up contacts. The relevant equations can be established for the contact pair and hence to supplement and improve the algorithm. This part of work will be carried out in the future.

   To be clear and concise, the relevant theory and method in the literature will be introduced in the following section, taking the 2D model as an example. The mechanical model is shown in Figure 12.8, in which a crack S is present in a continuous elastic medium. The upper side and lower side of S are denoted by $S+$ and $S-$, respectively. $S+$ and $S-$ are assumed to coincide at the initial moment. S is assumed to be smooth so that the normal vector n and the tangential vector t of S exist everywhere (Fig. 12.8). Under dynamic loading, $S+$ and $S-$ may separate from, collide or contact with each

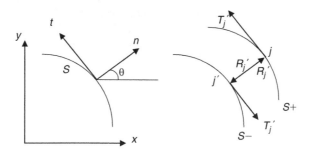

Figure 12.8 Contact model.

other. When the crack surfaces contact with each other, normal and shear contact stresses of the same magnitude and opposite direction act on $S+$ and $S-$.

The continuous medium is discretized by the finite element method. The same distribution is set for nodes in the upper and lower sides of the crack, i.e. the nodes in $S+$ and $S-$ are in one-to-one correspondence. The force generated by dynamic contact of the crack surfaces appears in the form of node force. By the lumped mass finite element method, the motion equations for any node j can be expressed as follows:

$$M_j \ddot{U}_j + \sum_l K_{jl} U_l = F_j + R_j + \tau_j \tag{12.20}$$

where

$$R_j = \begin{cases} \overline{n_j^T} |R_j| & j \in S+ \\ -\overline{n_j^T} |R_j| & j \in S- \\ 0 & j \notin S \end{cases} \tag{12.21}$$

$$\tau_j = \begin{cases} \pm \overline{t_j^T} |\tau_j| & j \in S \\ 0 & j \notin S \end{cases} \tag{12.22}$$

$$\overline{n_j} = \{\cos \theta_j \quad \sin \theta_j\} \tag{12.23}$$

$$\overline{t_j} = \{-\sin \theta_j \quad \cos \theta_j\} \tag{12.24}$$

In Equations (12.20)–(12.24), M_j is the lumped mass of node j; K_{jl} is the stiffness matrix for node j and the adjacent node l. $\ddot{U}_j$ is the acceleration vector of node j; U_l is the displacement vector of node l; F_j is the known external load vector acting on node j; R_j and τ_j are the contact force vector of node j due to dynamic normal and shear contact stresses between crack surfaces, respectively; $|R_j|$ and $|\tau_j|$ are the norm of R_j and τ_j, respectively; θ_j is the angle between the normal direction of the crack surface S at node j and the x-axis (Fig. 12.8). The sign in Equation (12.22) depends on the location of node j ($S+$ or $S-$) and the motion state of the node.

When solving Equation (12.20) by the step-by-step integration method, as R_j and τ_j are related to the motion state at the current moment and the moment before, the displacement at the current moment cannot be determined directly. Hence, supplementary conditions are required. According to the idea presented by Liu and Wang (1995a, 1995b), the displacement can be divided into three parts:

$$U_j^{P+1} = \overline{U}_j^{P+1} + \Delta\overline{U}_j^{P+1} + \Delta V_j^{P+1}$$

$$\Delta U_j^{P+1} = \frac{\Delta t^2}{M_j} R_j^p$$

$$\Delta V_j^{P+1} = \frac{\Delta t^2}{M_j} \tau_j^p \tag{12.25}$$

where $\overline{U}_j^{P+1}$ is the displacement without considering the effect of dynamic contact of crack surfaces at the moment P; ΔU_j^{P+1} and ΔV_j^{P+1} are the additional displacements caused by the dynamic contact force R_j^p and τ_j^p at the moment P, respectively. As R_j^p and τ_j^p are unknown, the dynamic contact state of the crack has to be considered and supplementary contact conditions are required.

12.4.1.1 Determination of R_j^p and ΔU_j^{P+1}

Let i and i' the two corresponding nodes on crack surfaces $S+$ and $S-$. The dynamic contact forces on nodes i and i' are of same magnitude but opposite direction. The condition for occurrence of dynamic contact between i and i' at moment $P+1$ is:

$$\overline{n}_i(\overline{U}_{i'}^{P+1} - \overline{U}_i^{P+1}) \geq 0 \tag{12.26}$$

when the above formula is not satisfied, nodes i and i' are not in contact with each other and R_i^p and τ_i^p are zero. When the above formula is satisfied, nodes i and i' are in contact and R_i^p is non-zero. In this case, the actual motion of nodes i and i' meets the contact condition, i.e. the displacement compatibility condition:

$$\overline{n}_i(\overline{U}_{i'}^{P+1} - \overline{U}_i^{P+1}) = 0 \tag{12.27}$$

From Equations (12.25) and (12.27), we have

$$R_i^P = \frac{M_i M_{i'}}{(M_i + M_{i'})\Delta t^2} \overline{n}_i^T \Delta_{1i}$$

$$\Delta_{1i} = \overline{n}_i(\overline{U}_{i'}^{P+1} - \overline{U}_i^{P+1})$$

$$\Delta U_i^{P+1} = \frac{M_{i'}}{(M_i + M_{i'})} \overline{n}_i^T \Delta_{1i} \tag{12.28}$$

$$\Delta U_{i'}^{P+1} = -\frac{M_i}{(M_i + M_{i'})} \overline{n}_i^T \Delta_{1i} \tag{12.29}$$

12.4.1.2   Determination of τ_j^P and ΔV_j^{P+1}

When the corresponding nodes i and i' are in contact, the two nodes are in different motion state, i.e. static friction state and dynamic friction state. When the node is in dynamic friction state, τ_j^P is related to the normal stress R_j^P and the dynamic friction coefficient. When the node is in static friction state, the upper limit of τ_j^P can be given by R_j^P. Liu and Wang (1995a, 1995b) derived the following equations:

$$\Delta_{2i} = \bar{t}_i[\overline{U}_{i'}^{P+1} - \overline{U}_i^{P+1}) - (U_i^P - U_{i'}^P)]$$

$$\tau_i^P = \frac{M_i M_{i'}}{(M_i + M_{i'})\Delta t^2}\bar{t}_i^T \Delta_{2i}$$

$$|\tau_i^P| = \frac{M_i M_{i'}}{(M_i + M_{i'})\Delta t^2}|\Delta_{2i}|$$

$$|\tau_i^P| \le f_S |R_i^P|$$

$$\left.\begin{aligned}
\Delta V_i^{P+1} &= \frac{M_i M_{i'}}{M_i + M_{i'}}\bar{t}_i^T \Delta_{2i} \\
\Delta V_{i'}^{P+1} &= -\frac{M_i M_{i'}}{M_i + M_{i'}}\bar{t}_i^T \Delta_{2i}
\end{aligned}\right\} \tag{12.30}$$

where f_S is the static friction coefficient, R_i^P is given by Equation (12.28). When the inequalities are not satisfied, the node is in dynamic friction state.

When the node is in dynamic friction state, the following equations shall be adopted:

$$\tau_i^P = \text{sgn}(\Delta_{2i})f_D|R_i^P| \tag{12.31}$$

$$\text{sgn}(x) = \begin{cases} 1 & x > 0 \\ 0 & x = 0 \\ -1 & x < 0 \end{cases} \tag{12.32}$$

At moment $P+1$, the necessary and sufficient condition for node being in dynamic friction state is:

$$\left.\begin{aligned}
\bar{t}_i(U_{i'}^{P+1} - U_i^{P+1}) &> (U_{i'}^P - U_i^P), \quad \text{when } \text{sgn}(\Delta_{2i}) = 1 \\
\bar{t}_i(U_{i'}^{P+1} - U_i^{P+1}) &< (U_{i'}^P - U_i^P), \quad \text{when } \text{sgn}(\Delta_{2i}) = -1
\end{aligned}\right\} \tag{12.33}$$

When Equation (12.33) is not satisfied, it means that the node has changed from dynamic friction state to static friction state. The relevant equations for static friction state need to be recalculated.

$$\left.\begin{aligned}
\Delta V_i^{P+1} &= \text{sgn}(\Delta_{2i})f_D\frac{\Delta t^2}{M_i}\bar{t}_i^T|R_i^P| \\
\Delta V_{i'}^{P+1} &= -\text{sgn}(\Delta_{2i})f_D\frac{\Delta t^2}{M_{i'}}\bar{t}_i^T|R_i^P|
\end{aligned}\right\} \tag{12.34}$$

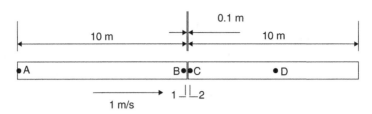

Figure 12.9 Numerical model.

Table 12.1 Parameters of model material.

	Elastic modulus (MPa)	Poisson's ratio	Density (kg/m³)
Left bar	10	0.25	1000
Right bar	10	0.25	1000

In short, according to the idea proposed by Liu and Wang (1995a, 1995b), the displacement is the summation of three parts. The contact pair is first searched, and the displacement without considering the dynamic contact between crack surfaces is calculated. The contact state between nodes is evaluated according to Equation (12.26). If the nodes are in contact, Equations (12.27–12.33) can be used to find the solution and the stress state can then be obtained.

12.4.2 Validation

In order to verify the accuracy of numerical simulation, a fundamental elastic collision problem is taken as an example in this section. The numerical model is shown in Figure 12.9 and the material parameters are listed in Table 12.1. The left bar strikes the right bar at a constant speed of 1 m/s. The right bar is static before the strike. Two boundary conditions of the right bar are considered, namely, free bar and right-end fixed cantilever bar. The model is assumed to be a plane strain problem. By comparing with the calculation results by ANSYS_LSDYNA, it is verified that the method can be applied to both contactable cracks and impact problems. In addition, the results are very close to ANSYS_LSDYNA results and of high accuracy. For simple notation, R.A. and L.A. are used to represent the average calculation results of key point A by RFPA and LSDYNA, respectively. R.A.m2 stand for the calculation results for key point A when $m = 2$. In the profile analysis, sections 1-1 and 2-2 are the vertical profiles on which the elements are in close proximity to the free surface (Fig. 12.9). R.m2.1-1 stands for the calculation results for section 1-1 by the RFPA program, and similarly for other notations.

12.4.3 Sample calculation 1: Impact response of homogeneous material

12.4.3.1 Impact with free bar

The left bar strikes the right bar at a constant speed. As the two bars have the same material properties, when the left bar stops, the displacement is about 0.2 m, as can

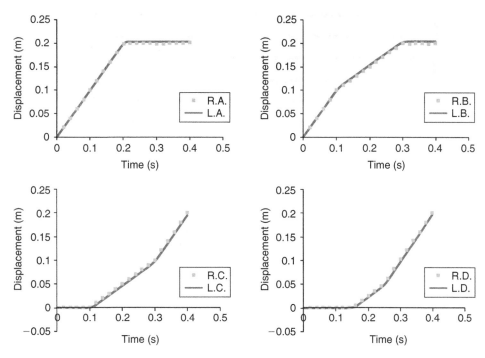

Figure 12.10 Displacement history of different observation points simulated by RFPA and ANSYS_LSDYNA for impact with free bar.

be seen from Figure 12.10. From the gradients of the displacement-time curves for key points B and C, the gradient at point C is 1 m/s before impact. Since elastic collision is considered in the sample calculation, the energy is fully transferred to the right bar. Hence, the right bar moves at the same constant speed eventually. The results obey momentum and kinetic energy theorem and prove that the simulation of collision by this method is accurate.

The distance between two bars is 0.1 m, and the left bar strikes the right bar at a constant speed of 1 m/s. Furthermore, the right bar is static before the strike. Hence, the first impact time is 0.1 s. As shown in Figure 12.11, the distance of wave propagation is still short at 0.106 s, and the maximal stress near the contact surfaces is higher than other area. At 0.293 s, as the time is close to the separation time, most energy has been transferred from the left bar to the right one, and the internal force becomes smaller gradually. When the time is 0.4 s, the two bars have already separated from each other. Because the two bars are homogeneous and have the same properties, the magnitude of contact force between them is the same. If a middle plane is defined as the plane passing the middle point between the two contact surfaces and parallel to the contact surfaces, as shown in Figure 12.11, it can be found that the maximal stress is symmetrical with respect to the middle plane during the impact.

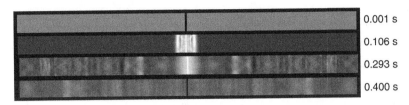

Figure 12.11 Distribution of maximal stress simulated by RFPA for impact with free bar.

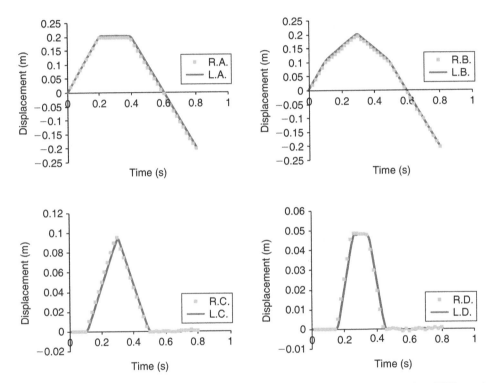

Figure 12.12 Displacement history of different observation points simulated by RFPA and ANSYS_LSDYNA for impact with cantilever bar.

12.4.3.2   Impact with cantilever bar

When the right bar is a right-end fixed cantilever bar, the displacement-time curve is shown in Figure 12.12, and distribution of the maximal stress is shown in Figure 12.13. Similarly, only elastic collision is considered.

As depicted in Figure 12.12, the impact course can be divided into two stages. At the first stage, when the left bar strikes the cantilever bar, compressive deformation is generated in the cantilever bar, and a part of the kinetic energy of the left bar is converted to the potential energy of the cantilever bar. Moreover, the potential energy further increases until the kinetic energy becomes zero. At the second stage, the potential energy is converted to the kinetic energy of the left bar, and the velocity

(a) 0.001 s

(b) 0.106 s

(c) 0.450 s

(d) 0.800 s

Figure 12.13 Distribution of maximal stress simulated by RFPA for impact with cantilever bar.

is opposite to the initial one. Therefore, the velocity of key points A and B, which corresponds to the curve gradient in Figure 12.12, can be used to judge the stage of energy transformation. For example, at the time of 0.106 s, the curve gradient of displacement history at key point B is positive. So it is in the first stage. Similarly, the state at 0.450 s can be judged to be in the second stage because the curve gradient of displacement history at key point B is negative.

Figure 12.13 shows the distribution of maximal stress for impact with a cantilever bar, and it can also depict the courses of contact and separation. Similar to impact with the free bar, the maximal stress is symmetrical about the middle plane before the wave propagates to the fixed end (Fig. 12.13b). However, when the wave is reflected by the fixed surface, the maximal stress is no longer symmetrical. It can be found that the stress near the right fixed surface is different from the relevant location in the left bar. A compressive stress wave is reflected as a tensile one on a free face, but as a compressive one on a fixed boundary. At the time of 0.8 s, the two bars have already separated from each other. As illustrated by the curve gradient in Figure 12.12, the left bar bounces back at the original speed, and the cantilever bar stays at its initial position at 0.8 s.

12.4.4   Sample calculation 2: Impact response of heterogeneous materials

Materials are not ideally homogeneous in nature. The stiffness and strength parameters at meso-scale are stochastic. For materials with distinct heterogeneity, such as rock, if they are simplified as homogeneous materials, the macroscopic nonlinearity cannot be reflected. Based on the previous studies, the RFPA program can reflect the stochastic features of a material by the Weibull distribution and the macroscopic nonlinearity can be reflected by the linearity at meso-scale. The method has been proved to be effective in solving the dynamic contact problems for impact between homogeneous materials in the earlier section. It is now imported into the RFPA program to analyze the impact problems for heterogeneous materials.

Material heterogeneity is considered for a two-bar impact case. Cases with $m = 2$, 5, 10 and 20 are calculated and the effect of different degrees of heterogeneity on impact response is analyzed. It can be seen from Figure 12.14, the left bar stops at a different location as compared with the homogeneous case. From the stress distribution plotted in Figures 12.17 and 12.21, the wavefront is not smooth due to the effect of material heterogeneity.

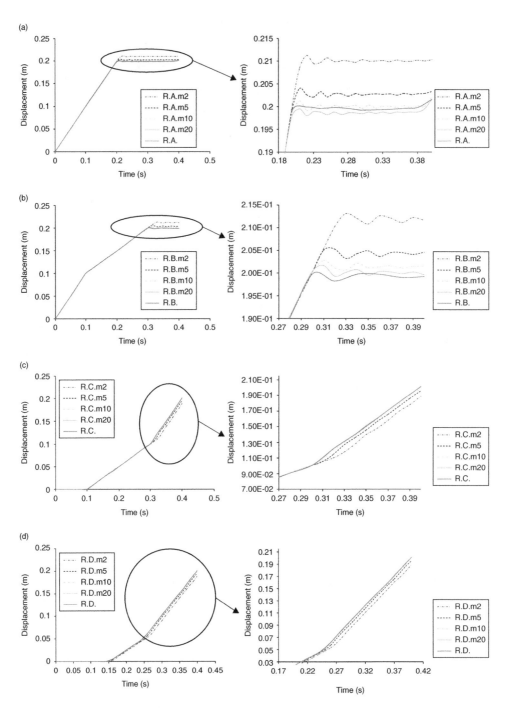

Figure 12.14 Displacement history of different observation points for different heterogeneity.

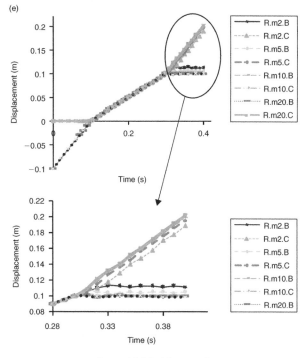

Figure 12.14 (Continued)

12.4.4.1   Impact with free bar

From the gradients of the displacement-time curves shown in Figures 12.14a–d, it can be seen that the stress wave propagation in heterogeneous materials is very different although the macroscopic speed is similar. The most prominent phenomenon is that the time from contact to separation is longer than that for homogeneous material, i.e. the more heterogeneous the material is (smaller m), the larger the final positive displacement of the left bar is, and the longer the time for the right bar to reach the peak speed. In order to better illustrate the entire process from contact to separation between the two bars, point C is taken as a reference and set as the origin. The displacement-time curves at points B and C for different heterogeneity coefficients are plotted in Figure 12.14e. For the cases with same heterogeneity coefficient, the displacements at points B and C are equal, which means that the two bars are in contact. As the initial distance between the two bars is 0.1 m for all models, the time for first contact is the same. However, the time for separation is different due to the effect of heterogeneity. The smaller the value of m is, the later the separation occurs.

In order to compare the impact response for models with different heterogeneity more comprehensively, two sections are selected for comparision of the major principal stresses. The major principal stresses at $t = 0.106$ s and 0.293 s in the contact phase are taken as examples, as the time corresponds to first contact and impending separation for homogeneous media. Taking sections 1-1 and 2-2 as examples, it can be seen

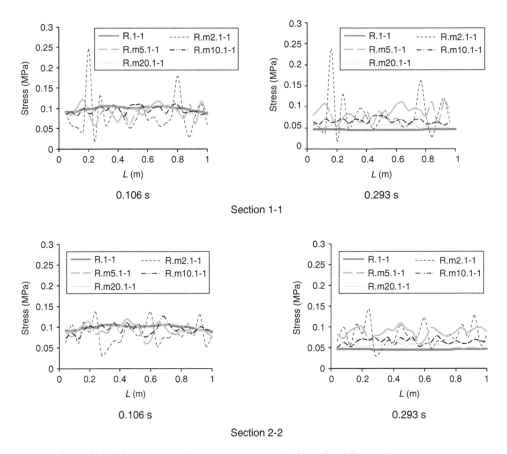

Figure 12.15 Major principal stresses in sectional planes for different heterogeneity.

from Figure 12.15 that the more heterogeneous the material is (smaller *m*), the more fluctuated the stress is, which is different from that for homogeneous material. From Figure 12.16, the stress curves at the two sections do not overlap with each other. Although the contact forces in the two bars are the same when they are in contact, the elastic moduli of the elements in sections 1-1 and 2-2 are different (considering the Weibull distribution). Hence the stress in the corresponding element varies. The smaller the value of *m* is (more heterogeneous), the more fluctuated the stress is. The larger the value of *m* is, the less fluctuated the stress is, which is closer to the homogeneous case.

Figure 12.17 shows the distribution of major principal stress for different heterogeneity. Taking heterogeneity into account, the property of the model is no longer similar in meso-scale, which leads to differences in wave propagation. Firstly, the action of impact is non-linear, as illustrated by the wavefront shown in Figure 12.17. Secondly, when the wave propagates in bars, the wave length and the peak stress are distinctive. In homogeneous media, the peak stress in every element is the same, but

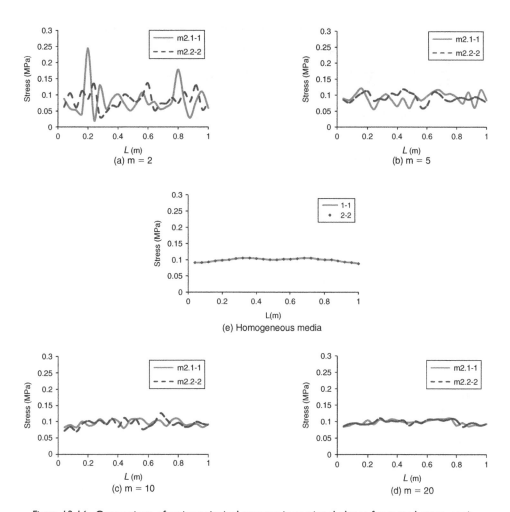

Figure 12.16 Comparison of major principal stresses in sectional planes for same heterogeneity.

with different arrival time. However, in heterogeneous media, the peak stress in some elements may be higher or lower than others. In general, the smaller m is, the more prominent the feature is. Furthermore, as shown in Figure 12.17 at 0.293 s, the smaller m is, the larger the area in compressive stress. That means the wavelength in medium with higher degree of heterogeneity is longer.

12.4.4.2 Impact with cantilever bar

Figures 12.18a–d show variations of displacement with time at key points for different material heterogeneity. Figure 12.18e shows the displacement-time curves at two contact points B and C. When the curves for points B and C coincide with each other, the two bars are in contact. It can be found from the displacements at key points

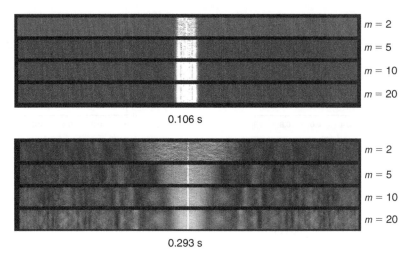

Figure 12.17 Distribution of major principal stress for different heterogeneity.

that smaller m leads to longer contact time. It also can be seen from the gradient of the displacement-time curves that the left bar bounces back at approximately the same speed and then separates from the right bar when it is back to the first contact position. However, the smaller m is, the later the rebound occurs.

Both impact and rebound are considered in section analysis, taking the representative moments $t = 0.106$ s and 0.45 s. When the material is homogeneous and elastic, the wavefront is smooth and the stresses in the two sections are consistent. Upon the first contact, the element stresses are about 0.1 MPa. When the material heterogeneity is considered, the element stresses in the two sections fluctuate and the stresses are not equal (Figs. 12.19 and 12.20). The smaller m is, the less smooth the wavefront is. The larger m is, the closer the results are to the case of homogeneous material. As the cantilever bar is fixed at the right end, the left bar eventually bounces back at the original speed. Although the macroscopic speed of the heterogeneous material is about the same, the location of stress wave propagation is different for the same moment, as shown in Figure 12.21. Figures 12.18e and f also indicate that smaller m leads to longer contact time.

Figure 12.20 compares the major principal stresses in two sections. The stresses corresponding to $t = 0.106$ s and 0.45 s are selected for illustration. Similar to impact with the free bar, the more heterogeneous the material (smaller m) is, the more fluctuated the stress is. The stress in the corresponding elements is equal (see Fig. 12.20e) when the homogeneous bars are in contact. It is also found that the stress curves for the two sections do not overlap for heterogeneous media. Although the contact forces are a pair of interaction forces in the two bars when they are in contact, the elastic moduli of the elements in sections 1-1 and 2-2 are different (considering the Weibull distribution). Hence the stresses in the corresponding elements are different. The smaller the value of m is (more heterogeneous), the more fluctuated the stress is. The larger the value of m is, the less fluctuated the stress is, which is closer to the homogeneous case.

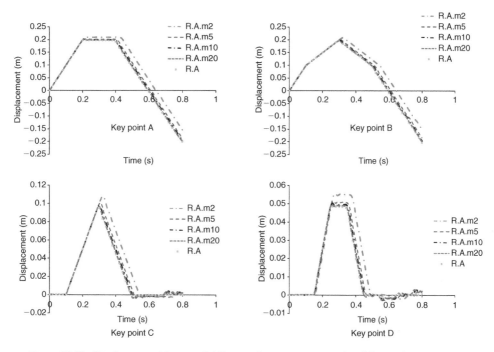

Figure 12.18 Displacement history of different observation points for different heterogeneity.

12.5 INFLUENCE OF STRESS WAVE AMPLITUDE ON ROCK FRACTURING PROCESS AND FAILURE PATTERN IN THE BRAZILIAN TENSILE TESTS

The dynamic mechanical response of rock materials is important to determine the required level of protection for important constructions such military structures. This is why some scientific researches in this field have focused on determination of the dynamic properties and development of accurate constitutive models and failure criteria at high strain rates (Zhao *et al.*, 1999).

One of the important features recognized for rocks or other brittle materials is their rate-dependence, i.e. their properties (ultimate strength, Young's modulus, fracture energy) are highly dependent on the loading rate. The general trend for rate effects is an increase in dynamic strength as the loading rate increases (Barpi, 2004). Much effort has been made to improve knowledge of the constitutive relationship for a wide range of strain rates by developing a more realistic material law.

Although such work has received considerable attention in the past decades (Cho, Ogata and Kaneko, 2003; Fourney, 1993; Donze, Bouchez and Magnier, 1997; Ma, Hao and Zhou, 1998; Sato *et al.*, 1999; Schmidt, Boade and Bass, 1979; Warpinski *et al.*, 1979; Zhao and Li, 2000; Zhu *et al.*, 2004), it still remains poorly understood because of the complexity associated with the dynamic response of rock that is

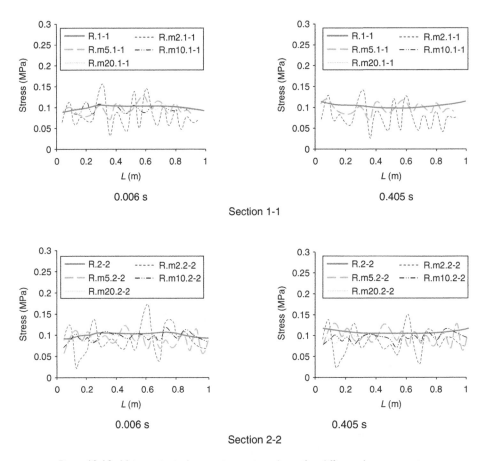

Figure 12.19 Major principal stress in section planes for different heterogeneity.

fully heterogeneous. Lots of facts demonstrate that this heterogeneity feature plays a significant role in fracture patterns of the rocks under dynamic loadings.

Experimental investigations provide good opportunities for examining the ultimate failure patterns of rock samples. However, it is found that sometimes analyzing the time history of failure is more important than simply examining the final outcome. Although most of the post-experimental observations can reveal fracture patterns and their relative proportions, they do not indicate the sequential order of the events or the conditions for fracture initiation, propagation and coalescence. Additionally, the post-experimental observations do not provide sufficient information about microfracture nor evolution of stress field.

The focus of this section is to numerically investigate the failure mechanisms and the fracture patterns of rocks using Brazilian samples under different stress wave amplitude. The study is conducted by using a RFPA-Dynamics code (Zhu *et al.*, 2004). This includes an assessment of how the pressure wave in heterogeneous rocks affects the dynamic fracture propagation and patterns. Different pressure stress waves in terms of peak values are employed to consider the waveform variation of the applied dynamic

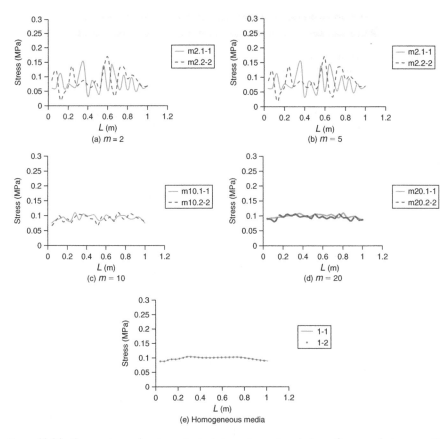

Figure 12.20 Comparison of major principal stress in sectional planes for same heterogeneity.

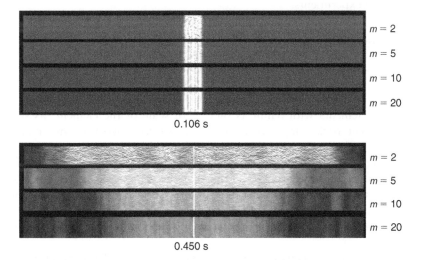

Figure 12.21 Distribution of major principal stress for different heterogeneity.

Figure 12.22 Numerical model of the sample and transmitter bars (400 × 15 elements with 5 mm length scale for the element).

loading. The code consists of a finite element model which allows both shear and tensile failure and fragmentation of the samples under dynamic loading.

One of the advantages of RFPA code is that it can take heterogeneity into account in the model. Smear method with small elements (SMSE for short) is proved to be suitable for simulating fracturing processes in rocks, since this heterogeneous material, in spite of the relatively small grain size, develops long fracture process zones due to bridging and interlocking of the heterogeneous local materials in the wake of the fracture. These process zones constitute the main energy dissipation mechanism for this kind of material (Tran, Kobayashi and White, 1999; Bower & Ortiz, 1991; Yu, Ruiz and Pandolfi, 2004). Indeed, RFPA code with SMSE applied to heterogeneous rock has successfully explained the dependency of some of their properties and loading methods on the heterogeneity, shape and size of the samples under static loading conditions (Tang *et al.*, 2000a, 2000b). Furthermore, SMSE is also feasible for handling the dynamic fracture appearing in rocks under dynamic loading.

Another feature of RFPA is the explicit treatment of fracture and fragmentation. It tracks individual fractures as they nucleate, propagate, branch and possibly link up to form fragments. It is incumbent upon the mesh to provide a rich enough set of possible fracture paths since the model allows fracture to occur within the elements only. However, almost no mesh dependency is expected as long as the element size is sufficiently small to resolve the fracture process zone of the rock. The simulations in this section give good prediction of the dependence of fracture initiation on the stress waveforms and come out with fracture patterns very similar to the actual ones observed in the experiments.

12.5.1 Numerical models

To evaluate the influence of heterogeneity on stress wave propagation, three samples with different homogeneity indices, $m = 2$, 5 and 10, subjected to a pressure wave input, were used for the simulations. The model with sample and bars is divided into 400 × 15 elements with 5 mm as the length scale of the element, as shown in Figure 12.22. The rock sample was positioned between two transmitter bars. The parameters and calculation conditions are listed in Table 12.2.

To investigate the influence of stress waveform in terms of peak value on fracture process and failure pattern of rocks, 2D Brazilian disc samples are used for the simulations, as shown in Figure 12.23. The radius of the disc is 80 mm. A section of the finite element layout for the disc is also illustrated in Figure 12.23. The model is divided into 25,600 square elements. The parameters and calculation conditions are listed in Table 12.3.

The modeling system consists of an incident boundary and a transmitter platen to transfer the incident compressive pulse that propagates toward the sample. The pulse

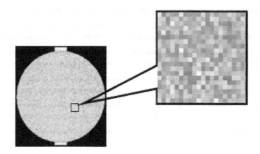

Figure 12.23 The numerical Brazilian disc sample and the loading conditions (the sample with 160 × 160 elements).

Table 12.2 Material properties for the models.

Setting	Sample	Bar
Young's modulus	60,000 MPa	210,000 MPa
Poisson's ratio	0.25	0.20
Homogeneity index	2, 5, 10, 100	200
Density	2.5e-6 kg/m^3	7.8e-6 kg/m^3

Table 12.3 Material properties for the models.

Setting	Sample		Bar
Young's modulus	37,500 MPa		210,000 MPa
Poisson's ratio	0.25		0.20
Compressive strength		205 MPa	
Tensile strength		18 MPa	
Homogeneity index	3		200
Density	2.5e-6 kg/m^3		7.8e-6 kg/m^3

is partially reflected at the border of the transmitter platen and partially transmitted through the sample.

12.5.2 Results and discussions

12.5.2.1 Influence of heterogeneity on stress wave propagation

Rock is a heterogeneous material, and the heterogeneity plays a significant role in the fracture process and the failure pattern. To demonstrate this influence, the stress wave in samples that did not fracture was simulated using the model shown in Figure 12.22. Figure 12.24 shows the stress wave propagation along the bars and the sample. The samples with $m = 2$, 5, 10 and 100, which correspond to relatively heterogeneous,

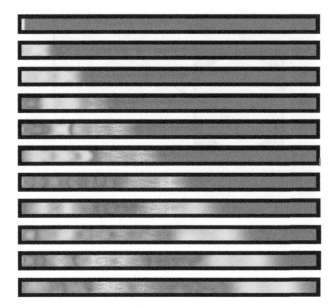

Figure 12.24 Numerically obtained stress wave propagation along the bars and the sample with $m = 2$.

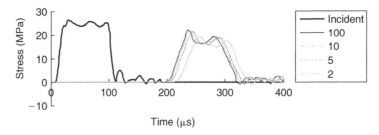

Figure 12.25 Numerically obtained stress wave–time curves in points A and B in the two bars.

medium homogeneous and relatively homogeneous rocks respectively, are used for the simulations.

The numerically obtained stress waves in the medium point in the two bars are shown in Figure 12.25. These compressive stress waves caused by the incident pressure reached point A in the first bar at 5 μs and reached point B in the second bar at 200 μs. Figure 12.25 shows that, after passing through the heterogeneous samples, the three stress waveforms differ largely. This implies that rock heterogeneity has a significant influence on dynamic stress wave propagation. After the stress wave travels through the sample, the peak value becomes lower for heterogeneous rock than for homogeneous rock. On the other hand, the pulse length becomes longer for heterogeneous rock than for homogeneous rock. This influence will surely cause the failure pattern to be different when a sample failure modeling is conducted.

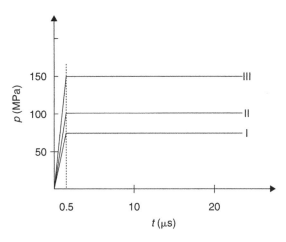

Figure 12.26 Applied pressure waveforms with three peak values of stress.

12.5.2.2   Influence of pressure stress wave amplitude on fracture process and failure pattern

The Brazilian samples shown in Figure 12.23 are used to investigate the influence of stress wave amplitude on fracture process and failure pattern. Figure 12.26 shows the applied pressure waveforms with three peak values of stress, with the rise time from t_0 up to the peak stress being constant. Figure 12.27 shows the stress history obtained in the transmitter platen during one of the simulations of a sample. The dashed line averages the oscillations in the plateau of the incident pulse.

Selected results of the numerically obtained fracture processes and failure patterns are presented in Figure 12.28. It should be carefully noted that in these plots, in order to aid visualization, displacements have been magnified by a factor of 5. Also shown in the figures are the level contours of stress magnitude, defined as the relative value of maximum shear stress. A fully fractured surface is shaded in black, whereas the zones that are intact or failed but not fully fractured remain in the color of level contours of stress magnitude.

The comparison between the three cases shown in Figure 12.28 displays graphically the good prediction of the dependency of dynamic fracture patterns on the transmitted stress wave amplitude.

Snapshots of the first row in Figure 12.28 show the failure pattern and shear stress evolution for the case of stress waveform with lower amplitude (peak value is 75 MPa). The fractures start nucleating and propagating at about $120\,\mu s$ when the peak value of the compressive pulse is reflected from the bottom and reaches the center of the sample. Distinct features such as the formation of double major fractures and branching are observed. Due to the heterogeneity, the fractures are well developed in the bottom area in the right side when the sample transmits the maximum load. The development of these fractures generates relief waves which temporarily halt the failure process. The stress waves subsequently travel from the left side towards the center of the disc, inducing a further fracture growth as well as some microfracturing in the center area

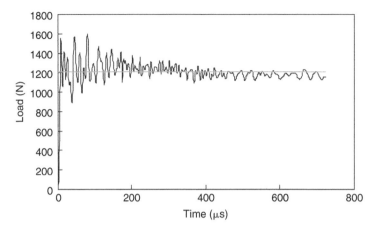

Figure 12.27 Stress history obtained in the transmitter platen during the simulation of sample with input peak stress 150 MPa.

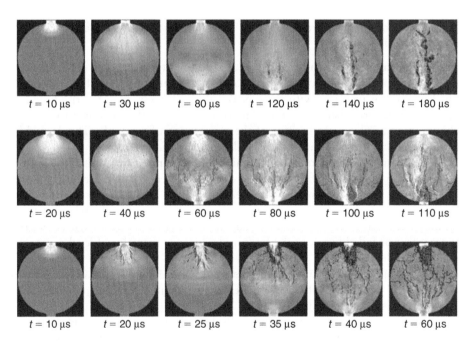

Figure 12.28 Fracture sequence and failure patterns for the three cases with different stress wave amplitude (with peak stress being 75 MPa, 100 MPa and 150 MPa, respectively).

part, which finally forms a main fracture. This through-fracture is clearly seen in the snapshots in the first row in Figure 12.28.

In contrast to the first case shown in the first row of Figure 12.28, the third case shown in Figure 12.28 reveals that, for the case of stress waveform with higher amplitude (peak value is 150 MPa), the fractures start nucleating and propagating at about $20\,\mu s$, which is much earlier than that in the first case, shown in Figure 12.28. A difference from the first case is that the fractures around the loading areas nucleate due to the incident wave, not the reflected wave as in the first case. In the two sides of the bottom support areas, however, the compressive stress waves are reflected as tensile waves when they reach the free surface, and are then superimposed upon the tail of the compressive waves. The superimposed stress waves developed an increasing amount of tension. When the tensile stresses are high enough, they induce opening fractures parallel to the surfaces. The process is also accompanied by the formation of a new wedge-shaped inclined fracture zone due to the intensity of the pressure stress waves near the loading area.

The predicted sequence of fracture patterns in terms of location closely followed what we expected based on our analysis. It showed that the fracture processes and the failure patterns are markedly affected by the stress wave amplitudes that are applied to the samples. The nucleation of fractures and failure patterns when $p_0 = 75$ MPa and 150 MPa shown in first line and third line in Figure 12.28 differed significantly. For lower p_0, the fractures start from the vicinity of the bottom compressive zone approximately after $t = 120\,\mu s$, whereas for higher p_0, the fractures occur almost immediately after the stress wave front enters into the sample, which is $100\,\mu s$ early than that for lower p_0.

It is seen from these snapshots that the failure patterns are enriched by the random distribution of the elemental stiffness and strength.

The snapshots of the third row in Figure 12.28 further reveal the development of profuse fracturing at the loading area which even leads to some fragmentation. Secondary fractures, parallel to the main diametric fracture, also appear as the load decreases, leading to the typical columnar failure of Brazilian tests (Yu, Ruiz and Pandolfi, 2004). For smaller stress wave amplitude, the model predicts the formation of the principal fracture that nucleates in the center of the sample and grows towards the bearing areas, as well as some secondary fracturing parallel to the main fracture and near the loading areas. For higher load stress wave, the simulation reports the formation of radial fractures starting in the circular border and growing to the center. The ability of RFPA-dynamics to account for such complex fracture patterns with relative ease is a remarkable feature of SMSE.

12.6 SUMMARY

The focus of this chapter is to introduce a FEM-based RFPA method which can be used to investigate elastic wave propagation and stress wave induced dynamic failure in brittle and heterogeneous rocks. First, the finite element method is briefly introduced. Two major problems in wave propagation are then discussed, namely, artificial boundary and impact dynamic contact. When it comes to transient wave propagation in infinite medium, the visco-elastic boundary is embedded into the RFPA2D, and the accuracy can meet the engineering requirements. The Lamb's problem is taken as an example to compare the modeling results of visco-elastic boundary and the extended solutions. In the model, the material heterogeneity is taken into consideration by assuming that the material properties of elements satisfy the Weibull distribution. Comparison of displacements at key points indicates that the visco-elastic boundary is also applicable for heterogeneous materials. In order to show the generality of comparison, a few cases with different Weibull distribution are simulated. It is shown that the smaller the m is, the more heterogeneous the material is, and the greater the peak displacement, and the more fluctuated the wavefront. It can be seen from the stress plot that the stress distribution is very different.

The dynamic contact method is imported into the collision contact problem and it is verified by comparison with ANSYS_DYNA modeling. The method is also valid for use for collision and contact between heterogeneous materials.

As an example, dynamic fracturing process analysis of rock material under the Brazilian tensile test condition is conducted. The analysis aims to investigate the influence of applied stress wave amplitude on the fracturing process and failure induced in the rock material. Influence of heterogeneity on stress wave propagation is discussed. Numerical simulations of three samples with different heterogeneity demonstrate that the heterogeneity of rock has significant influence on stress wave propagation. The dynamic failure process analyses are extended to investigate the influence of waveforms in terms of stress wave amplitude on failure modes. These simulations reveal that the failure modes are affected strongly by the stress wave amplitude. Stress wave of higher amplitude generates fracturing earlier. The effective dynamic behavior of samples under three different loading conditions is predicted as an outcome of the calculations. Numerical simulations show that at lower peak load of the stress wave, only a few larger microfractures form along the diametrical line under the applied load, which eventually causes the sample to split. The outcomes obtained from the simulations, which are very rich in information concerning fracture initiation and kinetics as well as the stress field observation, make this method an ideal candidate for the analysis of material failure under a fully dynamic framework. The simulations not only allow identification of model parameters but also explain the different failure mechanisms of rocks as a function of loading waveforms. It is seen that the model is suitable for simulating fracture processes and the failure patterns in rock materials.

REFERENCES

Barpi, F.: Impact behavior of concrete: a computational approach, *Engineering Fracture Mechanics* 71 (2004), pp.2197–2213.

Bower, A.F. and Ortiz, M.: The influence of grain size on the toughness of monolithic ceramics. *Trans ASME* 115 (1991), pp.228–236.

Chau, K.T., Zhu, W.C., Tang, C.A. and Wu, S.Z.: Numerical simulations of failure of brittle solids under dynamic impact using a new computer program DIFAR. *Key Engineering Materials* 261–263 (2004), pp.239–244.

Chen, J., Pan, J.Z. and Duan, Y.L.: A new method for solid contact analysis by transferring displacement DOF space. *Chinese Journal of Computational Mechanics* 23(1) (2006), pp. 101–106. (in Chinese)

Cho, S.H., Ogata, Y. and Kaneko, K.: Strain rate dependency of the dynamic tensile strength of rock. *Int J Rock Mech Min Sci* 40 (2003), pp.763–777.

Donze, F.V., Bouchez, J. and Magnier, S.A.: Modeling fractures in rock blasting. *Int J Rock Mech Min Sci* 34 (1997), pp.1153–1163.

Fang, Q. and Chen, Z.L.: Accuracy of explicit Newmark scheme for wave propagation problems. *Chinese Journal of Geotechnical Engineering* 15(1) (1993), pp.10–15. (in Chinese)

Fang, Q.: Studies on the accuracy of finite element analysis of implicit Newmark method for wave propagation problems. *Explosion and Shock Waves* 12(1) (1992), pp.45–53. (in Chinese)

Fourney, W.L.: Mechanisms of rock fragmentation by blasting[M]. In Hudson, J.A. (ed.). *Comprehensive Rock Engineering, Principles, Practice and Projects*. Oxford: Pergamon Press, 1993.

Fu, Y.F.: *Study on rock brittle failure process by numerical testing*[D]. Northeastern University, Shenyang, 2003. (in Chinese)

Lamb, H.: On the propagation of tremors over the surface of an elastic solid. *Philos Trans R Soc. London. Ser A* 203 (1904), pp.1–42.

Liao, Z.P. and Liu, J.B.: Finite element simulation of wave motion-basic problems and conceptual aspects. *Earthquake Engineering and Engineering Vibration* 9(4) (1989), pp.1–14. (in Chinese)

Liao, Z.P.: *Introduction to Wave Motion Theories for Engineering*[M]. Second edition. Beijing Science Press, 2002. (in Chinese)

Liao, Z.P. and Liu, J.B.: Elastic wave motion in discrete grids(I). *Earthquake Engineering and Engineering Vibration* 8(2) (1986), pp.1–16. (in Chinese)

Liao, Z.P. and Yang, G.: Finite element simulation of steady SH wave motion. *Acta Seismologica Sinica* 16(1) (1994), pp.96–105. (in Chinese)

Liu, J.B. and Wang, D.: A dynamic contact force model for contactable crack considering the effect of interface friction. *The latest developments in elastodynamics*. Beijing Science Press, 1995a. (in Chinese)

Liu, J.B. and Wang, D.: A dynamic contact force model for contactable crack considering the effect of interface friction. *The latest developments in elastodynamics*. Beijing Science Press, 1995b. (in Chinese)

Liu, J.B., Gu, Y. and Du, Y.X.: Consistent viscous-spring artificial boundaries and viscous-spring boundary elements. *Chinese Journal of Geotechnical Engineering* 28(9) (2006), pp. 1070–1074. (in Chinese)

Liu, J.B. and Liao, Z.P.: Elastic wave motion in discrete grids(II). *Earthquake Engineering and Engineering Vibration* 9(2) (1989), pp.1–11. (in Chinese)

Liu, J.B. and Liao, Z.P.: Elastic wave motion in discrete grids(III). *Earthquake Engineering and Engineering Vibration* 9(2) (1990), pp.1–10. (in Chinese)

Liu, J.B. and Lu, Y.D.: A direct method for analysis of dynamic soil-structure interaction based on interface idea. In: Ahang, C.H. and Wolf, J.P. (eds.). *Dynamic Soil-Structure Interaction*. Beijing: International Academia Publishers, 31(3) (1997), pp.258–273.

Liu, J.B. and Lu, Y.D.: A direct method for analysis of dynamic soil-structure interaction. *China Civil Engineering Journal* 31(3) (1998), pp.55–64.

Liu, J.B., Du, Y.X. and Yan Q.S.: Method and realization of seismic motion input of viscous-spring boundary in general FEM software. *Journal of Disaster Prevention and Mitigation Engineering* 27(supp) (2007), pp.37–42.

Ma, G.W., Hao, H. and Zhou, Y.X.: Modeling of wave propagation induced by underground explosion. *Computers & Geotechnics* 22(3–4) (1998), pp.283–303.

Ohte, S.: Finite Element Analysis of elastic contact problems. *The Japan Society of Mechanical Engineers* 16(95) (1973), pp.797–804.

Ou, H.A. and Gong, J.X.: An improved mixed finite element method of analyzing three-dimensional elastic contact problems. *Journal of Chongqing University* 1 (1988), pp.52–58. (in Chinese)

Sato, K., Hashida, T., Takahashi, H. and Takahashi, T.: Relationship between fractal dimension of multiple microcracks and fracture energy in rock. *Geotherm Sci Tech* 6 (1999), pp.1–23.

Schmidt, R.A., Boade, R.R. and Bass, R.C.: A new perspective on well shooting-the behavior of contained explosion and deflagrations[C] *Proceedings of the 54th Annual Conference SPE of AIME*, Las Vegas, Nevada (1979).

Tang, C.A. and Kaiser, P.K.: Numerical simulation of cumulative damage and seismic energy release in unstable failure of brittle rock-Part I. Fundamentals. *Int J Rock Mech Min Sci* 35(2) (1998), pp.113–121.

Tang, C.A., Liu, H., Lee, P.K.K., Tsui, Y. and Tham, L.G.: Numerical tests on micro-macro relationship of rock failure under uniaxial compression – Part II: constraint, slenderness and size effect. *Int J Rock Mech Min Sci* 37(4) (2000a), pp.571–583.

Tang, C.A., Liu, H., Lee, P.K.K., Tsui, Y. and Tham, L.G.: Numerical tests on micro-macro relationship of rock failure under uniaxial compression – Part I: effect of heterogeneity. *Int J Rock Mech Min Sci* 37(4) (2000b), pp.555–569.

Tang, C.A.: Numerical simulation of AE in rock failure. *Chinese Journal of Rock Mechanics and Engineering* 16(4) (1997a), pp.368–377. (in Chinese)

Tang, C.A.: Numerical simulation of progressive rock failure and associated seismicity. *Int J Rock Mech Min Sci* 34 (1997b), pp.249–262.

Tran, D.K., Kobayashi, A.S. and White, K.W.: Process zone of polycrystalline alumina. *Exp Mech* 39(1) (1999), pp.20–24.

Tworzydloa, W.W., Cecota, W., Odenb, J.T. and Yew C.H.: Computational micro- and macro-scopic models of contact and friction: formulation, approach and applications. *Wear* 220 (1998), pp.113–140.

Warpinski, N.R., Schmidt, R.A., Cooper, P.M., Walling, H.C. and Northrop, D.A.: High energy gas fracs: multiple fracturing in a well bore[C]. In Austin, T.X. (ed.). *Proceedings of the 20th US Symposium on Rock Mechanics*, 1979.

Wen, W.D. and Gao, D.P.: Progress and current situation of numerical analysis methods in contact problems. *Journal of Nanjing University of Aeronautics & Astronautics* 26(5) (1994), pp.664–675. (in Chinese)

Yu, R.C., Ruiz, G. and Pandolfi, A.: Numerical investigation on the dynamic behavior of advanced ceramics. *Engineering Fracture Mechanics* 71 (2004), pp.897–911.

Zhang B.Y., Chen H.Q. and Tu, J.: Improved FEM based on dynamic contact force method for analyzing the stability of arch dam abutment. *Journal of Water Resources Engineering* 10 (2004), pp.7–12.

Zhao, J. and Li, H.B.: Experimental determination of dynamic tensile properties of a granite. *Int J Rock Mech Min Sci* 37 (2000), pp.861–866.

Zhao, J., Zhou, Y.X., Hefny, A.M., Cai, J.G., Chen, S.G., Li, H.B., Liu, J.F., Jain, M., Foo, S.T. and Seah, C.C.: Rock dynamics research related to cavern development for ammunition storage. *Tunnelling and Underground Space Technology* 14 (1999), pp.513–526.

Zhu, W.C., Liu, J., Yang, T.H., Sheng, J.C. and Elsworth, D.: Effects of local rock heterogeneities on the hydromechanics of fractured rocks using a digital-image-based technique. *Int J Rock Mech Min Sci* 43 (2006), pp.1182–1199.

Zhu, W.C., Tang, C.A., Huang, Z.P. and Liu, J.S.: A numerical study of the effect of loading conditions on the dynamic failure of rock. *Int J of Rock Mech Min Sci* 41(3) (2004), pp.424–424.

Chapter 13

Discontinuum-based numerical modeling of rock dynamic fracturing and failure

Tohid Kazerani and Jian Zhao

13.1 INTRODUCTION

Benefitting from rapid advancements in computer technology, numerical methods have provided powerful tools in rock dynamics study. For example, numerical modeling has been used to simulate dynamic response of fractured rock masses (e.g. Chen *et al.*, 2000; Hildyard and Young, 2002), fracture propagation in rock and concrete under static and dynamic loading condition (e.g. Liang *et al.*, 2004; Zhu and Tang, 2006), wave propagation in jointed rock masses (e.g. Chen and Zhao, 1998; Lei *et al.*, 2006), and acoustic emission in rock (e.g. Hazzard and Young, 2000b). A large number of numerical methods have been applied to rock mechanics problems, such as the Finite Element Method (FEM), Finite Difference Method (FDM), and Discrete Element Method (DEM). These methods are classically categorized as continuum- and discontinuum-based (Jing, 2003). However, most of the attempts have been performed through adopting continuum-based models, which are not basically able to explicitly simulate fracture. To overcome this shortcoming, discontinuum-based models have been introduced. With regard to fracture and fragmentation purposes, the advantages of discontinuum- to continuum-based models can be summarized as follows.

– Discontinuum-based models are not engaged with the flow rule, potential function and complicated mathematical formulation needed by continuum-based ones in order to implement nonlinear analysis.
– They are capable of representing a crack as an explicit separation within material, whilst continuum-based ones have to simulate it indirectly through modifying material properties.
– While predictions by continuum-based models are restricted only to fracture initiation, the use of discontinuum-based methods make it possible to examine both the initiation and propagation of fracture over time by tracking consecutive separation of structural elements.

The dominant stream in discontinuum-based modeling is owned by the DEM. It has been widely used in underground works (e.g. Lemos, 1993; Souley *et al.*, 1997; Zhao *et al.*, 1999), laboratory test simulations and constitutive model development (e.g. Jing *et al.*, 1994; Min and Jing, 2003), rock dynamics (e.g. Cai and Zhao, 2000), wave propagation in jointed rock masses (e.g. Chen and Zhao, 1998; Zhao *et al.*, 2006), nuclear waste repository design and performance assessment (e.g. Jing *et al.*,

1995), rock fragmentation process (e.g. Gong and Zhao, 2007), and acoustic emission in rock (e.g. Hazzard and Young, 2000b).

This chapter reviews the key concepts of the DEM and DEM-coupled methods as well as the related studies to provide a picture of the current research state. Following this, a distinctive application of the coupled DEM/FDM is introduced and its different features as well as its predictions for rock compressive, tensile and fracture response are discussed. The present study aims to explore the micro-mechanisms underlying rock fracture and fragmentation through answering how rock micro-structure influences

- rock strength and failure in compression and tension,
- rock fracture behavior and dynamic fracture toughness,
- and fracture rate-dependent behavior observed in macroscopic scale.

13.2 DISCRETE ELEMENT METHOD

The key concept of the DEM is that the domain of interest is treated as an assemblage of rigid or deformable *blocks/particles/bodies* (Cundall, 1971). The DEM is capable of analyzing multiple interacting deformable continuous, discontinuous or fracturing bodies undergoing large displacements and rotations. Formulation and development of the DEM have progressed over a long period of time since the pioneering study of Cundall.

Contact detection and *contact interaction* are the most important aspects in the DEM, as the DEM is distinguished from other methods because of its ability to detect (create) new contacts during calculation. There are several contact detection algorithms aiming at efficiency of computation time and memory space. The details on this topic are provided by Munjiza (2004). Jing and Stephansson (2007) have extensively provided the fundamentals of the DEM and its application in rock mechanics. According to the solution algorithm used, the DEMs can be basically divided into two groups of explicit and implicit formulation.

13.2.1 Explicit DEM (distinct element method)

As the explicit formulation of the DEM, the distinct element method appeared in the early 1970s in a fundamental paper on progressive movement of rock mass as a 2D assemblage of rigid blocks (Cundall, 1971). It was further developed by Lemos *et al.* (1985), Cundall and Hart (1992), and Curran and Ofoegbu (1993). The most popular numerical representation of the explicit DEM has been implemented with the computer codes of PFC and UDEC (Itasca, 2009a, 2009b). Other developments were made behind or in parallel with these two. The distinct element method appears to be the main direction of the DEM implementations for rock mechanics problems, although the term discrete element method is more universally adopted.

One use of the explicit DEM is to represent grained materials as a dense packing of irregular-sized particles interacting at their boundaries. This method has been known as the Bonded Particle Method (BPM) (Potyondy and Cundall, 2004). Many researchers have employed the BPM with circular particles to capture different failure features of rock material and other grained media (Azevedo *et al.*, 2008; Cho *et al.*, 2007;

Hazzard *et al.*, 2000a; Jensen *et al.*, 1999; Potyondy, 2007; Schöpfer, 2009; Tan *et al.*, 2008, 2009; Yoon, 2007; Wanne and Young, 2008), and some others have made use of polygonal particles (Camborde *et al.*, 2000; Damjanac *et al.*, 2007; Kazerani and Zhao, 2010; Kazerani *et al.*, 2010).

13.2.2   Implicit DEM (discontinuous deformation analysis)

As the DEM implicit formulation, the Discontinuous Deformation Analysis (DDA) was proposed by Shi (1988). The DDA is somewhat similar to the FEM, but accounts for the interaction of independent blocks along discontinuities in fractured and jointed rock masses. The DDA is typically formulated as a work-energy method, and can be derived using the principle of minimum potential energy or Hamilton's principle.

The applications of the DDA are mainly in tunneling, caverns, earthquake effects, and fracturing and fragmentation processes of geological and structural materials (e.g. Hatzor *et al.*, 2004; Hsiung and Shi, 2001; Zhang *et al.*, 2007). The DDA developments include discretizing the blocks with finite elements (Shyu, 1993), handling the contacts as stiff joints and removing penetration criteria to improve the efficiency and to accelerate the convergence (Cheng, 1998), coupled stress-flow problems (Kim *et al.*, 1999), 3D block system analysis (Jiang and Yeung, 2004), higher order elements (Hsiung, 2001), and more comprehensive representation of the fractures (Zhang and Lu, 1998).

13.2.3   Coupled DEM with continuum-based methods

Continuum-based and discontinuum-based methods are non-ideal for modeling, respectively, the post-failure and pre-failure behavior of rock. A combination of both methods will enhance rock mechanics applications, including the prediction of formation and interaction of fragments. Coupled methods can benefit from the advantages of each method while avoiding the disadvantages. Creating fractured zones with a discontinuum-based method and intact zones with a continuum-based one forms a direct simple coupled methodology. Examples of this kind of method include hybrid DEM/BEM models (Lorig *et al.*, 1986), combinations of the DEM, DFN and BEM approaches (Wei and Hudson, 1988), and hybrid DEM/FEM models (Chen and Zhao, 1998; Munjiza *et al.*, 1999; Ariffin *et al.*, 2006; Morris *et al.*, 2006; Cai, 2008; Karami and Stead, 2008).

As a hybrid DDA/FEM, the Numerical Manifold Method (NMM) was developed by Shi (1991). This method employs two sets of systems, one is mathematical and defines domain approximations, and the other is physical and defines integration fields. The main advantage of NMM is getting rid of meshing, and combining discontinuum and continuum problems into a single framework. Hence, the NMM is reported as suitable for fracture simulation (Chiou *et al.*, 2002; Zhang *et al.*, 1997). The Finite Cover Method (FCM) has been proposed by Terada *et al.* (2003) to enhance the NMM by using Lagrange multipliers for dealing with heterogeneous materials. Terada and Kurumatani (2005) and Terada *et al.* (2007) have recently developed the method for solving 3D problems.

13.3 COHESIVE FRAGMENT MODEL

According to the heterogeneous and grained texture of rock material, the discontinuum-based methods have been extensively employed to reproduce rock structure as an aggregate of particles connecting together by structural bonds. Predominantly, particles are taken as rigid random-sized discs or spheres depending on the 2D or 3D state of the modeling. Therefore, contact between the particles inevitably occurs at points where two particles are touching each other. However, the material constituting rock can be more realistically described as an assemblage of deformable and sharp-cornered fragments where contacts are not necessarily punctual and frictional contact planes can exist. This idealization is more analogous to the rock texture and reproduces its fragmentation pattern more accurately.

Referred to as Cohesive Fragment Model (CFM), the present numerical study is based on the DEM coupled with the Cohesive Process Zone (CPZ) theory, introduced by Dugdale (1960). The model assumes rock material as a collection of deformable irregular-sized triangular fragments interacting at their cohesive frictional boundaries.

13.3.1 Universal Distinct Element Code (UDEC)

The Universal Distinct Element Code (UDEC) is adopted to implement the CFM due to its helpful capabilities and relative ease of development (Itasca, 2009b). As a DEM/FDM coupled code, UDEC permits 2D plane-strain and plane-stress analyses. The CFM takes advantage of the particle/contact logic of UDEC to handle the fragments and boundaries in between. Using the 2D Delaunay triangulation (Du, 1996), a pre-processor program has been separately developed to generate arbitrary-sized triangular particles. They are then discretized into the constant-strain elements (CST) to provide deformability for the fragments.

Figure 13.1 presents a representative CFM particle assemblage used for the Brazilian test simulation, along with the configuration of the model-constructing particles and contacts.

Dynamic behavior is numerically represented by a time stepping algorithm in which time step duration is limited by the assumption that velocities and accelerations are constant within the time step.

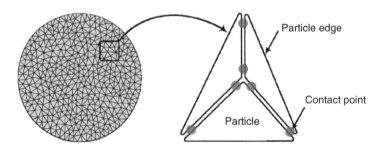

Figure 13.1 A representative CFM assemblage and configuration of particles and contacts.

The solution scheme is identical to that used by the explicit FDM for continuum analysis. The solving procedure in UDEC alternates between the application of a stress-displacement law at all the contacts and Newton's second law for all the particles. The contact stress-displacement law is used to find the contact stresses from the known and fixed displacements. Newton's second law gives the particles motion resulting from the known and fixed forces acting on them. The motion is calculated at the grid points of the triangular constant-strain elements within the elastic particle. Then, application of the material constitutive relations gives new stresses within the elements. Figure 13.2 schematically presents the calculation cycle in UDEC along with a brief review of the basic equations.

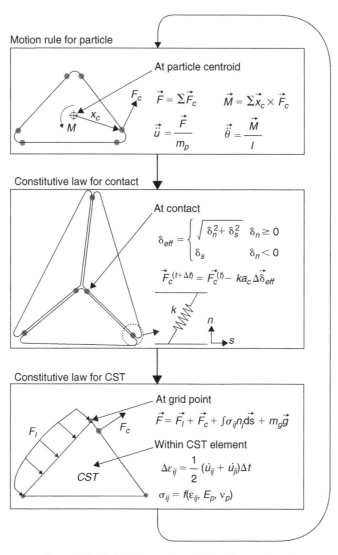

Motion rule for particle

At particle centroid

$$\vec{F} = \Sigma \vec{F}_c \qquad \vec{M} = \Sigma \vec{x}_c \times \vec{F}_c$$

$$\ddot{\vec{u}} = \frac{\vec{F}}{m_p} \qquad \ddot{\vec{\theta}} = \frac{\vec{M}}{I}$$

Constitutive law for contact

At contact

$$\delta_{eff} = \begin{cases} \sqrt{\delta_n^2 + \delta_s^2} & \delta_n \geq 0 \\ \delta_s & \delta_n < 0 \end{cases}$$

$$\vec{F}_c^{(t+\Delta t)} = \vec{F}_c^{(t)} - ka_c\Delta\vec{\delta}_{eff}$$

Constitutive law for CST

At grid point

$$\vec{F} = \vec{F}_l + \vec{F}_c + \int\sigma_{ij}n_j d\vec{s} + m_g\vec{g}$$

Within CST element

$$\Delta\varepsilon_{ij} = \frac{1}{2}(\dot{u}_{ij} + \dot{u}_{ji})\Delta t$$

$$\sigma_{ij} = f(\varepsilon_{ij}, E_p, \nu_p)$$

Figure 13.2 Calculation cycle in UDEC (Itasca, 2009b).

13.3.2 Orthotropic cohesive contact model

The model failure behavior is controlled by the contact constitutive law. Hence, the failure characteristics of rock, i.e. anisotropy, brittleness and rate-dependency, must be appropriately reflected in the contact model. Therefore, the model developed adopts an orthotropic behavior for fracture, and follows a decaying stiffness in the contact pre-failure stage in order to represent the material damaged in the fracture process zone. For this purpose, the stress σ applied on the contact surface is defined as

$$\sigma = \sigma(\delta_{eff}, k_t, k_s, t_c, c_c, \phi_c, D) \tag{13.1}$$

where δ_{eff} is the contact effective displacement, and k_t and k_s denote the contact initial stiffness coefficients in tension and shear, respectively. The parameters t_c, c_c, and ϕ_c characterize the strength of contact. They are respectively referred to as contact tensile strength, contact cohesion, and contact friction angle. D is the contact damage variable. In mixed-mode separation, i.e. concurrent existence of normal and shear displacements of contact, δ_{eff} is defined as

$$\delta_{eff} = \begin{cases} \sqrt{\delta_n^2 + \delta_s^2} & \delta_n \geq 0 \\ \delta_s & \delta_n < 0 \end{cases} \tag{13.2}$$

where δ_n and δ_s are the normal separation and shear sliding over the contact surface. δ_n is assumed to be positive where the contact undergoes opening (tension).

13.3.2.1 Tensile behavior of contact

Contact cohesive stress in tension is expressed as

$$\sigma = \begin{cases} k_t \delta_{eff} \exp(-\delta_{eff}/\delta_{ct}) & \delta_{eff} \leq \delta_{ct} \\ t_c(1 - D) & \delta_{eff} = \delta_{max} \\ k_{red} \delta_{eff} & \delta_{eff} < \delta_{max} \end{cases} \begin{array}{l} \\ \delta_{ct} < \delta_{eff} \leq \delta_{ut} \\ \\ \delta_{eff} > \delta_{ut} \end{array} \tag{13.3}$$

In the hardening stage ($\delta_{eff} \leq \delta_{ct}$), the governing equation is the exponential traction-separation law described by Xu and Needleman (1995). δ_{ct} is the critical tensile displacement of contact beyond which cohesive softening happens, and δ_{ut} is the ultimate tensile displacement of contact, at which contact entirely loses its cohesive strength. In this stage, the stress-displacement behavior of contact is elastic, i.e. the unloading and reloading paths are the same and no energy dissipation occurs within contact.

In the softening stage ($\delta_{ct} < \delta_{eff} \leq \delta_{ut}$), contact is permitted to release energy during unloading-reloading cycles. δ_{max} is then defined as the maximum effective displacement that contact has undergone (Fig. 13.3). δ_{max} is δ_{eff}, when contact is increasingly opened, and held fixed as it undergoes unloading or reloading, unless δ_{eff} again reaches δ_{max}. Substituting $\sigma = t_c$ and $\delta_{eff} = \delta_{ct}$, and solving for δ_{ct}, it is obtained as follows where $e = \exp(1)$ is the base of the natural logarithm:

$$\delta_{ct} = e\frac{t_c}{k_t} \tag{13.4}$$

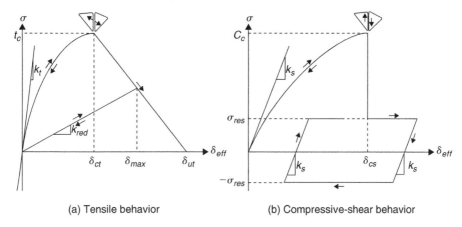

(a) Tensile behavior   (b) Compressive-shear behavior

Figure 13.3 Stress-displacement behavior of cohesive contact model.

The damage variable is defined as follows:

$$D = \frac{\delta_{max} - \delta_{ct}}{\delta_{ut} - \delta_{ct}} \tag{13.5}$$

As contact undergoes softening, D irreversibly increases from 0 to 1 or remains constant, even if multiple unloading-reloading cycles happen.

In unloading-reloading cycles $(\delta_{eff} < \delta_{max})$, contact follows a linear stress-displacement path. k_{red} is then defined as the secant stiffness at the point where the effective displacement equals δ_{max} (see Fig. 13.3a).

13.3.2.2 Compressive-shear behavior of contact

When contact is sheared under compression, the stress-displacement law is described as

$$\sigma = \begin{cases} k_s \delta_{eff} \exp(-\delta_{eff}/\delta_{cs}) & \delta_{eff} \leq \delta_{cs} \\ \sigma_{res} = -k_t \delta_n \tan(\phi_c) & \delta_{eff} > \delta_{cs} \end{cases} \tag{13.6}$$

Similarly, the critical shear displacement of contact is calculated as follows:

$$\delta_{cs} = e \frac{c_c}{k_s} \tag{13.7}$$

The unloading-reloading path of contact is linear as demonstrated in Figure 13.3b, where the contact stress increment (or decrement) is calculated through

$$\Delta\sigma = \begin{cases} k_s \Delta\delta_{eff} & \sigma < \sigma_{res} \\ 0 & \sigma = \sigma_{res} \end{cases} \tag{13.8}$$

In each solution iteration, $\Delta\sigma$ is calculated and added to the current value of the contact stress to update it. Finally, the normal and shear components of the contact force are obtained as follows, where a_c is the contact surface area, which is defined based on the contact length:

$$F_n = \begin{cases} -\sigma \dfrac{\delta_n}{\delta_{eff}} a_c & \delta_n \geq 0 \\ -k_t \delta_n a_c & \delta_n < 0 \end{cases} \tag{13.9}$$

$$F_s = -\sigma \frac{\delta_s}{\delta_{eff}} a_c \tag{13.10}$$

13.3.2.3 Contact fracture energy

The area under the curve in Figure 13.3a represents the energy needed to fully open the unit area of contact surface. Since contact is the numerical representation of fracture, the area under the curve should be equal to the Griffith's fracture energy, G_f:

$$G_f = \int_0^{\delta_{ut}} \sigma \, d\delta_{eff} = t_c \delta_{ct}(e-2) + t_c \frac{\delta_{ut} - \delta_{ct}}{2} \tag{13.11}$$

13.4 SIMULATION OF COMPRESSIVE AND TENSILE RESPONSE OF ROCK MATERIALS

The parameters involved in modeling are classified under the term *micro-parameter*. Table 13.1 lists them along with the analogous material properties.

Since any CFM simulation is controlled by the micro-parameters, they must be properly set such that the model reproduces a response similar to that of the physical material. To reach this purpose, the relation between the micro-parameters and the model behavior should be investigated. This purpose is fulfilled by establishing analytical and statistical relations, which provide physical interpretation for each micro-parameter in terms of the model macroscopic response.

13.4.1 Contact initial stiffness coefficients

As shown by Kazerani (2011), the contact initial stiffness coefficients can be expressed in terms of material mechanical properties. In plane-stress analysis, they are formulated as

$$k_t = \beta \frac{3E^2 \sigma_t}{K_{IC}^2} \quad \text{and} \quad k_s = \beta \frac{3E^2 \sigma_t}{2(1+v)K_{IIC}^2} \tag{13.12}$$

where β is a constant suggested as 0.25. In plane-strain analysis, those are defined as

$$k_t = \beta \frac{3E^2 \sigma_t}{(1-v^2)K_{IC}^2} \quad \text{and} \quad k_s = \beta \frac{3E^2 \sigma_t}{2(1-v^2)(1+v)K_{IIC}^2} \tag{13.13}$$

Table 13.1 Material properties and CFM micro-parameters.

Material property	Model micro-parameter
Young's modulus (E)	Particle Young's modulus (E_p)
Poisson's ratio (ν)	Particle Poisson's ratio (ν_p)
Fracture toughness in Mode-I (K_{IC})	Contact initial tensile stiffness coefficient (k_t)
Fracture toughness in Mode-II (K_{IIC})	Contact initial shear stiffness coefficient (k_s)
Brazilian strength (σ_t)	Contact tensile strength (t_c)
Internal cohesion (C)	Contact cohesion (c_c)
Internal friction angle (ϕ)	Contact friction angle (ϕ_c)

13.4.2 Particle elastic properties

The model global stiffness of a rock material is governed by the particle stiffness of the contacts together. Kazerani (2011) showed for a variety of brittle materials that the contact initial stiffness coefficients, obtained from Equations (13.12) and (13.13), are one to three orders greater than particle stiffness. This means contacts do not have any considerable effect on the model global elasticity. Therefore, the Young's modulus and the Poisson's ratio of the particles are assumed equal to those of the material:

$$E_p = E \quad \text{and} \quad \nu_p = \nu \tag{13.14}$$

13.4.3 Calibration process

The rest of the micro-parameters, i.e. t_c, c_c, and ϕ_c, are calculated by means of a calibration process in which the model responses are directly compared to the observed responses of the physical material (note that given G_f and k_t, δ_{ut} is related to t_c and calculated through Equation (13.11)). The calibrated micro-parameters should be unique and result in the best quantitative and qualitative agreement between the model response and that of tested rock in terms of the Brazilian tensile strength, uniaxial compressive strength, internal cohesion and internal friction angle. Note that these four parameters are dependent on each other. In other words, if having three of them for a typical material, the fourth is predictable by the Mohr-Coulomb equations. Therefore, the tensile strength, the internal cohesion, and the internal friction angle are considered as the parameters characterizing material mechanical response.

Since the Augig granite has been extensively tested by the Laboratory for Rock Mechanics (LMR) at EPFL, it is selected as the representative rock to perform calibration. As a coarse aggregate rock, it is composed of minerals ranging from 2 to 6 mm (4 mm in average). Its mechanical properties are the Young's modulus, $E = 25.8\,\text{GPa}$, the Poisson's ratio, $\nu = 0.23$, the Brazilian tensile strength, BTS $= 8.8\,\text{MPa}$, the uniaxial compressive strength, UCS $= 122\,\text{MPa}$, the internal cohesion, $C = 21\,\text{MPa}$, the internal friction angle, $\phi = 53°$, the fracture toughness in Mode-I, $K_{IC} = 1.5\,\text{MPa}\sqrt{\text{m}}$, and in Mode-II, $K_{IIC} = 3.0\,\text{MPa}\sqrt{\text{m}}$.

According to the specimen geometry and condition in laboratory, a plane-strain (axisymmetric) and a plane-stress analysis are respectively adopted for the compressive and Brazilian models. The compressive cylindrical sample is $80 \times 160\,\text{mm}$, and the

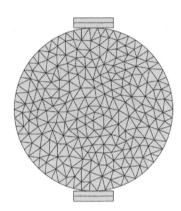

Figure 13.4 Model geometry for simulation of compressive and Brazilian tensile tests.

Brazilian specimen is an 80×80 mm disk (Fig. 13.4). The friction angle between the loading bars and the samples is assumed to be $5°$.

The particle assemblage is generated arbitrarily to capture the material heterogeneity and diverse fracture patterns. Both samples consist of irregular triangular particles with an average edge size of $d_p = 4.0$ mm, corresponding to 1122 and 452 particles for the compressive and tensile samples. d_p has been chosen according to the granite grain size.

Using Equation (13.14), the Young's modulus and Poisson's ratio of the particle are held fixed at $E_p = 25.8$ GPa and $v_p = 0.23$. Considering Equations (13.12) and (13.13), the tensile and shear initial stiffness coefficients of contact are obtained as $k_t = 1.95 \times 10^6$ MPa/mm and $k_s = 1.98 \times 10^5$ MPa/mm for the Brazilian sample and $k_t = 2.06 \times 10^6$ MPa/mm and $k_s = 2.10 \times 10^5$ MPa/mm for the compressive one.

As explained earlier, UDEC basically works with a dynamic algorithm. The loading rate is set to 10 mm/sec, and a high numerical damping, i.e., 0.85% of the critical damping, is applied to secure the quasi-static equilibrium.

13.4.3.1 Parametric study

A parametric study is carried out to determine how the model response is influenced by the micro-parameters. As a starting point, contact cohesion and friction angle are

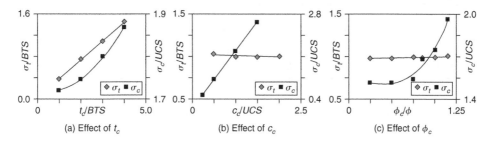

(a) Effect of t_c (b) Effect of c_c (c) Effect of ϕ_c

Figure 13.5 Tensile and compressive strength of the model versus contact tensile strength, cohesion and friction angle.

assumed as the rock UCS and internal frictional angle, respectively, i.e. $c_c = 122$ MPa, and $\phi_c = 53°$.

Based on the behavior of a CFM system, one would recognize that the model global strength is dependent on t_c. Figure 13.5a confirms that and indicates a linear relation between t_c and the tensile strength predicted by the model, σ_t. The relationship for the uniaxial compressive strength of the model, σ_c, is nonlinear.

By establishing a linear regression fit to the data, t_c is predicted to be 24.15 MPa to fit the tensile strength of the rock. Repetition of the simulation with this prediction gives $\sigma_t = 8.73$ MPa, which is satisfactorily close to the Augig granite tensile strength. Having $t_c = 24.15$ MPa, the sensitivity of the model to c_c and ϕ_c are examined as presented in Figures 13.5b and 13.5c, where the values of c_c and ϕ_c are normalized to their reference values, i.e. the uniaxial compressive strength and the internal friction angle of the rock, respectively.

The results show that the model Brazilian strength is independent of both c_c and ϕ_c, while the predicted uniaxial compressive strength is highly influenced by them. Figure 13.5c illustrates that the model uniaxial compressive strength will not change with ϕ_c any more, when ϕ_c goes below a certain threshold, i.e. about 0.75ϕ. All these results yield the fact that the model tensile strength depends only on the contact tensile strength. Therefore, the predicted $t_c = 24.15$ MPa is the target value of the contact tensile strength. However, the relationship between c_c and ϕ_c with the model response is not explicit yet and needs more investigation.

13.4.3.2 Response surface method

As a statistical discipline, the Response Surface Method (RSM) provides quantitative relations between a simulation response and its input factors (NIST/SEMATECH, 2003). It begins with the definition of the responses and the selection of the input variables. In this study, the unknown micro-parameters, i.e. c_c and ϕ_c, are chosen as the factors; and the assemblage macroscopic responses, in terms of the internal cohesion C, and internal friction angle ϕ, are considered as the responses.

The RSM provides a series of suggestions for the micro-parameters, using which the uniaxial and triaxial compressive tests are simulated. Each run results in a set of C and ϕ as the model response. Ultimately, the relation between the individual responses

Table 13.2 Experimental properties of Augig granite versus CFM predictions.

Property	E (GPa)	v	σ_t (MPa)	σ_c (MPa)	C (MPa)	ϕ (°)
Experimental value	25.8	0.23	8.8	122.0	21.0	53.0
Numerical prediction	25.2	0.24	8.7	125.4	20.9	53.5

and the micro-parameters is evaluated using the Fischer test in the quadratic form of (Park and Park, 2010)

$$\begin{cases} C = \alpha_0 + \alpha_1 c_c + \alpha_2 \phi_c + \alpha_3 c_c \phi_c + \alpha_4 c_c^2 + \alpha_5 \phi_c^2 \\ \phi = \beta_0 + \beta_1 c_c + \beta_2 \phi_c + \beta_3 c_c \phi_c + \beta_4 c_c^2 + \beta_5 \phi_c^2 \end{cases} \tag{13.15}$$

where α_0 to α_5 and β_0 to β_5 are the multipliers provided by the RSM.

Applying this method to the Augig granite samples, and solving the obtained equations for $C = 21$ MPa and $\phi = 53°$ of Augig granite, $c_c = 74.57$ MPa, and $\phi_c = 48.76°$ are finally obtained as the target micro-parameters. The details of this process are published by Kazerani (2011).

13.4.4   Solution verification

The CFM predictions using the target macro-parameters are listed in Table 13.2, which show fair agreement with the experimental measurements.

Comparisons between the curves of axial stress versus axial and lateral strain for the laboratory test and the simulation are presented in Figure 13.6. Note that some special aspects of rock behavior such as closure of initial flaws and pores are not captured by the CFM. This causes stress-strain curves in the simulation to be slightly different from those of the laboratory tests, particularly where the initial nonlinearity is not reflected in the modeling.

As shown by Paterson (1987), rocks exhibit higher ductility under triaxial circumstances than uniaxial. Figures 13.6c and 13.6d reveal that this phenomenon fairly captured by the CFM. The simulation gives further yielding and plastic deformation with confinement, whilst an abrupt softening at post-peak region is observed in the laboratory results.

Figure 13.7 qualitatively compares the CFM predictions with the laboratory observations in terms of failure mode. Good similarities are met, where the predicted compressive failure shows the typical cleavage happening in the laboratory test. For the Brazilian tests, the failure features, in terms of the major fault induced into the sample and the wedge-shaped zone created at the contact points with the platens, are also fairly well captured.

A series of simulations composed of particles with an average edge size ranging from 2 to 7.5 mm are designed. The results, presented in Figure 13.8, indicate that although the compressive strength has nearly no change, the tensile one decreases with the particle size decrease.

These results yield to the conclusion that the CFM's particles do not need to be extremely small. They must in fact be sufficiently small to allow the model to exhibit the actual failure patterns and processes of the rock, particularly in terms of the frequency

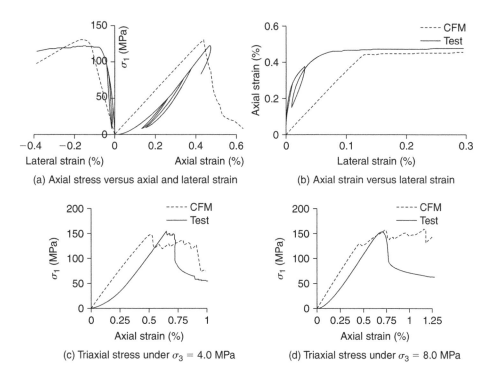

Figure 13.6 Comparison of predicted stress-strain curves with those obtained in laboratory tests.

of the dominant cracks controlling the failure procedure. For grained media, particle size is chosen mainly by the material texture and its average grain size.

13.5 SIMULATION OF DYNAMIC FRACTURE RESPONSE OF ROCK MATERIALS

The most critical parameter characterizing fast crack propagation is dynamic fracture toughness. It refers to the resistance of a material against fracture under high-rate loading. Fracture toughness tests on rock usually resort to compression-induced tension in order to avoid pre-mature failure due to gripping in purely tensile testing.

A disconcerting point regarding fracture toughness is that different test methods result in different measurements. The reason is often attributed to the undesired influences of a specimen's geometry, boundary condition and loading nature on the test results. How to obtain a unique dynamic fracture toughness as a reliable material property is still subject to open discussions.

Since numerical methods have provided a powerful tool to study dynamic fracture, they can potentially aid the experiments in terms of measurement verification or performing appropriate corrections (e.g. Maigre and Rittel, 1995). It will be shown how the CFM is able to evaluate the validity of the dynamic fracture toughness measured in the laboratory. As a representative case, the Semi-Circular Bend (SCB) test introduced by Chen *et al.* (2009) is examined.

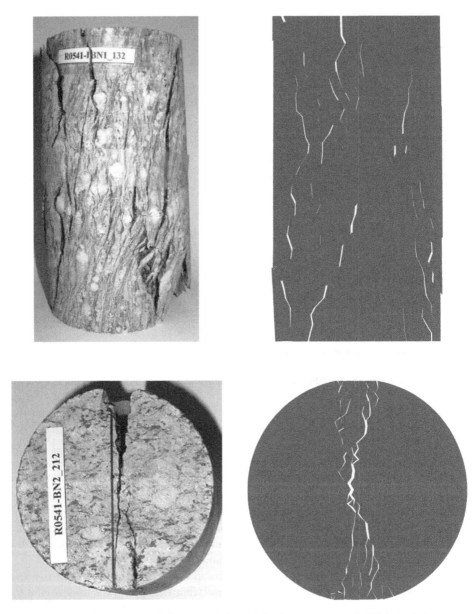

Figure 13.7 Comparison of laboratory failure of the Augig granite with CFM predictions.

13.5.1 Semi-Circular Bend (SCB) dynamic fracture toughness test

To make the SCB specimens, rock cores (40 mm nominal diameter) were drilled from the Laurentian granite blocks, and sliced into discs with an average thickness of 16 mm. The mechanical properties of the specimens used are the Young's

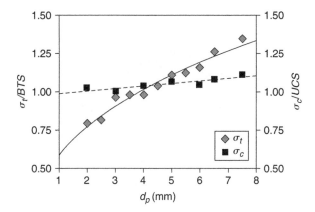

Figure 13.8 Variation of the compressive and the Brazilian strength versus particle size.

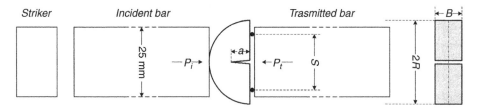

Figure 13.9 Schematics of the SHPB test setup and the SCB sample (Chen *et al.*, 2009).

modulus, $E = 92.0\,\text{GPa}$, the Poisson's ratio, $\nu = 0.21$, the Mode-I fracture toughness, $K_{IC} = 1.52\,\text{MPa}\sqrt{\text{m}}$, the Brazilian tensile strength, $\text{BTS} = 13.2\,\text{MPa}$, and the density, $\rho = 2630\,\text{kg/m}^3$. The dominant constituents of the Laurentian granite are feldspar (60%) and quartz (33%). The mineral grain size of the granite varies from 0.2 to 2 mm with average grain sizes of 0.5 and 0.4 mm for quartz and feldspar, respectively.

The SCB specimen is tested by a split Hopkinson pressure bar (SHPB) whose schematic configuration is presented in Figure 13.9, where R, B, and a respectively denote the radius, thickness, and depth of the initial notch of the sample. Two steel pins, spanning S, are placed between the transmitted bar and the specimen to minimize the disturbances that the specimen surface friction may make. More explanations about the test and the experimental setup are published by Dai *et al.* (2008, 2009).

The stress intensity factor for the Mode-I fracture in the SCB specimen is obtained as

$$K_{Id}(t) = \psi \frac{SP(t)}{BR^{1.5}} \tag{13.16}$$

where $P(t)$ is the time-varying loading force. $\psi = 0.96$ is a dimensionless factor, which depends on the specimen geometry. Since the specimen is in dynamic equilibrium, $P(t) = P_i = P_t$. The dynamic fracture toughness, K_{ICd}, corresponds to the maximum loading force, $P_{\max}$. The test results suggested that $K_{Id}(t)$ evolves with a nearly linear

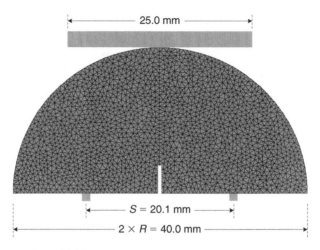

Figure 13.10 Model geometry and boundary condition.

trend, as its slope can be used to represent the average loading rate as follows, where t_d is when $K_{Id} = K_{ICd}$.

$$\dot{\kappa} = \frac{K_{ICd}}{t_d} \tag{13.17}$$

13.5.2 Simulation of the SCB dynamic fracture toughness test

Figure 13.10 presents the geometry and boundary condition of the model, which are the same as in the test. The model contains a 4 mm long slit along the centerline. Particle size is taken as the rock average mineral size, i.e. $d_p = 0.5$ mm. As the area in contact between the specimen and the support pins is about 1 mm^2, the pins are estimated by two fixed steel squares with 1 mm edge length.

The SHPB incident bar is simulated by the upper steel plate. It moves down to model the bar dynamic load, where its time-dependent velocity is

$$v(t) = \begin{cases} v_d \dfrac{t}{t_0} & t \leq t_0 \\ v_d & t > t_0 \end{cases} \tag{13.18}$$

v_d is the applied dynamic velocity and t_0 is the *arise time* to reach the applied velocity. It is assumed to be 20 μs for all the simulations. Equation (13.18) suggests that the applied velocity gradually increases to v_d. This is to help the specimen reach stress equilibrium. For this purpose, t_0 should be at least five times longer than the time needed for wave transmission through the specimen. The time step is taken at a small enough size at 5×10^{-10} sec, which secures the analysis stability (Kazerani, 2011).

Given Equation (13.12) for plane-stress, the tensile and shear stiffness coefficients of contact are calculated as $k_t = 3.63 \times 10^7$ MPa/mm and $k_s = 1.50 \times 10^7$ MPa/mm.

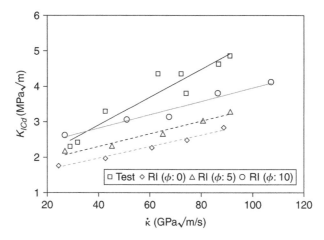

Figure 13.11 Variation of dynamic fracture toughness versus loading rate and specimen boundary friction.

To obtain the other micro-parameters, the calibration was repeated using the Laurentian granite properties, which led to $t_c = 66$ MPa, $c_c = 135$ MPa, $\phi_c = 46°$, and $\delta_{ut} = 7.54$ μm.

13.5.2.1   Calculation results

Three groups of simulation are designed, which are labeled as RI (ϕ: 0), RI (ϕ: 5), and RI (ϕ: 10). The values in the parentheses indicate the friction angle assumed for the specimen boundary surface, i.e. the interface of the loading plate and the support pins with the specimen. Each group includes five runs with different applied dynamic velocities as $v_d = 200, 400, 600, 800$ and 1000 mm/s. In all the simulations, no numerical damping is applied. This is to restrict the model to release energy only through contact failure but not particle viscosity.

As presented in Figure 13.11, the difference between the fracture toughness predicted by the model and that measured by the test is apparent. However, the rate-sensitivity observed in the experiment is partly captured, where the predicted K_{ICd} increases with the increase of loading rate. Nevertheless, the fitted lines to the CFM data are all less steep than the one fitted to the test results.

Figure 13.11 also suggests that the boundary surface friction greatly increases the CFM results. This signifies that the friction between the SCB specimen and the loading bars can potentially alter the measurements. The reason can be explored in the specimen motion mode. When the incident wave strikes the specimen, it splits in half, and each part laterally slides over the pins. Therefore, the slip friction between the pins and the specimen is mobilized, which disturbs the fracture opening process, and consequently increases the sample resistance against fracturing. As seen, the fitted lines to the data of the different groups have the same slope. This implies that the specimen surface friction does not control the model sensitivity to the applied loading rate.

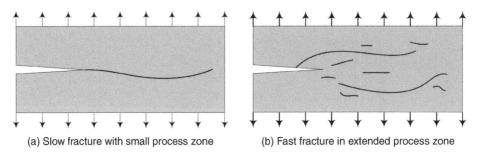

(a) Slow fracture with small process zone   (b) Fast fracture in extended process zone

Figure 13.12 Schematic formation of micro-cracks at fracture process zone.

13.5.2.2 Discussion

The presented results reveal that the rate-independent cohesive law with constant G_f cannot satisfactorily reproduce the experimental measurements. This means that structural inertia alone cannot explain the velocity-toughening effect.

Review of experimental observations on dynamic fracture (e.g. Shioya *et al.*, 1995; Shioya and Zhou, 1995) can help explore the reason. As schematically illustrated in Figure 13.12, the experimental observations on brittle material fracture suggest that the crack propagates straight forward at low crack velocities. Micro-cracking is not then significant, and thus the fracture surface is smooth. On the contrary, many micro-cracks are developed within the process zone when the fracture propagates at high speed. Consequently, dynamic fracture process happens in an expanded damage zone that causes a larger amount of energy to be dissipated. Hence, the velocity-toughening phenomenon can be attributed to these microscopic deformations and damage mechanisms happening at the fracture process zone.

One may argue that the correct macroscopic behavior of crack propagation would be captured by incorporating the details of the fracture process zone into the model, by adopting an extremely fine mesh e.g. 0.01 mm. This solution is presently impossible due to several reasons. First, crack propagation will not then be limited to the process zone and instead spreads over the whole specimen. Second, such a CFM simulation will contain millions of degrees of freedom, and is most likely too huge to be implemented by current computer facilities. Third, deformation mechanisms in the process zone, e.g. large deformation, nonlinear hardening, visco-plasticity, and thermal softening are so complex that their full numerical simulation seems out of reach. An alternative solution is to develop a rate-dependent model to implicitly introduce these effects into the calculation.

13.5.3 Rate-dependent cohesive model

In the adopted rate-dependent model, the contact opening speed, $\dot{\delta}_{eff} = \partial \delta_{eff}/\partial t$, is assumed as the factor controlling the energy release process within the crack-tip zone. As shown by Kazerani (2011), for PMMA plates undergoing fast fracture, the change in the fracture energy of contact, which represents the material fracture energy, G_f, can be then expressed through

$$G'_f = G_f \left[1 + \frac{\dot{\delta}_{eff}}{r_G - (1 - \alpha)\dot{\delta}_{eff}} \right] \tag{13.19}$$

where G'_f is the rate-dependent evaluation of G_f, r_G is a constant named reference opening speed of contact, and α is the ratio of the fracture terminal velocity to the Rayleigh surface wave speed. For PMMA, α is measured as 0.75 (Zhou et al., 2005).

Assuming contact peak strength as fixed, and since G_f is the product of the contact strength and displacement, the rate-dependent ultimate displacement of contact, δ'_{ut}, is estimated to follow an expression similar to Equation (13.19). Therefore, in general form,

$$\delta'_{ut} = \delta_{ut} \left[1 + \left(\frac{\dot{\delta}_{eff}}{r_\delta - (1 - \alpha)\dot{\delta}_{eff}} \right)^\eta \right] \tag{13.20}$$

where η and r_δ are called *rate-dependency parameters*.

Since any development of the contact model is needed to be consistent with the original numerical methodology outlined in Section 3.2, the linear-decaying irreversible cohesive law is still used. The rate-dependent tensile stress of cohesive contact is then expressed through

$$\sigma = \begin{cases} k_t \delta_{eff} \exp(-\delta_{eff}/\delta_{ct}) & \delta_{eff} \leq \delta_{ct} \\ t_c(1 - D) & \delta_{ct} < \delta_{eff} \leq \delta'_{ut} \\ 0 & \delta_{eff} > \delta_{fin} \end{cases} \tag{13.21}$$

The damage variable, D, is determined in the same manner as described in Section 3.2. The unloading-reloading cycles are also handled based on the suggestions outlined there. Compared to the rate-independent (RI) model, the rate-dependent (RD) one can be therefore illustrated as follows.

In the rate-dependent simulation, the tensile peak strength of contact is fixed at $t_c = 66$ MPa, while its ultimate tensile displacement is momentarily updated through Equation (13.20). Once the contact displacement exceeds δ'_{ut}, its displacement is recorded as the final displacement, δ_{fin}. From this instant on, the contact will carry no tension.

13.5.3.1 Influence of rate-dependency parameters

A series of CFM situations are arranged to evaluate the sensitivity of the model predictions to r_δ and η. For this purpose, r_δ is changed from 100 to 1000 m/s and η from 1 to 3. The specimen surface friction is assumed $5°$, and v_d is fixed at 1000 mm/s. The results, plotted in Figure 13.14, show that increasing η or decreasing r_δ leads to increasing K_{ICd}. In addition, increasing η makes the CFM predictions more sensitive to r_δ. If the loading rate, calculated by Equation (13.17), was plotted for all the points

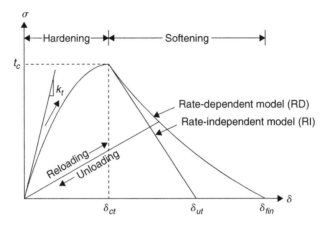

Figure 13.13 Rate-dependent cohesion law of contact.

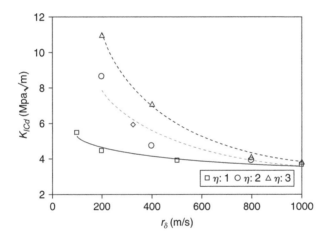

Figure 13.14 Variation of dynamic fracture toughness versus rate-dependency parameters.

of Figure 13.14, it would be seen to vary between 90 and 100 GPa$\sqrt{m}$/s where experimental $K_{ICd} \approx 5$ MPa$\sqrt{m}$ (Fig. 13.11). Figure 13.14 suggests $\eta = 2$ and $r_\delta = 400$ m/s as the best combination to get the dynamic fracture toughness of 5 MPa$\sqrt{m}$.

13.5.3.2 Reproduction of experimental results

Given $\eta = 2$ and $r_\delta = 400$ m/s, the SCB specimen is again simulated. The obtained results, labeled as RD (ϕ: 5), are compared to the experimental data and those of the RI model in Figure 13.15. The results clearly show that the RD model reproduces the test data much better than the RI model did. As seen, the slope of the fitted line to the RD model data is nearly as steep as the one fitted to the test data. This means that the RD model, unlike RI, is able to predict the actual rate-varying fracture toughness of the SCB specimens.

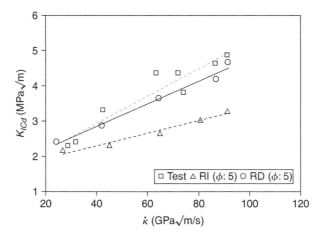

Figure 13.15 Variation of dynamic fracture toughness versus loading rate for RI and RD models.

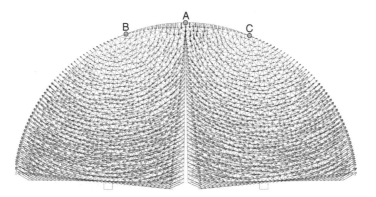

Figure 13.16 Nodal displacement vectors of particle assemblage.

Figure 13.16 depicts the nodal displacement vectors of the particle assemblage under $v_d = 1000$ m/s. As seen, the centre of rotation of each specimen half is located neither at the pins nor at the contact point with the incident bar (point A). It is at points B and C, which have no displacement. The plot shows that both the halves are sliding over the support pins. This demonstrates that the slip friction between the specimen and the pins is an unavoidable matter of the SCB test.

To explore the friction effects from an energy point of view, an energy analysis is performed, where the total boundary loading work supplied to the system, W, the current strain energy stored in the assemblage, U_c, the current kinetic energy of the system, U_k, the total dissipated energy through the specimen surface friction, W_j, are continuously calculated in the course of loading. The total energy released by fracture propagation will therefore be

$$W_f = W - U_c - U_k - W_j \tag{13.22}$$

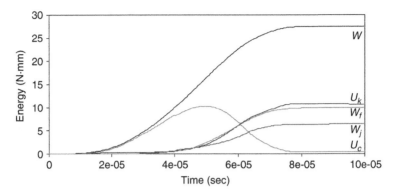

Figure 13.17 Variation of different components of energy versus time for $v_d = 1000$ mm/s.

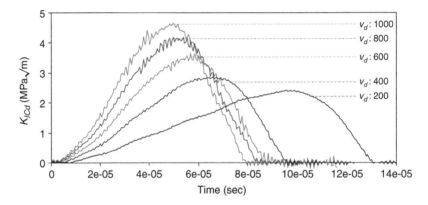

Figure 13.18 Variation of stress intensity factor under different values of applied dynamic velocity.

Figure 13.17 offers the energy analysis output for the RD model under $v_d = 1000$ m/s, and assuming 5° surface friction angle. It indicates that the amount of energy dissipated through the specimen surface friction (≈ 6.40 N·mm) is rather comparable to the total fracture energy (≈ 9.96 N·mm), although the friction angle is assumed to be very small. This again emphasizes how significantly the frictional effects influence the SCB test results.

Variation of the stress intensity factor versus time is plotted in Figure 13.18. Comparison of the results when $v_d = 1000$ m/s with Figure 13.17 indicates that the model strain energy as well as the stress intensity factor reaches their peaks at 50 μs.

However, Figure 13.19b demonstrates that the fracture just starts propagating 25 μs after the loading bar touches the specimen, when $K_{Id} = 2.33$ MPa√m. This means that the peak of K_{Id} (=4.67 MPa√m), which is experimentally reported as the dynamic fracture toughness, does not correspond with the instant of fracture initiation. The simulation suggests that the actual toughness value, regarding fracture initiation, is much lower than the laboratory measurement. The same is observed under $v_d = 200$ mm/s,

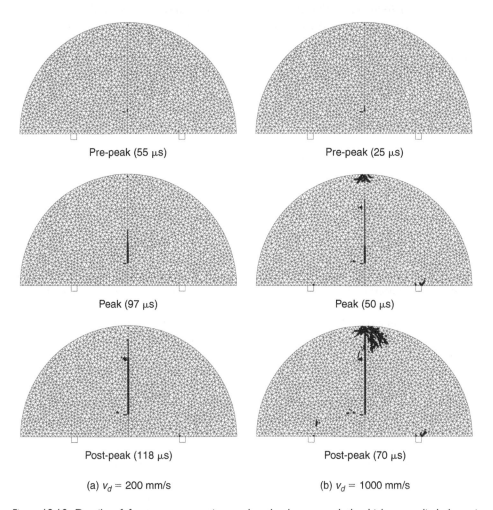

Pre-peak (55 µs)   Pre-peak (25 µs)

Peak (97 µs)   Peak (50 µs)

Post-peak (118 µs)   Post-peak (70 µs)

(a) $v_d = 200$ mm/s   (b) $v_d = 1000$ mm/s

Figure 13.19 Details of fracture propagation under the lowest and the highest applied dynamic velocities.

where the fracture starts propagating at 55 µs when $K_{Id} = 1.40$ MPa$\sqrt{\text{m}}$, but the peak value (=2.40 MPa$\sqrt{\text{m}}$) happens some time later at 97 µs.

Under $v_d = 1000$ mm/s, the fracture propagation ends at about 70 µs. From 80 µs on, the specimen does not bear any more load, and all the energy components become nearly constant (Fig. 13.17). Note that since the specimen is still in contact with the pins, the friction work (W_j) slightly evolves and consequently the kinetic energy decays a little.

The details of fracture propagation are illustrated in Figure 13.19, where the contacts fully damaged are colored in brown thick lines, and those in cohesive softening stage are plotted by thin red lines. It reveals that the fracture propagates much faster under high-rate than low-rate loading. Moreover, the cohesive zone, ahead the crack tip, broadens as the loading rate increases.

Figure 13.19b indicates that the high-rate loading may cause local damage around where the specimen touches the loading bars. As the particle generation is based on a random procedure, the arrangement of contacts in each half of the specimen is different from that in the other one. That is why the failure patterns of the specimen halves are not the same with each other, and thus the demonstrated examples are unsymmetrical.

13.6 CONCLUSIONS

This chapter demonstrated the capability and effectiveness of the discontinuum-based numerical modeling for rock dynamic fracture research. Besides the results presented, the study specifically provides some critical suggestions, which are believed to be necessary for any discontinuum-based modeling of rock fracture and fragmentation.

13.6.1 Particle size

Physical interpretation of particles varies with the microstructural characteristics of the simulated material. For rocks, particle size is basically a matter of their physical texture. Simulation of coarse-grained rocks needs bigger particles than that of fine-grained ones. Since rock fracture is expected to pass through a mineral interface, the weakest grains or mineral cement (if any), the rock fragments are at least as large as one or several minerals. Therefore, the model particles, which numerically represent the fragmented pieces of failed rock, do not need to be smaller than the rock grains size.

For isotropic homogeneous material, e.g. brittle polymers, particle size should respect the considerations suggested by continuum models, e.g. it should be a reasonable fraction of the stress wavelength or the length of the fracture process zone. However, there is also a lower boundary for particle size. Since the contact stiffness represents the stiffness of material existing in the fracture cohesive zone, only one particle is allowed to be placed within the cohesive zone thickness. Otherwise, the material located in this zone would be modeled less stiffly than reality because there would exist several contacts over the zone thickness, which act as springs in series. Therefore, particle size must not be smaller than the thickness of the fracture cohesive zone.

13.6.2 Necessity of a representative contact model

Since macroscopic response of any bonded particle assemblage is dominantly controlled by the contact (or bond) constitutive model, this model must be appropriately adopted to allow the assemblage to simulate material physical response. Therefore simulation of rock, as an anisotropic brittle material, needs a contact model providing these qualities. That is why the CFM contact model is orthotropic, cohesive, frictional, and rate-dependent. All these items are necessary for rock failure modeling. If contact orthotropy is neglected, the assemblage is no longer able to follow different compressive and tensile behaviors as physical rock does. Contact cohesiveness is required to provide real rock fracture energy, and contact friction is required to present real slip behavior of broken fragments past each other. However, the contact model should meet all these needs by using the fewest micro-parameters possible.

13.6.3 Numerical process of fracture energy release

No numerical damping must be applied to dynamic fracture simulations. We believe that the key for any successful simulation of fracture is that the adopted numerical model must be able to reproduce the actual amount of energy released in the material fracturing process. Very importantly, this issue must not be disturbed or manipulated at all by the application of any artificial (numerical) damping, e.g. local or Rayleigh damping. In other words, all the energy dissipation in a particle assemblage must occur only through the rupture of bonding elements (contacts).

13.6.4 Necessity of rate-dependent model for fracture in micro-scale

The thesis results supported the use of the rate-dependent contact (crack) model for dynamic fracture simulation. This approach is to let contact appropriately reproduce the actual amount of energy released during the material fracturing process. Otherwise, numerical energy released through contact failure is held constant whatever the applied loading rate is. There are two points that we should pay attention to in rate-dependent simulation: physical interpretation of rate dependency parameters and scale of modeling.

The results obtained from dynamic fracture simulation suggest that the rate dependency parameters somehow express the sensitivity of dynamic fracture response to loading rate. However, the essence of this sensitivity is not clear yet.

Although the rate-dependent model is needed in micro-scale simulation, its necessity for molecular-scale modeling is not approved. As mentioned, the key point in accurate simulation of fracture is that the simulation properly reproduces the actual amount of energy released through the material fracturing process. If current computational facilities allow us to establish a particle assemblage with a particle size in the order of material molecules, the assemblage would be able to reproduce all the micro-cracking events happening in the fracture process zone. Since this model produces more fracture energy than the CFM does, it could possibly match the actual amount of energy released in the fracture process zone without the rate-dependent assumption.

REFERENCES

Ariffin, A.K., Huzni, S., Nor, M.J.M. and Mohamed, N.A.N.: Hybrid finite–discrete element simulation of crack propagation under mixed mode loading condition. *Fracture and Strength of Solids* Vi, Pts 1 and 2 (2006), pp.306–308, 495–499.

Azevedo, N.M., Lemos, J.V., *et al.*: Influence of aggregate deformation and contact behavior on discrete particle modeling of fracture of concrete. *Engineering Fracture Mechanics* 75(6) (2008), pp.1569–1586.

Cai, J.G. and Zhao, J.: Effects of multiple parallel fractures on apparent attenuation of stress waves in rock masses. *Int. Journal of Rock Mechanics and Mining Sciences* 37(4) (2000), pp.661–682.

Cai, M.: Influence of intermediate principal stress on rock fracturing and strength near excavation boundaries—Insight from numerical modeling. *Int. Journal of Rock Mechanics and Mining Sciences* 45(5) (2008), pp.763–772.

Camborde, F., Mariotti, C., *et al.*: Numerical study of rock and concrete behavior by discrete element modeling. *Computers and Geotechnics* 27(4) (2000), pp.225–247.

Chen, R., Xia K., *et al.*: Determination of dynamic fracture parameters using a semicircular bend technique in split Hopkinson pressure bar testing. *Eng. Fract. Mech.* 76(9) (2009), pp.1268–1276.

Chen, S.G. and Zhao, J.: A study of UDEC modeling for blast wave propagation in jointed rock masses. *International Journal of Rock Mechanics and Mining Sciences* 35(1) (1998), pp.93–99.

Chen, S.G., Cai, J.G., Zhao, J. and Zhou, Y.X.: Discrete element modeling of an underground explosion in a jointed rock mass. *Geotechnical and Geological Eng.* 18(2) (2000), pp.59–78.

Cheng, Y.M.: Advancements and improvement in discontinuous deformation analysis. *Computers and Geotechnics* 22(2) (1998), pp.153–163.

Chiou, Y.J., Lee, Y.M. and Tsay, R.J.: Mixed mode fracture propagation by manifold method. *International Journal of Fracture* 114(4) (2002), pp.327–347.

Cho, N., Martin, C.D. and Sego, D.C.: A clumped particle model for rock. *International Journal of Rock Mechanics and Mining Sciences* 44(7) (2007), pp.997–1010.

Cundall, P.A.: A computer model for simulating progressive, large scale movements in blocky rock systems. *Proceedings of the International Symposium on Rock Fracture*, Nancy (1971).

Cundall, P.A. and Hart, R.D.: Numerical modeling of discontinua. *Eng Comp.* 9 (1992), pp.101–113.

Curran, J.H. and Ofoegbu, G.I.: Modeling discontinuities in numerical analysis. In: Hudson, J.A. (ed.): *Comprehensive Rock Engineering*, vol. 1. Pergamon Press, Oxford (1993), pp.443–468.

Dai, F., Xia K., *et al.*: Semicircular bend testing with split Hopkinson pressure bar for measuring dynamic tensile strength of brittle solids. *Review of Scientific Instruments* 79(12) (2008), pp.123903–123906.

Dai, F., Chen R., *et al.*: A semi-circular bend technique for determining dynamic fracture toughness. *Experimental Mechanics* (2009).

Damjanac, B., Board, M., *et al.*: Mechanical degradation of emplacement drifts at Yucca Mountain – A modeling case study: Part II: Lithophysal rock. *International Journal of Rock Mechanics and Mining Sciences* 44(3) (2007), pp.368–399.

Delaunay, B.: Sur la sphère vide. *Izvestia Akademii Nauk SSSR, Otdelenie Matematicheskikh i Estestvennykh Nauk* 7 (1934), pp.793–800.

Du, C.: An algorithm for automatic Delaunay triangulation of arbitrary planar domains. *Advances in Engineering Software* 27(1–2) (1996), pp.21–26.

Dugdale, D.S.: Yielding of steel sheets containing slits. *Journal of the Mechanics and Physics of Solids* 8(2) (1960), pp.100–104.

Gong, Q.M. and Zhao, J.: Influence of rock brittleness on TBM penetration rate in Singapore granite. *Tunnelling and Underground Space Technology* 22(3) (2007), pp.317–324.

Hatzor, Y.H., Arzi, A.A., Zaslavsky, Y. and Shapira, A.: Dynamic stability analysis of jointed rock slopes using the DDA method: King Herod's Palace, Masada, Israel. *International Journal of Rock Mechanics and Mining Sciences* 41(5) (2004), pp.813–832.

Hazzard, J.F. and Young R.P.: Micromechanical modeling of cracking and failure in brittle rocks. *Journal of Gheophysical Research* 105 (2000a), pp.16683–16697.

Hazzard, J.F. and Young, R.P.: Simulating acoustic emissions in bonded-particle models of rock. *International Journal of Rock Mechanics and Mining Sciences* 37(5) (2000b), pp.867–872.

Hildyard, M.W. and Young, R.P.: Modeling seismic waves around underground openings in fractured rock. *Pure and Applied Geophysics* 159(1–3) (2002), pp.247–276.

Hsiung S.M.: Discontinuous deformation analysis (DDA) with nth order polynomial displacement functions. *Rock Mech. in the National Interest, Swets & Zeitlinger Lisse* (2001), pp.1437–1444.

Hsiung, S.M. and Shi, G.: Simulation of earthquake effects on underground excavations using discontinuous deformation analysis (DDA). *Rock Mech. in the National Interest* (2001), pp.1413–1420.

Itasca: *PFC (Particle Flow Code). Theory and Background Volume.* Minneapolis, MN: ICG, Itasca Consulting Group Inc. (2009a).

Itasca: *UDEC (Universal Distinct Element Code). Theory and Background Volume.* Minneapolis, MN: ICG, Itasca Consulting Group Inc. (2009b).

Jensen, R.P., Bosscher, P.J., Plesha, M.E. and Edil, T.B.: DEM simulation of granular media–structure interface: effects of surface roughness and particle shape. *Int. J. Numer. Anal. Met. Geomechanics* 23 (1999), pp.531–547.

Jiang, Q.H. and Yeung, M.R.: A model of point-to-face contact for three-dimensional discontinuous deformation analysis. *Rock Mechanics and Rock Engineering* 37(2) (2004), pp.95–116.

Jing, L.: A review of techniques, advances and outstanding issues in numerical modeling for rock mechanics and rock engineering. *Int. J of Rock Mech. Mining Sciences* 40(3) (2003), pp.283–353.

Jing, L. and Stephansson, O.: *Fundamentals of Discrete Element Methods for Rock Engineering, Theory and Application.* Elsevier (2007).

Jing, L., Nordlund, E. and Stephansson, O.: A 3-D constitutive model for rock joints with anisotropic friction and stress dependency in shear stiffness. *International Journal of Rock Mechanics & Mining Sciences* 31(2) (1994), pp.173–178.

Jing, L., Tsang, C.F. and Stephansson, O.: DECOVALEX – an international co-operative research project on mathematical models of coupled THM processes for safety analysis of radioactive waste repositories. *Int. J of Rock Mechanics & Mining Sciences* 32(5) (1995), pp.389–398.

Karami, A. and Stead, D.: Asperity degradation and damage in the direct shear test: A hybrid FEM/DEM approach. *Rock Mechanics and Rock Engineering* 41(2) (2008), pp.229–266.

Kazerani, T. and Zhao, J.: Micromechanical parameters in bonded particle method for modeling of brittle material failure. *Int. J Num. Analit. Meth. in Geomech.* (2010), DOI: 10.1002/nag.884.

Kazerani, T., Zhao, J. and Yang, Z.Y.: Investigation of failure mode and shear strength of rock joints using discrete element method. *Proc. EUROCK 2010*, Lausanne, Switzerland (2010), pp.235–238.

Kazerani, T.: *Micromechanical Study of Rock Fracture and Fragmentation under Dynamic Loads Using Discrete Element Method.* Ph.D. thesis, École Polytechnique Fédérale de Lausanne (EPFL), Lausanne, Switzerland (2011).

Kim, Y.I., Amadei, B. and Pan, E.: Modeling the effect of water, excavation sequence and rock reinforcement with discontinuous deformation analysis. *International Journal of Rock Mechanics and Mining Sciences* 36(7) (1999), pp.949–970.

Lei, W.D., Teng, J., Hefny, A.M. and Zhao, J.: Transmission ratio (T–n) in the radian direction normal to joints in 2-D compressional wave propagation in rock masses. *Journal of University of Science and Technology Beijing* 13(3) (2006), 199–206.

Lemos, J.V.: Numerical modelling of fractured media applied to underground gas storage. *Short Course on Underground Storage of Gases, LNEC*, Lisboa (1993), pp.122–139.

Lemos, J.V., Hart, R.D. and Cundall, P.A.: A generalized distinct element program for modeling of jointed rock masses. *Proc. Int. Symp. Fundamentals of Rock Joints*, Bjorkliden, Sweden (1985), pp.335–343.

Liang, Z.Z., Tang, C.A., Li, H.X. and Zhang, Y.B.: Numerical simulation of the 3D failure process in heterogeneous rocks. *International Journal of Rock Mechanics and Mining Sciences* 41(3) (2004), pp.419–419.

Lorig, L.J., Brady, B.H.G. and Cundall, P.A.: Hybrid distinct element–boundary element analysis of jointed rock. *Int. Journal of Rock Mechanics and Mining Sciences* 23(4) (1986), pp.303–312.

Maigre, H. and Rittel, D.: Dynamic fracture detection using the force–displacement reciprocity: application to the compact compression specimen. *International Journal of Fracture* 73(1) (1995), pp.67–79.

Min, K.B. and Jing, L.R.: Numerical determination of the equivalent elastic compliance tensor for fractured rock masses using the distinct element method. *International Journal of Rock Mechanics and Mining Sciences* 40(6) (2003), pp.795–816.

Morris, J.P., Rubin, M.B., Block, G.I. and Bonner, M.P.: Simulations of fracture and fragmentation of geologic materials using combined FEM/DEM analysis. *International Journal of Impact Engineering* 33(1–12) (2006), pp.463–473.

Munjiza, A.: *The Combined Finite–Discrete Element Method.* John Wiley and Sons Ltd., University of London (2004).

Munjiza, A., Andrews, K.R.F. and White, J.K.: Combined single and smeared crack model in combined finite–discrete element analysis. *International Journal for Numerical Methods in Engineering* 44(1) (1999), pp.41–57.

NIST/SEMATECH: E–handbook of statistical methods. Retrieved on 8/11/2010 from www.itl.nist.gov/div898/handbook (2003, 6/23/2010).

Park, H.J. and Park, S.H.: Extension of central composite design for second-order response surface model building. *Communic. in Statis. – Theory and Methods* 39(7) (2010), pp.1202–1211.

Paterson, M.S.: *Experimental Rock Deformation, the Brittle Field.* Springer–Verlag (1978).

Potyondy, D.O.: Simulating stress corrosion with a bonded-particle model for rock. *International Journal of Rock Mechanics and Mining Sciences* 44(5) (2007), pp.677–691.

Potyondy, D.O. and Cundall P.A.: A bonded-particle model for rock. *International Journal of Rock Mechanics and Mining Sciences* 41(8) (2004), pp.1329–1364.

Schöpfer, M.P.J., Abe, S., *et al.*: The impact of porosity and crack density on the elasticity, strength and friction of cohesive granular materials: Insights from DEM modeling. *International Journal of Rock Mechanics and Mining Sciences* 46(2) (2009), pp.250–261.

Shi, G.H.: *Discontinuous Deformation Analysis, a New Numerical Model for the Statics and Dynamics of Block Systems.* PhD thesis, Univ. of California, Berkeley, Berkeley, Calif. 1988).

Shi, G.H.: Manifold method of material analysis. *Transactions of the 9th Army Conference on Applied Mathematics and Computing* U.S. Army Research Office, Minneapolis, MN (1991), pp.57–76.

Shioya, T. and Zhou, F.: Dynamic fracture toughness and crack propagation in brittle material. *Constitutive Relation in High/Very High Strain Rates* (1995), pp.105–112.

Shioya, T., Zhou, F. and Ishida, R.: Micro-cracking process in brittle crack propagation. *DYMAT J* 2 (1995), pp.105–118.

Shyu, K.: *Nodal-based Discontinuous Deformation Analysis.* PhD thesis, UC Berkeley (1993).

Souley, M., Homand, F. and Thoraval, A.: The effect of joint constitutive laws on the modeling of an underground excavation and comparison with in situ measurements. *International Journal of Rock Mechanics and Mining Sciences* 34(1) (1997), pp.97–115.

Tan, Y., Yang, D., *et al.*: Study of polycrystalline Al_2O_3 machining cracks using discrete element method. *International Journal of Machine Tools and Manufacture* 48(9) (2008), pp.975–982.

Tan, Y., Yang, D., *et al.*: Discrete element method (DEM) modeling of fracture and damage in the machining process of polycrystalline SiC. *Journal of the European Ceramic Society* 29(6) (2009), pp.1029–1037.

Terada, K. and Kurumatani, M.: An integrated procedure for three-dimensional structural analysis with the finite cover method. *International Journal for Numerical Methods in Engineering* 63(15) (2005), pp.2102–2123.

Terada, K., Asal, M. and Yamagishi, M.: Finite cover method for linear and non-linear analyses of heterogeneous solids. *International Journal for Numerical Methods in Engineering* 58(9) (2003), pp.1321–1346.

Terada, K., Ishii, T., Kyoya, T. and Kishino, Y.: Finite cover method for progressive failure with cohesive zone fracture in heterogeneous solids and structures. *Computational Mechanics* 39(2) (2007), pp.191–210.

Wanne, T.S. and Young, R.P.: Bonded-particle modeling of thermally fractured granite. *International Journal of Rock Mechanics and Mining Sciences* 45(5) (2008), pp.789–799.

Wei, L. and Hudson, J.A.: A hybrid discrete–continuum approach to model hydro–mechanical behavior of jointed rocks. *Eng Geol* 49 (1988), pp.317–325.

Xu, X.P. and Needleman, A.: Numerical simulations of dynamic interfacial crack growth allowing for crack growth away from the bond line. *Int. Journal of Fracture* 74(3) (1995), pp.253–275.

Yoon, J.: Application of experimental design and optimization to PFC model calibration in uniaxial compression simulation. *Int. J of Rock Mechanics and Mining Sciences* 44(6) (2007), pp.871–889.

Zhang, G.X., Sugiura, Y. and Hasegawa, H.: Manifold method and its applications to engineering. *Proceedings of the Second International Conference on Analysis of Discontinuous Deformation*, New York (1997).

Zhang, X.L., Jiao, Y.Y., Liu, Q.S. and Chen, W.Z.: Modeling of stability of a highway tunnel by using improved DDA method. *Rock and Soil Mechanics* 28(8) (2007), pp.1710–1714.

Zhang, X. and Lu, M.W.: Block-interfaces model for non-linear numerical simulations of rock structures. *International Journal of Rock Mechanics and Mining Sciences* 35(7) (1998), pp.983–990.

Zhao, J., Zhou, Y.X., Hefny, A.M., Cai, J.G., Chen, S.G., Li, H.B., Liu, J.F., Jain, M., Foo, S.T. and Seah, C.C.: Rock dynamics research related to cavern development for ammunition storage. *Tunnelling and Underground Space Technology* 14(4) (1999), pp.513–526.

Zhao, X.B., Zhao, J., Hefny, A.M. and Cai, J.G.: Normal transmission of S-wave across parallel fractures with Coulomb slip behavior. *J. of Eng. Mechanics–ASCE* 132(6) (2006), pp.641–650.

Zhou, F., Molinari, J.F., *et al.*: A rate-dependent cohesive model for simulating dynamic crack propagation in brittle materials. *Engineering Fracture Mechanics* 72(9) (2005), pp.1383–1410.

Zhu, W.C. and Tang, C.A.: Numerical simulation of Brazilian disk rock failure under static and dynamic loading. *Int. Journal of Rock Mechanics and Mining Sciences* 43(2) (2006), pp.236–252.

Chapter 14

Manifold and advanced numerical techniques for discontinuous dynamic computations

Gaofeng Zhao, Gen-Hua Shi and Jian Zhao

14.1 INTRODUCTION

With the improvement in computing power of modern computers, numerical methods have become extremely useful in scientific research. In addition to experimental methods, computer simulation using numerical methods has been proven as a powerful and effective tool for rock dynamics. There exist a large number of numerical methods, e.g. finite element method (FEM), finite difference method (FDM), finite volume method (FVM) and discrete element method (DEM). Generally, numerical methods used in rock mechanics are classified into continuum based method, discontinuum based method and coupled continuum/discontinuum method (Jing, 2003). The classical numerical methods, e.g. FEM, FDM and DEM have a few shortcomings when they are used for discontinuous dynamics computations. For example, directly using FEM to simulate dynamic cracking propagation problems is difficult due to the continuum assumption which leads to FEM being unsuitable for dealing with complete detachment and large-scale fracture opening problems (Jing, 2003). The DEM can well simulate the fracturing process of rock by the breakage of inter-block contacts or inter-particle bonds. However, it is not suitable for stress state analysis of the pre-failure stage.

In order to overcome these problems, some new numerical methods are proposed, e.g. smoothed particle hydrodynamics (SPH) (Monaghan, 1988; Randles and Libersky, 1996), numerical manifold method (NMM) (Shi, 1991), extended finite element method (XFEM) (Belytschko and Black, 1999), finite element method (FEM)/discrete element method (DEM) (Munjiza, Owen and Bicanic, 1995; Munjiza, 2004), discontinuous Galerkin method (DGM) (Reed and Hill, 1973), distinct lattice spring model (DLSM) (Zhao, Fang and Zhao, 2010) and multi-scale distinct lattice spring model (m-DLSM) (Zhao, 2010). These newly developed methods are not fully covered by the classical review paper on numerical methods in rock mechanics by Jing (2003). In this chapter, a comprehensive review on these newly developed numerical methods is presented including their basic principles, applications, advantages and disadvantages.

14.2 NUMERICAL MANIFOLD METHOD (NMM)

NMM was developed to integrate Discontinuous Deformation Analysis (DDA) and FEM. NMM employs two sets of cover system (Shi, 1991). One is the mathematical cover which is used to build approximation and is independent of the problem domain.

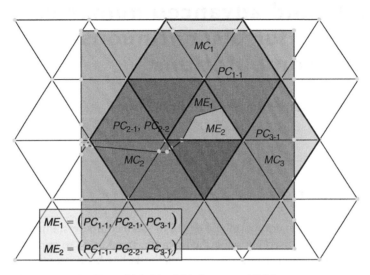

Figure 14.1 Manifold element in NMM.

Another is the physical cover which contains the geometry information of the problem domain and is used to define the integration fields. The advantages of NMM are releasing the task of meshing and combining continuum and discontinuum problems into one framework. The foundation of NMM is manifold approximation of the displacement function. The approximation method makes NMM suitable to deal with both the continuum and discontinuum problems. In this section, the method of NMM for building the approximation function will be presented in a concise way. The basic unit used in NMM is called the physical cover. It can be regarded as the intersection of the mathematic cover and the physical domain. The physical cover is equivalent to the FEM node used in the standard FEM. Degrees of freedoms (DOF) are defined in these physical covers to represent the deformation state of a small area. This representation of the physical cover is more suitable for description of discontinuous problems, as the influence domain can be separated naturally according to the topology of the studied body. The manifold element is made up from several physical covers (see Fig. 14.1). Details of how to construct these manifold elements can be found in the work of Shi (1991) and Ma *et al.* (2009).

The approximation function in NMM is given as follows. First, the displacement function is defined in the physical cover as

$$c_j(\mathbf{x}) = \sum_{i=1}^{n} b_{ji}(\mathbf{x}) \cdot u_{ji} \tag{14.1}$$

where $c_j(\mathbf{x})$ is the displacement function of the jth physical cover, u_{ji} is the general DOFs of the cover, $b_{ji}(\mathbf{x})$ is the basis of the displacement function and n is the number of DOFs. Finally the approximation function of the manifold element is written as

$$u^b(x) = \sum_{j=1}^{m} \phi_j(x) c_j(x) = \sum_{j=1}^{m} \phi_j(x) \sum_{i=1}^{n} b_{ji}(x) u_{ji} \tag{14.2}$$

where ϕ_j is the weight function of the cover and m is the number of physical covers of the manifold element. The weight functions should satisfy the partition of unity (PU), namely

$$\sum_{j=1}^{m} \phi_j(x) = 1 \tag{14.3}$$

The manifold elements can be called the three-cover element or six-cover element in order to distinguish them from the FEM. Equation (14.2) can be further written in a more familiar form as

$$u^b(\mathbf{x}) = \sum_{i=1}^{n \times m} N_i(\mathbf{x})u_i \tag{14.4}$$

where $N_i(\mathbf{x})$ is the shape function of ith general degree of freedoms. Now, the integration equations of NMM on elastic dynamics can be obtained through the weighted residual approach or the variational principle. Details of the integration equations and treatment methods for contacts can be found in the work of Shi (1991).

It should be mentioned that the NMM is proposed much earlier than the PU theory and other derived FEMs. Recently, it has also been called the cover-based generalized FEM (Terada et al., 2007). Actually, the solver in manifold code is very similar to that in the standard FEM. The distinct part of NMM is the manifold mesh generation technique which makes the method more suitable for describing continuum-discontinuum problems. For this reason, NMM has been used for fracture progress simulation (Chiou, Lee and Tsay, 2002; Ma et al., 2009; Terada et al., 2007) and dynamic fracturing computations. For example, the stress-based Mohr-Coulomb criterion and fracture mechanics based criterion are used by Chen et al. (2006) as the crack initiation law and the propagation law in NMM for modeling the two hole blasting problem (see Fig. 14.2 (a)). Another application of NMM is for dynamic large deformation simulation. An Eulerian form of explicit NMM is proposed by Okazawa et al. (2010) to solve problems involving dynamic boundary conditions and large deformation (see Fig. 14.2(b)). NMM is also used for stress wave propagation simulation. Non-reflection boundary conditions and more general integration method based Newmark assumptions are implemented into the standard NMM and have been used to model blasting wave propagation through jointed rock masses (Fig. 14.2 (c)).

NMM has been also used for other applications, e.g. stability analysis of ancient block structures under dynamic loading (Sasaki et al., 2009), modeling complex crack problems by using singularity cover (Ma et al., 2009), progressive failure in heterogeneous solids and structures with cohesive zone model (Terada et al., 2007). Even so, a few shortcomings also exist in NMM. For example, the *small* element will cause instability problems and rigid body rotation cannot be well modeled by NMM (Miki et al., 2009). Moreover, the implementation of NMM, especially for the 3D case, is very difficult and no commercial codes are available at present. Solving these shortcomings is necessary for its further application on discontinuous dynamic computations.

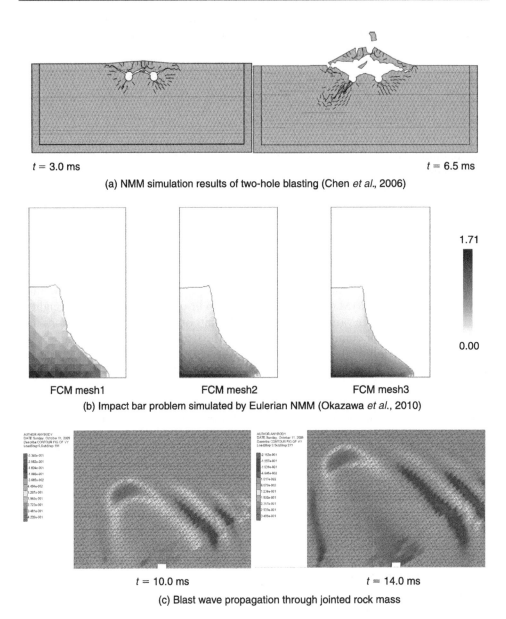

$t = 3.0$ ms $t = 6.5$ ms

(a) NMM simulation results of two-hole blasting (Chen *et al.*, 2006)

FCM mesh1   FCM mesh2   FCM mesh3

(b) Impact bar problem simulated by Eulerian NMM (Okazawa *et al.*, 2010)

$t = 10.0$ ms $t = 14.0$ ms

(c) Blast wave propagation through jointed rock mass

Figure 14.2 Applications of NMM on dynamic computations.

14.3 EXTENDED FINITE ELEMENT METHOD (XFEM)

The XFEM (Belytschko and Black, 1999) is based on the partition of unity method (PUM) (Babuska, 1997) which allows for addition of a priori knowledge about the solution into the approximation space of the numerical solution. XFEM focuses on crack propagation problems. Compared with the standard FEM, XFEM has several

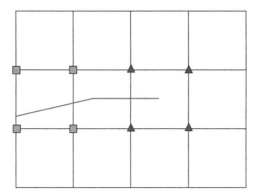

Figure 14.3 The enrichment strategy of XFEM.

advantages in aspect of mesh independence. The most important feature of XFEM is that it can perform extending crack without any remeshing. XFEM treats cracks at element level by using the level sets technique (Prabel *et al.*, 2007). Usually, the Heaviside function and asymptotic functions are used to deal with the discontinuity and singularity. The approximation function of the XFEM element is written as

$$u^b(\mathbf{x}) = \overbrace{\sum N_i(\mathbf{x})u_i}^{\text{Standard FEM}} + \overbrace{\sum N_i(x)H(\bar{f}(\mathbf{x}))b_i}^{\text{Discontious enriched}} + \overbrace{\sum N_i(x)\sum \gamma_k(x)a_i^k}^{\text{Crack tip enriched}} \qquad (14.5)$$

In this equation, the set of nodes whose support is cut by the crack and that contain the crack tip are enriched by discontinuous and singular functions (see Fig. 14.3).

Another technique used in XFEM is the level set method. It uses a scalar function within the domain with zero-level to represent the discontinuity. The domain is divided into two subdomains where the level-set function is negative or positive (see Fig. 14.4). The level-set function to describe the discontinuity in Figure 14.4 is given as

$$\phi(x, y) = \sqrt{x^2 + y^2} - r \qquad (14.6)$$

In practical, level-set functions are defined by values at the nodes in the domain by standard FEM shape functions, which are written as

$$\phi^b(\mathbf{x}) = \sum_{i \in I} N_i(\mathbf{x}) \cdot \phi_i \qquad (14.7)$$

To capture the dynamic evaluation of discontinuities, level set functions need to be correspondingly updated during calculation (Gravouil, Moes and Belytschko, 2002). When the level set is defined on an unstructured irregular mesh, a special algorithm for triangulation is used (Barth and Sethian, 1998). Another solution is to use auxiliary regular structured mesh for level set representation (Prabel *et al.*, 2007).

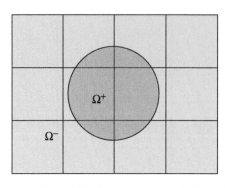

Figure 14.4 Represent discontinuity by level set method.

In XFEM, elements containing a crack are not required to conform to crack edges, and mesh generation is much simpler than in the classical FEM. Because of these advantages, XFEM was successfully used in the simulation of crack propagation (Stolarska *et al.*, 2001), dynamic crack propagation (Prabel *et al.*, 2007; Song, Wang and Belystchko, 2008) and three-dimensional crack propagation (Pedro and Belytschko, 2005; Sukumar, Chopp and Moran, 2003).

Examples of dynamic crack propagation are shown in Figure 14.5. It was reported that the simulation results of XFEM compared well with the experimental results. Recent development of XFEM includes dealing with cohesive fracturing (Asferg, Poulsen and Nielsen, 2007), explicit formulation of XFEM (Menouillard *et al.*, 2006), anisotropic XFEM (Asadpoure, Mohammadi and Vafai, 2006) and considering contact between crack surfaces (Khoei and Nikbakht, 2007; Ribeaucourt, Baietto-Dubourg and Gravouil, 2007).

XFEM has the advantage of mesh independence and being able to deal with weak or strong discontinuities efficiently. These merits make it suitable for fracturing process analysis. Nevertheless, it also has its own disadvantages. For example, the global stiffness matrix will become singular if the crack passes a very tiny part of the XFEM element (Markus, 2005). This is an existing problem for all derived FEMs including NMM. Implementation of XFEM into available commercial FEM code is also difficult (Stéphane *et al.*, 2006) since additional degrees of freedom are introduced. There are methods to reduce the singularity caused by *small element*, but with the price of sacrificing the description of discontinuity inside enriched elements. In spite of these drawbacks, XFEM is still the most promising method for dynamic discontinuous computation. This is mainly attributed to the success of the standard FEM idea and its inherent merits, e.g. robust and easy to deal with complex geometry, various loading and material conditions.

14.4　SMOOTHED PARTICLE HYDRODYNAMICS (SPH)

In recent years, a large family of meshless methods with the aim of getting rid of mesh constraints has been developed. Their requirements for model generation are

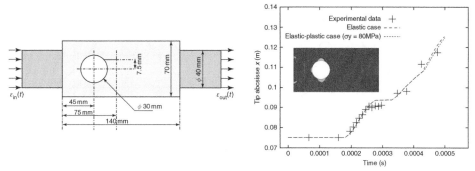

(a) Dynamic crack propagation of PMMA plate under compression (Prabel *et al.*, 2007)

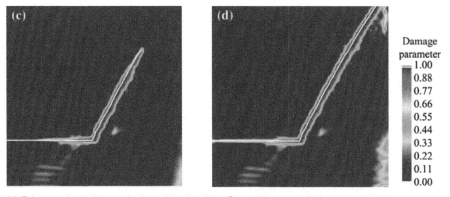

(b) Edge-cracked plate under impulsive loading (Song, Wang and Belystchko, 2008).

Figure 14.5 XFEM for dynamic fracturing computations.

only the generation and distribution of discrete nodes without fixed element-node topological relations as in FEM. Compared to mesh generation, it is relatively simple to establish a point distribution and adapt it locally. SPH is the oldest meshless method which was first invented to deal with problems in astrophysics (Monaghan, 1988) and later extended to elastic problems (Libersky and Petschek, 1991). In SPH, a local approximation function for the PDEs is built based on points grouped together in 'clouds'. A kernel interpolation method is used to estimate a function $f(\mathbf{x})$, and its differential form $\nabla f(\mathbf{x})$ at point i can be represented by the discretized points following the support domain as (see Fig. 14.6)

$$\langle f(\mathbf{x})\rangle_{x=x_i} \cong \sum_{j=1}^{N} m_i f_i W_{xi}(|\mathbf{x}_i - \mathbf{x}|/h)/\rho_i \tag{14.8}$$

$$\langle \nabla f(\mathbf{x})\rangle_{x=x_i} \cong -\sum_{j=1}^{N} m_i f_i \nabla W_{xi}(|\mathbf{x}_i - \mathbf{x}|/h)/\rho_i \tag{14.9}$$

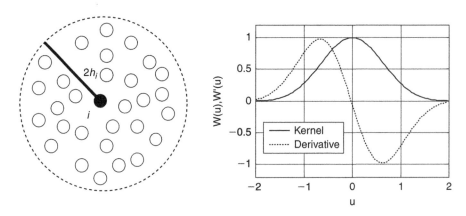

Figure 14.6 SPH neighborhood and its kernel function.

where h is named the smoothing length, which is the supporting domain of the particle, and $W_{xi}(|\mathbf{x}_i - \mathbf{x}|/h)$ is the smooth kernel function. A typical kernel function and its derivative are shown in Figure 14.6.

For dynamic solid mechanics, the mass and momentum equations can be expressed as

$$\frac{d\rho_i}{dt} = \rho_i \sum_{j=1}^{N} \frac{m_j}{\rho_j}[(v_\alpha)_i - (v_\alpha)_j]\frac{\partial W_{ij}}{\partial(x_\alpha)_i} \tag{14.10}$$

$$\frac{d(v_\alpha)_i}{dt} = \sum_{j=1}^{N} m_j \left(\frac{(\sigma_{\alpha\beta})_i}{\rho_i^2} + \frac{(\sigma_{\alpha\beta})_j}{\rho_j^2} + \Pi_{ij}\delta_{\alpha\beta}\right)\frac{\partial W_{ij}}{\partial(x_\beta)_i} \tag{14.11}$$

where v_α is the velocity, $\sigma_{\alpha\beta}$ is the stress tensor, α and β are used to denote the coordinate directions and Π_{ij} is the artificial viscosity to get stabilized solution (Gingold and Monaghan, 1977). An explicit time integration method is normally used in SPH to calculate $\sigma_{\alpha\beta}$ from the constitutive model through the strain-rate tensor. Details can be found in Monaghan (1988) and Randles and Libersky (1996). Due to its meshless property, the applications of SPH are mainly in fragmentation analysis and blasting simulation. For example, Figure 14.7(a) shows how concrete fragmentation under impact loading is simulated by SPH (Rabcauk and Eibl, 2003) and Figure 14.7(b) gives one example of blasting simulation (Randles and Libersky, 1996). Other applications include dynamic fragmentation in brittle elastic solid (Benz and Asphaug, 1995), high distortion impact computations (Johnson, Stryk and Beissel, 1996), formation of cracks around magma chambers (Gray and Monaghan, 2004) and strain rate effect for heterogeneous brittle materials (Ma, Wang and Li, 2010). There are two advantages of SPH over other methods as it is easy to be implemented and meshless. However, there are also a few shortcomings, e.g. SPH exhibits an instability problem called the tensile instability and a problem known as the zero-energy mode. Both of them need special

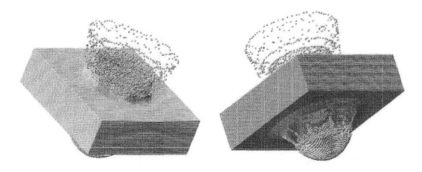

(a) Fragmentation of concrete slab by SPH (Rabczuk and Eibl, 2003)

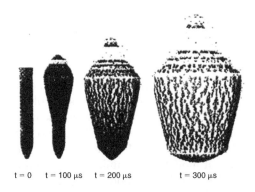

t = 0   t = 100 μs   t = 200 μs   t = 300 μs

(b) Thick-wall bomb simulation (Randles and Libersky, 1996)

Figure 14.7 Examples of SPH on dynamic fragmentation and blasting problems.

treatment in order to produce stable and accurate results (Dyka, Randles and Ingel, 1997). Furthermore, the kernel function of SPH has great influence on the simulation results (Fulk and Quinn, 1996), and its accuracy is not as good as FEM. Overall speaking, SPH has the advantage in simulation of dynamic fragmentation and is easy to be implemented. However, the accuracy, computational time and contact treatment are still problematic in SPH. Solving these problems will strengthen its abilities in dynamic discontinuous computations.

14.5 FEM/DEM METHOD

The continuum-based methods are unsuitable to capture the post-failure discontinuous stage, while the discontinuum-based methods are unsuitable to capture the pre-failure stage of rock. A combination of continuum and discrete methods is required in many rock mechanics applications, such as predicting the formation and interaction of fragments for projectile penetration into rock (Morris *et al.*, 2006). Coupled continuum and discontinuum methods can take advantages of the strength of each method while

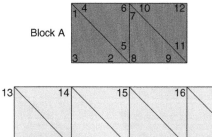

Figure 14.8 Blocks in FEM/DEM method.

avoiding their disadvantages. For fracturing simulation, a coupled method is required to be able to capture both the pre-failure and the post-failure behavior after collapse occurs (Darve *et al.*, 2004). To develop continuum-discontinuum coupled methods, most researchers incline to couple FEM with DEM. The review in this subsection will be limited to the FEM/DEM method (Munjiza, Owen and Bicanic, 1995; Munjiza, 2004). This method aims at modeling dynamic failing, fracturing and fragmenting of solids. In the FEM/DEM method, each body is represented by a single discrete element that interacts with other discrete elements that are close to it (see Fig. 14.8). In addition, each discrete element is divided into FEM elements. For a breakable block (block A in Fig. 14.8), the FEM elements have separate nodes and the block can be further broken into smaller blocks during calculation. One of the most distinct aspects of FEM/DEM method is the contact treatment method. As reported in the book by Munjiza (2004), the classical contact treatment cannot preserve kinematic energy balance. In the FEM/DEM method, a potential contact force concept is used to overcome this shortcoming.

As shown in Figure 14.9, the contact force is adopted for two blocks in contact, one is named as the contactor and another is called the target. An infinitesimal contact force is given by the following contact force equation:

$$df = [grad\varphi_c(P_c) - grad\varphi_t(P_t)]dA \qquad (14.12)$$

where df is the infinitesimal contact force by the overlap dA defined by the overlapping area belonging to both the contactor and target. In FEM/DEM, blocks are discretized into tri-node triangle elements for 2D and four-node tetrahedron elements for 3D. Special equations and procedures for potential contact force equations are provided in FEM/DEM. Details can be found in the book by Munjiza (2004).

In the FEM/DEM method, the explicit time integration method is used to solve the system equations. Due to the continuum-discontinuum property of the method, it has

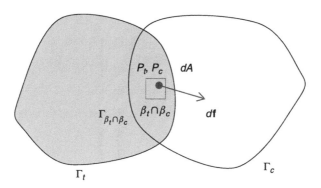

Figure 14.9 Potential contact force between two bodies.

been widely used to simulate dynamic fracturing processes. For example, Figure 14.10 shows examples of the blasting process of a square block with an explosive charge and the fracturing process of Brazilian disc under dynamic loading. Other applications of FEM/DEM method include the investigation of effect of explosive and impact loading on geological media (Morris *et al.*, 2006), the crack propagation under mixed mode loading (Karami and Stead, 2008) and the processes of joint surface damage and near-surface intact rock tensile failure (Ariffin *et al.*, 2006). The FEM/DEM method is a powerful method to solve the fracturing process problems. However, implementation of this method into a computer code needs complex skills and extensive efforts.

14.6   DISCONTINUOUS GALERKIN METHOD (DGM)

DGM is a generalization of the Galerkin method which is the basis of the standard FEM. The name of DGM was first proposed by Lesaint and Raviart (1974) to link separate domains in a weak form. Recently, DGM has been used for dynamic discontinuities computations (Mergheim, Kuh and Steinmann, 2004). When DGM is used for dynamic fracturing simulation, a splitting mesh scheme (see Fig. 14.11) is always used. Interface elements are inserted between two adjacent elements. This treatment is similar with that in the FEM/DEM method. The difference is that there is no dynamic contact detection used in DGM. Moreover, the cohesive law is always used in DGM to model dynamic failure. DGM produces block-diagonal matrices as it uses a discontinuous approximation, which makes it highly parallelizable (Levy, 2010). In DGM, dynamic computation is also taken as finite element discretization and the mass matrix and stiffness matrix are given as

$$\mathbf{M} = \int_{\Omega} \rho \mathbf{N}^T \mathbf{N} d\Omega \tag{14.13}$$

$$\mathbf{K} = \int_{\Omega} \mathbf{B}^T \mathbf{D} \mathbf{B} d\Omega \tag{14.14}$$

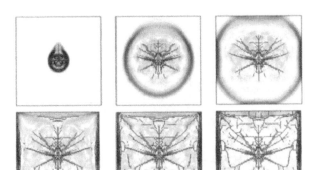

(a) Explosive simulation (Munjiza, 2004)

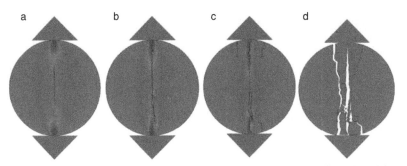

(b) Failure process of Brazilian disc under dynamic loading (Mahabadi *et al.*, 2010)

Figure 14.10 Examples of FEM/DEM method on discontinuous dynamic computation.

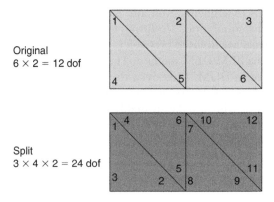

Original
6 × 2 = 12 dof

Split
3 × 4 × 2 = 24 dof

Figure 14.11 Splitting of a mesh in DGM.

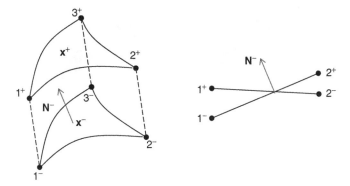

Figure 14.12 Interface elements for 3D and 2D cases in DGM.

where ρ is the density, $\mathbf{N}$ and $\mathbf{B}$ are respectively the interpolation matrix of displacement and strain, $\mathbf{D}$ is the elastic matrix.

The most important part of DGM is the interface element (see Fig. 14.12) which enables the method to well handle non-local behavior and failure behavior. The use of average numerical fluxes and introduction of appropriate quadratic terms makes the interface element keep both consistency and stability. The expression of force between two interfaces is given (Noels and Radovitzky, 2008) as

$$\mathbf{f}_{ia}^{\pm} = \pm \int_{\partial_1 B_{0h}} \langle P_h \rangle \cdot \mathbf{N}^- N_a dS \pm \int_{\partial_1 B_{0h}} \left[\left\langle \frac{\beta}{h_s} \mathbb{C} \right\rangle : [\![\mathbf{x}_a]\!] \otimes \mathbf{N}^- \right] \mathbf{N}^- N_a dS \qquad (14.15)$$

where $[\![x_a]\!] = x_a^+ - x_a^-$ are the jumps in the local coordinates, P_h is the first Piola-Kirchhoff stress tensor, $\mathbf{N}^-$ is the unit surface normal, N_a is the conventional shape function, $\mathbb{C}$ is the elastic tangent moduli for stabilization, h_s is the element size, $\beta > 0$ is the stabilization parameter. For dynamic computation, an explicit time integration is used. Details of the implementation of interface elements and dynamic DGM can be found in the work of Noels and Radovitzky (2008).

Due to the discontinuous nature of DGM, it has significant advantage for modeling dynamic fracturing processes and being efficient for parallel implementation. Figure 14.13 shows one example of DGM in modeling three dimensional fragmentation (Levy, 2010). Moreover, the DGM is also regarded as having other advantages, such as stable, high-order conservative, precisely tracking stress wave (Radovitzky et al., 2010) and modeling discontinuities involving irregular meshes (Castillo et al., 2001; Oden, Babuska and Baumann, 1998). The author considers that the DGM method is very similar to DDA (Shi, 1988), the difference between them should be the treatment of interaction between elements (blocks) and the time integration method used to solve system equations (DGM using the explicit scheme with DDA using the implicit algorithm). The shortcoming of DGM is the increasing of total DOFs (see Fig. 14.11) and the contact detection is not fully considered. These limitations may constrain the method which can be further applied in complex fracturing process problems, e.g. dynamic compressive failure and large deformation.

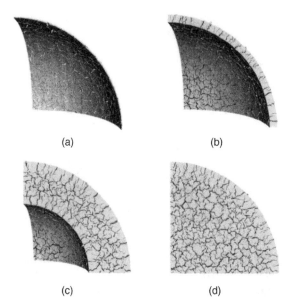

Figure 14.13 Dynamic fragmentation simulation of a hollow sphere under dynamic loading by DGM (Levy, 2010).

14.7 MULTI-SCALE DISTINCT LATTICE SPRING MODEL (M-DLSM)

14.7.1 Distinct Lattice Spring Model (DLSM)

The classical elasticity theory could provide an adequate description of the macroscopic mechanical response of most materials, even though they are actually heterogeneous when viewed at the microscopic level. However, dynamic fracturing of heterogeneous materials such as rock and concrete cannot be modeled realistically without appealing to their microstructures. This requires that a successful numerical method must be capable of considering not only the elastic stage, but also the formulation and evolution of micro-discontinuities. Lattice models (Bažant *et al.*, 1990; Ostoja-Starzewski, 2002) represent material by a system of discrete units (e.g. particles) interacting via springs, or, more generally, rheological elements. These discrete units are much coarser than the true atomic ones and may represent larger volumes of heterogeneities such as grains or clusters of grains. Lattice models are close relatives of the common finite element method (FEM) when dealing with elastic problems. Yet, due to their discrete nature, lattice models are known to be more suitable for complex fracturing simulation. For example, lattice models have been successfully applied to investigate the spatial cooperative effects of crack formation and heterogeneities in elastic-plastic (Buxton, Care and Cleaver, 2001) and elastic-brittle (Ostoja-Starzewski, Sheng and Jasiuk, 1997) systems. However, for lattice models composed of normal springs transmitting central forces only, it is known that the modeled Poisson's ratio approaches, in the limit of an infinite number of particles, a fixed value e.g. 1/4 in three-dimensional cases. Recently, an alternative 3D dynamic lattice spring model, DLSM (Zhao, Fang and Zhao, 2010), has been proposed to overcome the restriction on the Poisson's ratio while preserving the rotational invariance.

The DLSM is a microstructure-based numerical model based on the Realistic Multidimensional Inter Bond (RMIB) model (Zhao, 2010), which is an extension of the Virtual Multidimensional Inter Bond (VMIB) model (Zhang and Ge, 2005). In DLSM, materials are discretized into mass particles linked through distributed bonds (see Fig. 14.14(a)). Whenever two particles are detected in contact, they are linked together through bonds between their centre points. Based on Cauchy-born rules and the hyper-elastic theory, the relationship between the micromechanical parameters and the macro-material constants can be obtained as follows (Zhao, Fang and Zhao, 2010):

$$k_n = \frac{3E}{\alpha^{3D}(1 - 2v)}$$ (14.16)

$$k_s = \frac{3(1 - 4v)E}{\alpha^{3D}(1 + v)(1 - 2v)}$$ (14.17)

where k_n is the spring normal stiffness, k_s is the shear stiffness, E is the Young's modulus, v is the Poisson's ratio and α^{3D} is the microstructure geometry coefficient, which is obtained from

$$\alpha^{3D} = \frac{\sum l_i^2}{V}$$ (14.18)

where l_i is the original length of the ith bond and V is the volume of the geometry model. The particles and springs comprise a whole system, which represents the material. For this system, the equation of motion is expressed as

$$[\mathbf{K}]\mathbf{u} + [\mathbf{C}]\dot{\mathbf{u}} + [\mathbf{M}]\ddot{\mathbf{u}} = \mathbf{F}(t)$$ (14.19)

where $\mathbf{u}$ represents the particle displacement vector, $[\mathbf{M}]$ is the diagonal mass matrix, $[\mathbf{C}]$ is the damping matrix and $\mathbf{F}(t)$ is the vector of external forces on particles. The motion equations of the particle system are solved through an explicit central finite differences scheme. The calculation cycle is illustrated in Figure 14.14(b). The details of the implementation and verification of DLSM can be found in the paper by Zhao, Fang and Zhao (2010) and Zhao (2010).

Due to the explicit considerations of the discontinuous microstructure of material, the model has advantages in modeling of material failure behavior. For example, Kazerani, Zhao and Zhao (2010) used DLSM to model dynamic cracking propagation of PMMA by implementation of a rate-dependent cohesive law (Fig. 14.15 (a)), and Zhu et al. (2011) applied the DLSM on modeling wave propagation through jointed rock masses (Fig. 14.15 (b)). Moreover, the DLSM is also used in analysis of the cutting process of coal under a single blade (Zhao and Zhao, 2010), and dynamic fracture toughness of granite (Zhao, 2010).

In DLSM, there is no need to form the global stiffness matrix and only a local interaction is considered during calculation. This is very suitable for large scale parallel computing implementation (Zhao, 2010). The DLSM can be viewed as a meshless method like EFG and FPM, but more closer to DEM methods. Compared with the

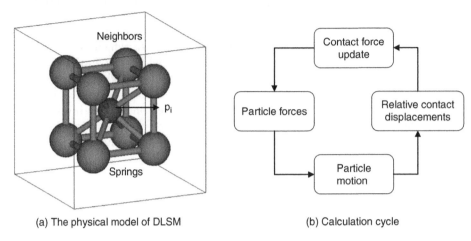

(a) The physical model of DLSM   (b) Calculation cycle

Figure 14.14 The physical model and the calculation cycle of DLSM.

particle-based DEM, the DLSM can directly use macroscopic parameters without a calibration process. This is regarded as the main advantage over other discrete element based methods. Moreover, DLSM also have advantages over existing meshless methods, e.g. EFG, FPM and SPH, on stability, no integration requirement and ease in dealing with heterogeneity problems. Due to the meshless and natural discrete properties of DLSM, it is suitable for dynamic discontinuous computations. However, as a newly developed numerical model, lots of work needs to be done, for example, calibrating the DLSM modeling results with experiments, developing multi-physical code, developing GPU-based high performance DLSM code, implementing complex constitutive models etc.

14.7.2 Multi-scale DLSM coupled with PMM

Multi-scale modeling is regarded as an exciting and promising methodology due to its ability to solve problems which cannot directly be handled by microscopic methods for the limitation of computing capacitance (Guidault *et al.*, 2007; Hettich, Hund and Ramm, 2008; Xiao and Belytschko, 2003). For this reason, the macro-material response can be directly obtained based on the micro-mechanical properties through multi-scale modeling. This advantage is extremely useful and essential in the study of material properties based on their microstructure information. It is well known that classical elasticity theory can only provide an adequate description of macroscopic mechanical response for most materials. It would be an unsuitable theory when facing the micro-mechanical response of these materials which are actually heterogeneous at microscopic level. Therefore, the microscopic modeling is necessary (Darve and Nicot, 2005). As has been mentioned above, directly building a microscopic model is usually impossible due to the limitation of computing resources. In this case, the multi-scale modeling provides a good choice. The most direct way to build a multi-scale numerical model is to combine two different scale methods. This methodology has been widely used in the coupling of MD with continuum mechanics models (Mullins

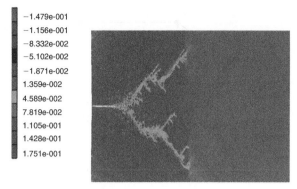

-1.479e-001
-1.156e-001
-8.332e-002
-5.102e-002
-1.871e-002
1.359e-002
4.589e-002
7.819e-002
1.105e-001
1.428e-001
1.751e-001

(a) Dynamic crack propagation through PMMA plate
(Kazerani, Zhao and Zhao, 2010)

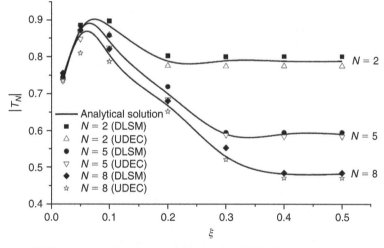

(b) Wave propagation through multi-joints by DLSM (Zhu *et al.*, 2011).

Figure 14.15 Examples of DLSM method on discontinuous dynamic computation.

and Dokainish, 1982; Tadmor, Ortiz and Phillips, 1996; Hasnaoui, van Swygenhoven and Derlet, 2003; Ma *et al.*, 2006).

In this section, a multi-scale model is developed to couple DLSM (Zhao, Fang and Zhao, 2010) and NMM (Shi, 1991). The reason for choosing NMM is that it is an advanced FEM and the background mesh used in the manifold method is independent of the physical model. Also, the DLSM is close to FEM due to the DOFs for each particle being the same as those of FEM nodes. These properties make it very suitable to couple these two different methods. A three layer structure is used to combine DLSM and NMM. The Particle based Manifold Method (PMM) is proposed to bridge between DLSM and NMM. The PMM element simplifies the contact detection between the particle in the DLSM model and NMM model and also serves as the cushion layer. The basic idea of the PMM element will now be introduced. The PMM element is realized by replacing the physical domain of the manifold element by the particle-based

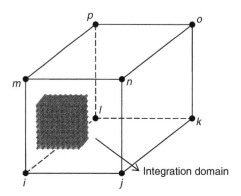

Figure 14.16 PMM element in m-DLSM.

DLSM model. The 3D PMM element used in m-DLSM is illustrated in Figure 14.16. The eight-node FEM element is used as the mathematic element and the DLSM model is used as the physical domain.

As the explicit integration method and lumped mass matrix are used in m-DLSM, the mass matrix of PMM element is taken as the 1/8 of the DLSM model included in the element:

$$\mathbf{M}_i^{\mathrm{PME}} = \frac{1}{8} \sum_{j=1}^{m_i} m_{ij}^p \tag{14.20}$$

where $\mathbf{M}_i^{\mathrm{PME}}$ is the mass matrix of PMM element, m_i is the number of particles included in the PMM element and m_{ij}^p is the mass of the particle. The stiffness matrix of the PMM element has to be obtained from a distinct method. As the deformation energy of the DLSM model is stored on the network of bonds between particles, the integration domain of the PMM element is neither 2D nor 3D. Actually, as the discrete natural property of the lattice network, the integration is realized through a summarizing operation as

$$\mathbf{K}_i^{\mathrm{PME}} = \sum_{j=1}^{n_i} \mathbf{K}_{ij}^b \tag{14.21}$$

where $\mathbf{K}_i^{\mathrm{PME}}$ is the stiffness matrix of the PMM element, n_i is the number of bonds included in the PMM element and $\mathbf{K}_{ij}^b$ is the stiffness matrix contributed by each lattice bond (a pair of normal and shear springs).

The work flow of the coupled calculation cycle in m-DLSM is shown in Figure 14.17. The DLSM and NMM computations are performed in parallel. Interactions between them are finished by the PMM model. Information exchange only happens at the beginning and the end of each cycle. The mapping of unbalance force from particles to PMM element computation is realized by using the following equation

$$\mathbf{F}_i^{\mathrm{ME}} = \mathbf{N}_{ij} \mathbf{F}_{ij}^{\mathrm{LS}} \tag{14.22}$$

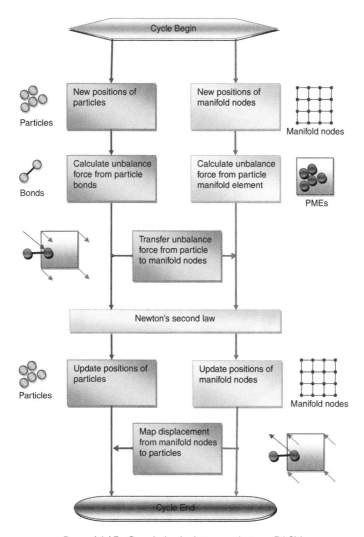

Figure 14.17 Coupled calculation cycle in m-DLSM.

where $\mathbf{F}_i^{ME}$ is the transferred force to the ith PMM element, $\mathbf{N}_{ij}$ is the interpolation matrix of displacement at the linked particle and $\mathbf{F}_{ij}^{LS}$ is the calculated unbalance force on the particle. After obtaining the unbalance force on particles and manifold nodes, new positions of these particles and manifold nodes can be obtained by using Newton's second law. Then, the displacement of NMM model is mapped to the particles which fall in the PMM model. The mapping operation is given as

$$\mathbf{u}_{ij}^{LS} = [\mathbf{N}_{ij}]^T \mathbf{u}_i^{ME} \qquad (14.23)$$

where $\mathbf{u}_{ij}^{LS}$ is the mapped displacement from PMM model to the linked particle and $\mathbf{u}_i^{ME}$ is displacement vector of the PMM element. The interaction between PMM and

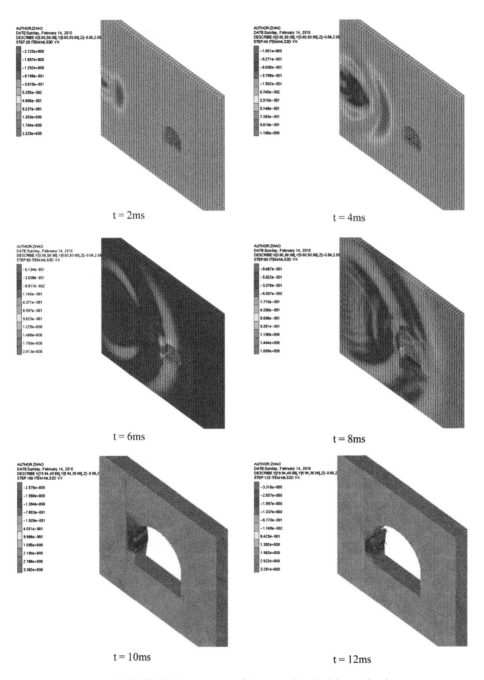

Figure 14.18 The failure process of the tunnel under blasting loading.

DLSM is realized through the interaction of the DLSM particle with the PMM particle. The interaction between PMM and NMM is realized by sharing common manifold nodes. The PMM model is used as the midst scale layer of the m-DLSM to realize coupling of the DLSM and the NMM. The implementation details of this method are given in Zhao (2010).

As the multi-scale model can largely reduce computing time required by the micro-numerical model, it is possible to deal with some large scale problems which cannot be handled by the micro model. As a newly developed numerical method, a few examples are given in Zhao (2010). For example, the blasting wave propagation through rock mass and the influence of discontinuities on the failure pattern of tunnels under a blasting wave is simulated by m-DLSM (see Fig. 14.18), in which the left side of the tunnel is broken under blast loading. For pure DLSM model, more than two million particles are needed to build this computational model. It means that information on more than ten million bonds needs to be stored, which is surely an inaccessible problem for the normal PC. However, only about half a million particles are used for the m-DLSM model which can easily be run on a normal PC. The m-DLSM can take advantages of both DLSM and NMM and further reduce computational resources. However, the implementation of the code is more complex and its functionality is mainly dependent on DLSM which still needs further development.

14.8 CONCLUSIONS

In this chapter, a few newly developed numerical methods for discontinuous dynamic computations are reviewed. It can be seen that developing a new numerical method is based on a new approximation method for the displacement function, e.g. NMM, SPH and DGM, or a new microscopic based model for elasticity, e.g. DLSM, or coupling between different methods, e.g. m-DLSM. These newly developed methods provide alternative choices for us when modeling discontinuous dynamic problems. However, each of them has its own advantages and demerits, thus for real applications, selection of a suitable numerical method should depend on the properties of the problem itself, i.e. microscopic or macroscopic, mechanism oriented or engineering oriented, precision requirement etc. Even if there are lots of numerical methods, development of new numerical methods is always needed to solve challenges in computational science (de Borst, 2008), i.e. to explicitly and accurately model dynamic crack propagation problems, multi-scale analysis and multi-physics analysis.

REFERENCES

Ariffin, A.K., Huzni, S., Nor, M.J.M. and Mohamed, N.A.N.: Hybrid finite-discrete element simulation of crack propagation under mixed mode loading condition. *Comput Meth Appl Mech Eng* 195 (2006), pp.4579–4593.

Asadpoure, A., Mohammadi, S. and Vafai, A.: Modeling crack in orthotropic media using a coupled finite element and partition of unity methods. *Finite Elem Anal Des* 42 (2006), 1165–1175.

Asferg, J.L., Poulsen, P.N. and Nielsen, L.O.: A consistent partly cracked XFEM element for cohesive crack growth. *Int J Numer Meth Eng* 72 (2007), 464–485.

Babuska, M. J. M. The partition of unity method. *Int J Numer Meth Eng* 40 (1997), pp.727–758.

Barth, T.J. and Sethian, J.A.: Numerical schemes for the Hamilton-Jacobi and level set equations on triangulated domains. *J Compul Phy* 145 (1998), pp.1–40.

Bažant, Z.P., Tabbara, M.R., Kazemi, M.T. and Pijaudier-Cabot, G.: Random particle model for fracture of aggregate or fiber composites. *J Eng Mech* 116(8) (1990), pp.1686–1705.

Belytschko, T. and Black, T.: Elastic crack growth in finite elements with minimal remeshing. *Int J Numer Meth Engng* 45 (1999), pp.601–620.

Benz, W. and Asphaug, E.: Simulations of brittle solids using smooth particle hydrodynamics. *Comp Phys Comm* 87 (1995), pp.253–265.

Buxton, G.A., Care, C.M. and Cleaver, D.J.: A lattice spring model of heterogeneous materials with plasticity. *Modell Simul Mater Sci Eng* 9(6) (2001), pp.85–97.

Castillo, P., Cockburn, B., Perugia, I. and Schötzau, D.: An a priori error analysis of the local discontinuous Galerkin method for elliptic problems. *SIAM J Numl Analy* 38(5) (2001), pp.1676–1706.

Chen, P., Huang, T., Yang, J. and Zhang, G.X.: Numerical simulation of rock fracture under dynamic loading using manifold method. *Key Eng Mat* 324–325 (2006), pp.235-238.

Chiou, Y.J., Lee, Y.M. and Tsay, R.J.: Mixed mode fracture propagation by manifold method. *Int J Fract* 114 (2002), pp.327–347.

Darve, F. and Nicot, F.: On incremental non-linearity in granular media: phenomenological and multi-scale views (Part I), *Int J Numer Anal Meth Geomech* 29 (2005), pp.1387–1409.

Darve, F., Servant, G., Laouafa, F. and Khoa, H.D.V.: Failure in geomaterials: continuous and discrete analyses. *Comput Methods Appl Mech Engrg* 193 (2004), pp.3057–3085.

de Borst, R.: Challenges in computational materials science: Multiple scales, multi-physics and evolving discontinuities. *Comput Mat Sci* 43(1) (2008), pp.1–15.

Dyka, C.T., Randles, P.W. and Ingel, R.P.: Stress points for tension instability in SPH. *Int. J. Numer. Methods Eng* 40 (1997), pp.2325–2341.

Fulk, D.A. and Quinn, D.W.: An analysis of 1-D smoothed particle hydrodynamics kernels. *J Comput Phys* 126 (1996), pp.165–180.

Gingold, R.A. and Monaghan, J.J.: Smooth particle hydrodynamics: theory and application to non-spherical stars. *Monthly Notices of the Royal Astronomical Society* 181(2) (1977), pp.375–389.

Gravouil, A., Moes, N. and Belytschko, T.: Nonplanar 3d crack growth by the extended finite element and level sets: Part II: level set update. *Int J Numer Meth Engng* 53 (2002), pp.2569–2586.

Gray, J.P. and Monaghan, J.J.: Numerical modelling of stress fields and fracture around magma chambers. *J Volcanol Geoth Res* 135 (2004), pp.259–283.

Guidault, P.A., Allix, O., Champaney, L. and Navarro, J.P.: A two-scale approach with homogenization for the computation of cracked structures. *Comput Struct* 85 (2007), pp.1360–1371.

Hasnaoui, A., van Swygenhoven, H. and Derlet, P.M.: Dimples on nanocrystalline fracture surfaces as evidence for shear plane formation. *Sci* 300 (2003), pp.1550–1552.

Hettich, T., Hund, A. and Ramm, E.: Modeling of failure in composites by X-FEM and level sets within a multiscale framework. *Comput Meth Appl Mech Eng*, 197 (2008), pp.414–424.

Jing, L.: A review of techniques, advances and outstanding issues in numerical modelling for rock mechanics and rock engineering. *Int J Rock Mech Min Sci* 40 (2003), pp.283–353.

Johnson, G.R., Stryk, R.A. and Beissel, S.R.: SPH for high velocity impact computations. *Comput Meth Appl Mech Eng* 139 (1996), pp.347–373.

Karami, A. and Stead, D.: Asperity degradation and damage in the direct shear test: A hybrid FEM/DEM approach. *Rock Mech Rock Eng* 41 (2008), pp.229–266.

Kazerani, T., Zhao, G.F. and Zhao, J.: Dynamic fracturing simulation of brittle material using the Distinct Lattice Spring Model (DLSM) with a full rate-dependent cohesive law. *Rock Mech Rock Eng* (2010), DIO: 10.1007/s00603-010-0099-0.

Khoei, A.R. and Nikbakht, M.: An enriched finite element algorithm for numerical computation of contact friction problems. *Int J Mech Sci* 49 (2007), pp.183–199.

Lesaint, P. and Raviart, P.: On a finite element method for solving the neutron transport equation. In: de Boor C. (ed.): *Mathematical Aspects of Finite Element Methods in Partial Differential Equations*, Academic Press, New York (1974), pp.89–123.

Levy, S.: *Exploring The Physics Behind Dynamic Fragmentation Through Parallel Simulations*. PhD thesis. École Polytechnique Fédérale de Lausanne (EPFL), Lausanne, Switzerland (2010).

Libersky, L.D. and Petschek, A.G.: Smooth particle hydrodynamics with strength of materials. *Lecture notes with physics* (1991), pp.248–257.

Ma, G.W., An, X.M., Zhang, H.H. and Li, L.X.: Modeling complex crack problems with numerical manifold method, *Int J Frac* 156 (1) (2009), pp.21–35.

Ma, G.W., Wang, X.J. and Li, Q.M.: Modeling strain rate effect of heterogeneous materials using SPH. *Rock Mech Rock Eng* 43 (2010), pp.763–776.

Ma, J., Lu, H., Wang, B., Hornung, R., Wissink, A. and Komanduri, R.: Multiscale simulation using generalized interpolation material point (GIMP) method and molecular dynamics (MD). *Comput Model Eng Sci* 14 (2006), pp.101–117.

Markus, P.K.H.: Numerical aspects of the eXtended Finite Element Method. *PAMM* 5(1) (2005), pp.355–356.

Menouillard, T., Réthoré, J., Combescure, A. and Bung, H.: Efficient explicit time stepping for the eXtended Finite Element Method (X-FEM). *Int J Numer Meth Eng* 63 (2006), pp.911–939.

Mergheim, J., Kuhl, E. and Steinmann, P.: A hybrid discontinuous Galerkin/interface method for the computational modelling of failure *Com Numl Meth Eng* 20 (2004), pp.511–519.

Miki, S., Sasaki, T., Koyama, T., Nishiyama, S. and Ohnishi, Y.: Development of coupled discontinuous deformation analysis and numerical manifold method (NMM-DDA) and its application to dynamic problems, *Proceedings of ICADD09* (2009), pp.255–263.

Monaghan, J.J.: An introduction to SPH. *Comput Phys Com* 48 (1988), pp.89–96.

Morris, J.P., Rubin, M.B., Block, G.I. and Bonner, M.P.: Simulations of fracture and fragmentation of geologic materials using combined FEM/DEM analysis. *Int J Impact Eng* 33 (2006), pp.463–473.

Mullins, M. and Dokainish, M.A.: Simulation of the (001) Plane crack in alpha-iron employing a new boundary scheme. *Philos Mag A* 46 (1982), pp.771–787.

Munjiza, A., Owen, D.R.J. and Bicanic, N.: A combined finite-discrete element method in transient dynamics of fracturing solids. *Eng Comput* 12 (1995), pp.145–174.

Munjiza, A.: *The combined finite-discrete element method*, Wiley (2004).

Noels, L., Radovitzky, R.: An explicit discontinuous Galerkin method for non-linear solid dynamics: Formulation, parallel implementation and scalability properties. *Int J Num Meth Eng* 74 (2008), pp.1393–1420.

Oden, J.T., Babuska, I. and Baumann, C.E.: A discontinuous hp finite element method for diffusion problems. *J Comput Phy* 146 (2) (1998), pp.491–519.

Okazawa, S., Terasawa, H., Kurumatani, M., Terada, K. and Kashiyama, K.: Eulerian finite cover method for solid dynamics. *Int J Comput Meth* 7 (2010), pp.33–54.

Ostoja-Starzewski, M., Sheng, P.Y. and Jasiuk, I.: Damage patterns and constitutive response of random matrix-inclusion composites. *Eng Fract Mech* 58(5–6) (1997), pp.581–606.

Ostoja-Starzewski, M.: Lattice models in micromechanics. *Appl Mech Rev* 55(1) (2002), pp.35–59.

Pedro, M.A.A. and Belytschko, T.: Analysis of three-dimensional crack initiation and propagation using the extended finite element method. *Int J Numer Meth Eng* 63 (2005), pp.760–788.

Prabel, B., Combescure, A., Gravouil, A. and Marie, S.: Level set X-FEM non-matching meshes: application to dynamic crack propagation in elastic-plastic media. *Int J Numer Meth Engng* 69 (2007), pp.1553–1569.

Rabczuk, T. and Eibl, J.: Simulation of high velocity concrete fragmentation using SPH/MLSPH. *Int J Numer Methods Eng* 56 (2003), pp.1421–1444.

Radovitzky, R., Seagraves, A., Tupek, M. and Noels, L.: A scalable 3D fracture and fragmentation algorithm based on a hybrid, discontinuous Galerkin, cohesive element method *Comput Meth Appl Mech Eng* (2010), doi:10.1016/j.cma.2010.08.014.

Randles, P.W. and Libersky, L.D.: Smoothed particle hydrodynamics: Some recent improvements and applications. *Comput Meth Appl Mech Eng* 139 (1996), pp.375–408.

Reed, W.H. and Hill, T.R.: Triangular mesh methods for the neutron transport equation, *Tech. Report LA-UR-73-479*, Los Alamos Scienti?c Laboratory, USA (1973).

Ribeaucourt, R., Baietto-Dubourg, M.C. and Gravouil, A.: A new fatigue frictional contact crack propagation model with the coupled X-FEM/LATIN method. *Comput Meth Appl Mech Eng* 196 (2007), pp.3230–3247.

Sasaki, T., Hagiwara, I., Sasaki, K., Yoshinaka, R., Ohnishi, Y., Nishiyama, S. and Koyama, T.: Stability analysis of ancient block structures by using DDA and manifold method. *Proceedings of ICADD09* (2009), pp.265–272.

Shi, G.H.: *Discontinuous deformation analysis, a new numerical model for the statics and dynamics of block systems*. PhD thesis, Univ. of California, Berkeley, Berkeley, Calif., USA (1988).

Shi, G.H.: Manifold method of material analysis. *Transactions of the 9th Army Conference on App Math and Comput*, U.S. Army Research Office, Minneapolis, MN, (1991), pp.57–76.

Song, J.H., Wang, H. and Belystchko, T.: A comparative study on finite element methods for dynamic fracture *Comput Mech* 42 (2008), pp.239–250.

Stéphane, B., Phu, V.N., Cyrille, D., Hung, N.D. and Amor, G.: An extended finite element library. *Int J Numer Meth Eng* 2 (2006), pp.1–33.

Stolarska, M., Chopp, D.L., Moes, N. and Belyschko, T.: Modelling crack growth by level sets in the extended finite element method. *Int J Numer Meth Eng* 51 (2001), pp.943–960.

Sukumar, N., Chopp, D.L. and Moran, B.: Extended finite element method and fast marching method for three-dimensional fatigue crack propagation. *Eng Fract Mech.* 70 (2003), pp.29–48.

Tadmor, E.B., Ortiz, M. and Phillips, R.: Quasicontinuum analysis of defects in solids. *Philos Mag A* 73 (1996), pp.1529–1563.

Terada, K., Ishii, T., Kyoya, T. and Kishino, Y.: Finite cover method for progressive failure with cohesive zone fracture in heterogeneous solids and structures. *Comput Mech* 39 (2007), pp.191–210.

Xiao, S.P. and Belytschko, T.: A bridging domain method for coupling continua with molecular dynamics. *Comput Meth Appl Mech Eng* 193 (2003), pp.1645–1669.

Zhang, Z.N. and Ge, X.R.: Micromechanical consideration of tensile crack behavior based on virtual internal bond in contrast to cohesive stress. *Theor Appl Fract Mech* 43 (2005), pp.342–59.

Zhao, G.F., Fang, J. and Zhao, J.: A 3D distinct lattice spring model for elasticity and dynamic failure. *Int J Numer Anal Meth Geomec* DOI: 10.1002/nag.930 (2010).

Zhao, G.F.: *Development of Micro-macro Continuum-discontinuum Coupled Numerical Method*. PhD thesis. École Polytechnique Fédérale de Lausanne (EPFL), Lausanne, Switzerland (2010).

Zhao, S.F. and Zhao, G.F.: Modeling the cutting process of coal under single blade through Distinct Lattice Spring Model (DLSM). *3rd International Conference on Advanced Computer Theory and Engineering, Chengdu, China* (2010), pp, 241–245.

Zhu, J.B., Zhao, G.F., Zhao, X.B. and Zhao, J.: Validation study of the distinct lattice spring model (DLSM) on P-wave propagation across multiple parallel joints. *Computers and Geotechnics* (2011), DOI:10.1016/j.compgeo.2010.12.002.

Chapter 15

Earthquakes as a rock dynamic problem and their effects on rock engineering structures

Ömer Aydan, Yoshimi Ohta, Mitsuo Daido, Halil Kumsar, Melih Genış, Naohiko Tokashiki, Takashi Ito and Mehdi Amini

15.1 INTRODUCTION

An earthquake is an instability problem of the Earth's crust and it is a subject of geoscience and geo-engineering. An earthquake is caused by varying crustal stresses and it is a product of rock fracturing and/or slippage of major discontinuities such as faults and fracture zones.

Earthquakes are known to be one of the natural disasters resulting in huge losses of human lives as well as of properties, as experienced in the 1999 Kocaeli earthquake. Therefore, the prediction of earthquakes is an important field of research. Some earthquake prediction projects such as the Tokai Earthquake Project in Japan, the Parkfield earthquake prediction project in USA have been recently undertaken. The most important item in earthquake prediction is how to assess the variations of the stress state of the Earth's crust with time. If both the stress state and the strength of the Earth's crust are known at a given time, one should be able to predict earthquakes with the help of some mechanical, numerical and instrumental tools.

When rock starts to fail, the stored mechanical energy in the rock tends to transform itself into different forms of energy. Experimental studies by Aydan and his group (Aydan, Minato and Fukue, 2001a; Aydan *et al.*, 2003a) showed that rock exhibits distinct variations of various measurable parameters such as electric potential, magnetic field, acoustic emission, and resistivity, besides load and displacement which are called multi-parameters, during deformation and fracturing processes. Furthermore, some in situ monitoring schemes were developed for structural safety of tunnels, abandoned lignite mines and historical underground structures as well as for earthquake prediction studies (Aydan *et al.*, 2005a, 2005b, 2006a; Aydan, Ohta and Tano, 2010). These variations may be useful in predicting the failures of rock structures in geoengineering as well as earthquakes in geoscience (Aydan, 2008a).

The dynamic responses of geo-materials during fracturing have not received any attention in the fields of geo-engineering and geo-science. However, these responses may be very important in the failure phenomenon of engineering structures (i.e. rockburst, squeezing, sliding) and the high ground motions induced by earthquakes (Aydan, 2003a; Aydan *et al.*, 2007). It is also known that the ground motions induced by earthquakes could be higher in the hanging-wall block or mobile side of the causative fault, as observed in the 1999 Kocaeli earthquake and the 1999 Chi-chi earthquake (Aydan *et al.*, 2007) as seen in Figure 15.1.

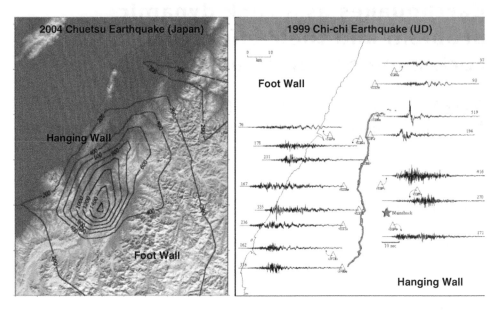

Figure 15.1 The foot-wall and hanging-wall effects on the maximum ground accelerations.

These surface ruptures and permanent ground deformations may cause the failure of foundations of super-structures such as bridges, dams, viaducts and pylons. The recent large earthquakes caused severe damage to pylons and the foundations of viaducts and bridges. Therefore, it is an urgent issue how to assess the effects of possible surface ruptures in potential earthquakes and how to minimize their effect on structures.

The 1999 Chi-chi, 2004 Chuetsu, 2005 Kashmir, 2008 Wenchuan and 2008 Iwate-Miyagi earthquakes caused many rock slopes and rockfalls. These slope failures and rockfalls, in turn resulted in the destruction of railways, roadways, housing and vehicles. The assessment of the stability of natural rock slopes against earthquakes is very important and one of our urgent issues is how to address it and how to devise the methods and technology for mitigation. It should be also noted that the failure forms induced by earthquakes might involve passive modes. The estimation of travel distance of natural slopes upon failure and their effect on engineering structures as well as on the natural environment is also of great importance. The travel distance may be of tremendous scale and it may cause severe damage to settlements and structures. Although this issue is well known and some simple methods are available, the present numerical methods are still insufficient to model post-failure motions.

It is well known that underground structures such as tunnels and powerhouses are generally resistant against earthquake-induced motions. However, they may be damaged when permanent ground movements occur in/along the underground structures. There are several examples of damage to tunnels due to permanent ground movements during the 1930 Tanna, 1978 Izu-Oshima-Kinkai, 1995 Kobe, 1999 Düzce-Bolu, 1999 Chi-chi, 2004 Chuetsu, 2005 Kashmir and 2008 Wenchuan earthquakes (Aydan *et al.*,

2010b,c). Most tunnels have non-reinforced concrete linings. Since the lining is brittle, the permanent ground movements may induce the rupture of the linings and falling debris may cause disasters with tremendous consequences to vehicles passing through. Therefore, this current issue must be urgently addressed. It should also be noted that the same issue is valid for the long-term stability of high-level nuclear waste disposal sites.

This article is concerned with earthquakes as a rock dynamic problem and their effects on rock engineering structures, which may be man-made or natural. In the first part, the responses observed in the multi-parameters experiments in the laboratory are presented so that they may be used to understand source characteristics of earthquakes and possible utilizations in the prediction of earthquakes. In the second part, the effects of earthquakes on rock engineering structures are described together with some laboratory model experiments and analytical studies, and issues to be addressed are pointed out and discussed.

15.2 MULTI-PARAMETER RESPONSES OF ROCKS DURING FRACTURING AND SLIPPAGE OF DISCONTINUITIES

15.2.1 Multi-parameter responses of rocks during deformation process and fracturing

Experimental studies for understanding of multi-parameter variations including electric potential, electrical resistivity, magnetic field, and acoustic emissions during deformation and fracturing process of geomaterials, which ranges from crystals, fault gauge-like materials to rocks under different loading regimes and environment have been taken by Aydan and his group (Aydan, Minato and Fukue, 2001a; Aydan et al., 2003a, 2005a, 2005b, 2007; Aydan, Ohta and Tano, 2010; Aydan, Tano and Ohta, 2010) (Fig. 15.2). Recently, the experiments have been repeated using an entirely manually operated loading system in order to eliminate the possible electric noise on the system (Aydan, Ohta and Tano, 2010). Furthermore, the dynamic responses of geo-materials during fracturing have not received any attention so far. The recent advances in measurement, monitoring and logging technologies enable us to measure and to monitor the dynamic responses of geo-materials during fracturing and slippage.

The applied load and induced displacement were automatically measured and stored on the hard disk of a laptop computer through an electronic logger. Electric potentials induced during the deformation of samples were measured through two electrodes attached to the top and bottom of samples using a voltmeter and logger unit and data were simultaneously stored on the hard-disk of the laptop computer. The electrodes were isolated from the loading frame with the use of isolators. Electric potentials were measured either as DC and/or AC. When electrical resistivity is measured, a function generator was used to produce electric current with a given amplitude. In some of experiments, magnetic field was also measured. In addition to the above measurement system, acoustic emissions (AE) devices and sensors and temperature sensors were used to measure the acoustic emissions as well as temperature variations during fracturing and sliding of samples. The acceleration responses of the samples during fracturing were measured by using an accelerometer, which can measure three components of

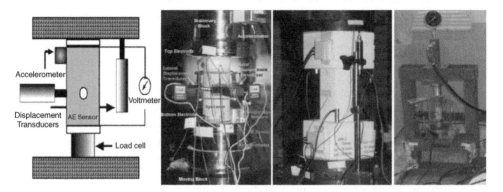

Figure 15.2 Experimental set-up and views of uniaxial compression experiments.

accelerations up to 10 g with a frequency range of 0–160 Hz. Various rock samples were tested. Although some of rock samples (e.g. granite, quartzite, sandstone etc.) contain piezo-electric minerals, some rock samples were selected such that they do not contain any piezo-electric substances, such as aragonite crystal, limestone, rocksalt, soapstone and marls.

Various responses measured during some of experiments are shown in Figures 15.3–15.4. The detailed discussions can be found in previous articles (Aydan, Minato and Fukue. 2001; Aydan *et al.*, 2003a, 2005a, 2005b, 2007; Ohta, Aydan and Tokashiki, 2008). Nevertheless, one can easily notice the distinct variations of multi-parameters during the deformation and fracturing of rocks. As seen from the experimental results, the deformation and fracturing of rock cause the distinct variations of electric potential, electrical resistivity, magnetic field and acoustic emissions in addition to conventional parameters such as displacement (strain) and force (stress). Despite the possibility of electrical noise from the loading devices in experimental results reported and discussed by Aydan, Minato and Fukue (2001) and Aydan *et al.* (2003a) previously, the same statements regarding the electric potential responses can be quoted herein:

1 Electric potential responses are closely related to strain response and they resemble the associated axial strain responses.
2 Bay-like variations of electric potential are distinctly observed before the rupture of samples.
3 Seismic electrical signals (SES) are generally observed before the rupturing of samples. These signals generally coincide with axial splitting type fracturing. However, seismic electric signals are less apparent for non-piezoelectric materials as compared with samples containing piezo-electric substances.
4 During a step-like loading path, which may resemble multi-stage creep tests, the measured responses of electric potential of various rock samples are closely related to the loading paths. In particular, the rate of electric potential development during load increment is very high and tends to decrease as the load is kept constant. However, it becomes asymptotic to an electric potential level greater than that induced

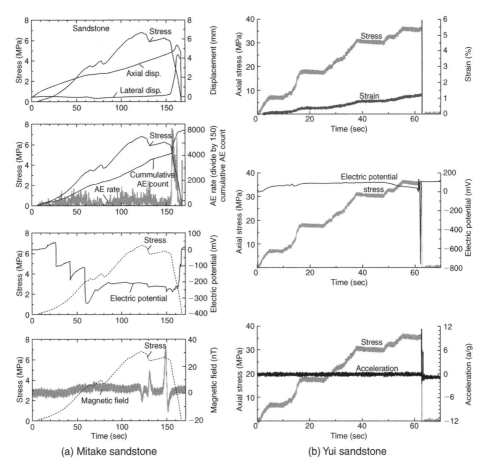

Figure 15.3 Multi-parameter response of some rocks.

in the previous load step. As the strain rate is quite high during load increment and tends to decrease once the load is kept constant, the electrical potential response seems to be closely associated with the strain rate response of the sample.

Fundamentally, the observed acceleration responses are similar to each other. The acceleration responses start to develop when the applied stress exceeds the peak strength and it attains the largest value just before the residual state is achieved, as seen in Figure 15.5. This pattern was observed in all experiments. Another important aspect is that the acceleration of the mobile part is much larger than that of the stationary part. This is also a common feature in all experiments.

The amplitude of accelerations of the mobile part of the loading system is higher than that of the stationary part (Fig. 15.6). It is very interesting to note that the maximum acceleration increases as the work done increases. This feature has striking similarities with the strong motion records of nearby earthquake faults observed in

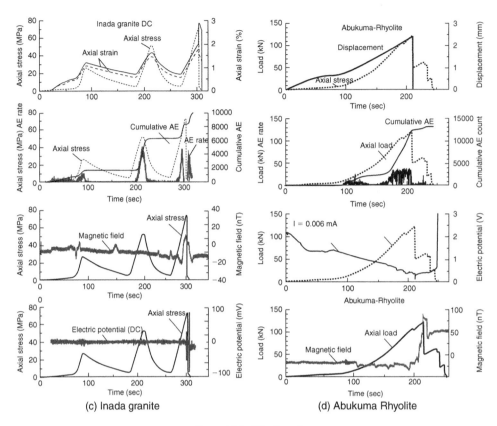

Figure 15.3 (Continued)

the recent large inland earthquakes. Furthermore, the waveforms of the acceleration records of the mobile part are not symmetric with respect to the time axis.

The amplitude of accelerations during the fracturing of hard rocks is much higher than that during the fracturing of soft rocks. This is directly proportional to the energy stored in samples before the fracturing. The chaotic responses in acceleration components perpendicular to the maximum loading direction may be observed. These may have some important implications for the procedures and interpretation of measurements for the short-term forecasting of failure events in geo-engineering and geo-science.

Figure 15.7(b) shows the frequency characteristics of measured accelerations of a sandstone sample shown in Figure 15.7(a). The integrated displacement and velocity responses using the EPS Method proposed by Ohta and Aydan (2007a, 2007b) are shown in Figure 15.8. It is important to note that the dominant frequency of the mobile part has a low frequency content compared to that of the stationary part (Ohta and Aydan, 2010).

The authors have also been performing some experiments on the elastic wave propagation velocity under creep loading regime. One example of experimental results is shown in Figure 15.9 for Oya tuff subjected to creep loading under a stress ratio

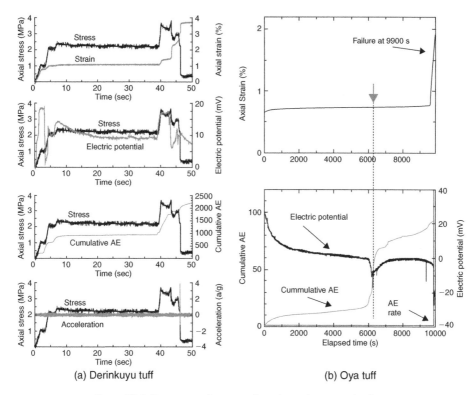

Figure 15.4 Response of some soft rocks under creep loading.

of 70%. The elastic wave velocity parallel to loading axis tends to decrease as the creep strain starts to increase during the tertiary creep phase. This observation was also reported in the previous studies (i.e. Toksöz, 1977).

Another important feature of rocks during deformation and fracturing is the change of their seepage characteristics. It is well known that the permeability decreases during the initial compression due to closure of pore space and micro-cracks, and tends to increase as the new micro-crack formation is initiated. The experiments reported by Kawamoto, Obara and Tokashiki (1981) using a servo-controlled testing machine indicated that the permeability of rocks becomes the largest when the peak strength is achieved and then it decreases to a value at residual strength level, which is greater than the initial permeability. This implies that the fluid flow rate will vary during formation of micro-cracks and their growth in relation to various stages of deformation process.

15.2.2 Multi-parameter responses of discontinuities during slippage

Tilting tests were carried out on samples containing a pre-existing discontinuity to study how electric potential is induced at the initiation of sliding and during the sliding process (Aydan *et al.*, 2003a). Tilting tests were performed using a wooden tilting device

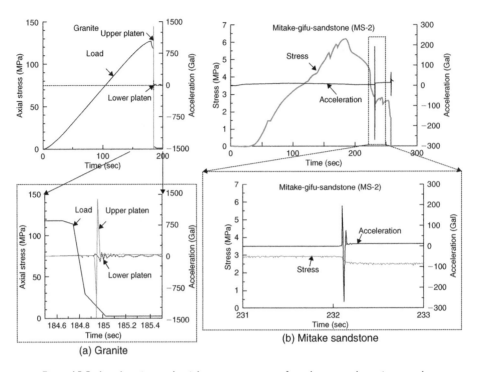

Figure 15.5 Acceleration and axial stress response of sandstone and granite samples.

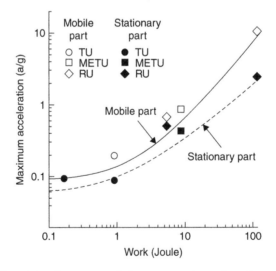

Figure 15.6 The relation between work done on samples and maximum acceleration.

to measure the electric potential variations produced during the sliding on existing discontinuities. The tilting was imposed manually. The electrodes were attached to the top and bottom sides of the assembled rock blocks containing a pre-existing discontinuity. The AE sensor was also attached to the top block and a laser displacement transducer

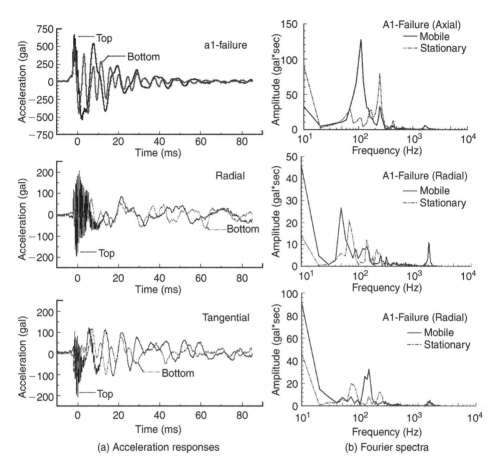

Figure 15.7 Acceleration response of a sandstone sample during fracturing and their Fourier spectra.

produced by KEYENCE was used to measure the translation displacement of the top block relative to the bottom block. Another laser displacement transducer was used to measure the amount of tilting from which the tilting angle was computed.

Figure 15.10 shows the responses of electric potential, displacement, stresses and AE measured during tilting tests of Ryukyu limestone (non-piezoelectric) and quartzite (piezoelectric) samples. Distinct seismic electric signals were observed during the tests on samples just before the initiation of sliding. Tests were carried out three times and very similar responses were obtained. The experiments further indicated that induced electrical potentials depend upon the rate of tilting.

15.2.3 Responses of discontinuities during slippage in stick-slip experiments

An experimental device consisted of an endless conveyor belt and a fixed frame (see Ohta and Aydan, 2011 for details). The inclination of the conveyor belt can be varied

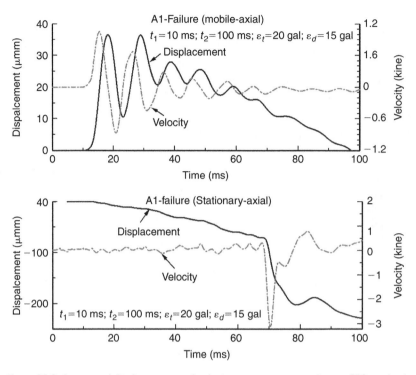

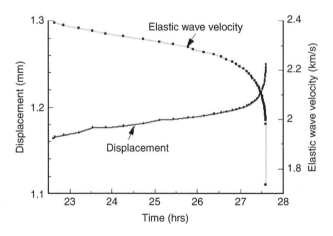

Figure 15.8 Integrated displacement and velocity responses according to EPS method.

Figure 15.9 Variation of elastic wave velocity in relation to creep response.

so that the tangential and normal forces can easily be imposed on the sample as desired. The belt itself was made of fiber reinforced rubber. In order to study the actual frictional resistance of interfaces of rock blocks, the lower block was stuck to the rubber belt while the upper block was attached to the fixed frame through

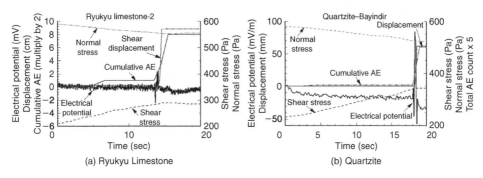

Figure 15.10 Multi-parameter responses during tilting experiments.

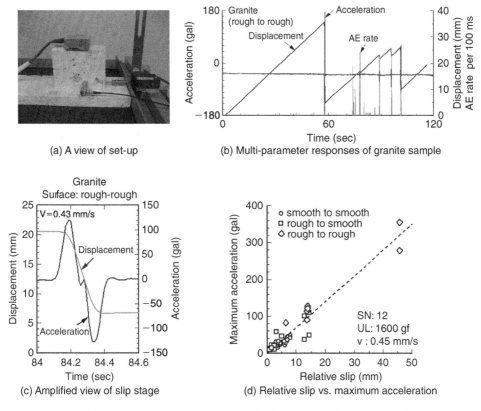

Figure 15.11 Multi-parameter responses of a discontinuity in granite during a stick-slip experiment.

a spring as illustrated in Figure 15.11(a). During experiments, displacement, acceleration and acoustic emissions were measured simultaneously. The stick-slip experiments were used to investigate the acceleration responses of the blocks of the experimental set-up as well as the recurrence periodicity of the stick-slip phenomenon, which are relevant to earthquakes. During the stick-slip experiments, the following conditions were investigated:

1   Variation of stiffness by either using different springs or weight of the upper block,
2   Variation of the velocity of the base plate and
3   Variation of the interface friction properties of blocks.

Stick-slip experiments were carried out on various natural rock blocks as well as other types of blocks made of foam, plastic, wood and aluminium. Responses of displacement, associated acceleration and acoustic emissions for stick-slip experiments on a granite sample are shown in Figure 15.11(b). There are many interesting observations in these responses, which are relevant to earthquake prediction as well as strong motions during earthquakes. Figure 15.11(c) shows an expanded view of responses of displacement and acceleration of a typical slip event, in which rise time, relative slip and stress drop can be clearly seen.

Figure 15.11(d) shows the relation between the amount of relative slip and maximum accelerations. It is noted that a linear relation holds and it should be also noted from Figure 15.11(b) that the induced acceleration waves are not generally symmetric with respect to the time axis.

15.2.4   Multi-parameter responses of rock during shock waves

Some experiments on the electric potential responses during shock waves were carried with the purpose of investigating co-seismic electrical potential variations during the passage of seismic waves. For this purpose, samples were sandwiched between two steel platens under a given small load and were isolated from the upper and lower platens by insulators. A hammer was used for creating shock waves by hitting the upper platen, and electric potential, accelerations and displacement responses were measured simultaneously. Piezo-electric and non-piezoelectric rock samples were used for experiments. Figure 15.12 shows two examples of experiments on gypsum and serpentine samples. The experimental results clearly show that the electric potential variations do occur during the passage of shock waves, distinctly.

15.2.5   Multi-parameter responses of discontinuities during seepage

It is also known that seepage through the ground induces electric field variations. Aydan and Daido (2002) carried out a series of experiments in order to investigate the fluid flow induced electric potential variations. A cylindrical container was used in seepage tests. The water head variations and acoustic emissions were measured. Silicious sand (No.7) was used and tap water as a seeping fluid through the pores of sand. Two copper electrodes having a length of 80 mm was inserted at two levels with a head difference of 10 cm. The induced electric potential was measured using a potentiometer and the electric current was measured as DC or AC. Figure 15.13(a) shows the electrical resistance variation under 0.1 mA electric current while Figure 15.13(b) shows the electric potential variation during seepage experiments. First a rapid increase in

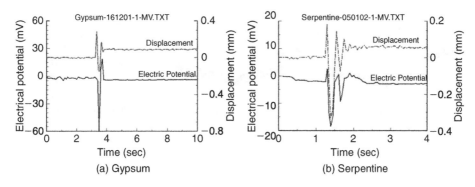

Figure 15.12 Electric potential and displacement responses during shock waves.

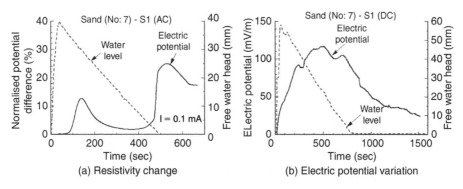

Figure 15.13 Electrical potential and resistance variations during seepage.

the electrical resistance occurs as the fluid flow front reaches the electrode level. Afterwards, the resistance starts to decrease for a while. Then a second peak, which is larger than the first one, is observed. And then, the gradual decrease of electrical resistance occurs. From this experimental result, it is clear that the fluid flow through pores of geomaterials increase their apparent resistance contrary to the common belief, that is, the fluid flow into induced pores during failure will decrease the apparent resistivity. Nevertheless, the fluid flow will cause large variations on the overall electrical resistance. Electric potential variation is proportional to the variation of seepage field with a certain time lag (Fig. 15.13 (b)). When unsaturated or partially saturated sand was filled with water rapidly, the induced electric potential increases first and then it starts to decrease with a negative rate. The seepage of fluid through pores and fractures of geomaterial induces an additional electrical potential. If the electric current is assumed to be equivalent to that of the electric current supplied into the ground, then the total electrical resistance of the ground will increase. Therefore, the apparent variation of electrical resistance of the ground will keep increasing. This may amplify the bay-like variation of the electrical resistance assumed in the literature. Furthermore, the variation of fluid rate during the formation of micro-cracks and their growth may enhance electric potential variations.

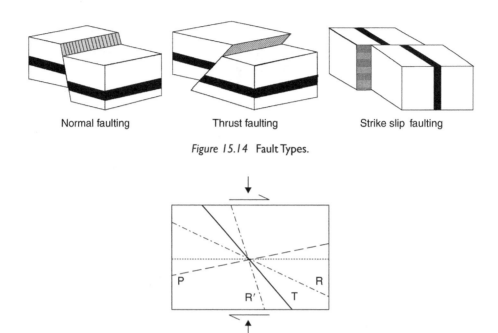

Normal faulting   Thrust faulting   Strike slip faulting

Figure 15.14 Fault Types.

Figure 15.15 Fractures in a shear zone or fault.

15.3   EARTHQUAKES AND THEIR PREDICTION

An earthquake is an instability problem of the Earth's crust. It is caused by varying crustal stresses and it is a product of rock fracturing and/or slippage of major discontinuities such as faults or fracture zones. Earthquake faults are geologically defined as a discontinuity in a geological medium along which a relative displacement took place. Faults are broadly classified into normal faults, thrust faults and strike-slip faults (Fig. 15.14). Their length may range from a few microns to thousands kilometers in a strict sense of definition. A fault is presumed to be geologically active if a relative movement took place in a period less than 2 millions years.

It is well known that a fault zone may involve various kinds of fractures as illustrated in Figure 15.16 and it is a zone having a finite volume (Aydan, Ito and Ichikawa, 1993; Aydan *et al.*, 1999a, 1999b; Aydan, Kumsar and Ulusay, 2002). In other words, it is not a single plane. Furthermore, the faults may have a negative or positive flower structure as a result of their trans-tensional or trans-pressional nature and the reduction of vertical stress near the earth surface as shown in Figure 15.16. For example, even a fault having a narrow thickness at depth may cause a quite broad rupture zones and numerous fractures on the ground surface during earthquakes due to flowering phenomenon.

Plate tectonics theory is often quoted to explain earthquake occurrences. However, the plate tectonics theory fails to explain why earthquakes occur within the plates. Therefore, a more broad concept must be introduced to explain the causes

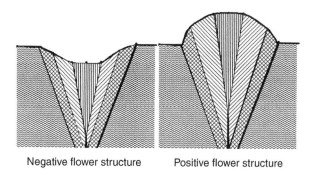

Negative flower structure   Positive flower structure

Figure 15.16 Negative and positive flower structures due to trans-tension or trans-pression faulting.

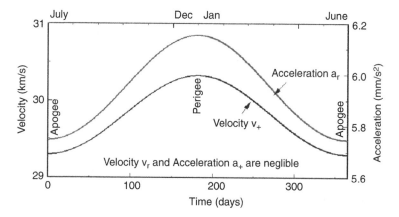

Figure 15.17 Acceleration and velocity fields of the earth during an entire year.

of earthquakes. It is well known that the earth travels around the sun in a given orbit. Furthermore, its distance with the sun varies during the entire year, which induces variations of its acceleration and velocity fields (Fig. 15.17). The 24 hours long rotation around its rotation axis with an inclination of 23.5° with respect to the orbit axis and the existence of the moon revolving around the earth with a period of aapproximately 28 days induces extra forces on the earth in addition to that of the sun.

Figure 15.18 shows how forces at the earth surface change during an entire day during a new and full moon. Furthermore, the surface area of the northern and southern hemispheres receiving sunlight varies during an entire year, which undoubtedly induces daily and year-long thermal stress cycles on the earth. All these additional factors must cause some cyclic variations of the stress state of the earth in addition to that caused by its gravitational system.

15.3.1 Stress conditions in the Earth's crust

Earthquakes are produced as a result of rupturing of the earth's crust caused by the stress state acting in the earth. It is generally believed that the so-called tectonic stresses

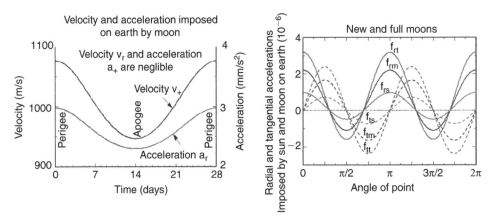

Figure 15.18 (a) Effect of moon on the velocity and acceleration of the earth per month, (b) effect of tidal forces of moon and sun on the earth per day.

are the principal actors. The plate tectonics theory has been presumed to be able to answer the causes of tectonic stresses and to explain why earthquakes occur along some certain locations. However, this theory is insufficient to explain intra-plate earthquakes since the theory is based on rigid body kinematics. The driving force for plate tectonics is assumed to be the mantle convection, which is thought to be resulting from non-uniform temperature distribution in the upper mantle caused by subducting plates. There is no doubt that such a temperature difference could cause the convection. The questions then are why the subduction of plates occurred and why the surface of the earth is divided into several plates. There are probably no answers to these questions in geophysics at present.

Aydan (1995) analyzed the stress state of the earth by modeling it as a spherical body consisting of layers exhibiting thermo-elasto-plastic behavior under pure gravitational acceleration. Figure 15.19 shows the distribution of radial and tangential stresses in the earth. From his study the following conclusions were drawn:

a) If the sphericity of the earth is taken into account, it is possible to explain why the horizontal stress is larger than the vertical stress near the ground surface. In other words, the large horizontal stress is due to gravity not due to presumed tectonic forces resulting from unknown sources.
b) For a spherical symmetric earth, the tangential stress (lateral stress) is the maximum principal stress and the radial stress (vertical stress) is the minimum principal stress irrespective of the mechanical behaviour of rocks.
c) The crust and the mantle are in a plastic state, which may have some important implications in rock mechanics and rock engineering as well as in geoscience. In other words, the earth is not an elastic body as presumed in many different studies.

The above approach provides very valuable information and it is a first approximation to the stress state of the earth. Although the gravitational pull is the governing

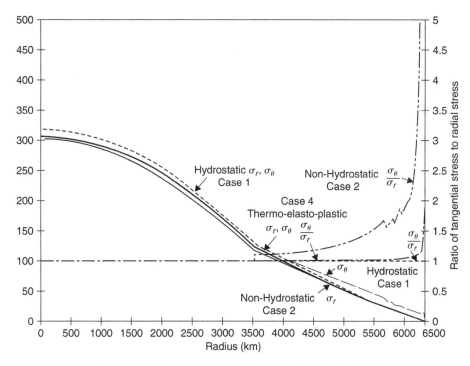

Figure 15.19 The stress state of the earth (from Aydan, 1995).

element in shaping the stress state of the earth, slight variations from that obtained from the gravitational model are caused by its rotation around its axis as well as its revolution around the sun, and mantle convection due to the non-uniformity in the thermal field resulting from the subduction of the cooler plates into the hot mantle. As the spherical symmetry condition does not strictly hold for the earth, better estimations may be done using actual geometry and distributions of materials constituting the earth. Nevertheless, it should be noted that our knowledge on the constitutive parameters of earth constituting materials are still insufficient.

In addition to the computational results given in Figure 15.19, rock stress measurements up to now showed that the earth's crust is under a compressive stress regime (Zoback *et al.*, 1985; Aydan and Paşamehmetoğlu, 1994; Aydan and Kawamoto, 1997). However, these measurements are restricted to depths less than 4 km while large earthquakes generally occur at depths greater than 10 km.

Aydan (2000a) proposed a new method to infer the crustal stresses from the striations of faults or other structural geological features, which may be quite useful in studying the stress state associated with past and current earthquakes. He recently advanced this method to infer the stress state of the earth's crust from focal plane solutions. The comparisons of inferred stress states with actual measurements confirmed the validity of the method (Aydan and Kim, 2002; Aydan, 2003a; Aydan and Tokashiki, 2003). An example of the method applied to Japan is shown in Figure 15.20.

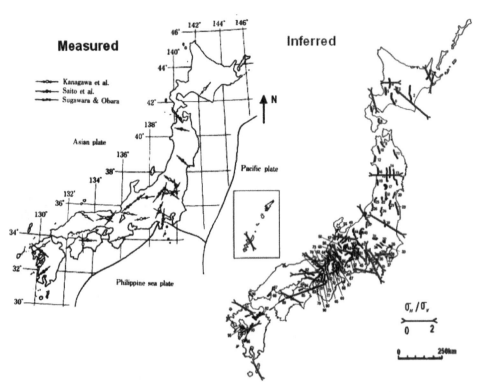

Figure 15.20 Directions and magnitudes of inferred maximum horizontal stresses with measurements (from Aydan, 2003d; Inset figure from Sugawara and Obara, 1993).

Figures 15.21 and 15.22 show an application of this technique to Turkey and a specific region along the North Anadolu Fault (NAF-KAF) of Turkey. This fault zone is about 1500 km long and produces very large earthquakes from time to time. Figure 15.22 shows the maximum horizontal stress normalized by the vertical stress (σ_H/σ_V).

15.3.2 Ground motions

Faulting during earthquakes results in the vibration of the surrounding medium upon the slippage, and it terminates as a result of re-distribution of crustal stresses. Furthermore, the crust deforms and some of the deformation is stored as elastic deformation while the rest of the deformation results in permanent ground deformations with/without relative slip during an earthquake. Some of relative slips are known as surface ruptures. Figure 15.1 shows the ground accelerations caused by earthquakes. The most important feature is that the ground accelerations are higher on the mobile side of the fault than that on its footwall side. This feature is particularly of great interest as it is very similar to the observations in the laboratory experiments on rocks presented in Section 2. Furthermore, the ground motions are also quite high at the ends of the earthquake fault.

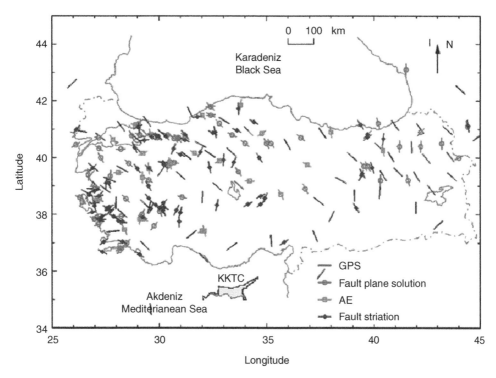

Figure 15.21 Directions of inferred maximum horizontal stresses with measurement.

The recent GPS and INSAR technologies provide us with ground deformations in a broad area following the earthquakes. Figure 15.23 shows the ground deformation vectors measured following the 1999 Kocaeli earthquake. It is natural to expect that these deformations induce permanent straining on the ground surface, which may impose tremendous forces on longer or larger structures such as elevated highways, dams, pipelines etc. Figure 15.24 shows computed principal strain variations associated with co-seismic deformations shown in Figure 15.23. The computation procedure is based on the method proposed by Aydan (2000b).

The authors have recently showed that the permanent ground deformations may be obtained from the integration of acceleration records. The erratic pattern screening (EPS) method proposed by the authors (Ohta and Aydan, 2007a, 2007b) can be used to obtain the permanent ground displacement with the consideration of features associated with strong motion recording. This method is applied to results of laboratory faulting and shaking table tests, in which shaking was recorded using both accelerometers and laser displacement transducers, simultaneously. Furthermore, the method was applied to strong motion records of several large earthquakes with measurements of ground movements by GPS as seen in Figure 15.25. The computed responses were almost the same as the actual recordings, implying that the proposed method can be used to obtain actual recoverable as well as permanent ground motions from acceleration recordings. Figure 15.26 shows the application of the EPS method to the strong

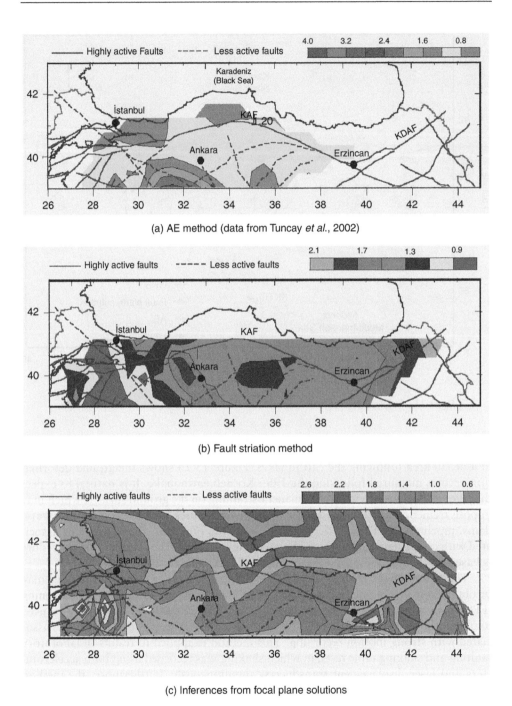

(a) AE method (data from Tuncay et al., 2002)

(b) Fault striation method

(c) Inferences from focal plane solutions

Figure 15.22 Comparison of normalized maximum horizontal stress obtained from different techniques along the North Anadolu Fault Zone of Turkey.

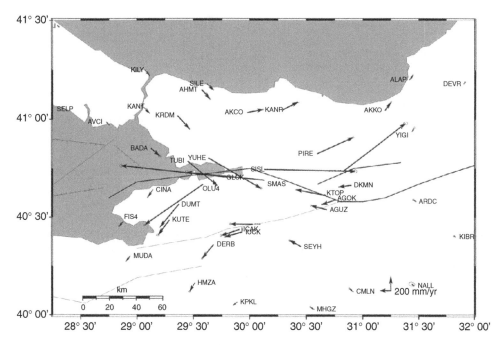

Figure 15.23 Permanent ground deformations induced by the 1999 Kocaeli earthquake (from Reilinger *et al.*, 2000).

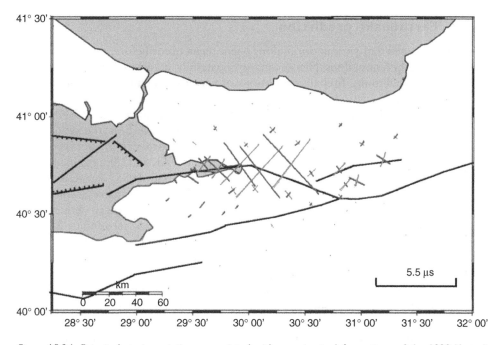

Figure 15.24 Principal strain variations associated with co-seismic deformations of the 1999 Kocaeli earthquake.

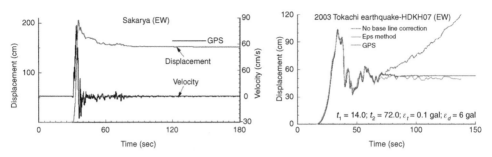

Figure 15.25 Comparison of the permanent ground deformation by the EPS method with measured GPS recordings (from Ohta and Aydan, 2007b).

motions records of the 2009 L'Aquila earthquake to estimate the co-seismic permanent ground displacements. These results are very consistent with the GPS observations. However, it should be noted the permanent ground deformations recorded by the GPS does not necessarily correspond to those of the crustal deformation. Surface deformations may involve crustal deformation as well as those resulting from the plastic deformation of ground due to ground shaking. The records at ground surface and 260 below the ground surface taken at IWTH25 during the 2008 Iwate-Miyagi earthquake clearly indicated the importance of this fact in the evaluation of GPS measurements (KIKNET, 2008).

15.3.3   Earthquake prediction

It is often reported that various anomalous phenomena occur before, during and after earthquakes. The anomalous phenomena are generally associated with the behavior of animals, lightning, fireballs, variations of various gas emissions, groundwater level, gravity, geomagnetic field and electric potential before, during and after earthquakes (Ikeya and Matsumoto, 1997; Ikeya *et al.*, 1997; Mizutani *et al.*, 1976). Some earth-scientists from the former USSR, China and Japan have been the pioneers in utilizing these phenomena as precursors of earthquakes in order to predict them. The earthquake prediction researches in Japan, USA, former USSR and China gained a considerable acceleration in the early 1970's (Toksöz, 1977; Mogi, 1985). Particularly, the successful prediction of the Haicheng earthquake in 1975 made many seismologists all over the world optimistic about the earthquake prediction. However, with the failure of predicting the 1976 Tangshan Earthquake, which killed more than 250,000 people, in the following year, many geo-scientists understood that it was still premature to predict earthquakes. This resulted in the disappearance of the enthusiasm for earthquake prediction studies and projects seen in the 1970's among scientists and politicians. Japan gave up the hope of success for earthquake prediction in 1997 after the Hyogo-ken Nanbu Earthquake, which devastated Kobe City and its close vicinity.

Although mankind is still premature in predicting earthquakes, it is believed that the accumulation of anomalous behaviors observed in each earthquake should be carefully documented for future generations, who might be successful in doing so.

Figure 15.26 Estimated permanent ground displacements by EPS method (from Aydan *et al.*, 2009c).

There is no doubt that the correct information on observation should provide some hints for such people to develop the methods for predicting earthquakes in spite of current pessimistic views.

Earthquakes are simply the products of rupture process of rocks composing the earth's crust. The stored mechanical work done on the earth's crust resulting from its deformation is transformed into various forms throughout its rupturing process if the energy conservation law of continuum mechanics holds. The forms of transformation of the mechanical work done can be observed as heat flux, electric current (magnetic current), kinetic energy and etc. as seen in Section 2. Without any doubt, these transformations will result in various phenomena, which may be called anomalous phenomena.

With the birth and advancement of rock mechanics in the 1960's, some physical backgrounds for various phenomena were established from laboratory tests on rocks, which are directly relevant to the rupturing process of the earth's crust. As experimentally shown in Section 2, the rock specimen starts to behave in a non-linear manner after a certain stress threshold. After this threshold, some fracturing starts to take place. Each time new fractures occur, various forms of transformation of the work done would take place and the imposed stress level increase would be stored as mechanical work done on the specimen. As experimentally shown in Section 2, these transformations may be seen as

a) sound waves (acoustic emissions),
b) electric potential (magnetic) variations and pulses,
c) increase in permeability and porosity implying decrease in pore pressure, and induced fluid flow,
d) temperature increase and heat flux,
e) gas emissions,
f) degradation of elastic properties, subsequently reduction in P and S wave velocities,
g) decrease in electrical resistivity and
h) creep.

It is experimentally known that the fracturing process becomes unstable after a certain stress level (i.e. Fig. 15.4(b)). From that level onward, the so-called secondary creep and subsequently tertiary creep processes take place and result in the failure of the specimen. Although the boundary conditions are different in actual earthquakes, the stages of the rupture process and transformation forms of the mechanical work done should be very similar to those observed on rock specimens tested under laboratory conditions.

The current earthquake prediction methods are mainly observational and they are basically too empirical. Figure 15.27 shows the basic concepts of the models for earthquake prediction adopted in the 1970's. The methods available may be categorized as follows:

a) Tilting or ground deformation anomaly method,
b) Creep method,
c) Ground water level anomaly method,
d) Elastic wave velocity anomaly method,
e) Electrical resistivity anomaly method,
f) Electric field anomaly method,
g) Magnetic field anomaly method,
h) Seismic gap method,
i) Gas emission anomaly method,
j) Gravity anomaly method, and
k) Anomalous animal behaviour method (Ikeya et al., 1997).

Although Nur (1972) tried to unify some of these methods into a dilatancy-diffusion method, there is presently no world-wide accepted approach, based on a sound universal theory. In many sites such as Parkfield in USA and Tokai region in Japan, some of these methods are simultaneously used.

In this section, the applicability of some of these techniques to earthquake prediction is presented with the consideration of the evaluations of the authors so far.

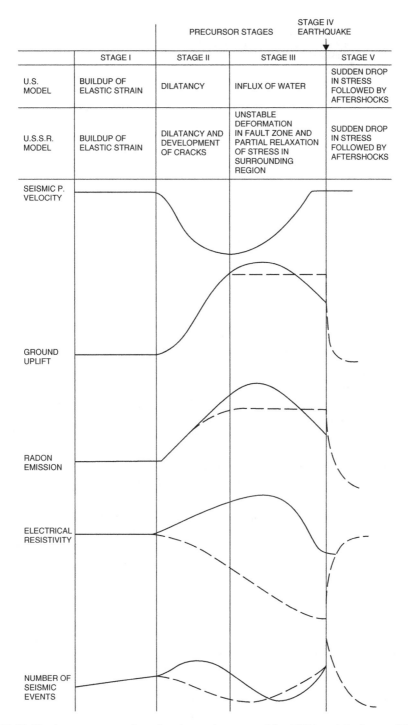

Figure 15.27 The basic concepts of earthquake prediction models of USA and the former USSR the 1970's (after Toksöz, 1977).

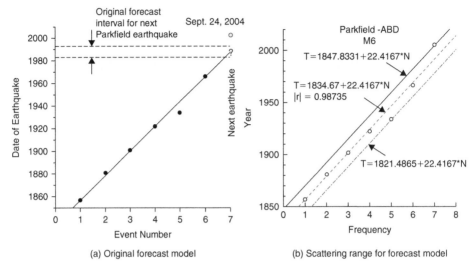

Figure 15.28 M6 class earthquake prediction models for the Parkfield region.

15.3.3.1 Stick-slip phenomenon and earthquake recurrence period concept

Earthquake prediction projects such as the Tokai Earthquake Project in Japan and the Parkfield Earthquake Project in USA have been initiated using the earthquake recurrence period model based on the stick-slip phenomenon. Although the stick-slip phenomenon is a valid concept, our experiments in the laboratory clearly showed that the recurrence period is not always constant even for apparently the same discontinuity and environmental conditions (see Ohta and Aydan (2009, 2010) for details and Figure 15.11). Multi-parameter observations have been undertaken for short-term prediction. The recent Parkfield earthquake on September 28, 2004 validated the prediction methodology based on the recurrence concept (Fig. 15.28 (a)). Nevertheless, the 2004 Parkfield earthquake implied that predictions should take into account the scattering range (Fig. 15.28(b)). Although this example clearly showed that it is difficult to make short-term prediction by this model, it can definitely be used for identifying regions with a high risk of earthquake occurrence provided that reliable past seismic data are available.

15.3.3.2 Effect of solar system on earthquake occurrence

An example of the effect of moon phases, the variation of yearly orbital acceleration of the earth around the sun, and the variation of heat input from the sun, on earthquake occurrence in Turkey for 1999 is given in this section. The values of functions for each parameter are subtracted from the average values and the resulting value is normalized by the average value as shown in Figure 15.29. Although the weighting of each parameter may be different, the total effect should be the sum of these parameters in addition to the sunspot index. A first glance at the figure implies that some co-relations exist between the earthquake occurrence and moon phases. It seems that the earthquake

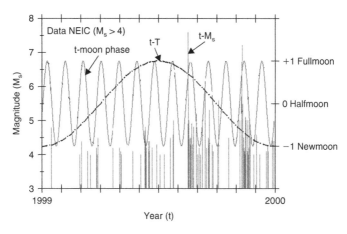

Figure 15.29 The relations between the earthquake magnitude and moon-phase, heat flux input.

occurrences, particularly those with great magnitude, are more likely during the new-moon and full-moon periods. In addition to the effect of moon phases, the variation of orbital acceleration of the earth and thermal stresses induced by the variation of heat flux input from the sun should have some effects. The effect of the variation of orbital acceleration of the earth around the sun should be minimum during summer and maximum during winter periods (see Fig. 15.17). The effect of the heat flux will have an effect on temperature distributions of the atmosphere and the crust, which should induce some cyclic thermal stresses. Although time variation of temperature distribution may be shifted from the theoretical heat input function due to the heat conduction characteristics of atmosphere, the resulting thermally induced stress variations must be compressive during hot periods and tensile during cold periods. Although some good co-relations exist among the parameters listed here, the weighting of each parameter on the overall stress state changes of the crust is still unclear.

15.3.3.3 GPS method

As stated previously, if the stress state and the yielding characteristics of the earth's crust are known at a given time, one may be able to predict earthquakes with the help of some mechanical, numerical and instrumental tools. The GPS method may be used to monitor the deformation of the earth's crust continuously with time. From these measurements, one may compute the strain rates and probably the stress rates. The stress rates derived from the GPS deformation rates can be effectively used to locate the areas with high seismic risk as proposed by Aydan, Kumsar and Ulusay (2000). Thus, daily variations of derived strain-stress rates from dense continuously operating GPS networks in Japan and the USA may provide high quality data to help understand the behaviour of the earth's crust preceding earthquakes.

First we describe a brief outline of the GPS method proposed by Aydan (2000b, 2004a, 2006b). The crustal strain rate components can be related to the deformation rates at an observation point (x, y, z) through the geometrical relations (i.e. Eringen, 1980) as given below:

$$\dot{\varepsilon}_{xx} = \frac{\partial \dot{u}}{\partial x}; \quad \dot{\varepsilon}_{yy} = \frac{\partial \dot{v}}{\partial y}; \quad \dot{\varepsilon}_{zz} = \frac{\partial \dot{w}}{\partial y};$$

$$\dot{\gamma}_{xy} = \frac{\partial \dot{v}}{\partial x} + \frac{\partial \dot{u}}{\partial y}; \quad \dot{\gamma}_{yz} = \frac{\partial \dot{w}}{\partial y} + \frac{\partial \dot{v}}{\partial z}; \quad \dot{\gamma}_{zx} = \frac{\partial \dot{w}}{\partial x} + \frac{\partial \dot{u}}{\partial z} \tag{15.1}$$

where $\dot{u}$, $\dot{v}$ and $\dot{w}$ are displacement rates in the direction of x, y and z respectively. $\dot{\varepsilon}_{xx}$, $\dot{\varepsilon}_{yy}$ and $\dot{\varepsilon}_{zz}$ are strain rates normal to the x, y and z planes and $\dot{\gamma}_{xy}$, $\dot{\gamma}_{yz}$, $\dot{\gamma}_{zx}$ are shear strain rates. The GPS measurements can only provide the deformation rates on the earth's surface (x (EW) and y (NS) directions) and it does not give any information on deformation rates in the z direction (radial direction). Therefore, it is impossible to compute normal and shear strain rate components in the vertical (radial) direction near the earth's surface. The strain rate components in the plane tangential to the earth's surface would be $\dot{\varepsilon}_{xx}$, $\dot{\varepsilon}_{yy}$ and $\dot{\gamma}_{xy}$. Additional strain rate components $\dot{\gamma}_{yz}$ and $\dot{\gamma}_{zx}$, which would be interpreted as tilting strain rate in this article, are defined by neglecting some components in order to make the utilization of the third component of deformation rates measured by GPS as follows:

$$\dot{\gamma}_{zx} = \frac{\partial \dot{w}}{\partial x}; \quad \dot{\gamma}_{zy} = \frac{\partial \dot{w}}{\partial y} \tag{15.2}$$

Let us assume that the GPS stations are re-arranged in a manner so that a mesh is constituted similar to the ones used in the finite element method. It is possible to use different elements as illustrated in Figure 15.30. Using the interpolation technique used in the finite element method, the displacement in a typical element may be given in the following form for any chosen order of interpolation function:

$$\{\dot{u}\} = [N]\{\dot{U}\} \tag{15.3}$$

where $\{\dot{u}\}, [N]$ and $\{\dot{U}\}$ are the deformation rate vector, shape function and nodal displacement vector of a given point in the element, respectively. The order of shape function $[N]$ can be chosen depending upon the density of observation points. The use of linear interpolation functions has already been presented elsewhere (Aydan, 2000b, 2003c). From Equations (15.1), (15.8) and (15.3), one can easily show that the following relation holds among the components of the strain rate tensor of a given element and displacement rates at nodal points:

$$\{\dot{\varepsilon}\} = [B]\{\dot{U}\} \tag{15.4}$$

Using the strain rate tensor determined from the Equation (15.4), the stress rate tensor can be computed with use of a constitutive law such as Hooke's law for elastic materials, Newton's law for viscous materials and Kelvin's law for visco-elastic materials (Aydan, 1997a; Aydan and Nawrocki, 1998). For simplicity, Hooke's law is chosen and is written in the following form:

$$\begin{Bmatrix} \dot{\sigma}_{xx} \\ \dot{\sigma}_{yy} \\ \dot{\sigma}_{xy} \end{Bmatrix} = \begin{bmatrix} \lambda + 2\mu & \lambda & 0 \\ \lambda & \lambda + 2\mu & 0 \\ 0 & 0 & \mu \end{bmatrix} \begin{Bmatrix} \dot{\varepsilon}_{xx} \\ \dot{\varepsilon}_{yy} \\ \dot{\gamma}_{xy} \end{Bmatrix} \tag{15.5}$$

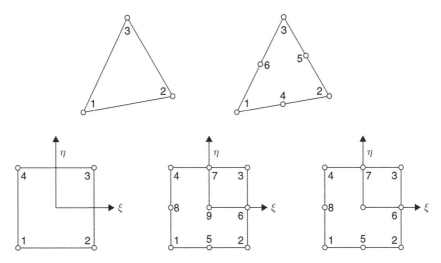

Figure 15.30 Finite elements for GPS method.

where λ and μ are Lame's constants, which are generally assumed to be $\lambda = \mu = 30 \, \text{GPa}$ (Fowler, 1990). It should be noted that the stress and strain rates in Equation (15.5) are for the plane tangential to the earth's surface. From the computed strain rate and stress rates, principal strain and stress rates and their orientations may be easily computed as an eigenvalue problem.

To identify the locations of earthquakes, one has to compare the stress state in the earth's crust at a given time with the yield criterion of the crust. The stress state is the sum of the stress at the start of GPS measurement and the increment from GPS-derived stress rate given as:

$$\{\sigma\} = \{\sigma\}_0 + \int\limits_{T_0}^{t} \{\dot\sigma\} dt \tag{15.6}$$

If the previous stress $\{\sigma\}_0$ is not known, a comparison for the identification of the location of the earthquake cannot be done. The previous stress state of the earth's crust is generally unknown. Therefore, Aydan, Kumsar and Ulusay (2000) proposed the use of maximum shear stress rate, mean stress rate and disturbing stress for identifying the potential locations of earthquakes. The maximum shear stress rate, mean stress rate and disturbing stress rate are defined below:

$$\dot\tau_{max} = \frac{\dot\sigma_1 - \dot\sigma_3}{2}; \quad \dot\sigma_m = \frac{\dot\sigma_1 + \dot\sigma_3}{2}; \quad \dot\tau_d = |\dot\tau_{max}| + \beta\dot\sigma_m \tag{15.7}$$

where β may be regarded as a friction coefficient. It should be noted that one (vertical) of the principal stress rates is neglected in the above equation since it can not be determined from GPS measurements. The concentration locations of these quantities may be interpreted as the likely locations of the earthquakes as they imply the increase in disturbing stress. If the mean stress has a tensile character and its value increases, it simply implies the reduction of resistance of the crust.

As Aydan, Kumsar and Ulusay (2000) have shown previously, the recent earthquakes in Turkey fall within the maximum shear stress concentration regions. Similarly close correlations exist between mean stress rate and disturbing stress rate concentrations and epicenters of the earthquakes. Therefore, the concentrations of maximum shear stress rate and disturbing stress rate may serve as indicators for identifying the location of potential earthquakes. The high mean stress rate of tensile character may also be used to identify likely earthquakes due to normal faults (Aydan, Kumsar and Ulusay, 2000). Figure 15.31 shows the contours of disturbing stress rate together with the epicenters of the earthquakes with a magnitude greater than 4 which occurred during 1995 and 1999 using the GPS data reported by Reilinger et al. (1997). Particularly, the epicenters of the 1999 Kocaeli, 1999 Düzce-Bolu, 2000 Orta-Çankırı and 2000 Honaz-Denizli earthquakes coincide with the regions of concentration of these stress rates. Therefore, the GPS method implies that it is possible to locate the earthquakes.

Aydan (2003c, 2004a) also showed that the time of occurrence of earthquakes in terms of weeks may be possible using the GPS measurements recorded during the 2003 Miyagi-Hokubu earthquake (Figs. 15.32 and 15.33). As noted from Figure 15.33, the stress rate components of Yamoto-Rifu-Oshika element indicated that remarkable stress variations started in October 2002. However, the strain rate components of the elements of Yamoto-Oshika-Onagawa, Yamoto-Onagawa-Wakuya and Yamoto-Wakuya-Miyagi-Taiwa started to change remarkably at the beginning of May, 2003 about 1 month before the M7.0 Kinkazan earthquake that occurred on May 26, 2003. The high rate of variations continued after the M7.0 earthquake and resulted in the July 26, 2003 Miyagi-Hokubu earthquakes. Variations seen in Figure 15.31 before the earthquake resembles those observed in creep tests. As the variations of disturbing stress rates were greater than those of mean and maximum shear stress rates, Aydan (2004a) concluded that the disturbing stress rate may be a good indicator of regional stress variations and precursors of following earthquakes. Therefore, the time of the earthquake may be obtained from the GPS measurements. However, the magnitude of the earthquake is still difficult to do so. Nevertheless, the area of stress rate concentrations with a chosen value may be used to determine the magnitude. As a result, the fundamental parameters of the earthquake prediction, i.e. location, time and magnitude, may be determined from the evaluation of GPS measurements. However, there are still some technical problems associated with GPS observations and artificial disturbances as pointed by Aydan (2003c, 2004a).

15.3.3.4   Multi-parameter method and its application to the 2003 Buldan earthquake

The first author and his co-workers (Aydan et al., 2005a; Kumsar et al., 2010) established a multi-parameter observation network in Denizli basin in order to investigate the relation between earthquakes and the changes within the earth-crust in Denizli region (Fig. 15.34). We show multi-parameter responses measured in relation to the 2003 July 23 Buldan earthquakes (Figs. 15.35–15.39).

There was earthquake activity in Buldan and surroundings of Denizli on 23rd of July 2003. An earthquake with a magnitude of 5.2 occurred at 07:56 a.m. on 23 July

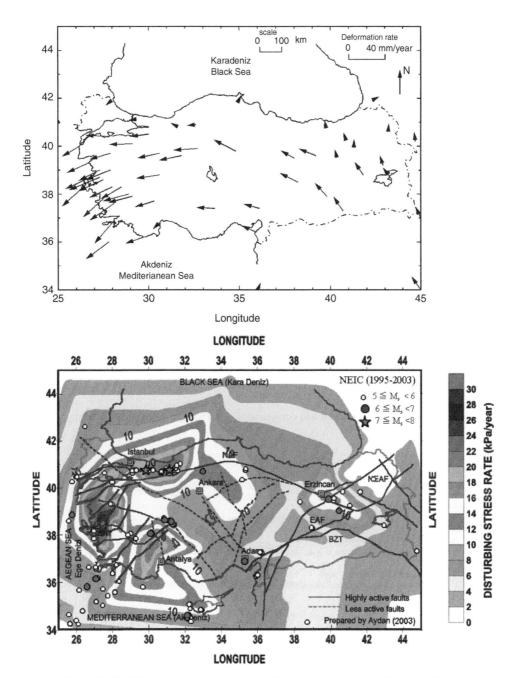

Figure 15.31 Displacement rate, computed disturbing stress rate and earthquakes.

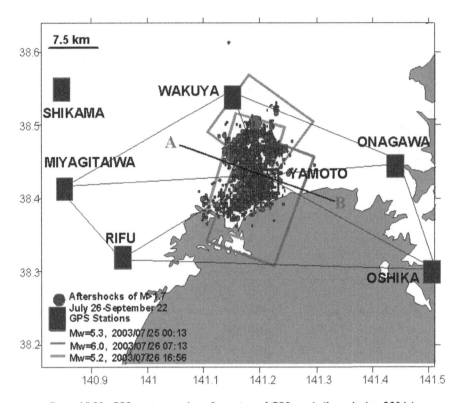

Figure 15.32 GPS stations and configuration of GPS mesh (from Aydan, 2004a).

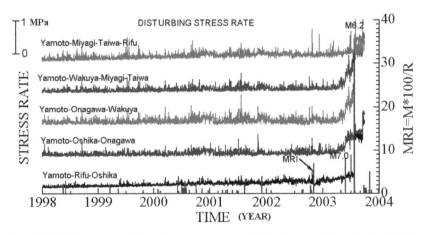

Figure 15.33 Time series of disturbing stress rates of GPS elements (from Aydan, 2004a).

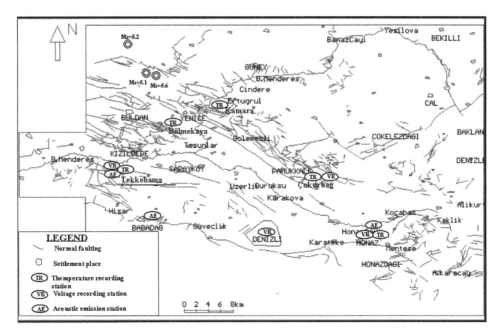

Figure 15.34 Multi-parameter measurement network in Denizli Basin.

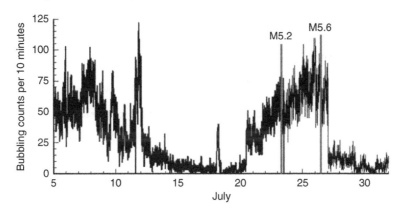

Figure 15.35 Bubbling (puf-puf) counts at Tekkehamam station.

2003. There were aftershocks with magnitudes of 4.1 on the following days. On 26th of July 2003, another earthquake with a magnitude of 5.6 happened near Buldan at 11:26 a.m. local time. This earthquake caused no life loss, but some damage to kerpiç and stonewall houses. The epicentre of the earthquakes was scattered north of Buldan. The nearest station to the earthquake activity was Tekkehamam station.

Bubble (Puf-Puf) count is defined as the number of bubbles hitting the sensor in a unit time in thermal spring mud. Normally the response of the Puf-Puf count is similar to the response of tidal waves. If there is an extraordinary difference on the Puf-Puf count, this can be related to an earthquake activity in the region. There was

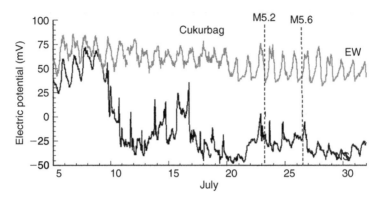

Figure 15.36 Electric potential variations at Cukurbag station.

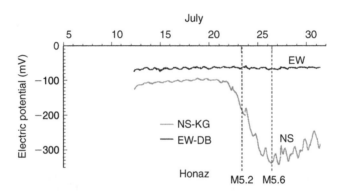

Figure 15.37 Electric potential variations at Honaz station.

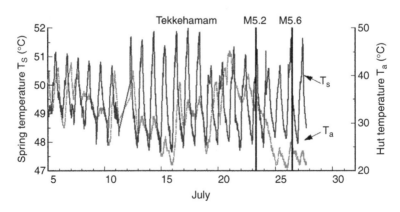

Figure 15.38 Temperature variations at Tekkehamam station.

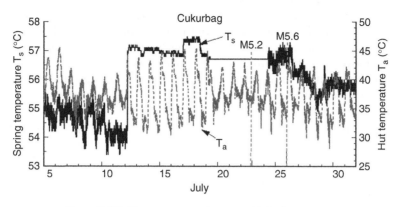

Figure 15.39 Temperature variations at Cukurbag station.

an important increase of the Puf-Puf counts 2 days before the M5.2 earthquake in Buldan. The activity of the Puf-Puf count continued as the earthquake activity was on. After 3 days, an earthquake with a magnitude of 5.6 occurred. When the magnitude of aftershocks became less than 3.5, the Puf-Puf count went down to a low level (Fig. 15.35). This shows that there is a correlation between the earthquake activity of the region and the Puf-Puf count changes.

At Çukurbağ station there are some electric potential variations, which may be related to the earthquakes with magnitudes of 5.2 and 5.6 (Figs. 15.36 and 15.37). The changes in the EW direction of Honaz station and the NS direction at Çukurbağ station started at the same time 2 days before the 5.2 magnitude. When the earthquake of 5.6 magnitude occurred, the electric potential value at the Honaz station in the NS direction was down to a minimum value and then the data plot started to increase. This type of change was also observed by Aydan, Minato and Fukue (2001) and Aydan *et al.* (2003a) in the laboratory experiments, some of which are also explained in Section 2. The changes at the Honaz and the Çukurbağ stations point out the possible changes of the regional crustal stresses.

Figures 15.38 and 15.39 show temperature changes of thermal water of hot-springs and atmosphere in the housing huts of the thermo recorder loggers at Çukurbağ and Tekkehamam. There were temperature changes of the thermal waters at Çukurbağ and Tekkehamam stations on 13th of July. There was a decrease at Tekkehamam between 20–25th of July. At Çukurbağ thermal spring, there was an increase of the temperature. Before the M5.2 earthquake, there was no temperature measurement for the thermal spring. Although the number of earthquakes and measurement period are still limited, a fairly good correlation exists with the crustal multi-parameter observations and earthquake activities, and these preliminary results are very promising.

15.4 EFFECTS OF EARTHQUAKES ON ROCK ENGINEERING STRUCTURES

It is well known that the Earth's crust is ruptured and contains numerous faults and various kinds of discontinuities, and it is almost impossible to find a piece of land without faults. During the construction of large or long structures such as tunnels, dams, power plants, roadways, railways, power transmission lines, bridges, elevated expressways etc., it is almost impossible not to cross a fault or faults. Therefore, one of the most important items is how to indentify which fault segments observed on ground surface will move or rupture during a future earthquake.

In addition to that, it is impossible to say the same ground breaks would re-rupture in the next earthquake in regions with thick alluvial deposits. Furthermore, the movements of a fault zone may be diluted if a thick alluvial deposit is found on the top of the fault. The authors will describe the model experiments and actual observations on the effects of earthquakes on rock engineering structures in this section.

15.4.1 Model experiments

Model experiments on rock engineering structures are one of the tools used in rock engineering since early times. If the design values are to be obtained directly from the model experiments, the similitude law between the model and actual structure is the most critical issue. As it is difficult to model an actual structure in a reduced scale of the geometry, stress conditions and constitutive parameters of materials, the model experiments should be used to validate the mathematical models for structural response under controlled conditions and material properties of the model experiments, and to study the mechanism of failure, which may be an extremely difficult task even in numerical analyses. The author and his co-workers prefer the second approach (Aydan et al., 2010a). Therefore, the model experiments presented in this section are intended to illustrate what we should expect under natural conditions and to understand the underlying mechanism of the response and stability of rock engineering structures subjected to earthquakes and earthquake faulting.

The authors used some model set-ups to investigate the effects of shaking or faulting due to earthquakes. Figure 15.40 shows an experimental device for investigating the effect of faulting under gravitational field. The orientation of faulting can be adjusted as desired. The maximum displacement of faulting of the moving side of the faulting

Figure 15.40 A view of the faulting experimental set-up.

experiments was varied between 25 and 100 mm. The base of the experimental set-up has a box of 780 mm long, 250 mm wide and 300 mm deep, and it can model rigid body motions of base rock. This experimental device is used to investigate the effect of forced displacement due to faulting on rock slopes and underground openings. The displacement and accelerations were measured simultaneously.

Dynamic testing of the models was performed in the laboratory by means of a one-dimensional shaking table, which moves along the horizontal plane (Fig. 15.41). The waveforms of the shaking table are sinusoidal: saw tooth, rectangular, trapezoidal or triangle. The shaking table has a square shape with 1000 mm long sides. It has a frequency interval between 1 Hz and 50 Hz, a maximum stroke of 100 mm, and

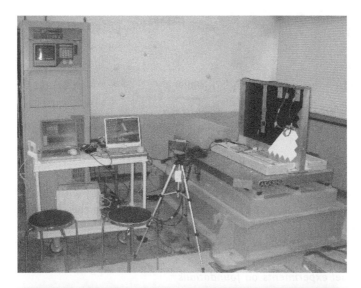

Figure 15.41 View of model experiments using the shaking table.

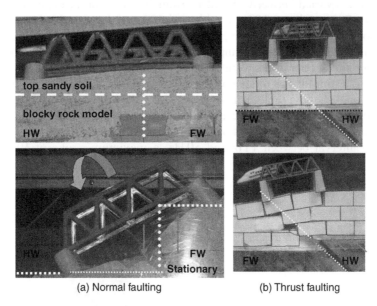

top sandy soil

blocky rock model

HW   FW

HW   FW

FW   HW

HW   FW   Stationary

FW   HW

(a) Normal faulting   (b) Thrust faulting

Figure 15.42 Views of the model truss bridge before and after experiments.

a maximum acceleration of 6 m/s^2 with a maximum load of 980.7 N. The model frame was fixed on the shaking table to receive the same shaking with that of the shaking table during the dynamic test. The accelerations acting on the shaking table and the models were recorded during the experiment, and saved on a data file as digital data. The displacement responses were recorded using laser displacement transducers.

15.4.1.1 Model experiments on foundations

Several experiments were carried out to investigate the effect of faulting on a bridge and its foundations. The bridge was a truss bridge just over the projected fault line. Figure 15.42 shows truss bridge models above the jointed rock mass foundation. Figure 15.42(a) shows views of the bridge model before and after the experiments, subjected to the forced displacement field of vertical normal faulting mode. Bridge foundations were pulled apart and tilted. The vertical offset was 0.37 times the bridge span. Similarly Figure 15.42(b) shows the bridge model before and after the experiments subjected to the forced displacement field of 45° thrust faulting mode. Bridge foundations were also pulled apart at the top and compressed at the bottom, and tilted as seen in Figure 15.42(b).

15.4.1.2 Model experiments on rock slopes

Model experiments on rock slopes were carried out by subjecting the model to shaking or forced displacement field due to faulting. Although some of these have not been completed yet, the experiments so far should be sufficient to explain the fundamental features of response of rock slopes under the assumed conditions. Model experiments

Figure 15.43 Failure modes of rock slope models with breakable material.

were carried out using either breakable blocks and layers or non-breakable blocks. Breakable blocks are made of BaSO$_4$, ZnO and Vaseline oil, which is commonly used in base friction experiments (Aydan and Kawamoto, 1992; Egger, 1979). Properties of block and layers are described in detail by Aydan and Amini (2009). When rock slopes are subjected to shaking, passive failure modes occur in addition to active modes (Aydan *et al.*, 2009a; 2009b; Aydan and Amini, 2009). Figures 15.43 and 15.44 show examples of slope failures consisting of breakable and non-breakable blocks and/or layers. The experiments also show that flexural toppling failure of passive type occur when layers (60°) dip into valleyside.

The authors have initiated a new experimental program on the effect of faulting on the stability and failure modes of rock slopes. The first series of experiments were carried out on rock slope models with breakable material under a thrust faulting action with an inclination of 45° (Fig. 15.45). When layers dip towards valleyside, the ground surface is tilted and the slope surface becomes particularly steeper. As for layers dipping into mountain side, the slope may become unstable and flexural or columnar toppling failure occurs. Although the experiments are still insufficient to draw conclusions yet, they do show that discontinuity orientation has great effects on the overall stability of slopes in relation to faulting mode. These experiments clearly show that the forced displacement field induced by faulting has an additional destructive effect besides ground shaking on the stability of slopes.

15.4.1.3 Model experiments on shallow undergound openings

The authors have been performing model experiments on underground openings for some time (Aydan, Shimizu and Karaca, 1994; Genis and Aydan, 2002). The first series

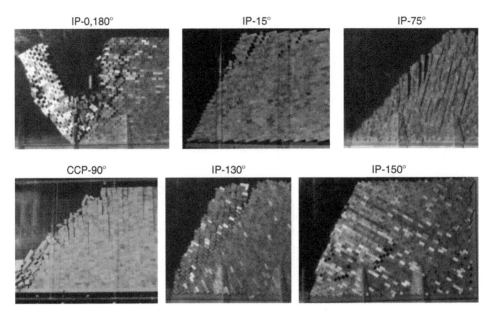

Figure 15.44 Failure modes of rock slope models with non-breakable material.

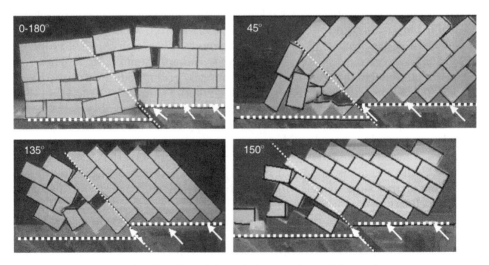

Figure 15.45 The effect of thrust faulting on the model rock slopes.

of experiments on shallow undergound openings in discontinuous rock mass using non-breakable blocks were reported by Aydan, Shimizu and Karaca (1994), in which a limit equilibirum method was developed for assessing their stability. These experiments have now been repeated using the breakable material following the observations of damage to tunnels caused by the 2008 Wenchuan earthquake. The inclination of continuous

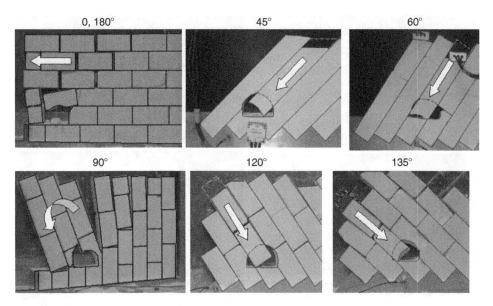

Figure 15.46 Failure modes of shalllow tunnels ajacent to slopes with breakable material.

discontinuity plane varied between 0° and 180°. Figure 15.46 shows views of some experiments. Unless the rock mass model failed itself, the failure modes were very similar to those of the model experiments using hard blocks. In some of experiments with discontinuities dipping into the mountain side, flexural toppling of the rock mass model occurred. The comparison of the preliminary experimental results with the theoretical estimations based on Aydan's method (Aydan, 1986; Aydan, Shimizu and Karaca, 1994) are remarkably close to each other (Ohta and Aydan, 2011).

The authors have also been performing some model experiments on the effect of faulting on the stability and failure modes of shallow undergound openings. Figure 15.47 shows views of some model experiments on shallow undergound openings subjected to the thrust faulting action with an inclination of 45°. Underground openings are assumed to be located on the projected line of the fault. In some experiments three adjacent tunnels were excavated. While one of the tunnels was situated on the projected line of faulting, the other two tunnels were located in the footwall and hanging wall side of the fault. As seen in Figure 15.47, the tunnel completely collapsed or was heavily damaged when it was located on the projected line of the faulting. When the tunnel was located on the hanging wall side, the damage was almost none in spite of the close approximity of the model tunnel to the projected fault line. However, the tunnel in the footwall side of the fault was subjected to some damage due to relative slip of layers pushed towards the slope. This simple example clearly shows the damage state may differ depending upon the location of tunnels with respect to fault movement.

There are many semi undergound openings or half tunnels in mountainous areas. Similar situations may also appear in the vicinity of cliffs next to the sea or rivers. Figure 15.48 shows examples of experiments on model half-tunnels in continuous and layered rock mass. The failure occurred due to bending.

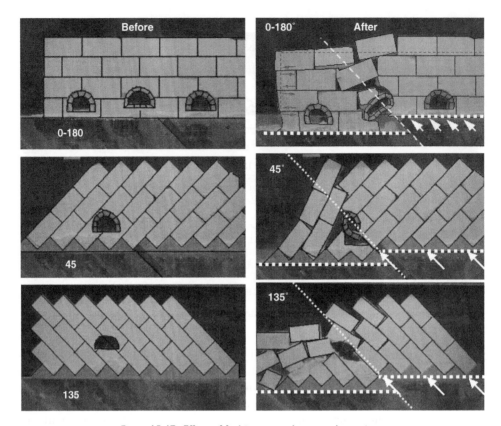

Figure 15.47 Effect of faulting on underground openings.

15.4.1.4 Model experiments on masonry structures

A series of model experiments on the seismic response and stability of masonry structures such as arches, castle walls, dams, retaining walls, bridges, pyramids, towers and buildings were performed using a shaking table (see Aydan *et al.*, 2003b for details). In these model studies, masonry structures were constructed using Ryukyu limestone blocks as building materials.

a) Experiments on arches

Five arch configurations denoted as Type A, B, C, D and Type-E, and four (Type A, B, D, E) of which are commonly used in Shuri Castle in Okinawa Island, Japan were tested (Fig. 15.49). The remaining arch form (Type-C) is quite common almost all over the world. The arches of Shuri Castle generally consist of two monolithic blocks in the form of a semi-circle or an ovaloid shape while the Type-C arch consists of several blocks and has a semi-circular shape. As the shaking table was uniaxial, the effect of the direction of input acceleration wave was investigated by changing the longitudinal axis of the arches. In this section, only some of experiments will be presented due to lack of space.

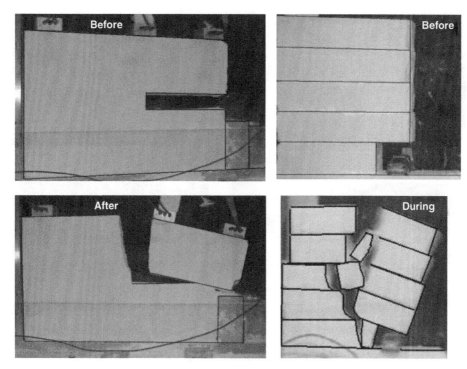

Figure 15.48 Effect of shaking on overhanging cliffs or half-tunnels.

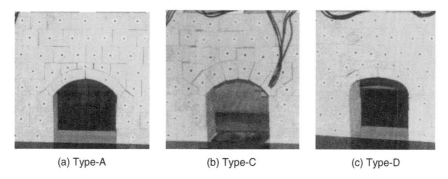

(a) Type-A   (b) Type-C   (c) Type-D

Figure 15.49 Model arch types.

Figure 15.50 shows the failure state of Type-A arch for shaking directions of 0°, 45° and 90°. The experimental results indicated that the common form of failure for all arch types for a shaking direction of 0° is sliding at abutments and inward rotational fall of arch blocks subsequently. As for 90° shaking, the arch failed in the form of toppling. The failure for 45° shaking was a combination of sliding and toppling. The experiments clearly indicated that the amplitude of acceleration waves, which cause failure, was the lowest for 90° shaking while it was the maximum for 0° shaking.

(a) Shaking direction of 0°

(b) Shaking direction of 45°   (c) Shaking direction of 90°

Figure 15.50 Failure modes of arch type-A.

Figure 15.51 Failure modes of 2D and 3D pyramids for horizontal shaking.

b) Experiments on pyramids

Both 2D and 3D models of pyramids consisting of limestone blocks were tested. In all experiments, the governing mode of failure was due to relative sliding among the layers of blocks as seen in Figure 15.51. If the motion of the blocky layer is not obstructed by the roughness of the block interfaces, the pyramids keep their original configuration during motion. However, if the inter-block sliding is obstructed by some asperities, then block separation starts to take place gradually as shown in Figure 15.51. The failure of pyramids is purely governed by the frictional properties of block interfaces.

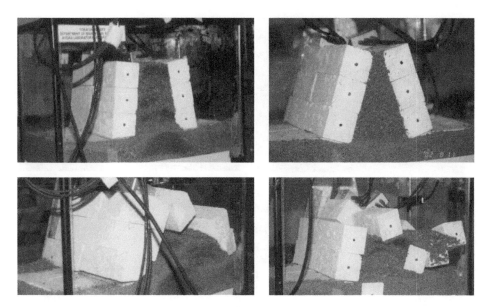

Figure 15.52 Castle walls before and after shaking.

c) Experiments on castle walls

The walls of many castles and historical cities all over the world are built either vertically or inclined. Furthermore, the outer shells of the walls are built with neatly placed blocks of stone while the core generally consists of rubble material. A series of experiments were carried out to see the effect of the inclination of the outer shell with the consideration of the basal inclination of the foundation blocks. The inclinations of the castle wall were 73°, 84° and 90° with 0° and 7° basal inclinations of the foundation block. Figure 15.52 shows the failure of castle walls with an inclination of 84° for with 0° and 7° basal inclinations of the foundation block. The experiments showed that the decrease in wall inclination results in higher resistance against shaking. Furthermore, walls with a 7° basal inclination have a higher resistance against shaking as compared with walls with a 0° basal inclination. The fundamental mode of failure is toppling. However, some inter-block sliding may be caused and the gap may be filled by backfill material, which may result in more unstable wall configuration after each wave passes through.

d) Experiments on retaining walls

In model tests, the aim was to investigate the height/width ratio, the inclination of retaining walls and back-filling material. In tests, the ratio of width to height was varied between 0.25 and 0.625. The wall inclinations were 73°, 84° and 90° with 0° and 7° basal inclinations. The back filling material was sand N7 and sandy gravel. Figure 15.53 shows some views of the model experiments on retaining walls for two types of back-filling materials. The fundamental model of failure was toppling (rotational failure) for walls with greater height/width ratios. After each shaking cycle, the interface gap was filled with a wedge of back-filling material, which resulted in an

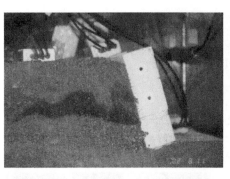
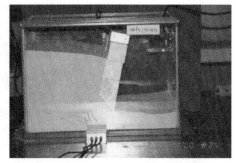

Figure 15.53 Failure modes of retaining walls.

unstable wall configuration. If the walls are inclined and/or have a 7° basal inclination, they can resist high acceleration amplitudes. This may also explain why such walls are more stable as compared with that of walls with 90° inclination during earthquakes.

e) Experiments on houses

The experiments were carried out on single story houses having a heavy roof. In experiments, plastic blocks were used. The model houses were built in such way that two sidewalls would be parallel to ground shaking while the other two walls would be subjected to out-of plane loading. Figure 15.54 shows some views of the experiments. The experiments clearly showed that the walls which are subjected to out-of plane loading, tend to collapse first while the sidewalls parallel to shaking tend to fail by inter-block sliding. The ground shaking causing total collapse of the sidewalls parallel to the direction of shaking should be such that the accumulated relative displacement of the inter-block sliding should exceed the half length of the block in the respective direction. The experiments also showed that the corners of the buildings are quite prone to fail first due to the concentration of two failure modes at such particular locations.

15.4.2 Effects of earthquakes on actual rock structures

The first author has been involved with earthquake reconnaissance since 1992 (Aydan and Hamada, 1992). In this section, the effects of earthquakes on various

Figure 15.54 Some views of experiments on model houses.

structures are described (see Aydan, 1997b, 2003b, 2004b, 2006a, 2008b; Aydan and Kawamoto, 2004; Aydan and Tokashiki, 2007; Aydan *et al.*, 1999a, 1999b, 2000, 2006a, 2009a, 2009b, 2009c; Aydan, Kumsar and Toprak, 2009; Aydan, Ohta and Hamada, 2009; Aydan, Tokashiki and Sugiura, 2008; Aydan and Kumsar, 2010; Kumsar, Aydan and Ulusay, 2000; Ulusay, Aydan and Hamada, 2002 for details).

15.4.2.1 Foundations and dams

Intraplate earthquakes may result in surface ruptures. Although it is difficult to estimate how surface ruptures will appear near the ground surface due to the reduction of vertical stress, they may induce negative or positive flower structures depending upon the sense of faulting. These surface ruptures may cause the failure of foundations of super-structures such as bridges, dams, viaducts and pylons (Aydan *et al.*, 1999a; Aydan, 2003b). The foundation may sometimes be catastrophic as observed in the foundation failure of Matsurube Bridge by the 2008 Iwate-Miyagi earthquake due to columnar toppling, and Shikang Gravity Dam by the 1999 Chi-chi earthquake by thrust faulting (Fig. 15.55). The earthquakes of 1995 Kobe, the 1999 Kocaeli, Chi-chi and Düzce-Bolu, and 2008 Wenchuan earthquakes caused severe damage to pylons (Fig. 15.56). The 1999 Kocaeli and Düzce-Bolu earthquakes damaged the foundations of viaducts and bridges.

(a) Matsurube bridge (toppling)   (b) Shikang gravity dam (faulting)

Figure 15.55 Damage to bridge and dam foundations (partly from Aydan, 2003b).

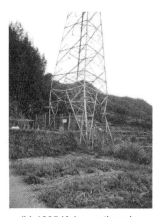

(a)1999 Kocaeli earthquake   (b) 1995 Kobe earthquake   (c) 1999 Chi-chi earthquake

Figure 15.56 Damage to pylons due to faulting (from Aydan, 2003b).

15.4.2.2 Slope failures and rockfalls

The 1999 Chi-chi, 2004 Chuetsu, 2005 Kashmir, 2008 Wenchuan and 2008 Iwate-Miyagi earthquakes caused many rock slopes and rockfalls (Fig. 15.58). These slope failures and rockfalls, in turn resulted in the destruction of railways, roadways, housing and vehicles. The failed rock mass body in the Aratozawa Dam landslide shown in Figure 15.57 by the 2008 Iwate-Miyagi earthquake luckily did not move into the reservoir, so that the incident similar to that of the Vaiont dam did not occur.

The assessment of the stability of natural rock slopes against earthquakes is very important and one of our urgent issues is how to address it and how to deploy the methods and technology for mitigation. It should also be noted that the failure forms induced by earthquakes might involve passive modes in addition to failures modes classified by Aydan, Shimizu and Ichikawa (1989).

(a) Chiufengershan (1999 Chi-chi Eq.)   (b) Shiraiwa (2004 Chuetsu Eq.)

(c) Hattian (2005 Kashmir Eq.)   (d) Beichuan (2008 Wenchuan Eq.)

(e) Aratozawa dam landslide   (f) Toppling slope failure (Isawa Valley)

Figure 15.57 Views of some slope failures caused by different earthquakes.

Rockfalls are generally observed in the areas where steep rock slopes and cliffs outcrop. Some rockfalls may induce heavy damage to structures. Nevertheless, rockfalls are the result of falls, after the initiation of failure, of individual rock blocks in the modes of sliding, toppling or combined sliding and toppling (Aydan, Shimizu and Ichikawa, 1989; Aydan, Shimizu and Kawamoto, 1992a). The fall of rock blocks due

(a) 2008 Wenchuan Eq.   (b) 2009 L'Aquila Eq.   (c) 2009 Pariaman Eq.

Figure 15.58 Rockfalls and their effects.

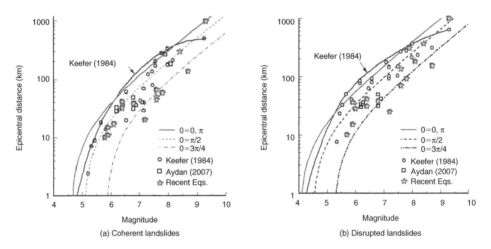

Figure 15.59 Comparison of empirical relations with observations.

Table 15.1 Parameters of equation (15.8) for disrupted and coherent landslides.

Condition	A	B
Disrupted	0.10	0.9
Coherent	0.08	0.9

to sliding mode occurs when the relative sliding exceeds their half width. As for toppling failure, the shaking should be large and long enough to cause the rotation of individual blocks. The height of block should generally be larger than the base width.

Keefer (1984) studied landslides in the USA and other countries, and he proposed some empirical bounds for landslides, which are classified as disrupted or coherent.

The empirical bounds of Keefer are not given specifically as a formula. The author compiled landslides (it would be better to name them as slope failures) caused by earthquakes according to Keefer's classifications and these are plotted in Figure 15.59. Besides the empirical bounds of Keefer, the following empirical equation is proposed for the maximum distance of disrupted and coherent landslides as a function of earthquake magnitude and fault orientation (Aydan, 2007; Aydan *et al.*, 2009a):

$$R = A * (3 + 0.5\sin\theta - 1.5\sin^2\theta) * e^{B \cdot M_w} \tag{15.8}$$

Constants A and B of Equation (15.8) for disrupted and coherent lanslides are given in Table 15.1. Since ground accelerations differ according to the location with respect to fault geometry, the empirical bounds proposed herein can provide some bases for the scattering range of observations.

Aydan (Aydan, 2006a; Aydan, Shimizu and Ichikawa, 1989) proposed a method based on the limit equilibrium approach to determine the limiting stable slope angle under the given seismic, geometrical and physical conditions. Figure 15.61 shows a plot of the lower slope angle of various rock slopes versus the inclination of the thoroughgoing discontinuity set whose strike is parallel or nearly parallel to the axis of the slope (please note that the overhanging slopes are not considered in this figure). Stable slopes are denoted by S and failed slopes by F. The plotted data include the data on presently stable natural rock slopes and rock slopes failed due to earthquakes. Most of the data are compiled by the authors. In the plots, the stability charts of a slope with a ratio of $t/H : 1/75$ for cross continuous and intermittent patterns $\xi = 26.5°$ for $\eta = 0.0$ are also included to have a qualitative insight rather than a quantitative comparison. The chosen value of t/H is arbitrary and may not correspond to the ratios of slopes plotted in the figure. The plotted cases confirm the qualitative tendency described in Figure 15.60. It is also interesting to note there are almost no failed slopes when the slope angle is less than 25–30° and most of the failed slopes have a slope angle greater than 25–30°. This is in accordance with the conclusion of Keefer (1984). Nevertheless, it is also noted that there are a great number of stable slopes having slope angle greater than 25–30°. This implies that slope angle and the height of slopes cannot be the only parameters determining the overall stability of natural rock slopes. Therefore, the orientations of discontinuity sets, their geometrical orientations with respect to slope geometry, and their mechanical properties and loading conditions must also play a great role in determining the stable angles of natural rock slopes. The results shown in Figure 15.60 may serve as guidelines for a quick assessment of the stability of natural rock slopes and how to select the slope-cutting angle in actual restoration of the failed slopes.

15.4.2.3 Underground structures

It is well known that underground structures such as tunnels and powerhouses are generally resistant against earthquake-induced motions. However, they may be damaged when permanent ground movements occur in/along the underground structures. There are several examples of damage to tunnels due to permanent ground movements during the 1930 Tanna, 1978 Izu-Oshima-Kinkai, 1995 Kobe, 1999 Düzce-Bolu, 1999 Chi-chi, 2004 Chuetsu, 2005 Kashmir and 2008 Wenchuan earthquakes (Aydan, Ohta and Tano, 2010; Aydan *et al.*, 2010a).

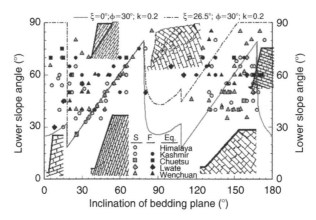

Figure 15.60 The relation between slope angle and bedding plane angle for stable (S) and failed (F) case histories in recent earthquakes.

Figure 15.61 Views of the collapsed section of Bolu tunnel and its surface depression.

The underground structures damaged by earthquakes worldwide are described in detail by Aydan *et al.* (2009a,b, 2010b). The damage to underground structures may be classified as:

a) shaking induced damage (Fig. 15.61)
b) portal damage (Fig. 15.62) and
c) permanent ground deformation induced damage (Fig. 15.63).

Permanent ground deformation induced damage is generally caused either by faulting or slope movements. Most tunnels have non-reinforced concrete linings. Since the lining is brittle, the permanent ground movements may induce the rupture of the linings and falling debris may cause disasters with tremendous consequences to vehicles passing through. Therefore, this current issue needs to be urgently addressed. It should be also noted that the same issue is valid for the long-term stability of high-level nuclear waste disposal sites.

Figure 15.62 Examples of damaged portals of tunnels.

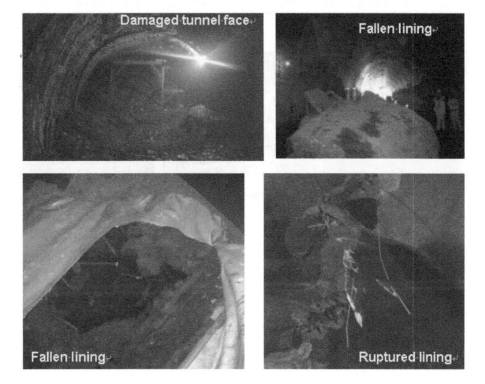

Figure 15.63 Earthquake damage at Jiujiaya Tunnel due to permanent deformations.

As suggested by Aydan *et al.* (2009b, 2010a), there may be two possible ways to deal with this issue. The first alternative would be to line the concrete lining with thin steel platens together with rockbolts. The other alternative may be to use fiber-reinforced polymers together with rockbolts to line the inner side of the concrete lining.

Earthquakes are also reported to cause some damage to mines. It is well known that the strike-slip faulting caused some damage in coalmines during the 1976 Tangshan earthquake. It is also reported that some collapse and damage to abandoned room and

Figure 15.64 Caving and sand boiling in Yamato town (from Aydan and Kawamoto, 2004).

Figure 15.65 Caving or damage to natural caves in L'Aquila earthquake.

pillar mines do occur from time to time. The 2003 Miyagi-Hokubu earthquake caused severe damage to abandoned mines in Yamoto town (Aydan and Kawamoto, 2004). The damage was in the form of caving (sink-hole) and water discharge. Figure 15.64 shows some examples of damage observed.

Karstic caves are geologically well-known to form along generally steep fault zones and fractures due to erosion as well as solution by ground water (i.e. Aydan, 2008b; Aydan and Tokashiki, 2007). Earthquakes may cause collapse or damage to natural caves. Aydan, Kumsar and Toprak (2009) and Aydan *et al.* (2009c) described some examples of damage to natural caves by the 2005 Nias earthquake and 2009 L'Aquila earthquake. Figure 15.65 shows examples of collapse or damage to natural caves caused by the 2009 L'Aquila earthquake.

In mountainous regions, half tunnels may have been excavated for various reasons. These half tunnels result in overhanging slope configurations. Similar situations are also observed along seashores and riverbanks. Rockfalls or collapses of half tunnels or overhanging slopes were observed in the 2005 Kashmir earthquake and 2008 Wenchuan earthquake. Figure 15.66 shows several examples observed in these earthquakes. Some rockfalls resulted in crushing vehicles and subsequently casualties.

Aydan *et al.* (2010a) compiled cases histories and developed databases for three different categories of damage, namely, faulting induced (18 cases histories), shaking

Figure 15.66 Rock falls from overhanging slopes or half tunnels.

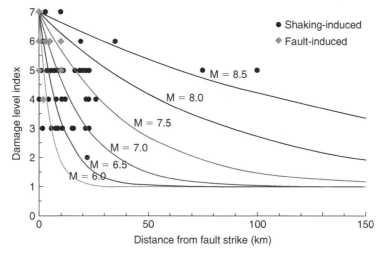

Figure 15.67 Relation between distance (Rf) from surface trace of the fault and damage level index (DLI).

induced (98 case history) and slope failure induced (47 cases histories). They also proposed a new classification of damage for underground openings, in which 7 levels of damage are defined. They proposed an empirical relation for delineating the damage levels as a function of distance and orientation from the fault, magnitude of earthquake and shear wave velocity of rock mass. Figure 15.67 shows a plot of the empirical function. As seen from the figure, the proposed relation by Aydan *et al.* (2010a) can closely estimate the observed damage level index of underground openings subjected to earthquakes.

15.5 NUMERICAL SIMULATIONS

In this section, numerical simulations of the response of underground structures subjected to ground shaking, fault propagation and its effect on structures are presented and discussed. The numerical simulations of some masonry structures are also presented.

Aydan and Mamaghani proposed the discrete finite element method (DFEM) for blocky systems, which is based on the principles of the finite element method (Aydan, Mamaghani and Kawamoto, 1996; Mamaghani *et al.*, 1994; Mamaghani, Aydan and Kajikawa, 1999; Tokashiki *et al.*, 1997a, 1997b). It is possible to handle deformable blocks and contacts that specify the interaction among them. Small displacement theory is applied to intact blocks while blocks can take finite displacement. Blocks are polygons with an arbitrary number of sides, which are in contact with neighboring blocks, and are idealized as a single or multiple finite elements. Block contacts are represented by a contact element, which has a finite thickness.

15.5.1 Simulation of post-failure motions of rock blocks and slopes

Aydan and Mamaghani (Aydan, Mamaghani and Kawamoto, 1996; Aydan *et al.*, 1997; Mamaghani *et al.*, 1994; Mamaghani, Aydan and Kajikawa, 1999; Mamaghani and Aydan, 2000) simulated the post-failure motions of rock blocks and slopes. We briefly describe some of the numerical simulations and details can be found in the referred articles.

First the dynamic stability of square and rectangular blocks on a plane with an inclination of 30° was analyzed by the DFEM. The rectangular block was assumed to have a height to breadth ratio $h/b = 1/3$. The friction angles for square and rectangular blocks are $\phi = 25°$ and $\phi = 35°$, respectively. Figure 15.68 shows computed configurations of the square block of size 4 m × 4 m and a rectangular block of size 12 m × 4 m. The square block slides on the incline (time step $\Delta t = 0.04$ sec) while the rectangular block topples (time step $\Delta t = 0.01$ sec). These predictions are consistent with the kinematic conditions for the stability of a single block in the previous example as well as with the experimental results reported by Aydan, Shimizu and Ichikawa (1989). It should, however, be noted that the discretization of the domain, mechanical properties of blocks and contacts, and time steps may cause superficial oscillations and numerical instability.

It is also worth noting that any hyperbolic type equation system requires a certain kind of damping (viscosity) to attain stationary solution, which requires information on the viscous characteristics of rocks and discontinuities. Since time-dependent characteristics of discontinuities and intact rocks are less studied and experimental data are still limited, the inertia term is neglected in the computations reported hereafter using the DFEM.

The next two examples are concerned with circular sliding and planar sliding of rock slopes. The material properties and boundary conditions were given in detail elsewhere (Aydan *et al.*, 1997). Figure 15.69 shows the configurations of the slope for each failure mode at each computation step. The displacement of the failure body for a circular sliding tends to become asymptotic to a certain value following the

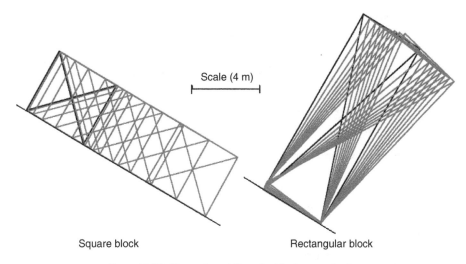

Square block   Rectangular block

Figure 15.68 Dynamic stability of a block on an incline.

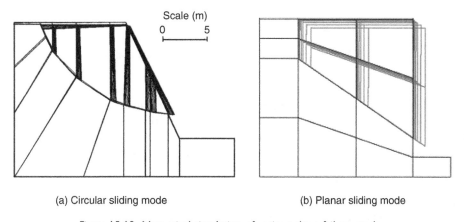

(a) Circular sliding mode   (b) Planar sliding mode

Figure 15.69 Numerical simulation of various slope failure modes.

rapid motion as the inclination of the sliding surface decreases after each computation step. As for planar sliding failure, a separation at the vertical discontinuity occurs and sliding along the inclined discontinuity takes place. The displacement of a sliding block increases as the computation step number increases as the failure plane remains the same.

The estimation of travel distance of natural slopes upon failure and their effect on engineering structures as well as on the natural environment is also of great importance (Aydan, 2006a; Aydan and Ulusay, 2002; Aydan and Kumsar, 2010; Aydan *et al.*, 2006b). The travel distance may be of tremendous scale and it may cause severe damage to settlements and structures. Although this issue is well known and some simple methods are available (i.e. Aydan, Shimizu and Kawamoto 1992b; Tokashiki

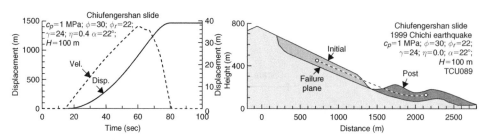

Figure 15.70 Comparisons of computed results for Chuifengershan landslide.

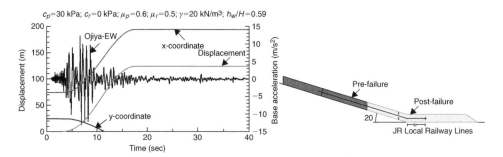

Figure 15.71 Comparisons of computed results for Shiraiwa landslide.

and Aydan, 2010), the present numerical methods are still insufficient to model post-failure motions. The accelerations records at a nearby station denoted TCU089 (CWB, 1999) were used in computations for the simulation of the post-failure motions of the Chiufengershan landslide. The computed response of displacement and velocity of the mass center and its path are shown Figure 15.70 together with the material and geometrical properties used in computation. As noted from the figure, the path of the mass center during motion is well estimated by the mathematical model.

Computations were carried out for Shiraiwa landslides (also called Myoken, Uragara or Yokowatashi). Figure 15.57(b) shows a view of the Shiraiwa landslide. The rock mass is andesitic tuff with a layered structure. The landslide took place in the form of plane sliding and hit the railway lines. The acceleration records at Ojiya, which is about 4 km south of the site, were used in computations. Figure 15.71 shows the absloute displacement, horizontal and vertical position of the landslide mass center in time space together with the EW component of Ojiya strong motion recorded by the K-NET(2004) and shear strength properties of the failure surface. The comparison of the computed position of the mass center with that obtained from the geometry of the landslide body is almost the same.

Another example of computation was carried out for the landslide in the reservoir of Aratozawa dam caused by the 2008 Iwate-Miyagi earthquake as shown in Figure 15.72. The maximum ground accelerations at Aratozawa dam were 865, 810 and 1024 gals for EW, NS and UD components, respectively. The landslide body was 1350 m long and 800 m wide. Although the thickness ranges from place to place,

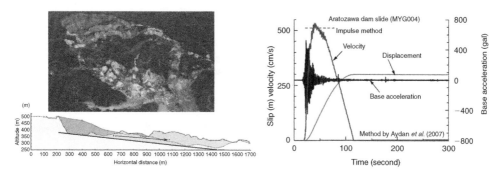

Figure 15.72 Comparisons of computed results for Aratozawa Dam landslide.

the highest landslide scarp was about 100 m. There are loosely cemented sandstone layers of volcanic origin, sandwiched by relatively impermeable marl-like mudstone. These sandstone layers played a major role in the landslide. The author carried out some laboratory tests on samples collected from the site. The computational results were based on the laboratory results and topographical maps before and after the landslides.

15.5.2 Numerical simulation of dynamic response of underground openings

A series of parametric numerical analyses on the shape of underground openings under different high in-situ stress regime and direction and amplitude of earthquake induced acceleration waves was carried out. The details of these numerical analyses can be found in publications by Genis (2002) and Genis and Gercek (2003). Figure 15.73 compares yield zone formations around circular and horseshoe-like tunnels subjected to in-situ hydrostatic stress condition ($P_o = 20$ MPa) under static and dynamic conditions. In the analyses, the rock mass behavior is assumed to be elastic brittle-plastic. The wave form used for dynamic analysis is the acceleration record taken at Erzincan during the 1992 Erzincan earthquake. The plastic zones around the underground openings are almost circular for the circular tunnels while it is almost elliptical for horseshoe-like tunnels although the acceleration record was uni-directionally applied.

Genis and Aydan (2007) carried out a series of numerical studies for the static and dynamic stability assessments of a large underground opening for a hydroelectric powerhouse. The cavern is in granite under high initial stress condition and approximately 550 m below the ground surface. The area experienced in 1891 the largest inland earthquake in Japan. In the numerical analyses, the amplitude, frequency content and propagation direction of waves were varied. We just present numerical results under the same frequency and amplitude of sinusoidal acceleration wave while changing its propagation direction as shown in Figure 15.74. The numerical analyses indicated that the yield zone formation is frequency and amplitude dependent (Genis and Aydan, 2007; Aydan, Ohta and Tano, 2010; Aydan *et al.*, 2010a). Furthermore, the direction of wave propagation has also a large influence on the yield zone formation around

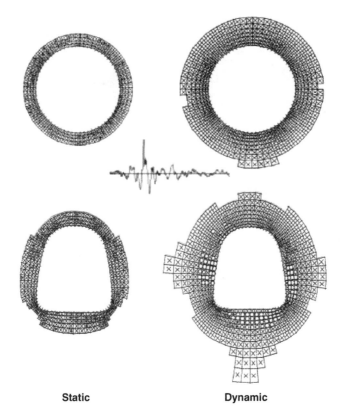

Static   **Dynamic**

Figure 15.73 Yield zone formations around a deep opening with different shapes.

the cavern. When maximum ground acceleration exceeds 0.6–0.7 g, it results in the increase of plastic zones around the opening. Thus, there will be no additional yield zone around the cavern if the maximum ground acceleration is less than these threshold values.

Genis and Aydan (2008) also reported some numerical studies on the stability of abandoned room and pillar mines under both static and dynamic loading conditions. They further showed the possible collapse mechanism of the abandoned mines next to a cliff along the Kiso River in Mitake town in Central Japan.

15.5.3 Fault propagation simulations

In this section, a pseudo-dynamic procedure of the DFEM is employed so that intact blocks and block contacts can behave elasto-plastically. The method of analysis is a pseudo time stepping incremental procedure. To model the elasto-plastic response of materials and contacts in numerical analysis, the initial stiffness method was employed together with the use of *Updated Lagrangian Scheme*.

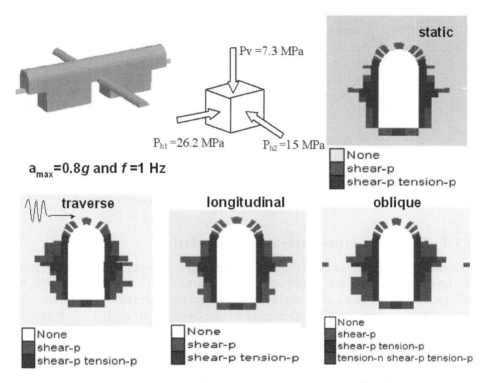

Figure 15.74 Yield zone formation around underground powerhouse for different directions of input ground motions (adapted from Genis and Aydan, 2007).

The details of computations can be found in the article by Aydan (2003b). The thickness of the fault plane was selected as 10 mm in view of past experiences. In the simulations, the fault plane was modelled through contact elements. The fault plane behaves elastically when the normal and shear stresses are below its yield strength. However, if yielding takes place, its behaviour is simulated as an elastic-perfectly plastic behaviour. In normal and thrust faulting the displacements having an amplitude of 10 cm are imposed at the boundary nodes as indicated in Figure 15.75 both in x and y directions. As for the strike-slip faulting simulation, the prescribed displacements of 10 cm are imposed only in the y-direction at selected points shown in Figure 15.75(b). Figures 15.76(a)–(c) show the deformed configurations of the model at computation steps 1 and 10. The computed results for Step 1 correspond to initial elastic responses after the prescribed displacement conditions being imposed. Results at computation Step 10 correspond to the fault propagation if yielding along the fault plane takes place. After a certain number of computation steps, the deformation of the fault tends to become stationary for the prescribed displacement boundary conditions. The propagation of faults starts at the bottom side and migrates towards the ground surface as expected. These three specific examples clearly show that the method used is capable of simulating the fault propagation processes for three different faulting modes.

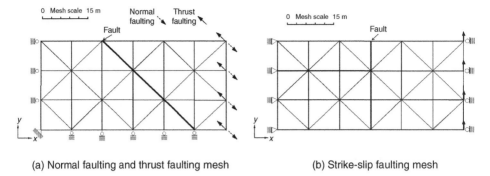

(a) Normal faulting and thrust faulting mesh   (b) Strike-slip faulting mesh

Figure 15.75 Finite element meshes and boundary conditions used in simulations.

15.5.4 Fault-Structure interaction simulation

The most important aspect in earthquake engineering is the interaction between structures and fault breaks. For this purpose, a truss structure straddling over the projected fault trace on the ground surface was considered, and normal faulting and thrust faulting conditions are imposed through prescribed displacement at selected points as in the previous computations. Figure 15.77 shows the finite element meshes and boundary conditions used in simulations (see Aydan, 2003b for details). Figure 15.78 shows the deformed configurations at computation steps 1 and 10 for normal and thrust type-faulting modes. In both cases, the truss structure tilts. While the thrust type faulting causes the contraction of trusses, the normal faulting condition results in the extension of trusses and separation of the supporting members fixed to the ground. These responses resemble to those shown in Figure 15.56.

15.5.5 Simulation of response of masonry structures

In this sub-section, we describe several applications of DFEM method and some analytical methods developed for retaining walls (Mamaghani, Aydan and Kajikawa, 1999; Aydan, Tokashiki and Sugiura, 2008).

Figure 15.79(a) shows a masonry arch bridge analyzed using the DFEM under static loading. The dimension of the blocks perpendicular to the xy plane is taken as $w = 1.0$ m. The arch is stable under its own weight. It is still stable when the distributed uniform traction load per unit length over the arch is less than 1.47 kN/m. However, if the traction load exceeds that level (step 27), the arch starts to become unstable.

A two-dimensional pyramid configuration was considered as shown in Figure 15.79(b). Two concentrated loads ($F_1 = 4.9$ kN and $F_2 = 127.4$ kN) were applied together with the application of gravity. As seen from the computed configurations, the inter-block sliding occurs and the pyramid tends to become unstable after each computation step.

The next three analyses were carried out under purely dynamic conditions. The wave forms shown in Figure 15.80 were used and the responses computed for each wave form are shown in the respective figures. Basically, a 12 m high and 4 m wide

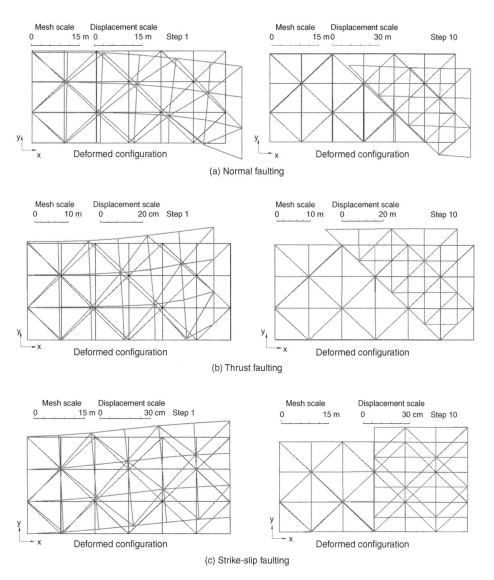

Figure 15.76 Deformed configurations at Steps 1 and for (a) normal faulting; (b) thrust faulting; (c) strike-slip faulting simulations.

masonry tower, a 12 m wide and 12 m high masonry wall, and a 12 m high masonry arch model were modelled. Figures 15.81, 15.82 and 15.83 show the computed responses for each structure. For plotting the deformed configurations, the displacement in the deformed configurations is amplified by 50 times to make the mode of failure (deformed configuration) more visible from the initial configuration. Although the masonry tower was stable at the end of shaking, there was relative sliding among blocks and the base. Furthermore, the separation of blocks occurred at the top. As for

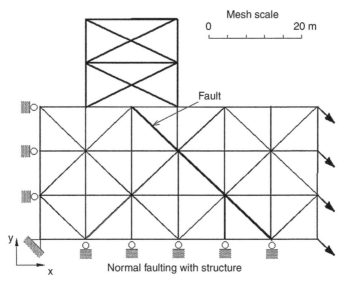

(a) Normal faulting mode

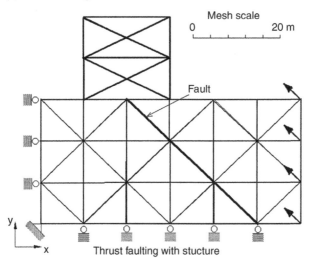

(b) Thrust faulting mode

Figure 15.77 Finite element meshes and boundary conditions for fault-structure interaction simulations.

the masonry wall, similar results were observed. In particular, block rotation at the top was quite visible.

Figure 15.83(b) shows that the arch slid at the base at the time step 23 under Acc. No. 1 and the crown blocks of the arch started to fall apart while the side columns were still stable. The columns slid relative to the base, and they tended to topple in two opposite directions. The blocks tended to separate within the side columns.

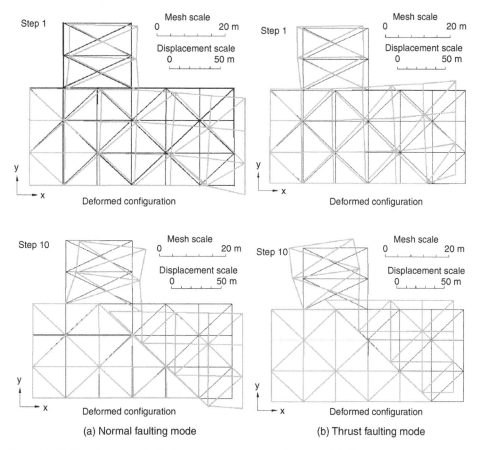

Figure 15.78 Simulations of the fault-structure interaction for normal faulting and thrust conditions.

As for Acc. No. 2 at time step 23, there is no slide at the base of the arch while the crown blocks are separated and tend to fall apart. At time step 23, the side columns of the arch exhibit relatively stable behavior under Acc. No. 2 as compared with Acc. No. 1. However, under Acc. No. 2 at time step 50 (10 seconds), the side columns of the arch slide at the base and the arching action disappears while the blocks start to fall apart. As expected, the toppling (failure) modes of the side columns of the arch differ, depending on the nature of the imposed form of acceleration waves, as shown in Figure 15.83. Figure 15.83(d) shows the displacement responses with time of a nodal point at the top right corner of the arch corresponding to Acc. No. 1 and Acc. No. 2. The results indicated that, as expected, the displacement of the side column of the arch with time is much more severe under Acc. No. 1 as compared with Acc. No. 2, especially in the early stage of loading. Under both of the imposed acceleration waves, the reaction of the toppled columns forces the crown block to move upward. This is because of the geometrically symmetric configuration of the structure and outward inclination of the crown block contact interfaces at the center of symmetry.

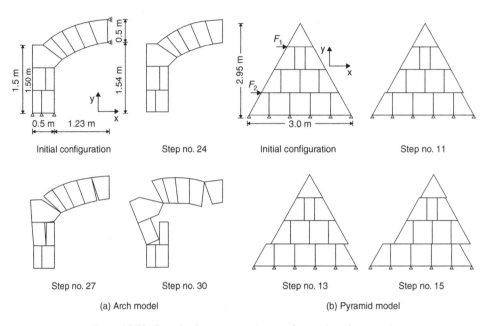

Figure 15.79 Pseudo-dynamic simulation of an arch and pyramid.

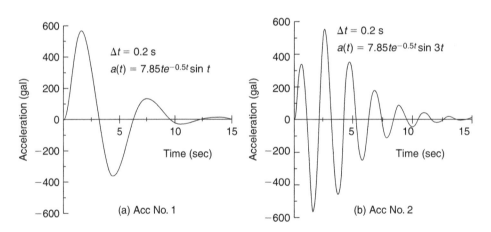

Figure 15.80 Wave forms used in computations.

The Kameyama earthquake occurred at 12:19 JST on April 15, 2007 and it had a magnitude (Mj) of 5.3 on the magnitude scale of the Japan Meteorological Agency. The earthquake injured 12 people and caused some structural damage (Aydan, Tokashiki and Sugiura, 2008). The earthquake caused the collapse of the northern corner of the Kameyama Castle (Fig. 15.84). The collapsed north wall of the castle is about 5 m high with an inclination of about 70° and block sizes of 50–60 cm. The block size of older parts of the castle walls is more than 100 cm and their inclination is about

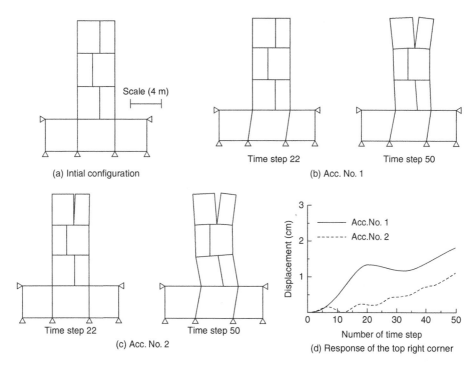

(a) Intial configuration

Time step 22

Time step 50

(b) Acc. No. 1

Time step 22

Time step 50

(c) Acc. No. 2

(d) Response of the top right corner

Figure 15.81 Initial and deformed configurations of the masonry tower and the displacement response of the top right corner.

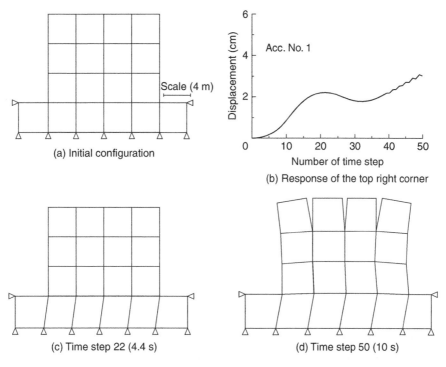

(a) Initial configuration

Scale (4 m)

Acc. No. 1

(b) Response of the top right corner

(c) Time step 22 (4.4 s)

(d) Time step 50 (10 s)

Figure 15.82 Initial and deformed configurations of the masonry wall and the displacement response of the top right corner.

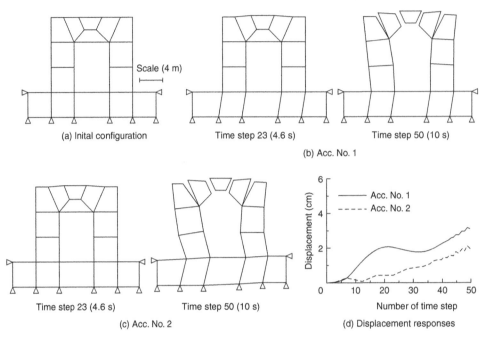

Figure 15.83 Initial and deformed configurations of the masonry arch and the displacement response of the top right corner.

(a) Eastward view of NE corner   (b) Southward view of NE corner

Figure 15.84 Views of intact and collapsed section of Kameyama Castle.

50°. Furthermore, the other walls of the castle are more than 10 m high. In the computations, the NS component of the acceleration records taken at Kameyama strong motion station of K-NET was used and the responses of the castle wall during shaking for sliding and toppling failure modes were computed. The computed displacement responses shown in Figures 15.85 and 15.86 correspond to those of the wall mass center. The sliding mode indicates that the wall would be displaced about 160 cm while

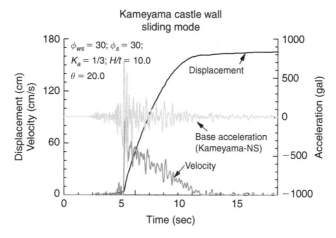

Figure 15.85 Computed displacement and velocity response of mass center for sliding mode of failure.

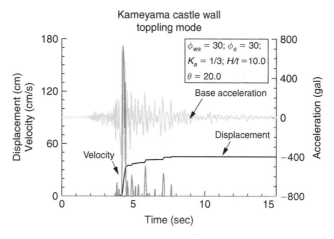

Figure 15.86 Computed displacement and velocity response of mass center for toppling mode of failure.

the toppling mode implies rotation of about 10° (45/250). These results imply that the earthquake shaking was sufficient to induce both sliding and toppling modes of failure. Nevertheless, the effect of sliding mode is more dominant. Since the displacement exceeds the wall width, it may be inferred that the failure of the castle wall was a natural consequence of earthquake shaking.

15.6 CONCLUSIONS

An earthquake is an instability problem of the Earth's crust caused by the varying crustal stresses. It involves rock fracturing and/or slippage of major discontinuities such as faults and fracture zones.

When rock starts to fail, the stored mechanical energy in the rock tends to transform itself into different forms of energy. As shown in Section 2, experimental studies by Aydan and his group showed that rock indicates distinct variations of various measurable parameters such as electric potential, magnetic field acoustic emission, resistivity etc. besides load and displacement, which are called multi-parameters, during deformation and fracturing processes (Fig. 15.1). These variations may be useful in predicting the failures of rock structures as well as earthquakes in geoscience.

The experimental results clearly indicate that the deformation, fracturing and sliding processes induce electric potential in geomaterials.

The magnitude of induced electric potential depends both upon the piezo-electric characteristics of minerals or grains, and the moment caused by the separation of electrons of minerals as a result of deformation and inter-crystal or inter-grain separation and/or sliding during dislocations as a result of fracturing or sliding.

The amplitude of accelerations of the mobile part of the loading system is higher than that of the stationary part. This feature has striking similarities with the strong motion records of nearby earthquake faults observed in the recent large inland earthquakes such as the 1999 Kocaeli earthquake of Turkey, 1999 Chi-chi Earthquake of Taiwan and the 2003 Miyagi Hokubu earthquake of Japan (see Fig. 15.1). Furthermore, the waveforms of the acceleration records of the mobile part are not symmetric with respect to the time axis.

The amplitude of accelerations during the fracturing of hard rocks is higher than that during the fracturing of soft rocks. This is directly proportional to the energy stored in samples before the fracturing.

The experimental results clearly indicate that the surface morphology of discontinuities greatly influences the periodicity of the stick-slip cycle and dynamic responses during the slip phases. However, the recurrence time is not constant even when the surface conditions of the discontinuity and loading conditions are almost the same. The maximum acceleration becomes larger as the system stiffness becomes higher for the same amount of slip. This may imply that the system stiffness has a great influence on the amplitude of the acceleration upon the slippage besides the effect of other parameters.

The recurrence period decreases as the friction angle decreases. Furthermore, the maximum acceleration increases in relation to the amount of relative slip. This may be related to the difference between static and kinetic friction angles of the interfaces in relation to the surface roughness.

It is geometrically possible to compute strain rate components on a plane tangential to the earth's surface from the variation of coordinates of stations of Global Positioning System (GPS) at a given time interval.

As seen in the previous applications to the GPS measurement in Turkey (Aydan et al., 2000c; Aydan, 2003c), the mean, maximum shear and disturbing stress rates can be quite useful for identifying the areas with a high seismic risk. They may be further useful for earthquake prediction in the near future (in the order of months). However, they cannot be used for very near future predictions, say, in the order of several hours to days unless the sensitivity of the measurements is substantially improved.

The major slope failures (landslides) are greatly influenced by the geological structure of the rock mass as well as the shaking characteristics of earthquakes. Specifically, the orientation and shear strength properties of bedding planes, schistosity and existing faults are of great importance. Therefore, the stability assessment of natural and cut slopes must strictly consider this fact in the risk assesments of slope failures due to earthquakes.

The scale and number of slope failures are much larger on the hanging wall side of the earthquake fault as compared with those on the footwall side. Higher ground motions on the hanging wall side and the permanent ground movements are probably the major causes for this observational fact.

The consideration of failures of natural rock slopes has received very little attention in earthquake engineering and regional seismic risk assessments. However, the recent earthquakes showed clearly that the scale of natural slope failures is much larger than that for cut slopes. Therefore, much attention must be given to the possibility of natural rock slopes with the due considerations of facts indicated in this chapter.

Model experiments on various underground openings showed that they are strong against shaking. Nevertheless, the existence of discontinuities makes them vulnerable to collapses particularly in the case of shallow underground openings. This may have some important implications for areas where shallow abandoned mines, underground shelters and new and old tunnels exist.

The GPS measurements of ground deformations during earthquakes (M > 6) clearly indicated that permanent ground deformations do occur. Permanent ground deformation may result from different causes such as faulting, slope failure, liquefaction and plastic deformation induced by ground shaking. They may cause tremendous forces on long and/or large underground structures such as tunnels, powerhouses and underground storage facilities for oil, gas and nuclear wastes.

Plastic zones around underground openings may form during the passage of seismic waves. The maximum amplitude and frequency content of the seismic waves may have some influence on the shape and size of plastic zone formations.

Case histories compiled by the authors indicated that the damage to underground structures might be classified as shaking induced damage, portal damage and permanent ground deformations induced damage. Permanent ground deformation induced damage is generally caused either by faulting or slope movements.

The relation proposed by Aydan *et al.* (2010b) for assessing the seismic damage to underground structures under various circumstances may serve as guidelines. However, further refinement may be possible with more data on the underground structures.

Underground openings crossing faults and fracture zones may be enlarged to accommodate relative slips along faults and fracture zones. The lining of the openings should be ductile to accommodate permanent ground deformations at such zones. Furthermore, the brittle linings of the existing underground structures should be lined with ductile thin plates or fiber-reinforced polymers together with rockbolts at fracture and fault zones, where permanent ground deformations may occur.

The Discrete Finite Element Method (DFEM) is used for the simulation of fault propagation and fault-structure interaction. In the simulations, fundamental faulting modes such as normal, thrust and strike-slip faulting mechanisms are considered. The computational results clearly demonstrated that the DFEM is capable of simulating both the fault propagation and fault-structure interaction, although a simple version of this method is used. Nevertheless, it is possible to use the other versions of the DFEM depending upon the available information on the characteristics of solid and fault planes.

ACKNOWLEDGEMENTS

The authors would like to sincerely thank Emeritus Professor T. Kawamoto of Nagoya University for encouraging the authors to undertake these studies, Assoc. Prof. Dr. Iraj H.P. Mamaghani of North Dakota University (USA) for his collaboration in developing the DFEM and help in some of the computations, Prof. Z. Hasgür of Istanbul Technical University, Prof. R. Ulusay of Hacettepe University, Prof. M. Hamada of Waseda University, Prof. G. Barla of Politecnico di Torino for joining the authors during many site investigations of worldwide earthquakes since 1992 and former and current undergraduate and graduate students of Aydan's laboratory in Tokai University for experimental studies in the laboratory. Finally, the authors dedicate this chapter to the late Prof. Dr. Rifat YARAR of Istanbul Technical University, Turkey for his continuous support and encouragement for the first author since 1992 until he passed away in 2004.

REFERENCES

Aydan, Ö.: *Stability of slope and shallow underground openings in discontinuous rock mass.* Dept. of Geotechnical Engineering, Nagoya University, Interim Report, 1986, 16 pages (unpublished).

Aydan, Ö.: *The stabilization of rock engineering structures by rockbolts.* Doctoral Thesis, Nagoya University, 1989, 204 pages.

Aydan, Ö.: The stress state of the earth and the earth's crust due to the gravitational pull. *The 35th US Rock Mechanics Symposium*, Lake Tahoe, 1995, pp.237–243.

Aydan Ö.: Dynamic uniaxial response of rock specimens with rate-dependent characteristics. *SARES'97*, 1997a, pp.322–331.

Aydan, Ö.: *Seismic characteristics of Turkish earthquakes.* Turkish Earthquake Foundation, TDV/TR 97-007, 41 pages, 1997b.

Aydan, Ö.: A stress inference method based on structural geological features for the full-stress components in the earth's crust, *Yerbilimleri* 22 (2000a), pp.223–236.

Aydan, Ö.: A stress inference method based on GPS measurements for the directions and rate of stresses in the earth' crust and their variation with time. *Yerbilimleri* 22 (2000b), pp.21–32.

Aydan, Ö.: An experimental study on the dynamic responses of geomaterials during fracturing. *Journal of School of Marine Science and Technology, Tokai University*, 1(2) (2003a), pp.1–7.

Aydan, Ö.: Actual observations and numerical simulations of surface fault ruptures and their effects engineering structures. *The Eighth U.S.-Japan Workshop on Earthquake Resistant Design of Lifeline Facilities and Countermeasures Against Liquefaction. Technical Report,* MCEER-03-0003, 2003b, pp.227–237.

Aydan, Ö.: *The earthquake prediction and earthquake risk in Turkey and the applicability of Global Positioning System (GPS) for these purposes.* Turkish Earthquake Foundation, TDV/KT 024-87 (2003c), pp.1–73 (in Turkish).

Aydan, Ö.: The Inference of crustal stresses in Japan with a particular emphasis on Tokai region. *Int. Symp. on Rock Stress*, Kumamoto, 2003d, pp.343–348.

Aydan, Ö. and Tokashiki N.: The Inference of crustal stresses in Ryukyu Islands. *3rd International Symposium on Rock Stress*, Kumamoto, 2003, pp.349–354.

Aydan Ö.: Implications of GPS-derived displacement, strain and stress rates on the 2003 Miyagi-Hokubu earthquakes, *Yerbilimleri*.30 (2004a), pp.91–102.

Aydan, Ö.: Damage to abandoned lignite mines induced by 2003 Miyagi-Hokubu earthquakes and some considerations on its possible causes. *J. of School of Marine Science and Technology, Tokai University,* 2(1) (2004b), pp.1–17.

Aydan, Ö.: Strong motions, ground liquefaction and slope instabilities caused by Kashmir Earthquake of October 8, 2005 and their effects on settlements and buildings. *Bulletin of Engineering Geology* 23 (2006a), pp.15–34.

Aydan, Ö.: The Possibility of Earthquake Prediction by Global Positioning System (GPS). *J. of School of Marine Science and Technology, Tokai University* 4(3) (2006b), pp.77–89.

Aydan, Ö.: Inference of seismic characteristics of possible earthquakes and liquefaction and landslide risks from active faults (in Turkish). *The 6th National Conference on Earthquake Engineering of Turkey,* Istanbul, 2007, Vol.1, pp.563–574.

Aydan, Ö.: New directions of rock mechanics and rock engineering: Geomechanics and Geoengineering. *5th Asian Rock Mechanics Symposium (ARMS5),* Tehran, 2008a, pp.3–21.

Aydan Ö.: Investigation of the seismic damage to the cave of Gunung Sitoli (Tögi-Ndrawa) by the 2005 Great Nias earthquake. *Yerbilimleri* 29(1), 2008b, pp.1–16.

Aydan, Ö. and Amini, M.: An experimental study on rock slopes against flexural toppling failure under dynamic loading and some theoretical considerations for its stability assessment. *Journal of Marine Science and Technology, Tokai University,* 7(2) (2009), pp.25–40.

Aydan, Ö. and Daido, M.: An experimental study on the seepage induced geo-electric potential in porous media. *Journal of School of Marine Science and Technology, Tokai University,* No.55, (2002), pp.53–66.

Aydan, Ö. and Hamada, M.: The site investigation of the Erzincan (Turkey) Earthquake of March 13, 1992. *The 4th Japan-US Workshop on Earthquake Resistant Design of Lifeline Facilities and Countermeasures Against Soil Liquefaction,* Honolulu, 1992, pp.17–34.

Aydan, Ö. and Kawamoto, T.: The stability of slopes and underground openings against flexural toppling and their stabilisation. *Rock Mechanics and Rock Engineering* 25(3) (1992), pp.143–165.

Aydan, Ö. and Kawamoto, T.: The general characteristics of the stress state in the various parts of the earth's crust. *Int. Symp. Rock Stress,* Kumamoto, 1997, pp.369–373.

Aydan, Ö. and Kawamoto, T.: The damage to abandoned lignite mines caused by the 2003 Miyagi-Hokubu earthquake and some considerations on its causes. *3rd Asian Rock Mechanics Symposium,* Kyoto, 2004, Vol.1, pp. 525–530.

Aydan, Ö. and Kim, Y.: The inference of crustal stresses and possible earthquake faulting mechanism in Shizuoka Prefecture from the striations of faults. *J. of the School of Marine Sci. and Technology, Tokai University* 54 (2002), pp.21–35.

Aydan, Ö. and Kumsar, H.: An Experimental and Theoretical Approach on the Modeling of Sliding Response of Rock Wedges under Dynamic Loading. *Rock Mechanics and Rock Engineering* (2010), 43, pp.821–830.

Aydan, Ö. and Nawrocki, P.: Rate-dependent deformability and strength characteristics of rocks. *Int. Symp. On the Geotechnics of Hard Soils-Soft Rocks,* Napoli, 1998, 1, pp.403–411.

Aydan, Ö. and Paşamehmetoğlu, A.G.: Dünyanın çeşitli yörelerinde ölçülmüş yerinde gerilimler ve yatay gerilim katsayısı. *Kaya Mekaniği Bülteni* 10 (1994), pp.1–17.

Aydan, Ö. and Tokashiki, N.: Some damage observations in Ryukyu Limestone Caves of Ishigaki and Miyako islands and their possible relations to the 1771 Meiwa Earthquake. *Journal of The School of Marine Science and Technology, Tokai University* 5(1) (2007), pp.23–39.

Aydan, Ö. and Ulusay, R.: Back analysis of a seismically induced highway embankment during the 1999 Düzce earthquake. *Environmental Geology* 42 (2002), pp.621–631.

Aydan, Ö., Shimizu, Y. and Ichikawa, Y.: The Effective Failure Modes and Stability of Slopes in Rock Mass with Two Discontinuity Sets. *Rock Mechanics and Rock Engineering,* 22(3) (1989), pp.163–188.

Aydan, Ö., Shimizu, Y. and Kawamoto, T.: The stability of rock slopes against combined shearing and sliding failures and their stabilisation. *Int. Symp. on Rock Slopes*, New Delhi, 1992a, pp.203–210.,

Aydan, Ö., Shimizu, Y. and Kawamoto, T.: The reach of slope failures. *The 6th Int. Symp. Landslides, ISL 92*, Christchurch, 1992b, Vol.1, pp.301–306.

Aydan, Ö., Shimizu, Y. and Karaca, M.: The dynamic and static stability of shallow underground openings in jointed rock masses. *The 3rd Int. Symp. on Mine Planning and Equipment Selection*, Istanbul, 1994, pp.851–858.

Aydan, Ö., Ito, T. and Ichikawa, Y.: Failure phenomena and strain localisation in rock mechanics and rock engineering: A phenomenological description. *Int. Symp. Assessment and Prevention of Failure Phenomena in Rock Engineering*, Istanbul, 1993, pp.119–128.

Aydan, Ö., Mamaghani, I.H.P. and Kawamoto, T.: Application of discrete finite element method (DFEM) to rock engineering structures, *NARMS '96*, 1996, pp.2039–2046.

Aydan, Ö., Kumsar, H., Ulusay, R. and Shimizu, Y.: Assessing limiting equilibrium methods (LEM) for slope stability by discrete finite element method (DFEM). *IACMAG*, Wuhan, 1997, pp.1681–1686.

Aydan, Ö., Ulusay, R., Hasgür, Z. and Taşkın, B.: *A site investigation of Kocaeli Earthquake of August 17, 1999"*. Turkish Earthquake Foundation, TDV/DR 08-49, 180p, 1999a.

Aydan, Ö., Ulusay, R., Hasgür, Z. and Hamada, M.: The behaviour of structures built in active fault zones in view of actual examples from the 1999 Kocaeli and Chi-chi Earthquakes. *ITU-IAHS International Conference on the Kocaeli Earthquake: A Scientific Assessment and Recommendations for Re-building*, İstanbul, 1999b, pp.131–142.

Aydan, Ö., Ulusay, R., Kumsar, H. and Tuncay, E.: *Site investigation and engineering evaluation of the Düzce-Bolu Earthquake of November 12, 1999*. Turkish Earthquake Foundation, TDV/DR 095-51, 307p, 2000.

Aydan, Ö., Dalgıç, S. and Kawamoto, T.: Prediction of squeezing potential of rocks in tunneling through a combination of an analytical method and rock mass classifications. *Italian Geotechnical Journal*, 34(1) (2000), pp.41–45.

Aydan, Ö., Kumsar, H. and Ulusay, R.: The implications of crustal strain-stress rate variations computed from GPS measurements on the earthquake potential of Turkey. *Int. Conf. of GIS on Earth Science and Application. ICGESA'2000*, Menemen, 2000(on CD).

Aydan, Ö., Minato, T. and Fukue, M.: An experimental study on the electrical potential of geomaterials during deformation and its implications in Geomechanics. *38th US Rock Mech. Symp.*, Washington, 2001, Vol.2, pp.1199–1206.

Aydan, Ö., Tokashiki, N., Shimizu, Y. and Mamaghani, I.H.P.: A stability analysis of masonry walls by Discrete Finite Element Method (DFEM). *10th IACMAG Conference*, Austin, 2001, pp.1625–1628.

Aydan, Ö, Kumsar, H., and Ulusay, R.: How to infer the possible mechanism and characteristics of earthquakes from the striations and ground surface traces of existing faults. *JSCE, Earthquake and Structural Engineering Division* 19(2) (2002), pp.199–208.

Aydan, Ö., Tokashiki, N., Ito, T., Akagi, T., Ulusay, R. and Bilgin, H.A.: An experimental study on the electrical potential of non-piezoelectric geomaterials during fracturing and sliding, *9th ISRM Congress*, South Africa, 2003a, pp.73–78.

Aydan, Ö., Ogura, Y., Daido, M. and Tokashiki, N.: A model study on the seismic response and stability of masonry structures through shaking table tests. *Fifth National Conference on Earthquake Engineering*, Istanbul, Turkey, 2003b, Paper No: AE-041(on CD).

Aydan, Ö., Daido, M., Tano, H., Tokashiki, N. and Ohkubo, K.: A real-time multi-parameter monitoring system for assessing the stability of tunnels during excavation. *ITA Conference*, Istanbul, 2005a, pp.1253–1259.

Aydan, Ö., Sakamoto, A., Yamada, N., Sugiura, K. and Kawamoto, T.: A real time monitoring system for the assessment of stability and performance of abandoned room and pillar lignite mines. *Post Mining 2005*, Nancy, 2005b(on CD).

Aydan, Ö., Daido, M., Ito, T., Tano, H. and Kawamoto, T.: Instability of abandoned lignite mines and the assessment of their stability in long term and during earthquakes. *4th Asian Rock Mechanics Symposium*, Singapore, 2006a, Paper No. A0355 (on CD).

Aydan, Ö., Daido, M., Ito, T., Tano, H. and Kawamoto, T.: Prediction of post-failure motions of rock slopes induced by earthquakes. *4th Asian Rock Mechanics Symposium*, Singapore, 2006b, Paper No. A0356 (on CD).

Aydan, Ö., Daido, M., Tokashiki, N., Bilgin, A. H. and Kawamoto, T.: Acceleration response of rocks during fracturing and its implications in earthquake engineering. *11th ISRM Congress*, Lisbon, 2007, Vol.2, pp.1095–1100.

Aydan, Ö., Tokashiki, N. and Sugiura, K.: Characteristics of the 2007 Kameyama earthquake with some emphasis on unusually strong ground motions and the collapse of Kameyama Castle. *Journal of The School of Marine Science and Technology, Tokai University* 6(1) (2008), pp.83–105.

Aydan, Ö., Hamada, M., Ito, J and Ohkubo, K.: Damage to Civil Engineering Structures with an Emphasis on Rock Slope Failures and Tunnel Damage Induced by the 2008 Wenchuan Earthquake. *Journal of Disaster Research* 4(2) (2009a), pp.153–164.

Aydan, Ö., Ohta, Y., Hamada, M., Ito, J. and Ohkubo, K.: The characteristics of the 2008 Wenchuan Earthquake disaster with a special emphasis on rock slope failures, quake lakes and damage to tunnels. *Journal of the School of Marine Science and Technology, Tokai University* 7(2) (2009b), pp.1–23.

Aydan, Ö., Kumsar, H. and Toprak, S.: The 2009 L'Aquila earthquake (Italy): Its characteristics and implications for earthquake science and earthquake engineering. *Yerbilimleri*, 30(3) (2009), pp.235–257.

Aydan, Ö., Ohta, Y. and Hamada, M.: Geotechnical evaluation of slope and ground failures during the 8 October 2005 Muzaffarabad earthquake in Pakistan. *Journal of Seismology* 13(3) (2009), pp.399–413.

Aydan, Ö., Kumsar, H., Toprak, S. and Barla, G.: Characteristics of 2009 L'Aquila earthquake with an emphasis on earthquake prediction and geotechnical damage. Geotechnical Damage. *Journal of Marine Science and Technology, Tokai University* 7(3) (2009c), pp.23–51.

Aydan, Ö., Ohta, Y. and Tano, H.: Multi-parameter response of soft rocks during deformation and fracturing with an emphasis on electrical potential variations and its implications in geomechanics and geoengnineering. *The 39th Rock Mechanics Symposium of Japan*, Tokyo, 2010, pp.116–121.

Aydan, Ö., Ohta, Y., Geniş, M., Tokashiki, N. and Ohkubo, K.: Response and Earthquake induced Damage of Underground Structures in Rock Mass, *Journal of Rock Mechanics and Tunnelling Technology* 16(1) (2010a), pp.19–45.

Aydan, Ö., Ohta, Y., Geniş, M., Tokashiki, N. and Ohkubo, K.: Response and stability of underground structures in rock mass during earthquakes. *Rock Mechanics and Rock Engineering*, 43(6), (2010b), pp.857–875.

Aydan, Ö., Tano, H. and Ohta, Y.: A multi-parameter measurement system at Koseto (Shizuoka, Japan) and its responses during the large earthquakes since 2003. *Proceedings of World Geothermal Congress 2010*, Bali, Indonesia, (2010d), P.N.1387 (on CD).

CWB: *Free field strong-motion data from the 1999 Chi-Chi earthquake*. Seismological Center, Central Weather Bureau, Taipei, Taiwan, 1999.

Egger, P.: A new development in the base friction technique. *Colloquium on Geomechnanical Models, ISMES*, Bergamo, 1979, pp.67–81.

Eringen, A.C.: *Mechanics of Continua*. R.E. Krieger Pub. Co., New York, 1980.

Fowler, C.M.R.: *The solid earth – An introduction to Global Geophysics*. Cambridge University Press, Cambridge, 1990.

Ikeya, M. and Matsumoto, H.: Reproduced earthquake precursor legends using a Van de Graaff electrostatic generator: candle flame and dropped nails. *Naturwissenschaften*, 84 (1997), pp.539–541.

Ikeya, M., Komatsu, T., Kinoshita, Y., Teramoto, K., Inoue, K., Gondou, M. and Yamamoto, T.: Pulsed electric field before Kobe and Izu earthquakes from Seismically-induced *Anomalous Animal Behaviour (SAAB)*. *Episodes*, 20(4) (1997), pp.253–260.

Genis, M.: *Investigation of the effects of geometrical design parameters on the dimensions of failure zone occurring around deep underground openings under static and dynamic conditions*. Ph.D. Thesis (in Turkish), Department of Mining Engineering, Zonguldak Karaelmas University, Zonguldak, Turkey, 2002, 352 p.

Genis, M. and Aydan, Ö.: Evaluation of dynamic response and stability of shallow underground openings in discontinuous rock masses using model tests. *Korea-Japan Joint Symposium on Rock Engineering*, Seoul, Korea, 2002, pp.787–794.

Genis, M. and Gerçek, H.: A numerical study of seismic damage to deep underground openings. ISRM 2003-Technology Roadmap for Rock Mechanics, *10th ISRM Congress*, South African Institute of Mining and Metallurgy, 2003, pp.351–355.

Genis, M and Aydan, Ö.: Static and dynamic stability of a large underground opening. In: N. Bilgin *et al.* (eds): *Proc of the 2th symposium on Underground excavations for Transportation* (in Turkish), TMMOB, Istanbul, 2007, 138, pp.317–326.

Genis, M. and Aydan, Ö.: Assessment of dynamic response and stability of an abandoned room and pillar underground lignite mine. In: *The 12th International Conference of International Association for Computer Methods and Advances in Geomechanics (IACMAG)*, Goa, India, 2008, pp.3899–3906.

Kawamoto, T. and Aydan, Ö.: A review of numerical analysis of tunnels in discontinuous rock masses. *International Journal of Numerical and Analytical Methods in Geomechanics* 23 (1999), pp.1377–1391

Kawamoto, T., Obara, Y. and Tokashiki, N.: Characteristics of deformation and permeability of fractured rock. In: *Proc. Int. Symp. on Weak Rock*, Tokyo, Japan, 1981, Vol. 1, pp.63–68.

Keefer, D.K.: Landslides caused by earthquakes. *Geological Society of American Bulletin 95* (1984), pp.406–421.

K-NET (2004): http://www.bosai.go.jp/K-NET

KIK-NET (2004, 2008): http://www.bosai.go.jp/KIK-NET

Kumsar, H., Aydan, Ö., and Ulusay, R.: Dynamic and static stability of rock slopes against wedge failures. *Rock Mechanics and Rock Engineering* 33(1) (2000), pp.31–51.

Kumsar, H., Aydan, Ö., Tano, H., Ulusay, R., Celik, S., Kaya, M., Karaman, M.Y.: An on-line monitoring system of multi-parameter changes of geothermal systems related to earthquake activity in Western Anatolia in Turkey. *Proceedings World Geothermal Congress 2010*, Bali, Indonesia, 2010, pp.25–29.

Mamaghani, I.H.P. and Aydan, Ö.: Stability analysis of slopes by discrete finite element method. *GeoEng 2000*, Melbourne, Australia, 2000(on CD).

Mamaghani, I.H.P., Baba, S., Aydan, Ö. and Shimizu, S.: Discrete finite element method for blocky systems, *Computer Methods and Advances in Geomechanics*, IACMAG, Morgantown, 1994, 1, pp.843–850.

Mamaghani, I.H.P., Aydan, Ö. and Kajikawa, Y.: Analysis of masonry structures under static and dynamic loading by discrete finite element method. *Journal of Structural Mechanics and Earthquake Engineering*. Japan Society of Civil Engineers, JSCE, No. 626/I-48 (1999), pp.1–12.

Mizutani, H., Ishido, T., Yokokura, T. and Ohnishi, S.: Electrokinetic phenomena associated with earthquakes. *Geophysical Res. Letters* 3(2) (1976), pp.365–368.

Mogi, K.: *Earthquake Prediction*. Academic Press, Orlando, Florida, 355 p, 1985.

Nur, A.: Dilatancy, pore fluids and premonitory variations of ts/tp travel times. *Bulletin of the Seismological Society of America* 62(5) (1972), pp.1217–1222.

Ohta, Y. and Aydan, Ö.: Integration of ground displacement from acceleration records. *JSCE Earthquake Engineering Symposium*, 2007a, pp.1046–1051.

Ohta, Y. and Aydan, Ö.: An integration technique for ground displacement from acceleration records and its application to actual earthquake records. *Journal of The School of Marine Science and Technology, Tokai University* 5(2) (2007b), pp.1–12.

Ohta, Y. and Aydan, Ö.: An experimental and theoretical study on stick-slip phenomenon with some considerations from scientific and engineering viewpoints of earthquakes. *Journal of The School of Marine Science and Technology, Tokai University* 8(3) (2009), pp.53–67.

Ohta, Y. and Aydan, Ö.: The dynamic responses of geomaterials during fracturing and slippage. Rock Mechanics and Rock Engineering, 43(6), (2010), pp.727–740.

Ohta, Y. and Aydan, Ö.: Failure modes of shallow tunnels induced by earthquakes and evaluation of their stability. *The 40th Rock Mechanics Symposium of Japan*, (in Japanese), 2011(on CD).

Ohta, Y., Aydan, Ö. and Tokashiki, N.: The dynamic response of rocks during fracturing and its implications in geo-engineering and earth science. *5th Asian Rock Mechanics Symposium (ARMS5)*, Tehran, 2008, pp.965–972.

Reilinger, R.E, McClusky, S.C., Oral, M.B., King, R.W., Toksöz, M.N., Barka, A.A., Kınık, I., Lenk, O. and Şanlı, I.: Global positioning system measurements of present-day crustal movements in the Arabia-Africa-Euroasia plate collision zone, *J. Geophysical Research* 102(B5) (1997), pp.9983–9999.

Reilinger, R.E., Ergintav, S.; Burgmann, R., McClusky, S., Lenk, O., Barka, A., Gürkan, O., Hearn, L., Feigl, K.L., Çakmak, R.; Aktug, B., Özener, H. and Toksöz, M.N.: Coseismic and postseismic fault slip for the 17 August 1999, M = 7.5, Izmit, Turkey Earthquake, *Science* 289 (2000), pp.1519–1524.

Sugawara, K. and Obara, Y.: *Measuring rock stress. In Comprehensive Rock Engineering*, Pergamon Press, Chapter 21, Vol. 3 (1993), pp.533–552.

Tokashiki, N. and Aydan, Ö.: Kita-Uebaru natural rock slope failure and its back analysis. *Environmental Earth Sciences*. (2010) DOI 10.1007/s12665-010-0492-8.

Tokashiki, N., Aydan, Ö., Mamaghani, I.H.P. and Kawamoto, T.: The stability of a rock block on an incline by discrete finite element method (DFEM), *IACMAG '97*, Wuhan, China, 1997a., pp.523–528.

Tokashiki, N., Aydan, Ö., Shimizu, Y. and Kawamoto, T.: The assessment of the stability of a very old tunnel by discrete finite element method (DFEM). *Numerical Methods in Geomechanics, NUMOG VI*, Montreal, 1997b, pp.495–500.

Toksöz, M.N.: Earthquake prediction research in the United States. Predicting Earthquakes, Panel for Earthquake Prediction of the Committee on Seismology, NRC,1977, pp.37–50.

Tuncay, E., Ulusay, R., Watanabe, H., Tano, E., Yüzer, E. and Aydan, Ö.: Acoustic Emission (AE) technique: A preliminary investigation on the determination of in-situ stresses by AE technique in Turkey, *Yerbilimleri*, 25 (2002), pp.83–98.

Ulusay, R., Aydan, Ö. and Hamada, M.: The Behavior of structures built on active fault zones: examples from the recent earthquakes of Turkey. *Structural Eng/Earthquake Eng, JSCE*, 19(2) (2002), Special Issue, pp.149–167.

Zoback, M.D., Moos, D., Mastin, L. and Anderson, R.N.: Well bore breakouts and in situ stress. *Journal of Geophysical Research* 90(B7) (1985), pp.5523–5530.

Chapter 16

Constraining paleoseismic PGA using numerical analysis of structural failures in historic masonry structures: Review of recent results

Yossef H. Hatzor and Gony Yagoda-Biran

16.1 INTRODUCTION

Seismic hazard is defined in terms of the probability of exceeding a certain ground motion in a specific area, and is typically discussed in terms of Peak Ground Acceleration (PGA). Predicted PGA values for specific regions are commonly reported in national seismic building codes and therefore PGA is used extensively in earthquake engineering practice throughout the world. A new method to constrain expected earthquake PGA values, by back analysis of finite block displacements in historic masonry structures, is presented here. To demonstrate the new approach two archeological masonry structures that exhibit seismogenic damage are used as illustrative examples: 1) a 2000 year old Nabatean (Roman Period) arch in which the keystone slid downward during an earthquake of an uncertain date (Fig. 16.1a), and 2) a 1400 year old Byzantine church in which a series of parallel granite and marble columns toppled down in the same direction, most probably due to an earthquake that struck the region in 749 AD (Fig. 16.1b). Both sites are located along the seismically active Dead Sea rift system.

The preserved and damaged structures are first mapped in great detail in the field, and on the basis of the acquired geometrical data a mesh of discrete blocks is generated numerically, representing the presumed un-deformed structural configuration. Then, the discrete block system is loaded numerically by dynamic input functions until the deformed configurations that best fit the preserved damaged configurations in the field are obtained. As a first order approximation harmonic, sinusoidal, input acceleration functions are used so that the most likely amplitude (PGA) as well as frequency and duration of the input motion are obtained directly as a result of the analysis. The analysis can be repeated with real earthquake records to further constrain the dynamic motion parameters.

We proceed with forward modeling performed for a simulated fracture pattern in an existing discontinuous rock slope – the upper terrace of King Herod's palace in Masada (Fig. 16.1c). Here we show how natural fracture patterns may be simulated to form a numerical block system mesh not restricted to any particular block shape. The generated mesh is loaded with a real earthquake record and the most likely failure modes that could result from earthquake loading are obtained as a result of the analysis.

The numerical approach used in this study is the implicit, discrete, Discontinuous Deformation Analysis (DDA) method developed by Dr. Gen-hua Shi (1993), the theory of which is briefly reviewed next.

(a)   (b)   (c)

Figure 16.1 Analyzed field case studies: (a) key stone displacement in a Roman arch, (b) toppling of a series of columns in a Byzantine church. (c) The highly fractured rock slope at the North face of Masada – Kind Herod's palace.

Following a brief summary of the theory, the accuracy of the DDA method is validated using analytical solutions before its application for real case studies is presented. We feel that by demonstrating the validity of DDA and its applicability for geotechnical earthquake engineering studies, it is now possible to safely proceed with more complicated tectonophysical studies for analyzing the kinematics of plate motion along tectonic boundaries.

16.2 BRIEF SUMMARY OF DDA THEORY

Discontinuous Deformation Analysis (DDA) is useful for investigating the kinematics of blocky rock masses. DDA models a discontinuous material as a system of individually deformable blocks that move independently with minimal amount of interpenetration. The formulation is based on dynamic equilibrium that considers the kinematics of individual blocks as well as friction along the block interfaces. The displacement and deformation of the discrete blocks are the result of the accumulation of small time steps. The equilibrium equations are derived by minimizing the total potential energy of the block system Π with respect to the displacement at block center D:

$$\frac{\partial \Pi_P}{\partial D} + \frac{\partial \Pi_V}{\partial D} + \frac{\partial \Pi_I}{\partial D} + \frac{\partial \Pi_E}{\partial D} + \frac{\partial \Pi_\sigma}{\partial D} + \frac{\partial \Pi_c}{\partial D} = 0 \qquad (16.1)$$

where P, V, I, E, σ, and C are the point load, body force, inertia force, elastic force, initial stress, and contact force, respectively. Application of an appropriate integral variation and expanding terms in the equation above results in the equations of motions:

$$M\ddot{D} + C\dot{D} + KD = F \qquad (16.2)$$

where M, C and K represent the mass, damping, and stiffness matrices of a geometrically non-linear system of equations subject to a time varying load F. The components of the matrices M, C, K, and F are extensively discussed in Shi (1993). In this study we use the original two dimensional version of DDA in which each block i in the general block system has six degrees of freedom, and the resulting displacement components (u, v) of an arbitrary point (x, y) in X and Y directions are derived using a first order approximation:

$$\begin{pmatrix} u \\ v \end{pmatrix} = \begin{pmatrix} 1 & 0 & -(y - y_o) & x - x_o & 0 & (y - y_o)/2 \\ 0 & 1 & (x - x_o) & 0 & (y - y_o) & (x - x_o)/2 \end{pmatrix} \begin{pmatrix} u_o \\ v_o \\ r_o \\ \varepsilon_x \\ \varepsilon_y \\ \gamma_{xy} \end{pmatrix} = [T_i][D_i] \qquad (16.3)$$

where the six degrees of freedom for block i are: rigid body translations (u_o, v_o), rigid body rotation angle (r_o) with a rotation center at $(x_o \ y_o)$, and the normal and shear strain components $(\varepsilon_x, \varepsilon_y, \gamma_{xy})$. In matrix form $[Ti]$ is the first order displacement function and $[Di]$ is the vector displacement variables of Block i. The algebraic equation for the increase in displacement is solved for each time increment by substituting the appropriate terms for acceleration and velocity, provided by a time integration formulation similar to Newmark direct integration method with parameters $\beta = 0.5$ and $\gamma = 1.0$, into the general equation of motion (Doolin and Sitar, 2004; Wang, Chuang and Sheng, 1996). The result is a system of equations for solving the dynamic problem which, after collecting terms on both sides, are typically expressed as:

$$\hat{K}D = \hat{F} \qquad (16.4)$$

Or in matrix form:

$$\begin{pmatrix} K_{11} & K_{12} & K_{13} & \dots & K_{1n} \\ K_{21} & K_{22} & K_{23} & \dots & K_{2n} \\ K_{31} & K_{32} & K_{33} & \dots & K_{3n} \\ \cdot & \cdot & \cdot & \cdot & \cdot \\ K_{n1} & K_{n2} & K_{n3} & \dots & K_{nn} \end{pmatrix} \begin{pmatrix} D_1 \\ D_2 \\ D_3 \\ \cdot \\ D_n \end{pmatrix} = \begin{pmatrix} F_1 \\ F_2 \\ F_3 \\ \cdot \\ F_n \end{pmatrix} \qquad (16.5)$$

where K_{ij} is a 6×6 coefficient sub-matrix, D_i is a 6×1 deformation matrix of block i and F_i is a 6×1 loading matrix of block i. Sub-matrices $[K_{ii}]$ depend on the material properties of block i and sub-matrices $[Kij]$ are defined by the contacts between blocks i and j.

The solution to the system of equations (16.5) is constrained by inequalities associated with block kinematics: the "*no penetration – no tension*" condition between blocks. The kinematic constraints on the system are imposed using the penalty method. The minimum energy solution is one with no tension or penetration. When the system converges to an equilibrium state, the energy of the contact forces is balanced by the penetration energy, resulting in inevitable, but very small, penetrations. The energy of the penetrations is used to calculate the contact forces, which are, in turn, used to calculate the frictional forces along the interfaces between blocks. Shear displacement along the interfaces is modeled using the Coulomb-Mohr failure criterion.

Three user-specified, numeric control parameters are required in DDA: The normal contact spring stiffness ($g0$), the time step size ($g1$), and the assumed maximum displacement per time step ratio ($g2$). The possible range for these control parameters in relation to the size of the problem is discussed in Shi's thesis (Shi, 1988), and optimal values for dynamic applications are reported by Tsesarsky, Hatzor and Sitar (2005). A user specified energy dissipation coefficient ($k01$) allows distinction between "dynamic" ($k01 = 1$) and "static" ($k01 = 0$) analyses. "Dynamic" analysis implies that at the beginning of a time step each block element inherits its terminal velocity at the previous time step, whereas "static" analysis implies that the initial velocity for each block element in every time step is always zero. Variations of the value of $k01$ in the range of $0.9 < k01 < 1$ allows introduction of up to 10% kinetic damping into the dynamic system as demonstrated for natural rock slopes (Hatzor *et al.*, 2004) and for shaking table experiments (Tsesarsky, Hatzor and Sitar, 2005). Implementation of more robust damping algorithms in DDA contact mechanics is the subject of intensive current research (Ohnishi *et al.*, 2005).

The validity and accuracy of DDA has been studied extensively over the past decade and a comprehensive review is presented by MacLaughlin and Doolin (2005); new DDA validations pertaining to the specific goals of this research are reported by Kamai and Hatzor (2008).

16.3 SEVERAL DYNAMIC DDA VALIDATIONS

Several DDA validation studies are performed for calibration purposes using existing or originally developed analytical solutions. The first case is for two-dimensional (2D) dynamic forward and backward sliding of a block which responds to induced cyclic motions in an underlying block; in the second case the 2D dynamic rocking of a column subjected to a sine impulse function applied at its centroid is studied; the third example is from a real rock slope problem where the "block slumping" failure mode is used for validation, and the fourth case is the study of the well-known problem of a block on an incline, in three dimensions.

16.3.1 Block response to induced displacements at foundation

DDA allows application of time-dependent displacements to "fixed" points in the block system, defined and positioned by the user. This feature may be used to simulate seismic ground motions at the foundation and to investigate the response of overlying stone structures. An analytical solution for the response of a single block resting on a block that is subjected to time-dependent displacement input function was developed (Kamai and Hatzor, 2008) to study the validity and applicability of this approach. The

Figure 16.2 The modeled DDA block system constraining the motion of Block 2 to horizontal sliding only. Block 0 is the foundation block, Block 1 receives the dynamic input motion (horizontal – cyclic), and Block 2 responds (after Kamai and Hatzor, 2008).

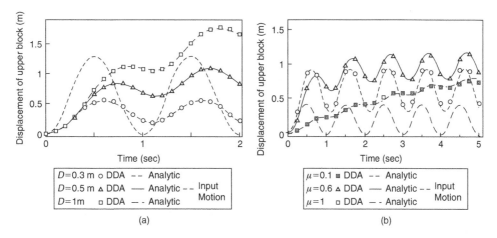

Figure 16.3 Comparison between analytical (line) and DDA (symbols) solutions for the response of a block to cyclic displacements at the foundation (after Kamai and Hatzor, 2008): (a) influence of cyclic displacement amplitude ($f = 1$ Hz and $\mu = 0.6$), (b) influence of friction coefficient along interface ($D = 0.5$ m, $f = 1$ Hz).

analyzed system consists of three blocks: a fixed foundation block (0), the activated block (1), and the responding block (2) as shown in Figure 16.2 above.

The displacement function for block 1 is in the form of a cosine function, starting from 0:

$$d_t = D(1 - \cos(2\pi\omega t)) \tag{16.6}$$

where D is the amplitude of the harmonic wave, and the corresponding response of block 2 is investigated.

In order to compare between DDA and the analytical solution, the mode of failure of the analyzed block in DDA had to be constrained to sliding in one direction only without rotation or vertical motions. This was achieved by modeling the responding block as a flat element minimizing all other degrees of freedom other than horizontal sliding, as in the analytical solution.

A sensitivity analysis for input amplitude and interface friction was performed and the results are presented in Figure 16.3 above. In the left panel the response of Block 2 to changing amplitudes of motion (D) under a constant input frequency of 1 Hz and friction coefficient of 0.6 is presented. The cumulative displacement is in direct proportion to the amplitude, as expected. Note that the three displacement curves

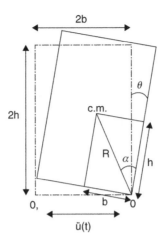

Figure 16.4 Free body diagram and sign convention for the rocking column analysis (after Makris and Roussos, 2000).

follow the periodic behavior of the input displacement function ($T = 1$ sec.), and that divergence between curves starts after 0.25 sec. where the displacement function has an inflection point. In the right panel the response of Block 2 to changing friction coefficients (μ) along the interface under constant displacement amplitude of 0.5 m and input frequency of 1 Hz is presented. The accumulating displacement is in direct proportion to the friction coefficient up to 0.5 sec., where the input displacement function changes direction. After that point the accumulating displacement for $\mu = 0.6$ is larger than for $\mu = 1$, since the high friction works in both directions: forward and backward. Note that curves for $\mu = 0.1$ and $\mu = 0.6$ follow the periodic behavior of the displacement function, whereas the curve for $\mu = 1.0$ is in delay of about 0.25 sec.

16.3.2 Dynamic rocking of a free standing column

Makris and Roussos (2000) studied the dynamic rocking of a column subjected to a sinusoidal input acceleration function for the free body diagram shown in Figure 16.4 above.

The solution for the dynamic rocking of a column subjected to an input loading function of a half-sine pulse is obtained in two stages:

1) Instantaneous response – dynamic motion which takes place simultaneously with application of the input acceleration function: $\ddot{u}_g(t) = a_p \sin(\omega_p t + \psi)$ from $t = 0$ to $t = 0.5$ sec., where ω is 2π ($f = 1$ Hz) and the phase angle (ψ) is $\psi = \sin^{-1}\left(\frac{a_g}{a_p}\right)$,

2) Consequent motion – rocking oscillations after pulse termination from $t = 0.5$ sec. and onwards. Naturally when the pulse terminates the input acceleration diminishes ($\ddot{u}_g(t) = 0$, hence $a_p = 0$) and the coefficients of integration are updated for changing rotation angle and angular velocity. Furthermore, following each impact (@ $\theta = 0$), the angular velocity and the coefficients of integration are recalculated as well.

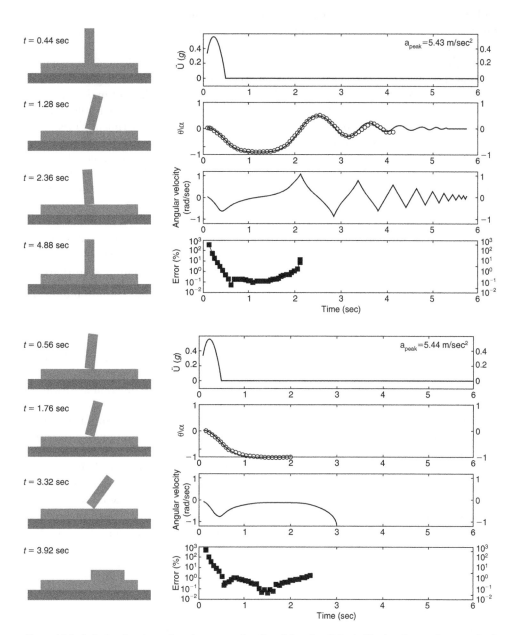

Figure 16.5 Solution for dynamic column rotation ($b = 0.2$ m, $h = 0.6$ m). (Top) a_p lower than required for toppling, (Bottom) a_p sufficient for column toppling. Solid line – analytical solution, Open circles – DDA results (after Yagoda-Biran and Hatzor, 2010).

The analytical and DDA solutions for column width and height of $b = 0.2$ m and $h = 0.6$ m are presented in Figure 16.5. In the top frame results obtained for amplitude $a_p = 5.43$ m/s² (0.5535 g), a value slightly lower than required for overturning, are shown, hence only column rocking is obtained. In the lower frame results obtained with an amplitude of $a_p = 5.44 \, m/s^2$ (0.5545 g), the minimum value required for overturning are shown, and indeed column overturning is obtained ~ 1.5 sec. after pulse termination.

In the DDA model the column rests on a fixed base and is subjected to dynamic input at its centroid. The friction angle along the interface is set to 89 degrees to avoid sliding, as in the analytical solution which ignores sliding. The optimal values for DDA numerical control parameters (g1, g2 in (Shi, 1993)) are obtained following an optimization study performed for a numerical shaking table experiment (Tsesarsky, Hatzor and Sitar, 2005). The optimal contact spring stiffness value (g0 in (Shi, 1993)) is selected by iterations, until the numerical computation returns the correct value of a_p necessary for column overturning as in the analytical solution. The numerical control parameters used here are: ($k01 = 1$, $g2 = 0.0075$, $g1 = 0.0025$ sec, $g0 = 83 * 10^6 N/m$, and $E = 3 \, GPa$, $\nu = 0.25$).

A remarkably good agreement between the analytical and numerical solutions is suggested by the data presented in Figure 16.5. The accuracy of the numerical solution with respect to the analytical solution can be assessed in terms of the relative numerical error $\left(\left| \frac{\theta_{anl.} - \theta_{num.}}{\theta_{anl.}} \right| \cdot 100\% \right)$. As can be seen in Figure 16.5 after initial perturbations the numerical error rapidly decreases below 1%. Note that the DDA solution deviates from the analytical solution as soon as the first impact between the rocking column and the fixed base occurs. This deviation may be explained by the way damping is addressed in the two solutions. While in the analytical solution the motion during impact is energetically damped due to conservation of angular momentum following the constant value of the coefficient of restitution which is used (Makris and Roussos, 2000), in DDA oscillations at contact points are restrained due to inherent algorithmic damping (see Doolin and Sitar, 2004; Ohnishi et al., 2005).

The same procedure is repeated to find the solution for a full sine input function, with $\omega = 2\pi$. The comparison between analytical and numerical solutions is presented in Figure 16.6 for column width and height of $b = 0.5$ m and $h = 1.5$ m, respectively.

Again in the top frame a_p is slightly lower and in the lower frame a_p is slightly higher than the peak acceleration required for overturning, as obtained from the analytical solution. The difference between the two values used for input in the two simulations is $0.003 \, m/s^2$ or $0.0001g$, suggesting an extremely high DDA accuracy. The spring normal stiffness found by optimization is different than for the previous run ($g0 = 64 * 10^6$ N/m), probably because of the difference in column stiffness due to the change in column geometry between the two cases. The other DDA input parameters are as in the half sine validation listed above. As in the previous validation for a half sine input the error remains very small until the first impact occurs, after which the error begins to increase. Naturally from the definition of relative error (see above), which depends on the actual value of θ at each time step, greater error is expected for very small values of θ, and vice versa.

16.3.3 Block slumping at the Snake path cliff, Masada

In a segment along the East face of the Masada mountain, locally known as the "snake path" cliff, a prismatic block resting on an easterly dipping bedding plane and separated from the cliff by two orthogonal "tension cracks", apparently exhibits a

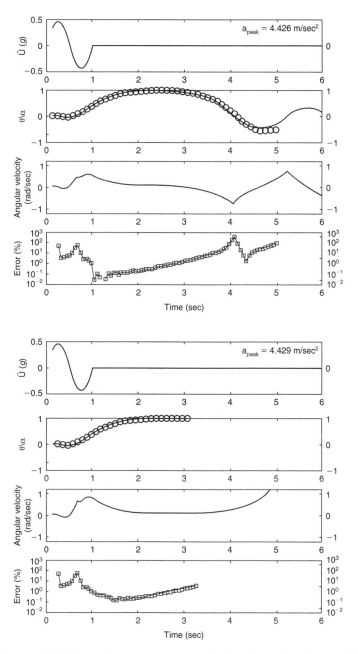

Figure 16.6 Analytical (solid line) and numerical (open symbols) solutions for column rocking under full sine input function ($b = 0.5$ m, $h = 1.5$ m): (Top) a_p slightly lower than required for toppling, (Bottom) a_p just sufficient for column toppling (after Yagoda-Biran and Hatzor, 2010).

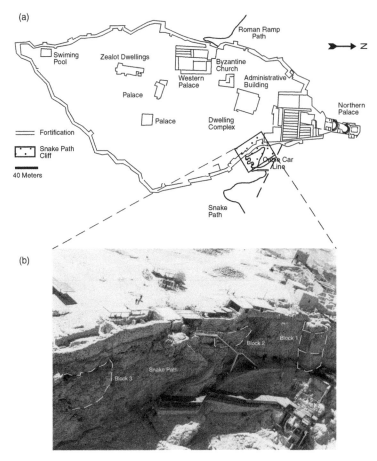

Figure 16.7 Plan (a) and bird eye (b) view of the top of the Masada monument. Block 1, located immediately above the old cable car station, is delineated (after Hatzor, 2003).

"block slumping" failure mode (Goodman and Kieffer, 2000). The block, 15 m high, 10 m wide, and weighs 13.7 MN (1400 ton), rests directly above the old cable car station and is situated along the path of a new bridge connecting between the new cable car station and the entrance gate to the monument at the mountaintop (Block 1 in Fig. 16.7). The block has clearly separated from the cliff over geologic and historic times by an accumulated displacement of up to 20 cm. Current displacement rates were monitored by Hatzor (2003) in connection with an overall dynamic stability analysis for the snake path cliff that culminated in the reinforcement of several removable blocks along the bridge path.

The tall and slender geometry of Block 1 makes it susceptible to the "block slumping" failure mode, initially proposed by Wittke (1965) and extended to multiple blocks by Kieffer (1998). Because the resultant weight vector trajectory of Block 1 acts on the steeply inclined plane (see free body diagrams in Fig. 16.8), sliding will commence by mobilizing shear strength along both the steep and the shallow inclined planes simultaneously. Thus, rotation around a center located outside of the block may take place – a failure mode defined as "Block Slumping" (Goodman and Kieffer, 2000).

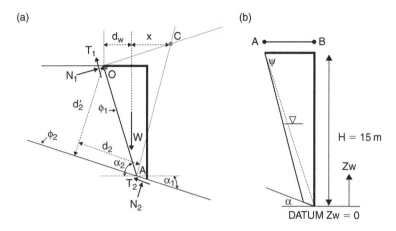

Figure 16.8 Block 1: (a) – Free body diagram, (b) – Actual geometry (after Hatzor, 2003).

It is intuitively clear that once slumping is initiated joint water pressures rapidly dissipate as a joint, with a wide base and sharp edge at the top, is formed behind the block at onset of motion.

The forces acting on a block that undergoes block slumping are shown in Figure 16.8a. Assuming the friction angles on the two sliding planes are equal ($\phi_1 = \phi_2$) three equilibrium equations are necessary for solution of the contact forces N_1 and N_2 and the mobilized friction angle $\phi_{\text{mobilized}}$:

$$\sum F_V = 0: W = N_1 \cos\alpha_1 + N_1 \tan\phi_1 \sin\alpha_1 + N_2 \cos\alpha_2 + N_2 \tan\phi_2 \sin\alpha_2 \quad (16.7)$$

$$\sum M_0 = 0: Wd_w + N_2 \tan\phi_2 d_2' = N_2 d_2 \quad (16.8)$$

$$\sum M_C = 0: Wx = N_2 \tan\phi_2 AC + N_1 \tan\phi_1 OC \quad (16.9)$$

where α_1 and α_2 are the inclinations of the sliding plane and the "tension crack" respectively. Simultaneous solution of the three equations for the geometry of Block 1 (Fig. 16.8b) yields a mobilized friction angle value of $\phi_{\text{mobilized}} = 22°$.

To test the validity of this solution DDA is employed. The exact two-dimensional geometry of Block 1 is studied under gravitational loading with different values of inter-face friction angle as the only varied parameter between simulations. The actual friction angle of Masada discontinuities was studied experimentally (Hatzor *et al.*, 2004) using tilt tests, tri-axial tests, and direct shear tests. The obtained failure envelopes are shown in Figure 16.9. The peak friction angle obtained from direct shear tests on rough surfaces is 41°. The residual friction angle, obtained from tri-axial tests performed on filled saw-cut planes is 23°. The analytical solution for block slumping indicates that for friction angle values lower than 22° Block 1 will exhibit back slumping by simultaneous shear along both interfaces. Therefore, for rough interfaces with available friction angle of 43° the block may be assumed to be stable. However, for interfaces

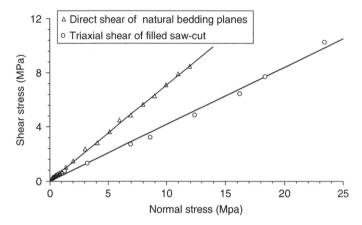

Figure 16.9 Failure envelopes for filled saw-cut (open circles) and rough (open triangles) discontinuities from Masada.

Figure 16.10 DDA results for Block 1 with interface friction angle of 20° and gravitational loading for $t = 0, 0.8, 1.6,$ and 2.5 sec.

possessing residual friction angle value of 23° the block may be considered at limit equilibrium considering the block slumping mode.

The original configuration of Block 1 is shown in the left panel of Figure 16.10. The block remains static until the input friction angle on the interfaces in DDA is reduced to 21° after which sliding ensues along both interfaces simultaneously. The dynamic deformation progress for interface friction angle of 20° is shown in Figure 16.10 where clearly the block slumping mode is obtained, confirming the analytical solution that requires a minimum friction angle of 22° for stability. It is important to note here that in DDA the failure mode is a *result* of the analysis and not a pre assumption.

16.3.4   Block on an incline in three dimensions

The 3D-DDA validation is performed with a simple model of a block on an incline using an existing analytical solution. The model is composed of two blocks: a base block of a triangular prism shape (10 m ∗ 10 m, 5 m width) with inclination angle $\beta = 45°$, and a sliding square prism shaped block (1 m ∗ 1 m ∗ 0.5 m) (see Fig. 16.11).

Figure 16.11 The 3-D model used for the 3D-DDA validation of a block on an incline.

The base block is fixed in space by 7 fixed points, therefore cannot move, and the sliding block is loaded by two loading points, for the 3rd step of the validation study.

16.3.4.1 Step one: gravitational loading

The first step of the validation study is subjecting the sliding block to gravity loading only. Down slope displacements are compared with displacements calculated by an analytical solution for the problem, presented below.

The analytical solution for this problem is as follows: The forces acting on the block on an incline are the gravitational force and the frictional force. The downslope destabilizing force can be expressed as $F_d = mg \sin \beta$ (where m is the block's mass and β is the inclination angle of the base block). The stabilizing force, i.e. the frictional force, can be expressed as $F_s = mg \cos \beta \tan \phi$ (where ϕ is the friction angle of the interface between the base block and the sliding block). Therefore the downslope acceleration, that is the resultant force acting on the block divided by the mass, is

$$a = g \sin \beta - g \cos \beta \tan \phi. \tag{16.10}$$

Double integration over time of the acceleration term will give the displacement (with zero initial velocity and displacement)

$$d(t) = \frac{1}{2}at^2 = \frac{1}{2}g(\sin \beta - \cos \beta \tan \phi)t^2 \tag{16.11}$$

The time step interval (g1) is set here to 0.001 sec, the maximum displacement per time step ratio (g2) to 0.002, and contact spring stiffness (g0) to $4 * 10^8$ N/m. The down slope displacement history is compared for three values of friction angle: 10, 20 and 30° (remembering the inclination angle of the slope is 45°).

The left panel of Figure 16.12 presents the results of the 1st step of the validation study. Note the good agreement between the analytical and numerical solutions. The right panel of Figure 16.12 presents the relative numerical error ($error = \frac{disp_{analy} - disp_{numer}}{disp_{analy}} \times 100\%$). After 0.2 sec the numerical error drops to values below 1% demonstrating good agreement between the two solutions.

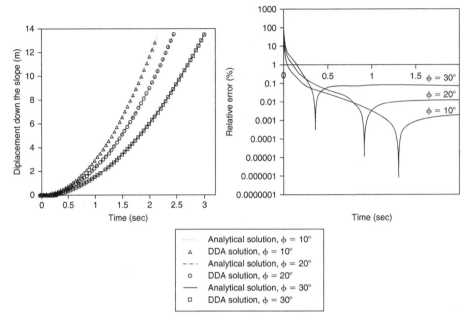

Figure 16.12 Left: Down slope displacement histories of a block on an inclined plane subjected to gravity alone. Legend: curves – analytical solution, symbols – DDA solution. Right: The relative numerical error.

16.3.4.2 Step two: gravitational loading and initial velocity

The next step of the validation study is applying initial velocity to the sliding block, and comparing the down slope displacements of the block to the ones computed by the analytical solution. Similar to the analytical solution presented in step one, double integration over time of the acceleration term is performed, this time remembering that there is initial velocity to the block, therefore the term for the downslope displacement is:

$$d(t) = \frac{1}{2}at^2 + v_0 t = \frac{1}{2}g(\sin\beta - \cos\beta\tan\phi)t^2 + v_0 t \qquad (16.12)$$

The model used in this validation step is identical to the previous one, with input numerical parameters: $g1 = 0.0001$ sec, $g2 = 0.002$, $g0 = 4 * 10^8$ N/m and input physical parameters: $\rho = 2.7$ g/cm^3, $E = 40$ GPa, $\nu = 0.18$, $\phi = 20°$.

The initial velocities are set in the horizontal (x) direction to three different values: 0.01 m/sec, 0.1 m/sec and 1 m/sec. Results for downslope displacement history are presented in the left panel of Figure 16.13. The agreement between the analytical and the numerical solutions is good for all three different velocities, as can be verified by the relative numerical error plotted in the right panel of Figure 16.13 – less than 1% after 0.5 sec of the analysis.

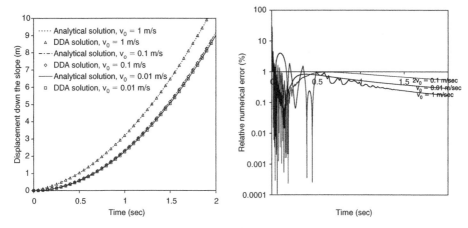

Figure 16.13 Left: Down slope displacement histories of a block on an inclined plane subjected to gravity and initial velocity. Legend: curves – analytical solution, symbols – DDA solution. Right: The relative numerical error.

16.3.4.3 Step three: gravitational and 1-D sinusoidal acceleration

The third step of the validation study is comparing the downslope displacements of the block computed by an analytical solution, with those computed by the DDA, when the sliding block is subjected to a 1-D horizontal sinusoidal acceleration in the form of $a(t) = A\sin(\omega t)$, as well as gravitational acceleration.

The analytical solution is as follows: as the friction angle of the sliding surface (50°) in this validation step is higher than the inclination angle (45°), block sliding will initiate only when the acceleration has reached the value of the yield acceleration and beyond. This type of analysis has been referred to as Newmark' type analysis (Goodman and Seed, 1966; Newmark, 1965). Newmark (1965) and Goodman and Seed (1966) have shown that in the case of a block on an incline, the yield acceleration is $a_{yield} = \tan(\phi - \beta)g$. Once the sinusoidal input acceleration has reached or exceeded a_{yield}, at time t_1, the block begins to gain downslope velocity and displacement. When the sinusoidal input acceleration drops again below the value of a_{yield}, the velocity decreases, as the block is restrained by the frictional force, until it reaches zero and the block stops. When the sinusoidal input acceleration exceeds a_{yield} again at time t_2, motion will again initiate and so on, as in Figure 16.14.

In order to obtain the Newmark displacement, one must perform double integration over time of the downslope acceleration term, which in this case is

$$a(t) = [A\sin(\omega t)\cos\beta + g\sin\beta] - \tan\phi[g\cos\beta - A\sin(\omega t)\sin\alpha] \qquad (16.13)$$

Double integration of this term yields (after Kamai and Hatzor, 2008):

$$d(t) = g\lfloor(\sin\beta - \cos\beta\tan\phi)(1/2t^2 - \theta t + 1/2\theta^2)\rfloor$$
$$+ \frac{A}{\omega^2}[(\cos\alpha + \sin\alpha\tan\phi)(\omega\cos(\omega\theta)(t - \theta) - \sin(\omega t) + \sin(\omega\theta))] \qquad (16.14)$$

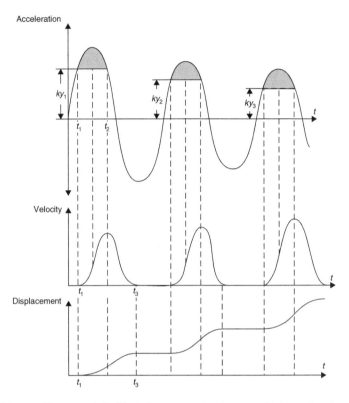

Figure 16.14 Newmark' type analysis. Shaded areas are the times at which acceleration exceeds a_{yield}. (after Goodman and Seed, 1966).

Since block movement initiates only once a_{yield} is reached or exceeded, double integration of the downslope acceleration is performed as long as the velocity is greater than 0.

The model used in this validation step is identical to the previous one, with the numerical parameters as follows: $g1 = 0.0001$ sec, $g2 = 0.002$, $g0 = 7 * 10^8$ N/m. Physical parameters: block density $= 2.7$ gr/cm^3, $E = 40$ GPa, $\nu = 0.18$, $\phi = 50°$.

Left panel of Figure 16.15 presents the downslope displacement history, calculated by the Newmark analysis and the DDA code. The agreement between the two is good, and can again be expressed in terms of relative error, presented in the left panel of Figure 16.15. During most of the analysis the error remains below 3%.

16.4 BACK ANALYSIS OF STONE DISPLACEMENTS IN OLD MASONRY STRUCTURES

16.4.1 Keystone displacement in a Roman masonry arch – Mamshit

Mamshit is the last Nabatean city built in the Negev on the trade route between Petra, Hebron, and Jerusalem (Negev, 1988). A unique structural failure is noticed in a tower

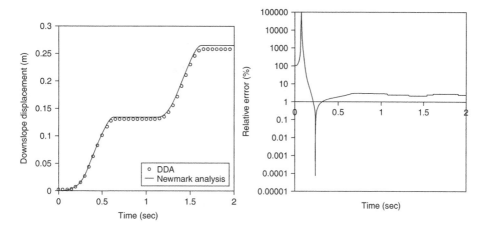

Figure 16.15 Left: Down slope displacement histories of a block on an inclined plane subjected to gravity and 1-D sinusoidal input function. Legend: curves – analytical solution, symbols – DDA solution. Right: The relative numerical error.

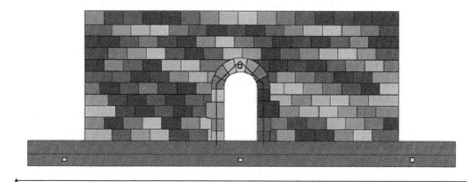

Figure 16.16 The DDA block system for the embedded arch at Mamshit. The modeled masonry wall rests on two fixed blocks. The lines intersecting the arch blocks represent material lines, and a measurement point (circle) is assigned for the keystone.

at the corner of the Eastern Church, where a key stone has slid approximately 4 cm downwards out of a still standing semicircular arch (Fig. 16.1a).

The numerical analysis of the arch at Mamshit is performed on a block system that contains the arch, with its accurate measurements from the field, confined by a uniform masonry wall. The arch is intersected by 'material lines' (Shi, 1993) which enable us to assign separate sets of mechanical parameters for the wall and the arch. These simulate the great difference between the hewn stones of the arch itself and the heterogenic confining wall material by assigning intact rock stiffness values to the arch stones ($E_{arch} = 17$ GPa) and soil-like stiffness values to the wall stones ($E_{wall} = 1$ MPa) because DDA does not yet contain specific features to model joint infilling. A measurement point is assigned at the keystone, and its vertical displacement is tracked versus time (Fig. 16.16).

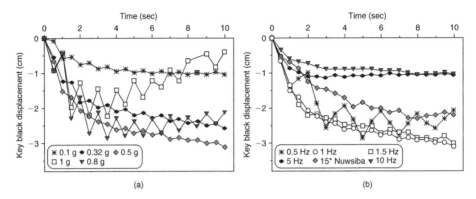

Figure 16.17 DDA results for the keystone at Mamshit (after Kamai and Hatzor, 2008): (a) Influence of input motion amplitude on vertical keystone displacement ($f = 1$ Hz), (b) Influence of input motion frequency on vertical displacement of keystone ($A = 0.5\,g$).

A very interesting and unique structural behavior is revealed through the sensitivity analyses of structural response to changing amplitude of motion. Instead of an intuitive proportion between induced motion amplitude and degree of damage, measured here by the magnitude of the displacement vector at the measurement point, there is apparently a structural "preference" to a certain range of amplitudes. The keystone at Mamshit exhibits the greatest displacement under specific amplitudes, not necessarily the largest. Figure 16.17a shows that the keystone of the arch at Mamshit exhibits the greatest downward displacement under an amplitude of $0.5\,g$, when everything else is kept equal. Downward displacement increases with acceleration amplitude up to an amplitude of $0.5\,g$. When the acceleration amplitude is greater than $0.5\,g$, the keystone response exhibits strong fluctuations and even a shift in displacement direction.

Structural response to frequency of motion is usually discussed in terms of the natural period of the structure, at which the structure will develop a resonance mode and collapse. Since the studied failures are local and not complete, each mode of failure will have its own natural period which can be different from that of the whole structure. The structural sensitivity to frequency, revealed in our sensitivity analyses, is significant considering that the common terminology for seismic risk evaluation uses mainly PGA (Peak Ground Acceleration) and largely ignores frequency.

In the case of the keystone at Mamshit a clear preference is detected for frequencies in the range of 1–1.5 Hz; only under those input frequencies is the downward displacement of the keystone continuous, and accumulates more than 3 cm of displacement, similar to the amount of displacement measured in the field (Fig. 16.17b). Lower or higher frequencies result in other modes of failure such as oscillations in the case of low frequencies (e.g. 0.5 Hz) and "locking" of the structure in the case of higher frequencies (e.g. 5–10 Hz in Fig. 16.17b).

A very important observation is made when a real earthquake record, that of Nuweiba 1995, is used as input motion (Fig. 16.18). The original record, deconvoluted to rock response (Hatzor *et al.*, 2004), is amplified by 15 in order for its PGA to reach the same amplitude of the synthetic motions used for the results which are plotted in Figure 16.17B, namely $0.5\,g$.

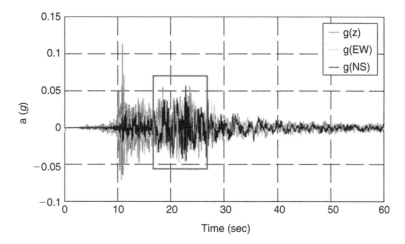

Figure 16.18 The Nuweiba 1995 record after de-convolution to rock response (Hatzor *et al.*, 2004). The rectangle marks the 10 seconds that were used for the analysis of the Mamshit block system.

Though the Nuweiba quake loads the structure with a wide range of frequencies and with two simultaneous components of motion (horizontal *and* vertical), the structural response to the natural quake is very similar to that of the simple sinusoidal ones, though more moderate (See Fig. 16.17b). This finding strongly suggests that the results of the sensitivity analysis, using synthetic records of horizontal motion only, are valid enough to be further discussed.

16.4.2 Pillar collapse in a Byzantine Church – Susita

The archeological site of Susita is located at the top of a diamond shaped plateau 350 m above the Sea of Galilee. Susita, or by its Greek name *Antiochia-Hippos*, was founded during the Hellenistic period, after 200 BC (Segal *et al.*, 2004). During the Hellenistic period it belonged to the Decapolis, a group of ten cities that were regarded as the centers of Greek culture in the region. The Southeastern church at Susita, the collapsed columns of which are analyzed in this paper, was built during the Byzantine period. The church columns are monolithic consisting of red and grey granites which were transported most likely from the Aswan region in Egypt (Segal *et al.*, 2004), as well as of some white and green marbles. One row of collapsed columns that originally supported the roof of the southeastern church can clearly be seen today (Fig. 16.1b). The collapsed columns rest parallel on the ground surface, all pointing in the same direction (Fig. 16.1b). Susita is believed to have been destroyed during the large earthquake of 749 AD (Amiran, 1996). This conclusion stems from several historical observations: 1) after 750 AD tax collection has moved from Susita to a different, nearby city, 2) no evidence for life in Susita has been found in archeological excavations after 750 AD, 3) the latest coins found in Susita excavations are dated to the Umayyad dynasty (early Arabic period) which ended at 750 AD.

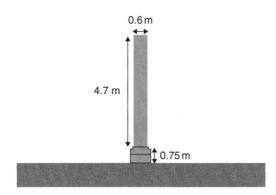

0.6 m

4.7 m

0.75 m

Figure 16.19 DDA block system used for a model column at Susita (after Yagoda-Biran and Hatzor, 2010).

Some of the columns that have collapsed are broken, most columns are displaced by some finite distance from their bases, and some columns have rolled on the ground after the collapse.

There are two fundamental assumptions in the following numerical treatment that must be declared in advance: 1) the numerical analysis is carried out for dynamic excitation of free standing columns. The results thus obtained pertain therefore to free standing columns and not for the entire structure which must have included some kind of roof. Although the roof must have been rather light, this assumption must be considered when regional seismic hazard is deduced from our results, 2) the dynamic analysis is carried out in two dimensions while the actual columns are three dimensional entities. At present we are conducting very preliminary investigations with 3D-DDA but the results reported here are only valid for two dimensions, for which an analytical solution also exists (Makris and Roussos, 2000).

The DDA block system used in this research for a typical Susita column is presented in Figure 16.19.

The input material properties for the column are: $E = 40\,\text{GPa}$, $\nu = 0.18$, $\rho = 2700\,\text{kg/m}^3$, and friction angle between column base and pedestal $\phi = 45°$. These values are obtained by engineering judgment and not by actual testing because it is not permissible to extract samples from the site. Nevertheless, the numerical analysis results do not seem to be very sensitive to material property variations, within reasonable bounds.

The input numerical control parameter values ($k01, g1, g2$ in (Shi, 1993)) are as in the validation study. The parameter for which the numerical analysis is most sensitive, the normal contact spring stiffness, is found by iterations using the analytical solution at the instantaneous response stage when only free oscillations are obtained under input peak acceleration value sufficiently small so as to avoid column toppling. An optimal contact spring stiffness of $g0 = 2 * 10^8$ N/m is thus obtained. The optimal $g0$ value was determined for very low values of θ to avoid sliding deformations which are not considered in the analytical solution but partake in the deformation both in reality and in DDA computations.

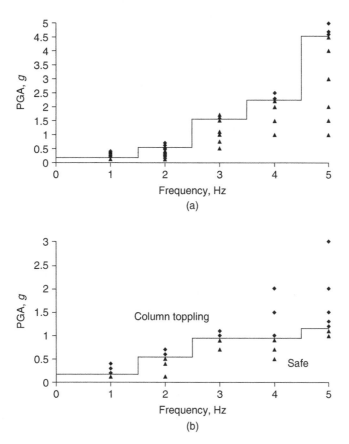

Figure 16.20 DDA results for the collapsed Susita columns: (a) required PGA for overturning under one input loading cycle. Solid triangles – stable columns, solid diamonds – overturned columns, solid line – stability boundary. (b) required PGA for overturning under three input loading cycles. Solid triangles – stable columns, solid diamonds – overturned columns, solid line – stability boundary (after Yagoda-Biran and Hatzor, 2010).

Once optimal numerical control parameters are found, forward DDA modeling can be performed for various input motion frequencies and amplitudes. DDA results for one and three cycles of a sinusoidal input function are presented in Figure 16.20.

Under a single loading cycle the required amplitude for overturning clearly increases with frequency (Fig. 16.20a). For example, the amplitude required for overturning at 5 Hz is 4.6 g, far greater than the expected PGA for this region, 0.3 g (S.I.I., 2004). Input loading frequencies of 1 Hz and 2 Hz return PGA values of 0.2 g and 0.6 g respectively, much closer to the expected PGA for this region. The required PGA for overturning also increases with frequency for three loading cycles, but at a smaller rate (Fig. 16.20b), and attain a "steady state" value of 1 g for 3 Hz frequencies and above.

Results of previous studies (Psycharis, Papastamatiou and Alexandris, 2000) suggest that for 1 Hz frequencies or below the required PGA for overturning is independent of the number of loading cycles, since the column will overturn during the first cycle or soon after it. For higher frequencies however, three loading cycles seem to better represent earthquake thresholds (Psycharis, Papastamatiou and Alexandris, 2000). It may seem more appropriate therefore to adopt the numerical results obtained for 3 cycles of input acceleration if only sinusoidal input motions are considered. The range of threshold PGA values thus obtained however is still too large: $0.2\,g < \text{PGA} < 1\,g$.

A possible way to constrain the obtained threshold PGA range is by assuming that the columns failed when rocking at their natural resonance frequency. However, it can be shown mathematically that a natural resonance frequency for a free standing column does not exist (Yagoda-Biran and Hatzor, 2010). Therefore, an alternative approach is needed for constraining PGA from back analysis of column overturning.

We overcome these two shortcomings by considering the frequency content of true earthquake records by subjecting the modeled column to real accelerograms recorded during strong earthquakes which occurred in tectonic settings similar to those as in the Dead Sea rift (DSR) system, namely where strike slip rather than normal or reverse faulting takes place. Possible candidates are the San Andreas Fault system in California as well as past strong earthquakes recorded along the DSR system.

We show below results obtained with such an approach for the modeled columns at Susita using real earthquake recordings with three component acceleration time histories. As our analysis is restricted for now to two dimensions, the analyzed accelerograms are applied twice for each earthquake, where in each simulation a different horizontal component (E-W or N-S) is aligned with the horizontal axis of the modelled column. In all simulations the vertical component is applied as well, so as to allow also for vertical motions in the analysis. The following earthquakes are modelled:

-   The 1995 Nueiba (Red Sea) earthquake, as recorded in Eilat, Israel (Hofstetter, Thio and Shamir, 2003). The recording station was about 60 km north of epicenter. Magnitude: 7.2, V-PGA: 0.11 g, H-PGA: 0.09 g, measured on fill.
-   The 1995 Nueiba (Red Sea) earthquake, as recorded in Eilat, Israel and deconvoluted for rock (Zaslavsky and Shapira, 2000). The recording station was about 60 km north of epicenter. Magnitude: 7.2. V-PGA: 0.11 g, H-PGA: 0.06 g (Fig. 16.18).
-   The 1989 Loma-Prieta earthquake, as recorded in Yerba Buena Island. The recording station was about 80 km north of epicenter. Magnitude: 6.9. V-PGA: 0.02 g, H-PGA: 0.05 g.
-   The 1940 Imperial Valley earthquake, as recorded by the 117 El Centro array #9. The recording station was about 11.5 km north of epicenter. Magnitude: 6.5. V-PGA: 0.2 g, H-PGA: 0.35 g.
-   San Francisco Bay area design earthquake. Magnitude 8, V-PGA: 0.55 g, H-PGA: 0.7 g (Law and Lam, 2003).

The modeled time histories are applied to "loading points" located at the centroids of the column and the pedestal (see Fig. 16.19), while the foundation block remains fixed. Each earthquake record is used twice as explained above, resulting in 10 different simulations. To determine the threshold PGA required for column overturning under the modelled earthquake the input records are up-scaled or down-scaled (by multiplying the entire record by a scalar), until column toppling is obtained. The

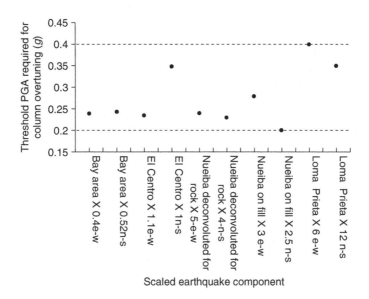

Figure 16.21 Threshold PGA required for column overturning obtained by subjecting the model to 10 different earthquake records. The original records were either down-scaled or up-scaled until column overturning was obtained (after Yagoda-Biran and Hatzor, 2010).

results of these analyses are plotted in Figure 16.21 below. Surprisingly, the threshold PGA values thus obtained remain constrained within relatively narrow bounds, between 0.2 g and 0.4 g, even though the records are from different earthquakes, measured on different subsurface conditions, and have different frequency contents. This approach narrows down quite significantly the range of possible threshold PGA values that could have been responsible for the detected damage at Susita. Furthermore, the expected PGA for this region of 0.3 g (S.I.I., 2004) falls nicely within this constrained PGA range obtained by our approach.

16.5 DYNAMIC DEFORMATION IN JOINTED AND FRACTURED ROCK SLOPES: THE CASE OF HEROD'S PALACE, MASADA

16.5.1 Geological and seismological setting

The top of Mount Masada consists of essentially bare hard rocks. The rocks are mainly bedded limestone and dolomite, with near vertical jointing. Structurally, the entire mountain is an uplifted block within the band of faults which forms the western boundary of the Dead Sea Rift, a seismically active transform (Garfunkel and Ben-Avraham, 1996; Garfunkel, Zak and Freund, 1981). A review of the tectonics and seismicity of the area is provided by Niemi *et al.* (1997). According to the Israel seismic building code (S.I.I., 2004) the Dead Sea valley has been classified as a region in which earthquake-induced peak horizontal ground acceleration (PGA) exceeding 0.2 g at the bedrock level is expected with a 10% probability within any 50 year window. This is analogous to a 475 year average recurrence interval for such acceleration.

Inspection of the historic earthquake record (Amiran, Arieh and Turcotte, 1994; Ben-Menahem, 1991) suggests that the strongest shaking events which have actually affected Mount Masada within the past two thousand years, were due to about ten identified earthquakes with estimated magnitudes in the range of $M = 6.0 \pm 0.4$ and focal distances probably in the order of several kilometers to a few tens of kilometers from the site. Under these assumptions it is highly likely that some of these earthquakes have caused at Mount Masada bedrock PGA's reaching and even exceeding $0.2\,g$, in general agreement with predictions for a 2000 year period based on the aforementioned building code assumptions.

One of the most notable historic earthquakes in this region occurred probably in the year 362 or 363, with a magnitude estimated at 6.4 (Ben-Menahem, 1991) or even 7.0 (Turcotte and Arieh, 1988). Reported effects included seismic seiches in the Dead Sea and destruction in cities tens of kilometers from the Dead Sea, both east and west. This is probably the earthquake identified by archeologists as "*the great earthquake which destroyed most of the walls on Masada sometime during the second to the fourth centuries*" (Netzer, 1991). The most recent of the major historic earthquakes near Mount Masada occurred on July 11th, 1927. This earthquake was recorded by tens of seismographs, yielding a magnitude determination of 6.2 and an epicenter location $30 \pm 10\,km$ north of Masada. It also caused seismic seiches in the Dead Sea and destruction in cities tens of kilometers away (Shapira and Eck, 1993). Recent findings on paleoseismicity in the region are reported by Salamon *et al.* (2007).

16.5.2 Documented historical stability as control for numerical simulations

The fortifications built by King Herod on Mount Masada about two thousand years ago included a casemate wall surrounding the relatively flat top of the mountain (Netzer, 1991). Clearly, because of its defensive function, the outer face of this wall was built so as to continue upward the face of the natural cliff, as much as possible. The outer wall was therefore founded typically on the flat top within several decimeters from its rim. Locally it was even founded slightly beyond the rim, on a somewhat lower ledge of rock. On the aforementioned three palace terraces, jutting at the northern tip of the mountain top, construction was again carried out up to the rim and beyond in order to achieve architectural effects and utilize fully the limited space. Thus, the remaining foundations effectively serve to delineate the position of the natural rim of the flat mountain top and associated northern terraces about 2000 years ago. Missing portions along such foundation lines indicate locations in which the rim has most probably receded due to rockfalls, unless the portions are missing due to other obvious reasons such as local erosion of the flat top by water or an apparent location of the foundation on fill beyond the rim.

Inspections of the entire rim of the top of Masada reveals that over almost the entire length of the casemate wall, which is about 1400 m long, the rock rim has not receded during the past two thousand years more than a few decimeters, if at all. Only over a cumulative total of less than 40 m, i.e. about 3% of the wall length, are there indications of rockfalls involving rim recessions exceeding 1.5 m, but not exceeding 4.0 m. Since the height of the nearly-vertical cliffs below the rim is in the order of tens of meters, these observations attest to remarkable overall stability in the face of the recurring earthquakes.

On King Herod's palace terraces there has been apparent widespread destruction, mostly of walls and fills which were somehow founded on the steep slopes. However, in the natural cliffs themselves there are few indications of rockfalls involving rim recessions of more than a few decimeters. Remarkably, most of the high retaining walls surrounding the middle and lower terraces are still standing, attesting to the stability of the rock behind them. In the upper terrace, on which this study is focused, there appears to be only one rockfall with depth exceeding several decimeters. It is a local rockfall near the top of the 22 m cliff, in the northeast, causing a rim recession of about 2.0 m. It is notable that this particular section of the terrace cliff was substantially modified by the palace builders, perhaps de-stabilizing the pre-existing natural cliff.

Rare aerial photographs of Mount Masada dated December 29th 1924, i.e. predating the 1927 earthquake were also inspected. Comparison with recent aerial photographs would have been capable of detecting rim recessions exceeding about one meter, if any had occurred in the northern part of the mountain. None were found, suggesting that the 1927 earthquake did not cause any significant rockfalls there (the southern part was less clear in the old photographs).

The information presented above essentially constitutes results of a rare rock-mechanics field-scale "experiment". Two thousand years ago the Masada cliff top was marked by construction. The mountain was later shaken by several major earthquakes, with deep bedrock accelerations certainly exceeding $0.1\,g$ and probably even exceeding $0.2\,g$. Observations at the present stage of the "experiment" show that all the cliffs surrounding the top of Mount Masada essentially withstood the shaking, with some relatively minor rockfalls at the top of the cliffs.

The above is a substantial result of a full-scale "experiment" on the real rock structure. Therefore, a fundamental test of any model of this structure is that it must essentially duplicate the above "experiment".

16.5.3   Rock mass properties

The rock in Masada is a massive and dense dolomite with low porosity (2%–8%) and density of $2{,}730\,kg/m^3$. The rock mass is bedded with local karstic voids between beds. The bedding planes are generally clean and tight, with crushed dolomite infilling in places.

Herod's palace, also known as the North palace, is built on three terraces at the north face of Masada. The rock mass structure at the foundations consists of two orthogonal, sub-vertical, joint sets striking roughly parallel and normal to the NE trending axis of the mountain, and a set of well developed bedding planes gently dipping to the north (Fig. 16.22b). The joints are persistent, with mean length of 2.7 m. The bedding planes, designated here as J_1, dip gently to the north with mean spacing of 60 cm. The two joint sets, J_2 and J_3, are closely spaced with mean spacing of 14 cm and 17 cm respectively (Fig. 16.22a).

The uniaxial compressive strength of intact rock samples exceeds 315 MPa, and typical values of Elastic modulus and Poisson's ratio are 40 GPa and 0.18 respectively. These strength and elasticity parameters are relatively high with respect to values determined experimentally for other dolomites and limestones in Israel (Hatzor and

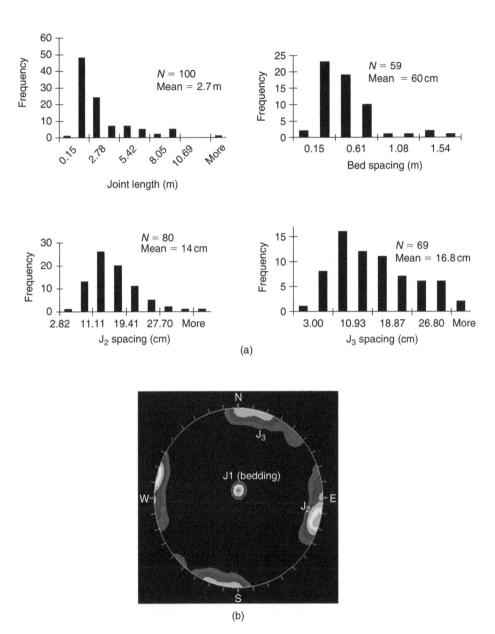

Figure 16.22 Rock mass structure at the Northern Palace – Masada (after Hatzor et al., 2004): (a) Joint length and spacing distribution, (b) Joint orientation (upper hemisphere projection of poles).

Palchik, 1997; Hatzor and Palchik, 1998). The shear strength of discontinuities was discussed above, and the experimental failure envelopes for rough and saw cut surfaces are presented in Figure 16.9, which indicated the Coulomb-Mohr failure criterion is nicely obeyed. While residual friction angle (23°) is used for stability analysis in the

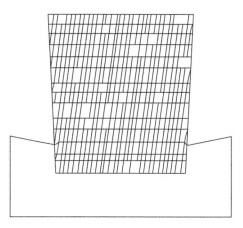

Figure 16.23 Synthetic trace map of the upper rock terrace of Herod's Palace in Masada using the statistical joint trace generation code (DL) of Shi (Shi, 1993).

"snake path" cliff where individual blocks are extremely large, in the North face of the mountain, where the average block size is much smaller it would be reasonable to base the design and the analysis on the available friction angle for rough joints, namely 41°.

16.5.4 Numerical generation of block mesh

The results of numerical analyses are extremely sensitive to: (a) the input mechanical and physical properties, (b) the geometrical configuration, namely the computed mesh, and (c) the input loading function. The geometrical configuration (b) is particularly important in distinct element methods where rock blocks and mesh elements are one and the same. In the previous section the determination of mechanical parameters was discussed. In this section the most suitable mesh configuration is discussed, followed by a discussion of the appropriate dynamic input motion.

Two principal joint sets and a systematic set of bedding planes comprise the rock structure at Herod's palace (Fig. 16.22). An E-W cross section of the upper terrace is shown in Figure 16.23, computed using the statistical joint trace generation code (DL) of Shi (Shi, 1993). It can be seen intuitively that while the east face of the rock terrace is prone to sliding of wedges, the West face is more likely to fail by toppling of individual blocks. Block theory mode and removability analyses (Goodman and Shi, 1985) confirm these intuitive expectations.

While it is quite convenient to use mean joint set attitude and spacing to generate statistically a synthetic mesh, the resulting product (Fig. 16.23) is quite unrealistic and bears little resemblance to the actual slope. The contact between blocks obtained this way is planar, thus interlocking between blocks is not modeled. Consequently the results of dynamic calculations may be overly conservative and the computed displacements unnecessarily exaggerated.

In order to analyze the dynamic response of the slope realistically a photo-geological trace map of the face was prepared using aerial photographs (Fig. 16.24a), and the joint trace lines were digitized. Then, the block-cutting (DC code) algorithm

Figure 16.24 (a) The upper terrace of Herod's palace, Masada, (b) Photo geological trace map.

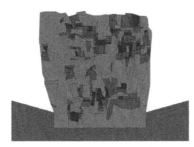

Figure 16.25 Deterministic joint trace map of the terrace prepared using the photogeological map of the upper terrace (Fig. 16.7) and the block cutting algorithm (DC) of Shi (Shi, 1993).

of Shi (Shi, 1993) was utilized in order to generate a trace map that represents more closely the reality in the field (Fig. 16.24b). Inspection of Figure 16.24 reveals that block interlocking within the slope is much higher and therefore the results of the forward analysis are expected to be less conservative and more realistic. The deterministic mesh shown in Figure 16.25 is used therefore in the forward modelling discussed below.

The trace lines obtained from the mapping were used as input to the block cutting program (DC) of Shi (Shi, 1993) to obtain the block system mesh, consisting of finite blocks for which all the information including area and contacts with adjacent blocks in the mesh is stored for further forward analysis (Fig. 16.25).

16.5.5   Selection of appropriate input motion

In this research we chose to use the recorded time history of the Mw = 7.1 Nuweiba earthquake which occurred in November 1995 in the Gulf of Eilat (Aqaba) with an epicenter near the village of Nuweiba, Egypt. The main shock was recorded at the city of Eilat where the tremor was felt by people, and structural damage was detected in houses and buildings. The city of Eilat is located 91 km north from the epicenter

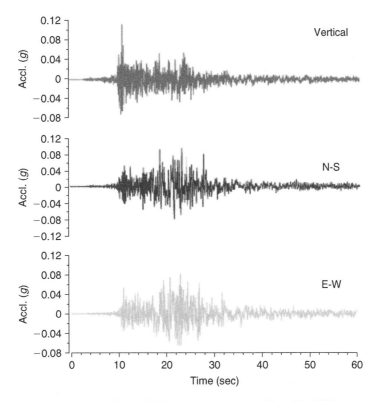

Figure 16.26 Time history of the Mw = 7.1 Nuweiba earthquake (Nov. 22, 1995) as recorded at the city of Eilat on a thick fill layer of Pleistocene alluvial fan deposits.

and 186 km south of Masada, on the northern coast of the gulf of Eilat (Aqaba). Figure 16.26 shows all three components of the accelerogram that were recorded in Eilat. The peak ground acceleration (PGA) of the Nuweiba record as measured in Eilat was 0.09 g.

The Eilat seismological station is situated on a thick fill layer of Pleistocene alluvial fan deposits. The recorded accelerogram therefore represents the response of a site situated on deep fill layer rather than on sound bedrock. Therefore, direct application of the original Eilat record for the case of the Masada rock site would be inappropriate. In order to obtain a "rock response" record for the Nuweiba event it would be necessary therefore to remove the local site effect of the fill layer, which typically amplifies ground motions, and to produce a corresponding "rock" response using an appropriate transfer function. This mathematical procedure is known as de-convolution.

In this research a one-dimensional multi-layer model for the fill was utilized with the key parameters being shear wave velocity, thickness, and density for the horizontal fill layers. The material and physical parameters were determined using both seismic refraction survey data and down-hole velocity measurements. The appropriate transfer function was developed by optimization of both theoretical and experimental results

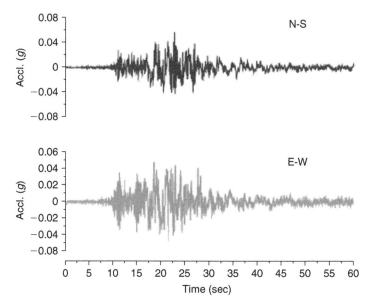

Figure 16.27 Deconvolution of the Eilat Fill record (Fig. 16.26) for bedrock response (after Hatzor et al., 2004).

(Zaslavsky and Shapira, 2000). The resulting de-convoluted record for rock response is shown in Figure 16.27.

Although the Masada site is situated directly on rock, a significant topographic effect was recorded in the field (Hatzor *et al.*, 2004) and it should therefore be considered in the development of the relevant input motion for the site. An empirical response function for Masada, developed on the basis of the field study discussed in Section 5.2 above, is shown in Figure 16.28. Three characteristic modes are found at 1.06, 3.8, and 6.5 Hz. The resulting time history is shown in Figure 16.29. The DDA forward modelling is performed using the modified input motion shown in Figure 16.29 below.

16.5.6 Forward DDA analysis

To test the hypothesis that the studied fractured rock slope withstood several episodes of shaking with PGA = 0.2 g, the modified Nuweiba record (Fig. 16.29) was up scaled to PGA = 0.2 g and was used as input loading function for the realistic block mesh shown in Figure 16.25 using the forward modelling code (DC) of DDA. The results of the numerical simulation are shown graphically in Figure 16.30 for the 10 most critical seconds of the event, from $t = 15$ s to $= 25$ s. Inspection of the graphical output reveals that under that level of excitation only minor damage should be anticipated, primarily in the form of block toppling in the west slope and local block sliding in the East slope.

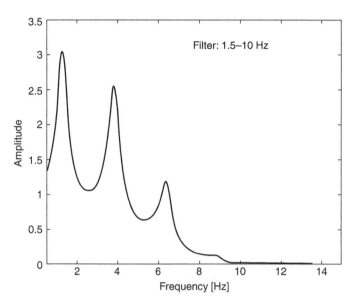

Figure 16.28 An empirical response function for the topographic site effect at Masada (after Hatzor et al., 2004).

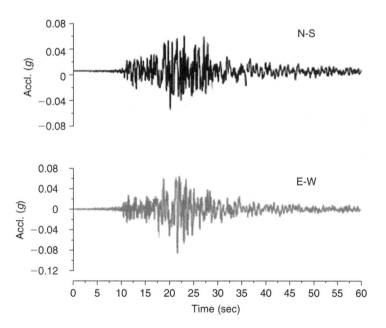

Figure 16.29 The Nuweiba record modified for rock response including local topographic site effect at Masada (after Hatzor et al., 2004).

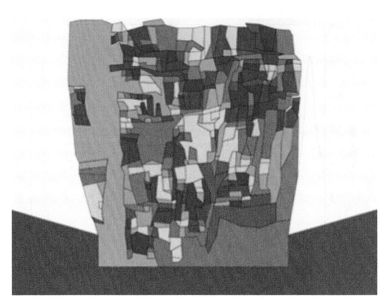

Figure 16.30 DDA prediction of fractured rock slope response to modified Nuweiba record scaled to PGA = 0.2 g.

These results are confirmed by the presence of intact structures at the rim of the cliff. The complete details of the numerical analysis are discussed by Hatzor *et al.* (2004).

16.6 SUMMARY AND CONCLUSIONS

In this paper a powerful numerical tool is discussed, the discrete discontinuous deformation analysis (DDA) method. Following a brief review of DDA theory several validations against analytical solutions are presented. These validations prove that DDA is suitable for dynamic applications for discrete systems such as masonry structures or fractured rock masses. Application of DDA for seismic hazard studies is demonstrated using firstly back analysis of finite block displacements in old masonry structures, and then by performing forward modelling of an existing fractured rock slope. The back analysis study allows us to constrain the peak ground acceleration of the seismic event that caused the observed damage. The forward modelling study demonstrates the ability of the method to predict damage patterns in discontinuous rock masses. DDA results in the study of deformed masonry structures are supported by PGA predictions of the seismic design code in Israel, values arrived at in a completely different and independent approach. DDA results for the fractured rock slopes are confirmed by field observations concerning mapped failure modes in the two opposite faces of the analyzed slope, as well as the still standing 2000 year old structures on the rim of the analyzed cliff.

Acknowledgment: This research is partially supported by the ISRAEL SCIENCE FOUNDATION, through grant No. 556/08.

REFERENCES

Amiran, D., Arieh, E. and Turcotte, T.: Earthquakes in Israel and Adjacent Areas: Macroseismic Observations since 100 B.C.E. *Isr. Explor J* 44(3–4) (1994), pp.260–305.

Amiran., D.: Earthquakes in the land of Israel. *Qadmoniot* 29 (1996), pp.53–61.

Ben-Menahem, A.: Four thousand years of seismicity along the Dead Sea rift. *J Geophys Res* 96(B12) (1991), pp.20,195-20,216.

Doolin, D.M. and Sitar, N.: Time integration in discontinuous deformation analysis. *Journal of Engineering Mechanics ASCE* 130(3) (2004), pp.249–258.

Garfunkel, Z. and Ben-Avraham, Z.: The structure of the Dead Sea basin. In : Dynamics of extensional basins and inversion tectonics. *Tectonophysics* 266 (1996), pp.155–176.

Garfunkel, Z., Zak, I. and Freund, R.: Active faulting in the Dead Sea Rift. *Tectonophysics* 80 (1981), pp.1–26.

Goodman, R.E. and Kieffer, D.S.: Behavior of rock in slopes. *Journal of Geotechnical and Geoenvironmental Engineering* 126(8) (2000), pp.675–684.

Goodman, R.E. and Seed, H.B.: Earthquake-induced displacements in sand embankments. *J. Soil Mech. Foundation Div, ASCE* 90(SM2) (1966), pp.125–146.

Goodman, R.E. and Shi, G.-H.: *Block Theory and its Application to Rock Engineering.* Prentice-Hall, Inc., Englewood Cliffs, New Jersey, 1985.

Hatzor, Y.H.: Keyblock stability in seismically active rock slopes – Snake Path Cliff, Masada. *Journal of Geotechnical and Geoenvironmental Engineering* 129(11) (2003), pp.1069–1069.

Hatzor, Y.H., Arzi, A.A., Zaslavsky, Y. and Shapira, A.: Dynamic stability analysis of jointed rock slopes using the DDA method: King Herod's Palace, Masada, Israel. *International Journal of Rock Mechanics and Mining Sciences* 41(5) (2004), pp.813–832.

Hatzor, Y.H. and Palchik, V.: The influence of grain size and porosity on crack initiation stress and critical flaw length in dolomites. *International Journal of Rock Mechanics and Mining Sciences* 34(5) (1997), pp.805–816.

Hatzor, Y.H. and Palchik, V.: A microstructure-based failure criterion for Aminadav dolomites. *International Journal of Rock Mechanics and Mining Sciences* 35(6) (1998), pp.797–805.

Hofstetter, A., Thio, H.K. and Shamir, G.: Source mechanism of the 22/11/1995 Gulf of Aqaba earthquake and its aftershock sequence. *Journal of Seismology* 7(1) (2003), pp.99–114.

Kamai, R. and Hatzor, Y.H.: Numerical analysis of block stone displacements in ancient masonry structures: a new method to estimate historic ground motions. *International Journal for Numerical and Analytical Methods in Geomechanics* 32 (2008), pp.1321–1340.

Kieffer, S.D.: *Rock Slumping – A Compound Failure Mode of Jointed Hard Rock Slopes* U. C. Berkeley, Berkeley, 1998.

Law, H.K. and Lam, I.P.: Evaluation of seismic performance for tunnel retrofit project. *Journal of Geotechnical and Geoenvironmental Engineering, ASCE* 129(7) (2003), pp.575–589.

MacLaughlin, M.M. and Doolin, D.M.: Review of validation of the Discontinuous Deformation Analysis (DDA) method. *Int J Numer Anal Meth Geomech* 30 (2005), pp.271–305.

Makris, N. and Roussos, Y.S.: Rocking response of rigid blocks under near-source ground motions. *Geotechnique* 50(3) (2000), pp.243–262.

Negev, A.: *The Architecture of Mampsis, Quedem.* monographs of the Institute of Archaeology. 26, The Hebrew University of Jerusalem, Jerusalem, 1988.

Netzer, E.: Masada III – the Yigael Yadin Excavations 1963–1965 Final Reports – The Buildings Stratigraphy and Architecture. *Israel Exploration Society* 665, 1991.

Newmark, N.: Effects of earthquakes on dams and embankments. *Geotechnique* 15(2) (1965), pp.139–160.

Niemi, T.M., Avraham, Z.B. and Gat, J.R.: The Dead Sea, The lake and its setting. Oxford Univ. Press Ltd, 1997.

Ohnishi, Y., Nishiyama, S., Sasaki, T. and Nakai, T.: The application of DDA to practical rock engineering problems: issues and recent insights. In: M.M.a.N. Sitar (ed.), *Proceedings of the 7th International Conference on the Analysis of Discontinuous Deformation*, Honolulu, Hawaii, 2005, pp.277–287.

Psycharis, I.N., Papastamatiou, D.Y. and Alexandris, A.P.: Parametric investigation of the stability of classical columns under harmonic and earthquake excitations. *Earthquake Engng Struct. Dynamics* 29 (2000), pp.1093–1109.

Salamon, A., Rockwell, T., Ward, S.N., Guidoboni, E. and Comastri, A. (2007), Tsunami hazard evaluation of the eastern Mediterranean: Historical analysis and selected modeling, *B Seismol Soc Am*, 97(3), 705–724.

S.I.I.: *Israel Building Code #413, 2nd amendment*. The Standards Institution of Israel, 2004.

Segal, A., Mlynarczyk, J., Burdajewicz, M., Schuler, M. and M.E.: *Hippos – Susita, fifth season of excavation and summary of all five seasons*, Zinman Institute of Archeology, University of Haifa, Haifa, Israel, 2004.

Shapira, A. and Eck, T.v.: Synthetic uniform hazard site specific response spectrum. *Natural Hazard* 8 (1993), pp.201–205.

Shi, G.: *Discontinuous Deformation Analysis – a new numerical method for the statics and dynamics of block system*. University of California, Berkeley, 1988.

Shi, G.: *Block System Modeling by Discontinuous Deformation Analysis*. Topics in Engineering, 11. Computational Mechanics Publication, Southhampton, UK, 1993.

Tsesarsky, M., Hatzor, Y.H. and Sitar, N.: Dynamic displacement of a block on an inclined plane: Analytical, experimental and DDA results. *Rock Mechanics and Rock Engineering* 38(2) (2005), pp.153–167.

Turcotte, T. and Arieh., E.: *Catalog of earthquakes in and around Israel*, Israel Electric Corp. LTD, 1988.

Wang, C.Y., Chuang, C.C. and Sheng, J.: Time integration theories for the DDA method with finite element meshes, *1st International Forum on Discontinuous Deformation Analysis (DDA) and Simulation of Discontinuous Media*, Berkeley, CA and Albuquerque, NM., 1996, pp. 263–287.

Wittke, W.: Methods to analyze the stability of rock slopes with and without additional loading (in German). *Rock Mech. and Eng. Geol*, Supp. II. (1965): 52.

Yagoda-Biran, G. and Hatzor, Y.H.: Constraining paleo PGA values by numerical analysis of overturned columns. *Earthquake Engineering and Structural Dynamics* 39 (2010), pp.462–472.

Zaslavsky, Y. and Shapira, A.: Questioning nonlinear effects in Eilat during MW=7.1 gulf of Aqaba earthqukae, *XXVII General Assembly of the European Seismological Commission (ESC)*, Lisbon, Portugal, 2000, pp.343–347.

Chapter 17

Explosion loading and tunnel response

Yingxin Zhou

17.1 INTRODUCTION

Damage to rock tunnels from shock loading resulting from explosions in rock is of great interest to the engineer designing for tunnel blasting and for protection of underground structures, including underground explosives storage safety. However, prediction of the explosion loading and the damage criteria to be used in design, can present some difficulties due to the highly variable nature of the rock mass and the varying conditions under which explosions can take place. Different sources of explosion loading may produce significantly different effects, because of the many factors at play. A review of the literature reveals large variations in the ground shock prediction equations and definitions of tunnel damage. It is thus important to have a rational approach to the prediction of the explosion loading and the analysis of tunnel response and damage. A clear understanding of the characteristics of the loading sources and how the various factors affect the explosion effects is essential to this approach.

A series of large-scale tunnel explosion tests were conducted between 2000 and 2003 to validate the safety design for underground ammunition storage, including response of tunnels. While design for measuring ground shock loading and tunnel response was relatively easy, initial design of the test tunnel presented some engineering challenges in the prediction of the dynamic loading, tunnel damage assessment, and dynamic rock support for the tunnel facility which was designed to last a few years under repetitive explosion loading. The design was made more difficult by the general lack of design guidelines on ground shock prediction, damage criteria, and dynamic rock support. This chapter discusses some of the design issues, lessons learned and results from the tests. Case studies of rock damage will also be discussed and compared with results of analytical solutions and large-scale tests. The effects of the various factors on damage will be discussed and quantified where possible.

A vigorous treatment of dynamic tunnel response, including the explosion process, will logically require the use of the dynamic properties of rock and the dynamic constitutive relations. However, such a treatise is beyond the scope of this chapter. Interested readers can refer to publications by Zhao (2000), Li, Zhao and Li (1999, 2000), Li *et al.* (2001), Zhao *et al.* (1999a), Zhao and Li (2000) and Zhao, Li and Cai (2000). This chapter will focus on the engineering aspects of the problem.

17.2   PREDICTION OF GROUND SHOCK LOADING

17.2.1   Sources of explosion loading and their characteristics

Ground shock can result from one of the following main sources: blasting in mining and civil engineering construction, conventional weapons, and accidental explosion of stored explosives. This section discusses the characteristics of these explosions. It will not discuss ground shock loading from nuclear explosions.

17.2.1.1   Tunnel and mine blasting

In rock blasting for tunnel excavation or mining, explosives are charged into drill holes, and detonated in multiple delays. In a typical tunnel face of 60–100 m², the maximum charge weight per delay is in the order of 100 kg. For larger caverns or open pit mining where benching blasting is used, the charge weight per delay can be higher, up to few hundred kg per delay. In special blasting applications (e.g. large chamber blasting), the charge weight can be as large as a few hundred tons.

Explosives used for commercial operations are generally of a lower strength (commonly ANFO and bulk emulsions) compared to TNT and other explosives used in weapons. Thus, consideration should be given to the equivalent charge weight if a ground shock equation based on different explosives is used. Blasting from mining or construction work is repetitive by nature, so analysis of their effects on adjacent structures must take into account possible fatigue effects. Allowable limits set for such blasting are usually much lower than those for storage purposes.

17.2.1.2   Conventional weapons

In military or civil defence applications, all protective design is based on an assumed threat of a certain weapon grade. Depending on the type of weapons, and the value of the facility as a target, the charge weight can range from a few kg to 2,000 kg. Penetration of weapons or cratering is also an important factor in ground shock prediction. Some weapons detonate on contact or with very limited penetration, while others are designed to penetrate deep into the ground before explosion. These will have strong effects on the ground shock wave generated in the ground. When a weapon penetrates deeply to the ground, the explosion is usually fully coupled and generates much higher ground shock than shallow or contact explosions, where a substantial amount of the explosion energy is transmitted to the air in the form of airblast. For the purpose of ground shock calculation, a scaled burial depth, expressed in $m/kg^{1/3}$, of 1.0 for hard rock and 1.2 for soft rock is considered sufficient to prevent any cratering (Department of the Army, 1961). Explosives of weapons grade generally have much higher explosion strength than commercial explosives.

17.2.1.3   Accidental explosion in explosives storage

In underground explosives storage, accidental explosion is often the design basis in terms of internal and external safety. The charge weight varies within a wide range and is typically a few tens to a few hundred tons. However, the explosives (or weapons) are usually stored in rock caverns with large empty space for operations and other technical installation, resulting in relatively low charge weight per unit volume (or loading density). This effect is called decoupling and can significantly reduce the ground shock generated in the ground. Accidental explosions are also extremely rare, typically with a probability of one in a few thousand years. The amount of explosives in an accidental explosion can vary significantly. As such, designing for explosives storage is generally based on an assumed maximum credible event (MCE), often expressed in equivalent quantities of TNT. Due to the extremely low probability of explosion, allowable limits for ground shock resulting from an accidental explosion in storage are generally much higher than those set for repetitive blasting work.

17.2.2 Ground shock equations

Explosions in rock generate a dynamic stress wave, or ground shock, that propagates through the geological media. This stress wave is typically represented by the time history of the acceleration or particle velocity. Figure 17.1 shows the time histories of acceleration, particle velocity, and displacement from a decoupled underground explosion in hard rock. The velocities and displacements were obtained from integration of the recorded acceleration time history.

For the purpose of engineering design, the dynamic load generated from an underground explosion in rock can be represented by the peak particle velocity (PPV), which has been shown by studies to be the most representative parameter when describing the ground motion and tunnel response (Dowding, 1984).

The PPV from a fully coupled explosion can be given in the following general form:

$$V = H\left(\frac{R}{Q^B}\right)^{-n} \tag{17.1}$$

where H and n are constants for a certain geology and explosion set up; $R =$ actual distance, m; $Q =$ charge weight, kg. The term $R/Q^{1/3}$ is the scaled range, expressed in $m/kg^{1/3}$.

The exponent B, representing the energy scaling law, is a function of the geometry of the charge and reflects the energy transmission from the explosive to the surrounding medium. Many mining applications tend to use ½, or the square root scaling, while most military applications tends to use 1/3 or cube root scaling. While the PPV equation developed from tests is essentially a curve fitting exercise, the exponent B should approximately follow the values indicated in Table 17.1. From an energy point of view, this argument is generally true for distances that are relatively small with respect to the dimensions of the charge. At a sufficiently large distance (compared to the explosive geometry), the geometry effect should decrease and the exponent should approach 1/3 (cube root scaling). For the purpose of this chapter, all discussions will use the cube root scaling.

In Equation (17.1), it can be said that H gives the initial shock magnitude transmitted to the ground, while n governs how the shock wave propagates, or attenuates, through the ground. For hard rock, most values of n are in the vicinity of 1.5, although there is a general trend for n to increase with decreasing rock quality. Table 17.2 shows some typical values of H, B, and n found in the literature.

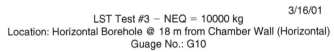

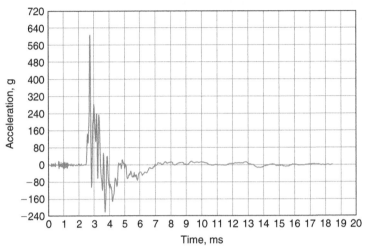

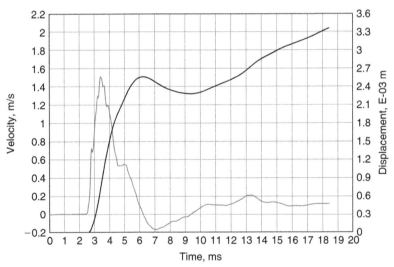

Figure 17.1 Time histories of acceleration, particle velocity and displacement.

Table 17.1 Scaling law and charge geometry.

Shape of Charge	Exponent B	Remarks
Spherical	1/3	Typical of explosives testing and tunnel blasting
Cylindrical	1/2	Multiple MS delay surface bench blasting
Plane	1	Long row of closely spaced holes detonating simultaneously

Table 17.2 Typical values for the constants H, B and n.

K value, mm/s	Exponent n	Exponent B	Remarks	Reference
1099	1.4	1/3	Fully coupled tests in granite	Zhou et al., 2000
21–804	0.88–2.8	1/3	Cube Root: China & Japan	Tao, 1979
1200	1.6	1/2	Square Root: Mining (USBM), Civil engineering	Nichols et al., 1971
1130	1.77	1/2	Hong Kong granite tunnel blasting	Smith & Mortan, 1986
700	1.5	0.467	Average Swedish bedrock	Holmberg and Persson, 1980
11430–12000	2.7–2.8	1/3	Fully coupled tests conducted in soft limestone and chalk using 500 kg of TNT	Johnson, Rozen and Rizzo, 1988
	2.8	1/3	Tests of de-decoupled explosions conducted in hard limestone (Magdalena tests) for explosives storage safety	Joachim, 1990
12700	1.77	1/3	Fully coupled tests in Klotz II	Hultgren, 1987
1800	2.5	1/3	Decoupled tests in hard limestone Linchburg Mine, Nex Mexicao, USA. Seismic velocities ranging from 3,636 m/s to 5,738 m/s. Loading density ranging from 10 to 48 kg/m^3	Murrel and Joachim, 1996
700	2.0	1/3	Granite rock	Hendrych, 1979
193–1930	1.6	1/2	Down hole bench blasting	Oriard, 1972
50–220	1.10	1/2	Coyote (large chamber) blasting	Oriard, 1972

The parameters H and n are generally a function of the soil/rock quality but can be affected by other factors such as types of explosives and scale of the explosion. For example, the initial value H has been shown to be related to the acoustic impedance coupling between the explosives and the rock wall for fully coupled explosions (Fig. 17.2). Likewise, data from Johnson, Rozen and Rizzo (1988) and Zhou and Ong (1996) show that the higher the acoustic impedance of the rock mass, the lower the attenuation coefficient, n.

A second point to note is the strength of explosives. Typically, commercial explosives have lower strengths compared to TNT and military explosives. However, in many PPV equations we reviewed, this point is seldom highlighted.

Finally, the scale of the explosion also has a strong influence on the PPV equation. Oriad (1972) presented several figures for PPV as a function of the scaled distance,

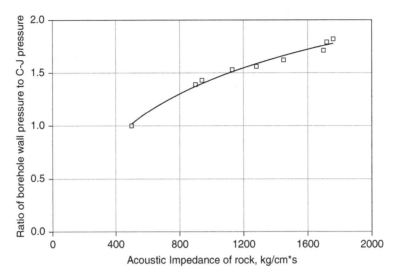

Figure 17.2 Coupling of blast pressure to rock wall for various types of rock. Data based on low density TNT (Density = 1,000 kg/m³, Velocity of Detonation = 4,850 m/s) in fully coupled explosions (Tao, 1979).

based on tens of thousands of data points from open pit blasting. The most important point of this chapter is the observation that at the same scaled range, PPV values for small charge amounts are much larger than those for coyote (large) blasts. Similar observations are also reported by Perret and Bass (1975) and Madshus and Langberg (1999).

As can be seen from Table 17.2, empirical PPV equations of the form in Equation (17.1) are inherently highly site-specific. Unless similar rock properties are present, the application of these equations may not be appropriate. Based on earlier work by Johnson, Rozen and Rizzo (1988), McMahon (1992), and Dowding (1985), Zhou and Ong (1996) developed a modified PPV equation that incorporates the rock mass density and seismic velocity.

$$V = \frac{0.5C^{2.17}}{\rho C} \left(\frac{R}{Q^{1/3}} \right)^{-n} \tag{17.2}$$

From the above equation, the initial value is:

$$H = \frac{0.5C^{2.17}}{\rho C} \tag{17.3}$$

where H = m/s; ρ = rock mass density, kg/m³; and C = seismic wave velocity, m/s.

The attenuation coefficient, n, is typically 1.5 for hard rock and can be more than 2.0 for softer rock. For softer rock with a $C < 1,500$ m/s, the following equations can be used to estimate n:

$$n = 2.31 e^{\frac{87390}{\rho C}} \tag{17.4}$$

17.2.3 Decoupled explosions

Decoupling occurs when the explosive charge does not fill up the volume completely, and at least some parts of the charge are not in direct contact with the rock. Here decoupling refers exclusively to volumetric decoupling.

Decoupling is the norm for underground explosive storage in rock caverns, where space is required for operations inside a chamber. In blasting, decoupling is often used to control rock damage, typically in controlled blasting or in the perimeter holes of tunnel face blasting, in the form of reduced charge density or alternative empty holes, to control. In some underground explosion tests, it is known that decoupled explosions were used to conceal the actual quantity of explosives in the test.

When decoupling occurs, the peak particle velocity generated by the same quantity of explosives at the same distance is substantially reduced.

Prediction of ground shock due to decoupled explosions is more complicated because of several factors affecting the transfer of explosion energy to the ground. Basically, two types of techniques can be used for predicting the peak particle velocity from decoupled explosions. The first method uses the wall or chamber pressure as the starting point and then predicts the ground shock based on some form of wave theory. The prediction cannot easily be done in analytical form unless significant simplifications are made. The second method applies a decoupling factor to the predicted ground motion based on a fully coupled explosion of the same quantity. The decoupling factor is often given as a function of the loading density, expressed in kg/m³.

17.2.3.1 Decoupling factor

Assuming the same attenuation coefficient for both coupled and de-coupled explosions, Zhou and Jenssen (2009) showed a decoupling factor as a function of the explosives density and attenuation coefficient n, as follows:

$$f_d = \frac{1}{\rho^{n/3}} (w)^{n/3} \tag{17.5}$$

where $w =$ loading density, kg/m³; and $n =$ ground shock attenuation coefficient, and $\rho =$ mass density of the explosives.

This equation satisfies the condition that the decoupling factor always approaches 1 as the loading density approaches the density of the explosives.

For $n = 1.5$ and an explosive density of 1600 kg/m³, the de-coupling factor becomes:

$$f_d = 0.025 (w)^{0.5} \tag{17.6}$$

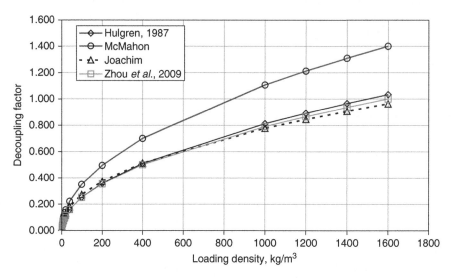

Figure 17.3 Comparison of decoupling factors.

Several decoupling factors are also found in the literature (Hultgren, 1987; McMahon, 1992; Joachim and Smith, 1988; Zhou and Ong, 1996; Zhou and Jenssen, 2009). Figure 17.3 shows a comparison of the various coupling factors. The de-coupling factor in Equation (17.6) is similar to the de-coupling factor derived by Hultgren (1987) for hard granite rock (uniaxial compressive strengths of 200 MPa and RQD values of 90–100%) and that of Joachim. It seems that these decoupling equations are primarily derived for hard rock. It is conceivable that the decoupling factor for softer rock may be different.

Several factors contribute to the difficulty with predicting the effects of decoupling. First, a truly decoupled explosion requires the explosives to be placed in the centre, with roughly equal distance to the rock walls. This, however, is almost never the case, at least for explosive storage, where the explosives are usually in contact with or much closer to the floor. The direct contact will result in a direct transmission of some shock energy to the floor. Second, data on the effects of loading density may not be fully representative as such tests are usually conducted in a fixed volume with varying charge weights, rather than with a fixed charge weight with varying volumes. Third, the geometry of the charge and that of the rock chamber and their relative dimension have a strong influence on the coupling of explosion energy. Finally, the effects of impedance coupling are normally not included. As discussed in the previous section, the impedance coupling of explosives and rock has been shown to be very important in energy transfer.

17.2.3.2 Chamber wall pressures

As discussed earlier, chamber pressure is sometimes used as a boundary condition for predicting ground shock without the use of decoupling factor, especially for relatively low loading densities in underground storage conditions. Typically, the chamber pressure is given as a function of the loading density, as shown in the following equation given by Strange, Dornbusch and Jr. Rooke (1995):

$$P = 2.25w^{0.72} \qquad (17.7)$$

Where P = peak chamber pressure in MPa; and w = loading density, kg/m^3.

This pressure can then be converted to the peak particle velocity by the following relationship:

$$V = P/(\rho C) \qquad (17.8)$$

This method has several shortcomings. First, variations of predicted chamber pressure are large (Strange, Dornbusch and Jr. Rooke, 1995; McMahon, 1994), and most equations for chamber wall pressure have been derived based on tests conducted in steel tubes which assume perfectly rigid and reflecting chamber walls. Numerous numerical simulations of large-scale tests using hydrocodes, which have been proven to be valid using data from shock tubes, tend to over-predict blast pressure by a factor of at least two. Various factors have been examined and the response of the rock chamber walls has been identified as a major contributing factor. Second, the conversion of pressure to peak particle velocity and the subsequent propagation must assume overly simplified relationships. Most analytical solutions reported in the literature so far use a one-dimensional elastic wave theory. Third, if the explosives are in contact with the floor of the storage chamber, the transmitted ground shock (or pressure) can never be measured and its value as compared with the chamber pressure is not known. Finally, the oxygen balance of the explosives has a strong effect on the degree of after-burning of the detonation products and thus the chamber pressure, depending on the loading density. For example, TNT will have a complete combustion with loading densities up to 0.5 kg/m^3. For loading densities higher than 0.5 kg/m^3, the peak chamber gas pressure as a function of the loading density will be lower due to the lack of oxygen in the chamber for after-burning. Figure 17.4 shows the peak chamber gas pressure as a function of the loading density for various types of explosives.

17.2.4 Correction for charge geometry

For close range explosions, the charge shape and geometry can have significant effects on the ground shock. For the same charge weight, a slender charge shape will produce less ground shock compared to a spherical charge at the same distance from the charge centre. The charge shape effect has been shown to be primarily a function of the distance from the charge and the aspect ratio of the charge (Ouchterlony et al., 1997; Hao and Wu, 2001). Where the charge length is significant compared to the distance, the effects of the charge shape should also be accounted for by applying a correction factor to the charge weight. According to Ouchterlony et al. (1997), the charge weight can be corrected by applying the following factor:

$$f = a\tan\left(\frac{H}{2R}\right)/(H/2R) \qquad (17.9)$$

where R = distance, H = charge length, f = is in radians.

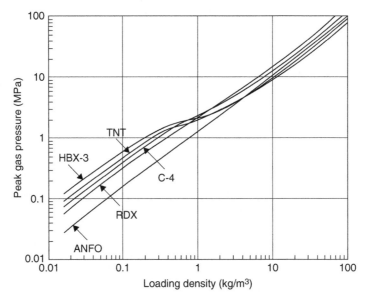

Figure 17.4 Average peak chamber gas pressure as a function of loading density for various explosives (NATO, 2006).

Figure 17.5 shows the charge weight correction factor as a function of the ratio of distance to the charge length based on Ouchterlony *et al.* (1997). As can be seen, the charge shape effect diminishes as the ratio of distance to charge length becomes greater than 1. For explosives storage, PPV equations have been derived directly as a function of the chamber geometry which shows an average PPV reduction of 20–30% for a storage chamber with a length-to-width ratio of 4:1 compared to a chamber with 2:1 ratio (DSTA, 2002).

17.3 TUNNEL RESPONSE

Essentially, rock damage can be in the form of crushing or spalling, depending on the proximity of the rock to the explosion source. The definition of spalling itself is also subject to discussion. In many discussions, the word spalling has been taken to mean the "peeling off" of loose rock pieces.

17.3.1 Spall analysis

In this section, a simple spall analysis is presented based on a 1-D elastic saw-tooth wave pulse (Zukas, 1982) travelling along a rock bar. While this is a simplified representation of reality, it nevertheless provides some good insights into the physical meaning of rock and tunnel damage.

The velocity of the first spall is given as follows:

$$V_{SP} = \frac{2\sigma_m - \sigma_{DT}}{\rho C} = 2ppv - \frac{\sigma_{DT}}{\rho C} \qquad (17.10)$$

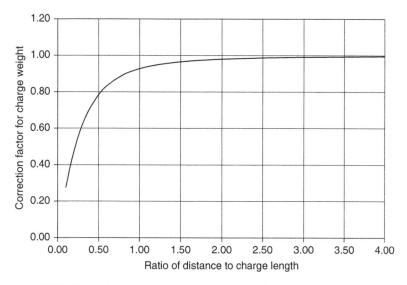

Figure 17.5 Correction factor for charge weight as a function of charge geometry.

where V_{SP} = velocity of the first spall; σ_m = magnitude of incipient stress; σ_{DT} = dynamic tensile strength of rock; ρ = rock mass density, kg/m³, or N/m³/(m/s²) or Ns²/m⁴; C = seismic wave velocity in rock, m/s.

The amplitude of the incipient stress can be related to the peak particle velocity by:

$$\sigma_m = ppv(\rho C) \tag{17.11}$$

The threshold PPV for spalling can be obtained by setting Equation (17.10) to zero:

$$ppv = \frac{1}{2}\frac{\sigma_{DT}}{\rho C} \tag{17.12}$$

The thickness of the first spall can be estimated using the following equation:

$$t_1 = \frac{\lambda}{2}\frac{\sigma_{DT}}{\sigma_m} = \frac{\lambda}{2}\frac{\sigma_{DT}}{ppv(\rho C)} \tag{17.13}$$

where λ = wavelength = C/f, and f = frequency.

The effect of frequency is obvious. From ground shock tests conducted in hard rock, typical frequencies of the stress wave range from 100 Hz to 500 Hz.

The maximum level of stress is reduced by σ_{DT} every time a spall is formed. The total number of spalls is given by the following equation:

$$n = \frac{1}{2} + \frac{\sigma_m}{\sigma_{DT}} = \frac{1}{2} + \frac{ppv(\rho C)}{\sigma_{DT}} \tag{17.14}$$

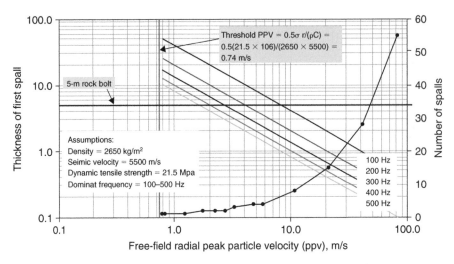

Figure 17.6 One-D spall calculations for a typical granite rock.

Assuming an average frequency of 250 Hz for granite, and dynamic tensile strength of 21.36 MPa, a mass density of 2650 kg/m³, and a seismic velocity of 5500 m/s, we can estimate the threshold ppv for spalling as:

$$ppv = \frac{21.36 \times 10^6}{2(2650)(5500)} = 0.733 \text{ m/s}$$

The thickness of the first spall is:

$$t_1 = \frac{21.36 \times 10^6(5500\text{ m/s}/250\text{ Hz})}{2(0.733)(2650)(5500)} = 22 \text{ m}$$

The above calculation shows that the incipient spalling ppv is 733 mm/s for a one-dimensional rock column. The spalling ppv threshold for in-situ rock mass may be higher due to the 3-D confining effects. For the given strength and ppv, the thickness of the 1-D rock column is 22 m, suggesting that the degree of damage is very limited, and from a practical point of view, tunnel damage would not be visible.

Figure 17.6 shows the thickness of the first spall and the total number of spalls possible calculated for a granite rock. It can be seen that the number of spalls will only increase to 2 at an incipient ppv close to 2–3 m/s.

17.3.2 Damage criteria

Assessing the response and damage of rock subjected to dynamic loading requires a set of damage criteria. It is extremely important to have a common understanding of the definition of damage, before any meaningful discussion of tunnel damage can be made. Unfortunately, there currently exist no established damage criteria for rock

Table 17.3 Damage criteria for hard Scandinavian bed rocks (Persson, 1997).

Pep, m/s	Tensile Stress (MPa)	Strain Energy (J/kg)	Typical Effect
0.70	8.7	0.25	Incipient swelling
1.00	12.5	0.5	Incipient damage
2.50	31.2	3.1	Fragmentation
5.00	62.4	12.5	Good fragmentation
15.0	187	112.5	Crushing

Table 17.4 Tunnel Damage Criteria for Unlined Rock Tunnels (Li & Huang, 1994).

Rock Parameters			Peak Particle Velocity, mm/s				
Rock Type	Unit Weight (g/cm³)	Comp. strength (MPa)	Tensile Strength (MPa)	No Damage	Slight Damage	Medium Damage	Serious Damage
Hard	2.6–2.7	75–110	2.1–3.4	0.27	0.54	0.82	1.53
Rock	2.7–2.9	110–180	3.4–5.1	0.31	0.62	0.96	1.78
	2.7–2.9	180–200	5.1–5.7	0.36	0.72	1.11	2.09
Soft	2.0–2.5	40–100	1.1–3.1	0.29	0.58	0.90	1.67
Rock	2.0–2.5	100–160	3.4–4.5	0.35	0.70	1.07	1.99

and rock support. Most of the damage definitions found in the literature are not well defined. Various terms have been found in the literature to describe damage, often with significant differences in definition and practical meaning. Rock damage at the micro-crack level may be interesting for numerical modelling, but for engineering applications such damage criteria are not very useful. Thus, the damage criteria discussed in this chapter typically refer to rock damage at the macro level.

Table 17.3 shows the damage criteria for Swedish hard rock by Persson (1997), which were developed based on a similar analysis as shown in Section 3.1, with observations of physical effects. These criteria suggest threshold damage (incipient damage) at a ppv of 1 m/s.

Li and Huang (1994) discussed damage criteria for rock tunnels, with the following damage definition:

- Slight damage – initial cracking
- Medium damage – partial collapse
- Serious damage – large-area tunnel collapse.

The respective values of ppv for the various types of rock mass are shown in Table 17.4. The ppv for slight damage seems to correspond to initial swelling described by Persson (1997).

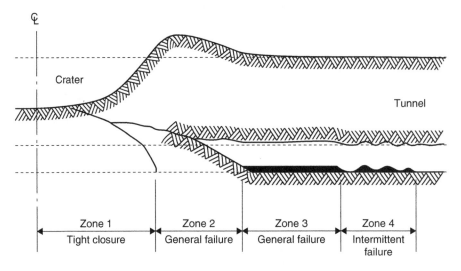

Figure 17.7 Tunnel damage zones (Hendron, 1977).

Table 17.5 Comparison of UET Tests (Hendron, 1977) and 1-D calculation.

Damage Zone	1	2	3	4
Damage	Tight closure	General failure	Local failure	Intermittent failure
Free-field radial strain	NA	40	13	3–6
Free-field ppv, m/s	NA	12	4	0.9–1.8
Calculated thickness of 1st spall, m		0.3–1.4	1–4.2	2–18.5
Calculated number of spalls		11	4	1

Damage threshold values from other studies can be found in Coates (1981), Dowding (1984), Kartuzov (1976), Oriad (1972), Philips *et al.* (1992), and Siskind (1997).

Perhaps the most comprehensive study of tunnel damage is the US Army's Underground Explosion Tests (UET), as reported in Hendron (1977). The study classifies tunnel damage into four damage zones (Fig. 17.7).

From Figure 17.7, it can be said that what is referred to as intermittent failure is most likely random spalling of loose rocks. Serious spalling (or damage) does not occur until the ppv reaches 4 m/s.

Using the methodology described in Section 3.1, a similar calculation has been done for the sandstone rock of the UET tests and shown in Table 17.5. From Table 17.5, the number of spalls corresponding to a ppv of 4 m/s for Zone 3 damage (local failure) is 4, which would seem to fit the physical description of tunnel damage quite well.

17.3.3   Damage of rock tunnel with support

Studies have been done to examine the potential damage of rock support due to blasting ahead of the tunnel face. It has been found that even very close to the blasting face,

Table 17.6 Explosive testing tunnel response (Dowding, 1984).

Type	Strain %	PPV, m/s
Unlined tunnel		
Joint movement, fall of loose stones		0.3
Intermittent failure	0.015	2.0
Local failure	0.04	3.6
Complete closure	0.1	
Lined tunnel		
Cracking of liner	0.02	1.0
Displacement of cracks		1.3
Local failure	0.15	7.4
Complete failure	0.8	40.0

damage to the rock bolt grout and shotcrete is negligible. Studies carried out by Stjern and Myrvang (1998) and Ortlepp and Stacey (1998) have shown that ppv up to 1 m/s will not cause any measurable damage of the rock support. For lined tunnels, Dowding (1984) suggested that the threshold PPV would roughly double that for unlined tunnels (Table 17.6). Again, it is important to make a distinction between tunnel support by rock bolts and shotcrete, and tunnel support by concrete liners.

17.3.4 Observations of tunnel damage

Based on the above one-dimensional analysis and comparisons with several studies of rock or tunnel damage, the following observations can be made:

– Most observations from field tests suggest initial tunnel damage is a result of falling loose rocks, rather than damage of the rock material created by the shock wave.
– The threshold damage based on theoretical cracking most likely will not result in any visible damage of a tunnel for engineering purposes.
– Visible tunnel damage in competent rock will not occur until the incipient ppv reaches about 1–2 m/s. This value will double for tunnels with lining support.
– For practical applications, the tunnel damage criteria defined in the UET tests seems to be the most realistic.

17.4 LARGE-SCALE TESTING

17.4.1 Test tunnel facility

From 2000 to 2001, several large-scale tests have been conducted in a rock tunnel facility in Älvdalen, Sweden, site of the existing Klotz Group tunnel (Chong et al., 2002). The tests were conducted to validate the safety design for underground explosives storage, including response of tunnels. While the design for measuring ground shock loading and tunnel response was relatively easy, the initial design of the test facility presented some engineering challenges in the prediction of the dynamic loading,

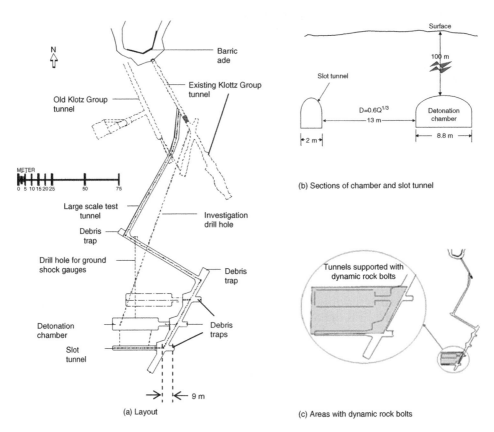

Figure 17.8 Test tunnel facility layout, chamber sections, and areas with dynamic rock bolts.

tunnel damage assessment, and dynamic rock support for the tunnel facility which was designed for repetitive explosion loading.

The test facility was constructed in a rock mass consisting of mostly red porphyry with some grey granitic intrusions. Fresh intact rock has uniaxial compressive strengths of 200–250 MPa. The rock mass quality is considered "good" with average Q values of 15–20. It consists of a detonating chamber connected by a series of tunnels. Adjacent and parallel to the chamber is a slot tunnel at criterion separation distance $(0.6Q^{1/3})$ to test and monitor response of an adjacent chamber (Fig. 17.8). The average rock cover over the chamber area is about 100 m. The chamber has a width of 8.8 m, a height of 4.2 m, and length of 33 m. The slot is 2-m wide and has the same height of the chamber. The main tunnel connecting the chamber and slot is 4.5 m. The actual separation between the chamber and slot tunnel is about 13 m based on a maximum Net Explosives Quantity (NEQ) of 10 tons.

As the tunnel was designed to last through four years of explosion testing, including fragment loading, dynamic rock bolts (Ansell, 1999) were used to support the detonation chamber and the areas within $0.6Q^{1/3}$ from the chamber wall, as shown by the shaded area in Figure 17.8c. The dynamic bolts have a smooth section which

Table 17.7 Test setup and objectives.

No.	NEQ (kg)	Charge Type	Objectives
1	10	Fully coupled bare charge	Ground shock calibration.
2	500	Bare charge	Bare charge test at loading density 0.5 kg/m^3 for airblast calibration
3	10,000	Bare charge	Bare charge test at loading density 10 kg/m^3 as reference test
4a	2,500	Bare charge	Bare charge test at loading density 2.5 kg/m^3 at intermediate loading density
4	10,000	Cased munitions	Effects of fragment loading and debris flow in tunnels at loading density 10 kg/m^3

allows them to detach from the grout and deform plastically under high dynamic loads. In the chamber, plain shotcrete was applied in two layers, with a wire mesh in between.

17.4.2 Test setup and test objectives

A total of 5 tests were carried out, with all detonations in the same chamber. Table 17.7 shows the test setup and test objectives. With the exception of Test #1, the charges for all other tests were evenly distributed in 10 tables in the chamber with the bottom raised to about 800–900 mm above the floor. Details of the tests and instrumentation are discussed in a separate paper (Chong *et al.*, 2002). This chapter will only discuss results of ground shock measurements and observations of tunnel response.

17.4.3 Ground shock instrumentation

Ground shock gauges were installed in the following locations (Fig. 17.9):

- A horizontal hole perpendicular to the chamber axis.
- A vertical hole above the chamber centre.
- Along the wall of the slot tunnel at 13-m from the chamber wall.

Strain gauges were also installed on two dynamic rock bolts installed along the middle of the slot tunnel wall.

Using ground shock prediction methods discussed earlier, the predicted incipient PPV for the slot tunnel wall was found to be between 0.75 and 1.49 m/s. Details of the ground shock prediction can be found in Chong *et al.* (2002). Table 17.8 shows a summary of the PPV calculations for the slot tunnel wall.

17.4.4 Results and observations of damage

Based on observations after the tests, the amount of damage sustained in the tunnel system was surprisingly less than anticipated. Figure 17.10 shows some photos of the chamber and the slot tunnel after the detonations.

For the 10-ton and 2.5-ton tests, ten craters of a similar size were created on the floor below the ten charges after each detonation. The ten craters of similar size also

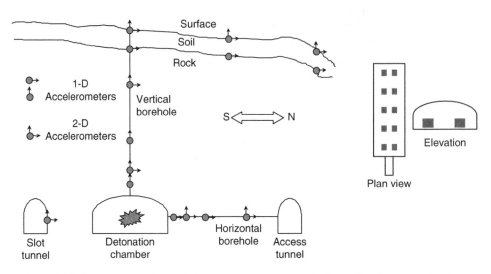

Figure 17.9 Section view of ground shock instrumentation and charge distribution in chamber.

Table 17.8 Summary of predicted PPV on slot tunnel wall.

Charge weight	10,000 kg
PPV for fully coupled explosion (13 m away from chamber wall)	$5000(R/Q1/3) - 1.5$ $= 5000(14/100001/3) - 1.5$ $= 10,760\,mm/s$
PPV correction for charge geometry (chamber length to width ratio 4:1)	0.6–0.8
Decoupling factor (loading density of 10 kg/m³)	0.116–0.23
Predicted PPV for slot wall (incipient)	$10,760 \times 0.6 \times (0.116 - 0.23)$ $= 748–1,485\,mm/s$

confirmed complete detonation of all charges. There was no rock fall from the roof or walls. In the 10-ton bare charge test, there were very few spots of observed spall of the shotcrete. For the 155-mm cased charges, there were more shotcrete spalls on the chamber walls due to the directional loading of the steel fragments. Even so, no bare rock was observed in the chamber wall, except in one location near the branch tunnel, where a piece of rock fell off. The surprisingly low damage in the chamber might be due to the "burned" shotcrete resulting from Test#3, which may have made the shotcrete into some form of sandy material. This sandy material could have acted as an energy-absorbing layer against the fragments of Test #4b, thereby limiting the tunnel wall damage.

In the slot tunnel, no damage of the shotcrete or bare wall was observed. The only exception was at a gauge hole, where the shotcrete had been damaged during drilling and subsequently dropped to the tunnel floor due to the ground shock loading. The loose material on the floor showed some movement from Test #3. All lights and fixtures

Chamber with craters after detonation (Test #3)

Slot tunnel showing soil movement after
Test #3

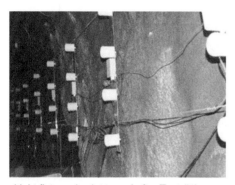

Light fixtures in slot tunnel after Test #4b   Chamber after detonation (Test #4b)

Figure 17.10 Photos of the detonation chamber and slot tunnel after Tests #3 and #4b.

in the slot tunnel remained fully functional after the tests. The relatively small damage of the rock separation is also confirmed by the low strains recorded on the rock bolts from Test #3 (Fig. 17.11). The calculated seismic velocities of the slot wall show an 8% reduction after Test #3 (from 4,636 m/s to 4,268 m/s) and subsequently remained unchanged (Table 17.9). The initial change was probably due to loosening of the joints, rather than new fractures created in the rock mass. Thus, it was concluded after Test #3 that the dynamic rock bolts were not necessary due to the relatively low dynamic loads.

For Tests #3 and #4b, the average PPV recorded on the slot wall was 1.4 m/s and 0.97 m/s, respectively (Fig. 12). Since the PPV's are recorded on the slot tunnel wall (perpendicular to it), their values represent the reflected ground shock wave. The

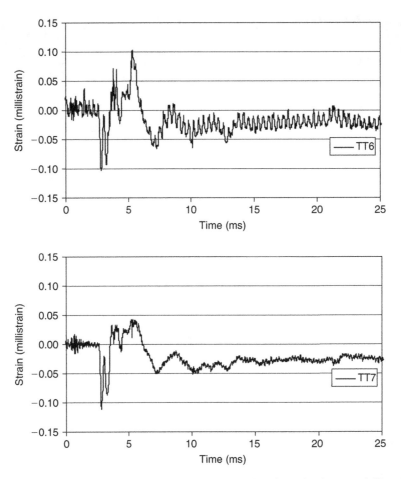

Figure 17.11 Dynamic strain recorded on rock bolts installed from the slot tunnel (Test #3).

Table 17.9 Calculated seismic velocities in slot tunnel wall for Tests 1–3.

Test and Charge	Peak Chamber Pressure, MPa	Average PPV on Tunnel Wall, mm/s	Time of Arrival, Ms	Calculated Seismic Velocity, m/s
Test 1–10 ton bare TNT	100	1390	3.07	4,636
Test 2–2.5 ton bare TNT		622	3.26	4,268
Test 3–10 ton TNT (1450 155 mm shells)	50	977	3.28	4,294
Ratio of Seismic Velocity after Test 2			—	0.93

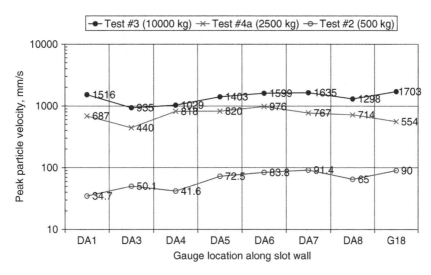

Figure 17.12 Comparison of measured PPV on the slot wall.

Table 17.10 Summary of PPV data on slot tunnel wall and calculated equivalent TNT.

Items	Test #3	Test #4b
Min PPV, m/s	0.94	0.62
Ratio of Min PPV	1.00	0.66
Max PPV, m/s	1.70	1.84
Ratio of Max PPV	1.00	1.09
Average PPV, m/s	1.39	0.98
Ratio of Avg PPV	1.00	0.70
Equivalent TNT Ratio	1.00	0.54

equivalent free field PPV at the same distance would be roughly one half of the recorded value. Test #4b produced a lower ground shock due to casing effects of the 155-mm shells with equivalent TNT of 0.54 based on ground shock measurements. This casing effect is also confirmed from blast pressure measurements, which showed equivalent TNT from 0.45 to 0.67 of bare TNT (Chong *et al.*, 2002). Table 17.10 shows the equivalent TNT calculated from the measured PPV values on the slot tunnel wall.

In the blast door niche area outside the detonation chamber and the main tunnel immediately next to it (located at $0.4Q^{1/3}$ from the chamber), there was no visible damage of the rock wall or roof, despite repeated blasts.

17.4.5 Effects of decoupling

Results of ground shock measurements are plotted in Figure 17.13, along with the PPV equation for fully-coupled detonation from earlier tests conducted at the same

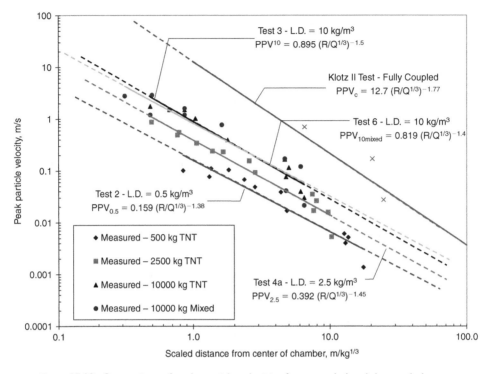

Figure 17.13 Comparison of peak particle velocities from coupled and de-coupled tests.

site (Hultgren, 1987). The distance for ground shock in Figure 17.13 is measured from the chamber centre.

This decoupling effect can be seen clearly from Figure 17.13, where at the same scaled range of $1 \, m/kg^{1/3}$, the average peak particle velocity was $15 \, m/s$, $1 \, m/s$, and $0.4 \, m/s$, for fully coupled and loading densities of $10 \, kg/m^3$ and $2.5 \, mg/m^3$, respectively. The respective decoupling factor for the PPV, for loading densities of $10 \, kg/m^3$ and $2.5 \, mg/m$ are 0.067 and 0.027. In other words, at a distance of about $22 \, m$, the peak particle velocity produced by a $10,000 \, kg$ explosion with a loading density of $10 \, kg/m^3$ will produce only about 7% of the PPV of a fully coupled charge of the same quantity.

In summary, based on results from large-scale tests in Sweden, for loading densities up to $10 \, kg/m^3$ in hard rock, the tunnel separation to prevent tunnel damage can be safely reduced from the current $1.0Q^{1/3}$ to $0.6Q^{1/3}$. Chambers sited at $0.6Q^{1/3}$ are expected to remain fully functional if access is available after an accidental explosion. For the main tunnel located at $0.4 \, Q^{1/3}$ from the end of a chamber, no significant damage is expected after the explosion of a designed quantity. It is clear that for loading densities up to $10 \, kg/m^3$, requirements for separation distances for hard rock based on the current codes are overly conservative.

17.5 CONCLUSIONS

In assessing tunnel damage from dynamic loading, accurate prediction of the ground shock loading is very important. The prediction must consider the type of explosion source and influencing factors such as the geology and rock mass properties, charge geometry, and loading density. The effects of explosion loading on tunnels are very much a function of the loading density. The same distance and same quantity of explosives, the peak particle velocity produced by an explosion of $10\,kg/m^3$ is less than 10% of that by an explosion of a fully coupled explosion. Assessment of tunnel damage for engineering purposes should follow a common understanding and definition of damage criteria.

Based on a literature review, theoretical analysis and the results of the tests, it can be concluded that damage of unlined tunnels in competent rock does not seem to begin until the incipient PPV reaches a value of at least 1–2 m/s. With the addition of tunnel support (such as rock bolts and fibre reinforced shotcrete), it is expected that the tunnel can sustain a much higher load, probably at least 2–4 m/s in ppv. For such load cases, normal static rock support is probably sufficient although the use of steel fibre reinforced shotcrete is recommended for its high energy capacity. The use of dynamic support in hard rock is not necessary unless the dynamic loading, as expressed by the incipient peak particle velocity, reaches more than 2–4 m/s.

REFERENCES

Ansell, A.: *Dynamically Loaded Rock Reinforcement.* TRITA-BKN. Bulletin 52, Doctoral Thesis, Royal Institute of Technology, 1999.

Chong, K., Zhou, Y., Seah, C.C. and Lim, H.S.: Large-scale tests – Airblast, Ground Shock, and Debris. *Proceedings of the International Symposium on Defence Construction, Singapore,* 17–18 April 2002.

Coates, D.F.: *Rock Mechanics Principles.* Monograph 874, Energy, Mines, and Resources Canada, 1981.

Defecne Science & Technology Agency (DSTA): Ground shock predictions for underground ammunition storage safety: technical background. Informal Working Paper, NATO Underground Storage Working Group. AC/258 SIN (ST) UGSWG IMP 2-2002. 2002.

Department of the Army: *Design of Underground Installations in Rock – Penetration and Explosion Effects.* TM 5-857-4. 1961.

Dowding, C.H.: Estimating earthquake damage from explosion testing of full-scale tunnels. *Adv. Tunnel Technology and Subsurface Use* 4(3) (1984), pp.113–117.

Dowding, C.H.: *Blasting Vibration Monitoring and Control.* Prentice-Hall, NJ, USA, 1985.

Hao, H. and Wu C.Q.: *Dynamic Response of Rock Tunnels Subjected to Blast Loads.* Report No.: NTU/PTRC/2001.04. Protective Technology Research Centre, Nanyang Technological University, 2001.

Hendron, A.J.: Engineering rock blasting in civil projects. In: Hall, W.J. (ed.): *Structural and Geotechnical Mechanics: A Volume Honouring Nathan M. Neumark.* Prentice Hall, 1977, pp. 242–277.

Hendrych, J. The dynamics of explosions and its use. Elsevier, Czechslovakia. 1979.

Holmberg, R. and Persson, P.A.: Design of tunnel perimeter blasthole patterns to prevent rock damage. *Trans. Inst. Min. Metall.,* London, Vol. 89 (1980), pp. A37–40.

Hultgren, S.: *Ground Motion – Measurements,* KLOTZ II. Report C4:87, Fort F, 1987.

Joachim, C.E. and Smith, D.R.: WES underground magazine model tests. In: *Proceedings of the 23rd Department of Defence Explosives Safety Seminar*, Department of Defence Explosives Safety Board, 1988.

Joachim, C.E.: *Shallow Underground Tunnel/Chamber Explosion Test Program – Summary Report*. TR SL-90-10, U.S. Army Engineer Waterways Experiment Station, Vicksburg, MS, 1990.

Johnson, W.J, Rozen A. and Rizzo P.C.: Explosions in soils: the effects of soil properties on shock attenuation. In: *Proceedings, 23rd Department of Defence Explosives Safety Seminar*, Atlanta, GA, USA, 1988, pp.1831–1841.

Kartuzov, M.I. *et al.*: Stability of underground mine workings with various types of supports when subjected to blasting. *Soviet Mining Science* 11(2) (1976), pp.99–103.

Li H.B., Zhao J. and Li T.J.: Triaxial compression tests of a granite at different strain rates and confining pressures. *Int J Rock Mech Min Sci* 36 (1999), pp.1057–1063.

Li H.B., Zhao J. and Li T.J.: Micromechanical modelling of mechanical properties of granite under dynamic uniaxial compressive loads. *Int J Rock Mech Min Sci* 37 (2000), pp. 923–935.

Li, H.B, Zhao J., Li T.J. and Yuan J. X.: Analytical simulation of the dynamic compressive strength of a granite using the sliding crack model. *International Journal for Numerical and Analytical Method in Geomechanics* 25 (2001), pp.853–869.

Li, Z. and Huang, H.: The calculation of stability of tunnels under the effects of seismic wave of explosions, In: *Proceedings of 26th Department of Defence Explosives Safety Seminar*, Department of Defence Explosives Safety Board, 1994.

Madshus, C. and Langberg, H.: Ground shock in rock – full scale tests in Norway (Part 1 and Part 2). In: Proceedings of the Joint Singapore-Norway Technical Workshop on Ground Shock, Lands & Estates Organisation, Singapore, 1999, pp.17–36.

McMahon, G.W.: Ground motions from detonations in underground magazines in rock. In: *Proceedings of 25th Department of Defence Explosives Safety Seminar*, Anaheim CA, USA, 1992, pp. 277–293.

McMahon, G.W.: Intermediate-scale underground magazine explosion tests – decoupled ground motion experiments. *Proceedings of 26th Department of Defence Explosives Safety Seminar*, Miami, FL, USA, 1994, pp. 1–12.

Murrel, D.W. and Joachim, C.E.: The 1996 Singapore Ground Shock Test. Department of the Army, Waterways Experiment Station, Corps of Engineers. 1996. 93 pp.

NATO: *AASTP-1 Part III (Edition 1, Change 2). NATO Safety Principles for the Storage of Military Ammunition and Explosives*, 2006.

Oriad, L.L.: Blasting effects and their control in open pit mining. *In Geotechnical Practice for Stability in Open Pit Mining, Society of Mining Engineers*, New York, 1972, pp.197–222.

Ortlepp, W.D. and Stacey, T.R.: Performance of tunnel support underground large deformation static and dynamic loading. *Tunnelling and Underground Space Technology* 13(1) (1998), pp.15–21.

Ouchterlony, N.S. *et al.*: Monitoring of large open cut rounds by VOD, PPV, and gas pressure measurements. *International Journal of Blasting and Fragmentation*, 1(1997), pp. 3–25.

Perret, W.R. and Bass, R.C.: *Free-field Ground Motion Induced by Underground Explosions*. SAND74-0252, Sandia Laboratories, Albuquerque, NM, 1975.

Persson, P.A.: The relationship between strain energy, rock damage, fragmentation, and throw in rock blasting. *International Journal of Blasting and Fragmentation* 1 (1997), pp.99–110.

Phillips, J.S., Luke, B.A., Long, J.W. and Lee, J.G.: *MISTY ECHO Tunnel Dynamics Experiment – Data Report, Volume 1*. Sandia National Laboratories Report SAND89-0972/1, 1992.

Siskind, D.E.: Vibration damages in industrial situations. *Journal of Explosives Engineering*, Nov/Dec/ 1997:40.

Smith, M.C.F. and Morton, D.G.: Construction and blasting methods at the Kornhill Site Formation works. In: *Proceedings of Rock Engineering and Excavation in an Urban Environment, Hong Kong*, 1986, pp. 365–380.

Stjern, G. and Myrvang A.: The influence of blasting on grouted rock bolts. *Tunnelling and Underground Space Technology* 3(1) (1998), pp.65–70.

Strange, J.N., Dornbusch W.K and Jr Rooke, A.D.: *Review and Evaluation of Technical Literature*. Contract Report SL-95-4, UAST-CR-94-002, US Army Corps of Engineers, Waterways Experiment Station, 1995.

Tao, S.L.: *Blasting Engineering*. Metallurgical Industry Publishing House, 1979. (in Chinese).

Zhao, J., Li, H.B., Wu, M.B. and Li, T.J.: Dynamic uniaxial compression tests of a granite. *Int J Rock Mech Min Sci* 36 (1999a), pp.273–277.

Zhao. J.: Applicability of Mohr-Coulomb and Hoek-Brown strength criteria to dynamic strength of brittle rock materials. *International Journal of Rock Mechanics and Mining Sciences* 37 (2000), pp.1115–1121.

Zhao, J. and Li H.B.: Experimental determination of dynamic tensile properties of a granite. *Int J Rock Mech Min Sci* 37 (2000), pp. 861–866.

Zhao, J., Li, H.B. and Cai, J.G.: *Dynamics Tests of a Cement Mortar*. Geotechnical Research Report NTU/GT/00-1, Nanyang Technological University, Singapore, 2000.

Zhou, Y. and Ong, Y.H.: Ground shock prediction methods – a critical appraisal. In: *Proceedings of the 1st Asia-Pacific Conference on Shock and Impact Loads on Structures*, 1996, pp.477–483.

Zhou, Y., Seah, C.C., Guah E.H., Foo S.T., Wu Y.K. and Ong P.F.: Considerations for ground vibrations in underground blasting. *International Conference on Tunnels and Underground Construction*, Singapore, 2000, pp.313–318.

Zhou, Y. and Jenssen, A. Internal separation distances for underground explosives storage in hard rock. *Tunnelling and Underground Space Technology*, 24 (2009), pp.119–125.

Zukas, J.A.: *Impact Dynamics*. John Wile, 1982.

Chapter 18

Rock support for underground excavations subjected to dynamic loads and failure

Charlie Chunlin Li

18.1 INTRODUCTION

In highly stressed rock masses, rock failure is an unavoidable matter after excavation. In certain circumstances, rockburst events occur, causing casualties and equipment damage. In shallow tunnels where rock stresses are low, the main objective of rock support is to stabilize the loosened rock blocks after excavation. The task of support elements, such as rock bolts and steel sets, is to prevent the loosened blocks from falling. This means that in shallow locations, the loading condition for the support element is the dead weight of the blocks. In this case, the strength of the bolts has to be larger than the dead weight of the potentially falling blocks. In other words, strong rock bolts have to be used in shallow locations. This obeys the principle of structural mechanics which states that a structure fails when it yields. In highly stressed rock masses, rock blocks seldom become loosened in underground openings. Instead, rock fails because of elevated stresses that are beyond the strength of the rock. In this case, the loading for the support system is no longer dead-weight controlled, but rather a displacement controlled process. The well-known Ground Response Curve (GRC) describes such a loading process (Carranza-Torres and Fairhurst, 2000). The more displacement is allowed, the less the need for support pressure to stabilize the ground. In high stress rock conditions, the rock support should be strong and also ductile, i.e. it should be able to absorb a good amount of deformation energy prior to failure.

In this chapter, we shall have an overview of the definitions of rockburst, current research results in dynamic rock support, available energy-absorbing rock bolts, principles of dynamic rock support, and dynamic test methods and facilities of support elements/systems.

18.2 ROCKBURST EVENTS

All rockburst events are related to high in-situ stresses in rock masses. Moreover, rockburst events usually occur in hard and brittle rocks. In accordance with the triggering mechanism, a rockburst event is either classified to *strain burst* or *fault-slip burst*. Strain burst refers to a burst event that is directly related to a stress concentration in the nearby field of an underground opening. After excavation, the tangential stresses in the superficial rock become elevated. In extreme cases, the stresses are so high that the rock is not capable of sustaining them. At this moment, the rock bursts out and

Figure 18.1 Rock pieces that burst down from the roof in a deep metal mine.

the elastic energy stored in the rock is released in a violent manner. The epicentre of a strain burst event is the place where the burst occurs. The superficial rock usually becomes finely fragmented with thin and knife-sharp slices of rock being spread all over the site. Figure 18.1 shows such a burst event that occurred in massive quartzite at a depth of approximately 1000 m in a mine. The quartzite rock mass was almost discontinuity free. Immediately after excavation blasting, the rock started to spall slice by slice from the roof surface. All the materials in the pile shown in the picture burst down from the roof after mucking.

Underground excavation results in a tangential stress concentration in the rock surrounding the opening, while on the other hand a reduction in the radial stress in the rock mass. The reduction in the radial stress will lead to a decrease in the normal stresses on some pre-existing faults in the rock mass and therefore a reduction in the shear resistance on the faults. Slippage may occur on the faults. Such fault slippage will induce strain/stress waves that propagate spherically outward from the epicentre where the slippage occurs, as shown in Figure 18.2. In the mining industry, this is called a seismic event. At great depths, the rock usually becomes fractured after excavation, owing to stress concentrations. When the strain waves released from the fault slippage reach the underground opening, a rockburst event may be triggered so that the fractured rock is burst down to a certain depth, which is the so-called fault-slip rockburst. Fault slippage usually releases a significant amount of energy. As a result, a fault-slip burst may cause more serious damage to underground infrastructures than a simple strain burst event. A rock pile from a fault-slip burst is composed of irregular rock pieces, ranging from finely fragmented pieces to large blocks. Figure 18.3 shows such a rock pile in a deep mine drift. The rockburst event was triggered by a fault-slippage located

Figure 18.2 Fault-slip rockburst event.

Figure 18.3 A fault-slip rockburst in a deep metal mine (Simser, 2001).

about 100 m from the mine drift. Fault-slip seismic events usually occur at a distance of tens to hundreds of metres from underground openings in deep mines.

The volume of rock excavation is usually on a very large scale in a mine. A mine stope, where ore is mined, can be tens to hundreds of metres in height and several hundred metres in length. Even though the mined-out space is backfilled with waste rocks and/or tailing sands in most cases, such a large mined-out space will disturb the in-situ stresses in a large portion of the rock surrounding the opening. Because of the large-scale stress disturbance, both strain burst and fault-slip burst could occur in deep mines. The excavation of tunnels and civil caverns usually disturbs the rock mass on a relatively small scale. For that reason, it is more common for strain burst than fault-slip burst to occur in civil excavations.

18.3 REVIEW OF PREVIOUS WORK

Many rockburst studies have been conducted in mining countries such as South Africa, Australia and Canada in the past years. Rockburst was thoroughly studied in The Canada Rockburst Research Program from 1990 to 1995. The program was established as a collaborative research project to investigate the problem of rockburst in Canadian mines after some serious rockburst incidents took place in Canada. The overall objective of the 5-year project was to obtain an understanding of how and why mines become susceptible to seismicity and rockburst problems, to determine what needs to be done in order to minimize the effect of rockbursts and how the damage caused by these seismic events is controlled. The results of the program are summarized in a six-volume handbook (CAMIRO, 1995) that covers topics of mining in burst-prone ground, rockburst support, seismic monitoring, numerical modelling, mining seismology and case histories.

In the handbook, rockburst damage mechanisms are classified into three categories: rock bulking due to fracturing, rock ejection due to seismic energy transfer and rockfall due to seismic shaking. *Rock bulking* due to fracturing usually results in an increase in volume. If the fracturing occurs in an unstable and violent manner, it is often referred to as a strain burst, which is perhaps the most common form of damage in both civil and mining excavations in burst-prone ground. *Rock ejection* due to seismic energy transfer occurs in a manner in which rock blocks are violently ejected from the periphery of an excavation, while a seismic event occurs nearby. The seismic energy is transferred to the blocks which then become kinetically active. This mechanism is considered to be a primary cause of rockburst damage in deep mines. *Rockfall* due to seismic shaking occurs when an incoming seismic wave accelerates a volume of rock that was previously stable under static conditions. The seismic shaking triggers the fall, although gravity is an important factor in this failure.

The severity of rockburst damage is represented by minor, moderate and major classes. *Minor damage* involves only a shallow skin of fractured or loosened rock less than 0.25 m in thickness. *Moderate damage* implies that the rock is heavily fractured and may have been violently displaced, and is generally characterized by fractured or loosened rock between 0.25 m to 0.75 m in thickness. *Major damage* involves the deep fracturing or presence of damaged rock to a depth of more than 0.75 m. Most ground support components would be broken or damaged when a drift sustains major damage.

In regard to rock support, a support element can provide one or more than one of the following three primary functions: to reinforce the rock mass, to retain broken rock and to hold the retaining elements (McCreath and Kaiser, 1992). *Reinforcing* the rock mass is to strengthen it and prevent a loss of strength, thus helping the rock mass to support itself. *Retaining* broken rock is required for safety reasons on the one hand, while on the other hand it prevents progressive failure that would lead to an unravelling of the rock mass.

Rock blocks may be ejected when subjected to a seismic wave. The ejection velocity can be obtained by performing back-calculations of the ejected rock mass (Tannant *et al.*, 1993):

$$v_e = d \sqrt{\frac{g}{2h\cos^2\theta + \sin\theta}} \tag{18.1}$$

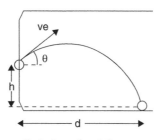

Trajectory of rock thrown
during a rockburst

where θ is the initial angle of motion measured upwards from the horizontal plane and g is the gravitational acceleration. Rockburst damage becomes evident when ejection velocities exceed 3 m/s. Back-calculations of ejected rock masses reveal that the ejection velocity of the ejected rock mass under a severe rockburst is approximately 5 m/s, even though the velocity of small rock pieces could be up to 10 m/s. A study by Yi and Kaiser (1993) confirmed an assumption that rock ejection velocity is approximately equal to the peak particle velocity (ppv) under typical mining and seismicity conditions. Their study also demonstrated that ejection velocities are greater than the peak particle velocity only when the ejected blocks are very small. Therefore, the kinetic energy E_k of an ejected block of rock can be determined by:

$$E_k = \frac{1}{2}m \times v_e^2 \approx \frac{1}{2}m \times ppv^2 \qquad (18.2)$$

where m is the mass of ejected rock, v_e is the ejection velocity and ppv is the peak particle velocity. The basic demand on a dynamic rock support system is that it can stop the motion of the ejected rock in order to bring the kinetic energy to zero. This demand means that the support elements should have the capability of absorbing the kinetic energy released from the potentially ejected rock.

18.4   PHILOSOPHY OF DYNAMIC ROCK SUPPORT

A large amount of elastic energy is stored in deeply located rock prior to excavation. The elastic energy stored in one cubic metre of rock, which is called energy density, is expressed by $\sum \sigma_i^2/2/E$, where σ_i ($i = 1$, 2 and 3) are the in-situ principal stresses and E is the Young's modulus of the rock. Taking a deep mine in Scandinavia as an example, the vertical rock stress at a depth of 1000 m is about 27 MPa, whereas the two horizontal principal stresses are roughly 2.3 and 1.4 times the vertical stress (Li, 2006). The energy density in the rock is then calculated to be 50 kJ/m^3, with an assumption that the Young's modulus of the rock is 60 GPa. There is no doubt that all the stored energy will be released when the rock is subjected to strain burst failure, though only a small portion of it, possibly a few to tens kJ, is dissipated for rock fracturing. The rest of it would be transformed to kinetic energy to eject broken rock pieces away from their original positions.

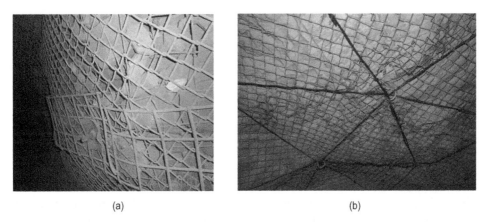

(a)   (b)

Figure 18.4 Examples of support elements to provide surface containment. (a) Finely shattered rock contained by mesh (Simser, 2001). (b) A combination of mesh and lacing to enhance the surface containing effect in a deep mine in South Africa.

In some cases, the stress concentration in the near field of an opening is not intense enough to create strain burst as opposed to only creating surface-parallel extension fractures. This is the so-called spalling/slabbing failure of rock. With such a failure mode, the fractured rock party does not completely lose its integrity; consequently, it is still able to carry some load and a certain amount of elastic energy is still stored in the rock slabs. When a fault-slip event occurs nearby, the seismic waves break the rock slabs and eject them into the opening. In this case, the kinetic energy of the ejected rock party is the sum of the elastic energy stored in the rock party and the seismic wave energy. The latter could be very powerful indeed. Because of this, a fault-slip rockburst could be more violent than a strain burst.

The analysis above indicates that the total energy released from the rock is dissipated for rock fragmentation and rock ejection when a rockburst event occurs. In order to prevent a rock party from being ejected, the support system has to be able to dissipate the energy released from the rock. It has been observed in the field that the ejected rock pieces are large in places where light rock support is applied, while in heavily supported places the ejected rock party is usually quite fragmented and contains a certain amount of fine powders. This implies that well contained rock would not only be subjected to fracturing but also grinding between rock pieces, which leads to fine powders. The grinding would dissipate enormous amounts of energy. The more the energy is dissipated by fracturing and grinding, the less or even no ejection that occurs. Figure 18.4a shows an example in which the mesh and mesh straps provided a surface containment to such a satisfactory degree that the rock was finely shattered behind the surface-containing mesh. Surface containment is usually achieved by mesh, strap, shotcrete, lacing or various combinations of them. Figure 18.4b shows an example of a surface containment using mesh and lacing in a South African mine. A satisfactory surface containment is a necessary measure for dynamic rock support.

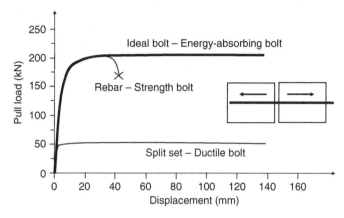

Figure 18.5 Strength bolt (rebar), ductile bolt (Split Set) and energy-absorbing bolt.

The elements in a support system, such as bolts, cables and meshes, are also able to absorb a certain amount of the released energy through their deformations. A satisfactory dynamic support system must be able to provide a good surface containment, while also being able to absorb a large amount of kinetic energy through the support elements.

Conventional rock bolts such as encapsulated rebar and a Split Set absorb little energy because of either a small deformation capacity (for rebar) or a small load-bearing capacity (for a Split Set). In recent years, efforts have been made to develop energy-absorbing rock bolts to enhance the energy absorption of hanging elements in dynamic rock support systems.

Fully-encapsulated rebar may be the most commonly used support element in both civil and mining engineering, and is widely adopted because of its high load-bearing capacity. The shortcoming of rebar is its small deformation capacity. Rebar is a suitable support element in cases where a dead weight, for instance, a loosened rock block, needs to be stabilized. In high stress rock masses, however, rock bolts have to be ductile in order to avoid premature failure of the bolt. A Split Set is a typical ductile support element which has been used for a long time to deal with squeezing/bulking rock conditions. A Split Set has a good ductility, although its load-bearing capacity is much smaller than rebar. In securing burst-prone rocks, the desired support elements should not only be able to accommodate large rock deformations (i.e. be ductile), but also be able to carry a high load. Such a type of element is called an energy-absorbing support element. Taking a rock bolt as an example, the load-bearing capacity of the ideal energy-absorbing rock bolt should be as high as, or close to, rebar, while at the same time being able to accommodate large deformations. In other words, it should be capable of absorbing a large amount of energy. Referring to Figure 18.5, three types of rock bolts are defined from an energy-absorbing point of view:

–   *Strength rock bolt:* This type has a high load-bearing capacity, but deforms little prior to failure. Rebar is a typical strength rock bolt.

- *Ductile rock bolt:* This type is able to accommodate a large rock deformation, though its load-bearing capacity is limited. A Split Set belongs to this category.
- *Energy-absorbing rock bolt:* This type has a high load-bearing capacity, and at the same time it is able to accommodate a large rock deformation. In other words, an energy-absorbing rock bolt is both strong and ductile.

When subjected to dynamic loading, the philosophy of ground support is no longer to balance the dead weight, but instead to help the rock to dissipate the dynamic energy. The dynamic force is not a constant, but rather a function of the stiffness of the support system. It is known from Newton's second law that the momentum of a mass, mv, is equal to the product of force and time, i.e. $mv = Ft$, when the momentum is transferred from one object to another in collision. m stands for the mass, v for the velocity, F for the force between the objects in collision and t for the acting time. For a given ejected rock mass at an ejection velocity, its momentum (mv) is a constant. Thus, a short acting time will result in a high collision force, while a long acting time leads to a low collision force. Dynamic ground supporting is a process of transferring momentum from the ejected rock to the support system. In the case of a stiff support system, the acting time for the momentum transfer is short so that the force induced on the support system would be high. In this case, the dynamic force may be beyond the strength of the support system, and the support system could be destroyed. In the case of a soft or ductile support system, the system would yield at a certain level of force, and deform together with the ejected rock to a certain extent. As a result, the acting time is longer than that in the case of a stiff support, and the kinetic energy of the ejected rock is dissipated during this process. The objective of a dynamic support system is to absorb the kinetic energy. Hence, ductility is crucial for a support system in burst-prone rock conditions.

Figure 18.6 shows a fault-slip induced rockburst event in a deep metal mine in Canada. The rock was already fractured prior to the occurrence of the burst event. The roof was shaken down when the fault-slip burst occurred. In the area, both conventional fully-encapsulated rebar bolts and energy-absorbing bolts were used for ground support. All the energy-absorbing bolts in the collapsed area survived the event, though the rebar failed in the manner of a brittle snap failure, see the close-up right-hand picture in Figure 18.6. Nevertheless, the energy-absorbing bolts did not prevent the collapse of the roof. This was due to a shortcoming of the support system in the surface containment. If the support system had a better surface containing function, the energy-absorbing bolts would help the roof rock to sustain the burst event without the occurrence of the roof collapse.

A dynamic support system is composed of surface support elements and holding elements, see Figure 18.7. The task of the surface support elements is to provide a good containment to the surface so that the surface rock does not disintegrate when it becomes damaged by the rockburst event. Therefore, a full coverage of the surface is necessary in order to achieve a satisfactory surface containment. The surface support should be ductile enough to accommodate large rock dilations. Wire mesh and fibre, or mesh reinforced shotcrete, are two types of surface support elements that have been used in many burst-prone mines. Ductile thin liners with a full surface coverage may be an even better solution for surface containment in the future. The holding support elements have two roles to play in a support system. The first is to limit rock dilation

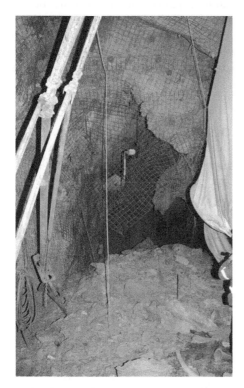
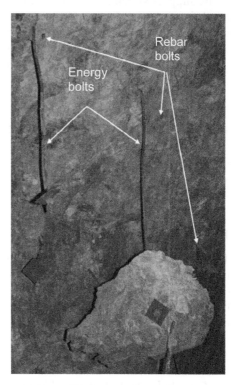

Figure 18.6 A fault-slip induced rockburst in a deep metal mine. The right picture is a close-up of the collapsed roof.

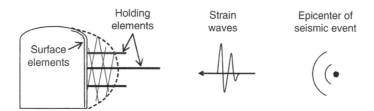

Figure 18.7 A sketch illustrating fault-slip burst waves and support elements in a dynamic support system.

and the second is to hang surface support elements so that the load on the surface elements is transferred into the rock mass through the holding elements. Rock bolts with a length of 2–3 m are the most commonly used holding elements in mines today. Long cable bolts (4–6 m long) can be added as a second layer of holding elements in mines where large rockburst events are expected.

All the support elements, particularly the holding elements (i.e. bolts and cables), in a dynamic support system must be energy absorbent. The energy absorption capacity of any dynamic support system is limited no matter what types of elements are used in the support system. When designing a dynamic support system, one must bear in mind that it is not the support system that absorbs all the burst energy. The aim of the support system is to contain the rock surface and limit the rock dilation. A well contained and dilation-limited rock party would absorb a large amount of energy through rock fragmentation and grinding/friction. The more fragmented the rock is, the more energy is dissipated.

18.5 ENERGY-ABSORBING ROCK BOLTS

Energy-absorbing rock bolts are the most important elements in a dynamic support system. A few of these bolts which are available so far are briefly introduced in this section.

18.5.1 Cone bolt

The cone bolt was invented in South Africa in the 1990s (Jager, 1992). It may well be the first energy-absorbing rock bolt designed to combat rockburst problems. The original cone bolt was designed for cement grout. A cone bolt consists of a smooth steel bar with a flattened conical flaring forged to one end, see Figure 18.8a. The smooth bar is coated with a thin layer of wax, so that it will be easily de-bonded from the grout under pull loading. The cone bolt was modified later for resin grout in Canada (Simser, 2001). The modified cone bolt (MCB) is similar to the cement cone bolt, although the difference is that an MCB has a blade at its far end for resin mixing, see Figure 18.8b.

Cone bolts are fully grouted in boreholes. The bolt is designed so that the deformation of the dilating rock is transferred to the bolt via the face plate, thus pulling the conical end through the grout, performing work and absorbing energy from the rock. The working mechanism of the bolt demands that the strength of the grout must be precisely as designed in order to achieve the designed performance. The reality is that it is not easy to control the strength of the grout, which means that the designed performance of the bolt is not guaranteed. Figure 18.9 shows the results of pull tests conducted in the field (Simser et al., 2006). The static load capacity varies over a relatively large range. The results of dynamic drop tests also vary over a large range,

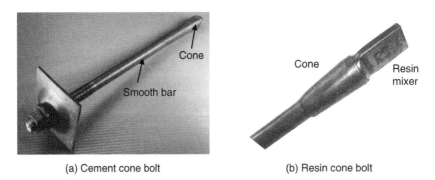

(a) Cement cone bolt   (b) Resin cone bolt

Figure 18.8 The cone bolt.

see Figure 18.10 (St-Pierre, 2007). The load capacity of the cone bolt depends on the mechanism of the bolt deformation. The bolt deforms either through a ploughing of the cone in the grout, an elongation of the bolt shank or a combination of them. The cone bolt is anchored in a two-point manner in the borehole, so there is a risk that the bolt will lose its reinforcement function if the face plate loses contact with the rock.

18.5.2 Durabar

Durabar is an element evolved from the cone bolt. The anchor of Durabar is a crinkled section of the smooth bar plus a smooth tail at the far end, see Figure 18.11. When the face plate is loaded, the anchor slips along a waved profile under the pull. The maximum displacement is equal to the length of the tail, which is approximately 0.6 m.

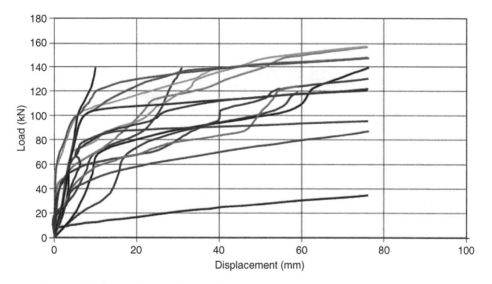

Figure 18.9 Static pull tests of MCB 33 mm resin bolts in the field (Simser *et al.*, 2006).

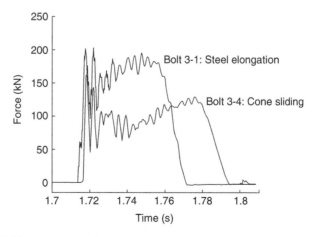

Figure 18.10 Impact load of the cone bolt in dynamic drop tests (St-Pierre, 2007).

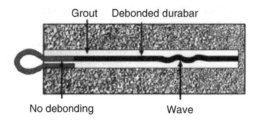

Figure 18.11 Durabar.

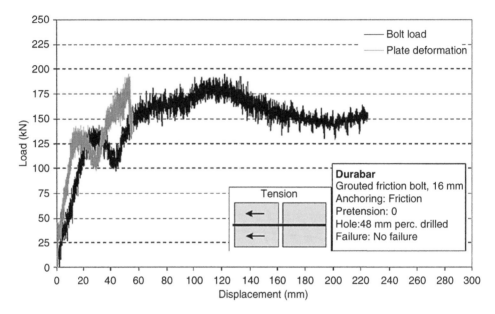

Figure 18.12 Static pull test result of a Durabar (Dahle and Larsen, 2006).

Similar to the cone bolt, Durabar is also a two-point anchored tendon. The static pull test results of a Durabar are shown in Figure 18.12. No results of dynamic tests are available for Durabar thus far.

18.5.3 Hybrid bolt

The hybrid bolt is composed of two bolt elements: a rebar resin-grouted in a Split Set, see Figure 18.13. Installation of the hybrid bolt is conducted in three steps. Step 1: Push a Split Set into the borehole and set epoxy cartridges (fast and low settings) in the tube of the Split Set; Step 2: Spin a rebar in the Split Set; Step 3: Tighten up the nut to apply a pre-load to the rebar.

Pullout tests show that the ultimate pull load of the hybrid bolt is up to 160 kN, and its displacement capacity is similar to a Split Set, see Figure 18.14. No results of dynamic tests are available for hybrid bolts thus far.

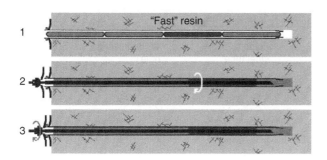

Figure 18.13 The hybrid bolt. (a) the process of installation, (b) a hybrid bolt in situ. (Mercier-Langevin and Turcotte, 2007).

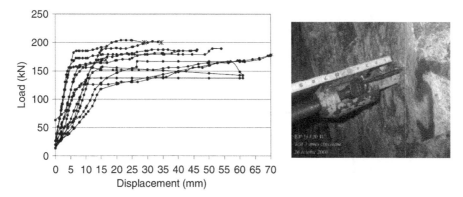

Figure 18.14 Results of pull tests done in the field on hybrid bolts (Mercier-Langevin and Turcotte, 2007).

18.5.4   Inflatable bolt

Swellex is a typical inflatable rock bolt that has been used in civil and mining engineering for a long time, shown in Figure 18.15. The bolt interacts with the rock mass through the friction between the cylindrical surface of the bolt and the wall of the borehole. The recently developed Mn24 version of the bolt has a much better energy absorption capacity than the standard Swellex (Charette, 2007). Figure 18.16 demonstrates the static and dynamic performances of the bolt. The tests show that the static load capacity of the bolt is higher than the dynamic load capacity. The tests by Charette (2007) showed that the energy absorption of Mn24 Swellex is 18–29 kJ.

18.5.5   Garford solid bolt

A Garford dynamic bolt consists of a steel solid bar, an anchor and a coarse threaded steel sleeve at the end, see Figure 18.17. The bolt is designed for use with resin grout. The threaded sleeve is used to mix resin substances. This bolt is characterized by its engineered anchor which allows the bolt to stretch by a large amount when the rock

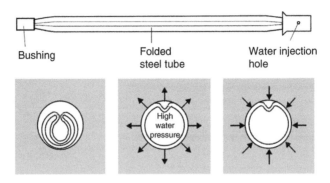

Figure 18.15 A sketch of the Swellex rock bolt (Li and Håkansson, 1999).

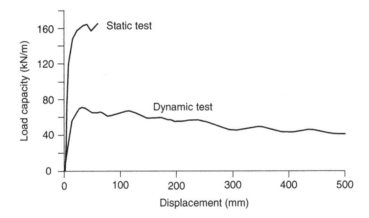

Figure 18.16 Static and dynamic performance of inflatable bolts. The static test specimen is Mn24 Swellex and the dynamic test specimen is Omega bolt. They are equivalent to each other. The embedment length of the specimen is 1 m. (Player, Villaescusa and Thompson, 2009)

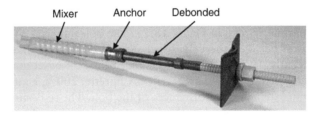

Figure 18.17 Garford dynamic solid bolt.

dilates. The anchor is a thick wall steel cylinder which is pressed to the solid steel bar at a position 350 mm from the far end of the bar. The diameter of the solid bar is reduced from its original size to a smaller one in the position of the anchor. The anchor is resin encapsulated in the borehole. When the rock dilates between the anchor and the face plate, the solid bar is pulled through the hole of the anchor so that an extruding process

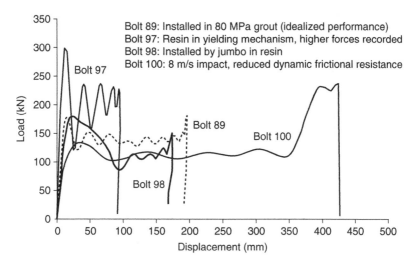

Figure 18.18 Dynamic force-displacement response for Garford solid bolts (Varden *et al.*, 2008).

Figure 18.19 Roofex rock bolt (Charette and Plouffe, 2007).

of the steel bar takes place. The extruding force remains at an approximately constant level when the bar is stretched. The bolt can accommodate an elongation of 390 mm prior to failure. Figure 18.18 shows the results of dynamic tests for the Garford bolt (Varden *et al.*, 2008).

18.5.6 Roofex

Roofex is a ductile bolt developed for dynamic rock support. It consists of an anchoring unit and a smooth bar, see Figure 18.19. The anchor is encapsulated with resin in the borehole. The smooth bar slips through the anchor, generating a constant frictional resistance of about 80 kN, see Figure 18.20. This yield load is designed slightly lower than the yield load of the bar material. The results of dynamic drop tests for different input energies are shown in Figure 18.21. The dynamic load of Roofex is approximately 60 kN.

18.5.7 D-Bolt

The D-Bolt is made of a smooth steel bar that has a number of integrated anchors spaced along its length, see Figure 18.22 (Li *et al.*, 2009; Li, 2010). The anchors are wider than the bar shank in order for the bar to be automatically centralized in the

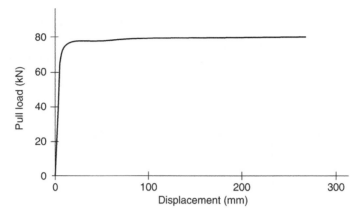

Figure 18.20 Redraw of the results of static pull tests for Roofex (Charette and Plouffe, 2007).

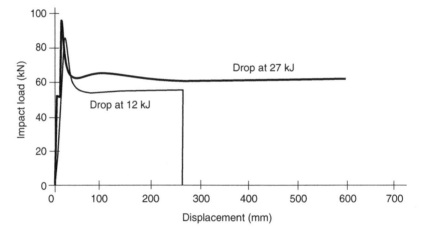

Figure 18.21 Redraw of the results of dynamic drop tests for Roofex at input energies (Charette and Plouffe, 2007).

borehole after installation. The bolt is fully encapsulated in a borehole, using either cement grout or resin. The anchors are firmly fixed in the grout, while the smooth bar sections between the anchors have no or a very weak bonding to the grout. When rock dilates between two adjacent anchors, the anchors will restrain the dilation so that a tensile load is induced in the smooth bar between the anchors. The section elongates a few millimetres elastically and then becomes yielded. After that, the bar section elongates plastically until the ultimate strain limit is reached. Both the strength and deformation capacity of the steel plays a role in this process. The ultimate strain of mild carbon steels, which are usually used for rebar bolts, is about 15–20% for a standard test length of approximately 200 mm. Figure 18.23 shows the pull test results of two D-Bolt sections that are 825 mm long. The bolt sections are elongated 100–120 mm at a load level of roughly 200 kN. Figure 18.24 shows the results of dynamic drop tests of

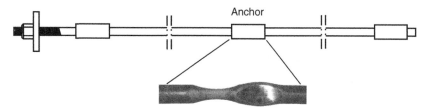

Figure 18.22 Layout of the D-Bolt.

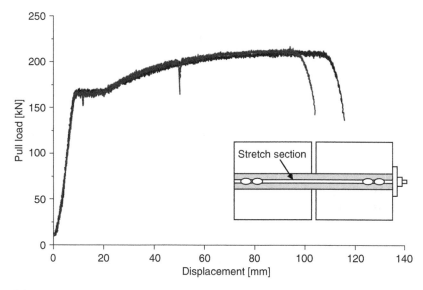

Figure 18.23 Pull test results of two D-Bolt sections, bolt diameter is 22 mm and the stretch length is 825 mm.

the bolt. The tested sections were 22 mm in diameter and 850 mm in length. The two samples were tested with input energies of 36 and 39 kJ, which corresponds to a drop mass of 2 452 kg and 2 675 kg, respectively, being dropped from a height of 1.5 m. The 850 mm long section absorbed about 39 kJ of dynamic energy prior to failure.

The D-Bolt absorbs deformation energy by fully mobilizing the strength and deformation capacity of the bolt steel. Every smooth section of the D-Bolt between two adjacent anchors works independently. Therefore, the failure of one section (or loss of one anchor) only has a local effect on the bolt's reinforcement capability. The other sections (or anchors) still provide reinforcement to the rock. This performance of the D-Bolt is a significant improvement in comparison to two-point anchored rock bolts. For example, if the anchoring function at the bolt plate is lost because of rock crushing under the plate, the D-Bolt only loses reinforcement in the short thread section, and the other sections are not affected at all. For two-point anchored rock bolts, however, the reinforcement would be lost after the loss of the plate.

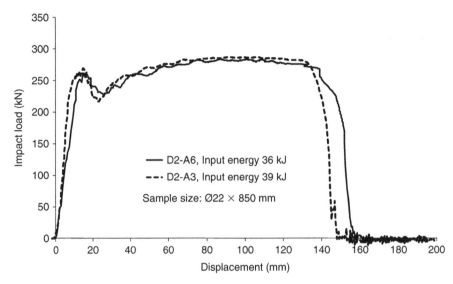

Figure 18.24 Dynamic drop test results of two D-Bolt sections, bolt diameter is 22 mm and the stretch length is 850 mm.

18.6 DYNAMIC SUPPORT PRINCIPLES USED IN SOME COUNTRIES

18.6.1 Australian support principle

The Australian support philosophy is to bolt the failed rock with tightly-spaced Split Sets (2.4 m long) and then nail the bolt-reinforced failed rock by the use of long cables to the deeply located competent rock strata in squeezing rock conditions. The rock surface is retained with a mesh, strap and mesh shotcrete. In burst-prone rock conditions, it is 2.4 m and 3 m long cone bolts which are used in conjunction with meshes or fibre/mesh shotcrete to construct a dynamic support system.

18.6.2 Canadian support principle

The Canadian support philosophy is to integrate the failed rock by short bolts (2.4 m long) in conjunction with meshes and sometimes fibre/mesh shotcrete. The types of bolts used are rebar, Split Sets, and cone bolts in cases of seismic rock conditions. Rebar is still preferred in many burst-prone mines because of its high load capacity, even though it is a stiff rock reinforcement element.

18.6.3 South African support principle

The idea of ductile support elements was initiated in South Africa in the 1990s. The South African rock support philosophy is to dissipate the dynamic energy within the rock that is reinforced with energy-absorbing bolts. The dynamic energy is partially absorbed by the energy-absorbing bolts and partially dissipated through fragmentation of the rock contained by the surface support elements. Lacing is quite often used in South African support systems.

In South Africa, drifts excavated in high stress rock are typically supported by energy-absorbing bolts such as cone bolts, Durabar, Split Set and cables (Durrheim, 2007). The primary bolting is a ring of 1.2 m bolts. The secondary bolting consists of a ring of 2.4 m bolts, mesh, 50 mm thick steel or polyester fibre shotcrete and lacing. The bolting pattern is usually 1 m × 1 m, and yielding props are used when needed.

In some South African gold mines, mining stopes are only about 1 m high. Preloaded yielding props are used to secure the stope, but short bolts are sometimes installed in the roof face. Preconditioning is conducted ahead of the advance face to mitigate the stress concentration to a far depth.

18.6.4   Scandinavian support principle

Similarly to the Canadian rock support, the Scandinavian support philosophy is to integrate the failed rock by short rebar bolts in conjunction with surface liners. Split Set is not used in Scandinavian mines. LKAB and Boliden are now testing energy-absorbing rock bolts. It is expected that energy-absorbing support elements will be soon introduced in some of their mines. The surface liner is either steel-fibre or plain shotcrete, with the former being used the most often.

18.7   DYNAMIC TESTING METHODS OF SUPPORT ELEMENTS

In the past decade, several dynamic testing facilities have been constructed for bolt testing in Canada, Australia and South Africa. An overview of dynamic testing of rock support was given by Hadjigeorgion and Potvin (2007). The testing principles of those facilities are similar. The dynamic load is applied by dropping a mass over a predefined distance onto a tendon installed in a simulated borehole. Thick wall steel tubes are usually used to simulate the borehole. The tendon to be tested is grouted in the hole of the tubes and a mass drops onto the target, either on the face plate of the tendon or on the lower tube. Three most reported testing facilities are briefly presented in this section.

18.7.1   CANMET dynamic test facility

The CANMET dynamic test facility is at the CANMET Mining and Mineral Sciences Laboratories, Ottawa, Canada, see Figure 18.25. The facility is primarily used to test rock bolts. Each test is conducted by dropping a known mass, from a known height, onto a plate connected to a tendon grouted inside a steel tube. The energy input is controlled by the drop height and the mass. The drop height can be varied from 0 to 2.1 m, with the maximum drop mass being 3 000 kg (Plouffe *et al.*, 2008). The maximum energy input of the facility is thus 62 kJ and the maximum impact velocity is 6.5 m/s. The mass is lifted with an electromagnet, which in turn is lifted by a pair of cranes mounted in parallel on top of the machine. By cutting the power to the magnet, the mass freely falls onto the sample.

Figure 18.25 The CANMET dynamic test facility.

18.7.2 **WASM dynamic test facility**

The WASM dynamic test facility is at the Western Australian School of Mines (WASM), Kalgoorlie, Australia, see Figure 18.26. The facility has a maximum impact velocity of 6 m/s and a drop mass of 2 000 kg, thus providing a maximum input energy of 36 kJ. Dynamic loading is applied to test samples through momentum transferring. For bolt testing, the bolt is installed in a split steel tube. The drop mass and bolt sample are lifted by a beam to a predefined height. The beam and the mass, as well as the sample, freely fall and the beam is then stopped by two pads at the bottom of the frame. The momentum of the falling package is then transferred to the bolt. This type of loading has some similarity to the dynamic loading of a tendon installed in situ.

18.7.3 **SIMRAC dynamic test facility**

The SIMRAC dynamic test facility is at the Savuka mine near Johannesburg in South Africa, see Figure 18.27. For bolt testing, the drop mass can be up to 2 706 kg and be dropped from a height of 3 m, corresponding to a maximum input energy of 80 kJ and an impact velocity up to 7.7 m/s. This facility can be also used to test other support elements such as props, mesh and shotcrete. Figure 18.28 shows a test arrangement for a prop support system. For those tests, a deck of 5.5×6.5 m is constructed. The central area of the deck, 3×3 m, consists of three Voussoir beams, each comprising 12 reinforced concrete blocks of $1 \times 1 \times 0.25$ m. These blocks are held together

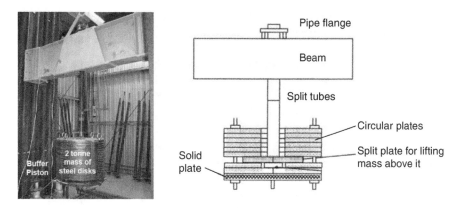

Figure 18.26 The WASM dynamic test facility.

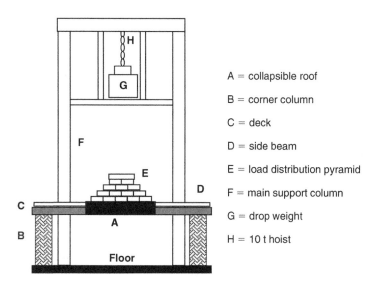

A = collapsible roof

B = corner column

C = deck

D = side beam

E = load distribution pyramid

F = main support column

G = drop weight

H = 10 t hoist

Figure 18.27 The SIMRAC dynamic test facility in South Africa.

under a tension load of 200 kN and props, mesh and shotcrete are set up under the deck. In Figure 18.28, the dead weight on the deck is 20 000 kg and the drop mass is 10 000 kg.

18.8 CONCLUSIONS

Rockburst events occur in highly stressed rock masses. There are two types of rockbursts: strain burst and fault-slip burst. Strain burst is directly related to stress

Figure 18.28 Dynamic testing of props on the SIMRAC test facility: the dead weight is 20 000 kg and the drop weight is 10 000 kg.

concentration in the roof and walls after excavation, while fault-slip rockburst is associated with seismicity in the near field of excavation. A fault-slip rockburst is usually more violent and damaging than a strain rockburst.

Rockburst damage mechanisms are classified into three categories: rock bulking, rock ejection and rockfall. Rock bulking due to fracturing usually results in an increase in volume. Rock ejection due to seismic energy transfer occurs in a manner in which rock blocks are violently ejected from the periphery of an excavation. Rockfall occurs when an incoming seismic wave accelerates a volume of fractured, but statically stable rock.

In the case of rock ejection, the support system has to be capable of absorbing the kinetic energy of the ejected rock. A satisfactory dynamic support system should not only have the capability to absorb deformation energies, but should also be able to contain the potentially ejected rock so that a portion of the energy is dissipated through rock fragmentation and grinding/friction under confinement.

Energy-absorbing rock bolts are needed in a dynamic support system. Ideally, a rock bolt should be not only strong, but also ductile. Among the available energy-absorbing rock bolts, the cone bolt, Durabar, the Garford solid bolt and Roofex are two-point anchored in the rock; the D-Bolt is multi-point anchored; and the hybrid bolt and inflatable bolt interact frictionally with the rock mass along their entire lengths.

The dynamic support philosophies are slightly different in different countries, but they all have accepted the use of energy-absorbing elements and recognized the importance of surface containment in a dynamic support system.

A dynamic support system may be comprised of three layers: a surface support layer, a short bolt layer and a long cable layer. The surface support layer provides a surface containment to avoid rock disintegration on the surface. The short bolt layer limits rock dilation. Rock fracture, fragmentation and friction occur in the dilation-limited volume so that a portion of the burst energy is dissipated in this process. The cable layer limits the movement of the bolt-reinforced rock party, and absorbs an extra amount of deformation energy.

All of the dynamic test methods of support elements are similar in their testing principle. The dynamic load is applied by dropping a mass onto the tendon installed in thick wall steel tubes. Three of the most widely reported test facilities are the CANMET dynamic test facility in Canada, the WASM facility in Australia and the SIMRAC facility in South Africa.

REFERENCES

CAMIRO: *Canadian Rockburst Research Program 1990–1995 – A comprehensive summary of five years of collaborative research on rockbursting in hard rock mines.* CAMIRO Mining Division. 1995.

Carranza-Torres, C. and Fairhurst, C.: Application of the convergence-confinement method of tunnel design to rock masses that satisfy the Hoek-Brown failure criterion. *Tunnelling and Underground Space Technology* 15(2000), pp. 187–213.

Charette, F. and Plouffe, M.: Roofex – results of laboratory testing of a new concept of yieldable tendon. In: Y. Potvin (eds): *Deep Mining 07. – proceedings of the 4th International Seminar on Deep and High Stress Mining.* Australian Centre for Geomechanics, Perth, 2007, pp. 395–404.

Charette, F.: Performance of Swellex rockbolts under dynamic loading conditions. In: Y. Potvin, J. Hadjigeorgiou and D. Stacey (eds.): *Challenges in Deep and High Stress Mining.* Australian Centre for Geomechanics, Perth, 2007, pp. 387–392.

Dahle, H. and Larsen, T.: *Full-scale pull and shear tests of 5 types of rock bolts.* SINTEF report, SBF55 F06033, 8p (2006).

Durrheim, R.J. 2007. The deep mine and future mine research programmes – knowledge and technology for deep gold mining in South Africa. In: Y. Potvin, J. Hadjigeorgiou and D. Stacey (eds): *Challenges in Deep and High Stress Mining.* Australian Centre for Geomechanics, Perth, 2007, pp. 131–139.

Hadjigeorgion, J. and Potvin, Y.: Overview of dynamic testing of ground support. In: Y. Potvin (eds): *Deep Mining 07. – proceeding of the 4th International Seminar on Deep and High Stress Mining.* Australian Centre for Geomechanics, Perth, 2007, pp. 349–371.

Jager, A.J.: Two new support units for the control of rockburst damage. In: P.K. Kaiser and D.R. McCreath (eds): *Proc. Int. Symp. on Rock Support.* Balkema, Rotterdam, 1992, pp. 621–631.

Li, C. and Håkansson, U.: Performance of the Swellex bolt in hard and soft rocks. In: E. Villaescusa, C.R. Windsor and A.G. Thompson (eds): *Rock Support and Reinforcement Practice in Mining – Proc. of Int. Conf. on Rock Support and Reinforcement Practice in Mining,* Kalgoorlie, Australia. Balkema, 1999, pp. 103–108.

Li, C.C.: Disturbance of mining operations to a deep underground workshop. *Tunnelling and Underground Space Technology* 21(2006), pp. 1–8.

Li, C.C.: A new energy-absorbing bolt for rock support in high stress rock masses. *Int. J. Rock Mech. Min. Sci.* 47(2010), pp. 396–404.

Li, C C, Doucet, C and Carlisle, S.: Dynamic tests of a new type of energy absorbing rock bolt – the D bolt. *3rd Canada-US Rock Mech Symp 20th Canadian Rock Mech Symp,* Toronto, Canada, 2009, Abstract pp. 199–200 (9p in CD).

McCreath, D.R. and Kaiser, P.K.: Evaluation of current support practices in burst-prone ground and preliminary guidelines for Canadian hard rock mines. P.K. Kaiser and D.R. McCreath (eds.): *Proc. Int. Symp. on Rock Support.* Balkema, Rotterdam, 1992, pp. 611–619.

Mercier-Langevin, F. and Turcotte, P.: *Evolution of ground support practices at Agnico-Eagle's LaRonde Division – Innovative solutions to high-stress yielding ground.* 8p, 2007.

Player, J.R., Villaescusa, E. and Thompson, A.G.: Dynamic testing of friction rock stabilisers. *3rd Canada-US Rock Mech Symp 20th Canadian Rock Mech Symp*, Toronto, Canada. 2009, Abstract pp. 123–124 (15p in CD).

Plouffe, M., Anderson, T. and Judge, K. 2008. Rock bolts testing under dynamic conditions at CANMET-MMSL. In: T.R. Stacey and D.F. Malan (eds.): *6th Int. Symp. on Ground Support in Mining and Civil Engineering Construction*, Cape Town, South Africa, 2008, SAMM symposium series S51, pp. 581–96.

Simser, B.: *Geotechnical review of the July 29th, 2001 West Ore Zone Mass Blast and the performance of the Brunswick/NTC rockburst support system.* Technical report, 46p, 2001.

Simser, B., Andrieux, P., Langevin, F., Parrott, T. and Turcotte, P.: Field Behaviour and Failure Modes of Modified Conebolts at the Craig, LaRonde and Brunswick Mines in Canada. *Deep and High Stress Mining*, Quebec City, Canada, 2006, 13p.

St-Pierre, L.: *Development and Validation of a Dynamic Model for a Cone Bolt Anchoring System.* Master thesis, McGill University, Canada, 2007.

Tannant, D.D., McDowell, G.M., Brummer, R.K. and Kaiser, P.K.: Ejection velocities measured during a rockburst simulation experiment. In: 3rd *Int Symp on Rockbursts and Seismicity in Mines*, A.A. Balkema, 1993, pp. 129–133.

Yi, X. and Kaiser, P.K.: Impact testing of rockbolt for design in rockburst conditions. *Int J Rock Mech Min Sci Geomech Absstr* 31(1993), pp. 671–685.

Varden, R., Lachenicht, R., Player, J., Thompson, A., and Villaescusa, E.: Development and implementation of the Garford Dynamic Bolt at the Kanowna Belle Mine. In: *10th Underground Operators' Conference*, Launceston, Australia, 2008, 19p.

List of symbols

A: the cross-sectional area
A_0: the initial area of the sample
A_p, A_s: the amplitudes of harmonic P- and S-waves, respectively
$\mathbf{B}$: the interpolation matrix of strain
C: the one dimensional longitudinal stress wave velocity
$[\mathbf{C}]$: the damping matrix
C_e: the effective velocity from EMM
C_{eff}: the effective velocity from existing analytical equations
c_d: the dynamic cohesion
$c_j(\mathbf{x})$: the displacement function of the jth physical cover
C_p, C_s: the P- and S-wave propagation velocities, respectively
D: the scalar damage variable
$\mathbf{D}$: the elastic matrix
d: the fracture closure
$[Di]$: the vector displacement variables of Block i
E: Young's Modulus
$E_{tra}, E_{ref}, E_{inc}$: the transmitted, reflected and incident wave energies
e_{tra}, e_{ref}: the transmitted and reflected energy rates
E_{vp}, E_{vs}: the normal and shear stiffness in a viscoelastic medium contributed by a rock joint, respectively
f_{c0}: the uniaxial compression strength
f_d: the de-coupling factor
f_p, f_s: the frequencies of P- and S-waves, respectively
f_S: the static friction coefficient
F_j: the known external load vector acting on node j
$\mathbf{F}(t)$: the vector of external forces on particles
G: the shear modulus of rock material
H: the Heaviside step function
h: the particle penetration
I: the amplitude of incident wave
k_{ne}: the effective normal fracture stiffness
K_I: the stress intensity factor
K_{jl}: the stiffness matrix for node j and the adjacent node l
K_n: the normalized normal stiffness
K_p, K_s: the normal and shear stiffness of a rock joint, respectively
l: the length of the input bar of SHPB

L: the length of the sample
m: the particle mass
$[\mathbf{M}]$: the diagonal mass matrix
M_j: the lumped mass of node j
n: the number of DOFs
N: the number of fractures
$\mathbf{N}$: the interpolation matrix of displacement
$N_i(\mathbf{x})$: the shape function of ith general degree of freedoms
P: peak chamber pressure
P_1: the dynamic force on the incident end of SHPB
P_2: the dynamic force on the transmitted end of SHPB
P_{max} is the maximum load
Q: charge weight
R: the reflection coefficient
r_{fc}^*: the residual strength
r_{ic}^*: the strength of the intact material
R_j: the contact force vector of node j due to dynamic normal contact stress between crack surfaces
$|R_j|$: the norm of R_j
S: the joint spacing
$S_{\alpha\beta}$: the traceless symmetric deviator stress
t: time
Δt: the time interval
$t_{tra}^1, t_{ref}^1, t_{inc}^1$: the final times of transmitted, reflected and incident waves
$t_{tra}^0, t_{ref}^0, t_{inc}^0$: the initial times of transmitted, reflected and incident waves
t_I, t_T: the time spots for the peak velocities of v_{I1} and v_{Te1}, respectively
T: the transmission coefficient
$|T_1|$: the magnitude of transmission coefficient across a single fracture
$|T_N|$: the magnitude of transmission coefficient across multiple parallel fractures
$|T_2|$: the magnitude of transmission coefficient across two parallel fractures
$[Ti]$: the first order displacement function of Block i
u: the displacement
$u^-(n, j+1)$: the displacement at time $j+1$ before the fracture at distance n
$u^+(n, j+1)$: the displacements at time $j+1$ after the fracture at distance n
u_{ji}: the general DOFs of the cover
$\ddot{U}_j$: the acceleration vector of node j
U_l: the displacement vector of node l
$\dot{u}, \dot{v}, \dot{w}$: the displacement rates in the direction of x, y and z respectively
$\{\dot{u}\}, [N], \{\dot{U}\}$: the deformation rate vector, shape function and nodal displacement vector of a given point in the element, respectively
V: the peak particle velocity
W: elastic energy carried by a stress wave
W_G: the total fracture energy
x_k: the length of the rock mass along the wave propagation path
z_k: the wave impendence and $z_k = \rho C_k$ ($k = p$ for P-wave, $k = s$ for S-wave)
Z_p: the P-wave impedance

α: the dimensionless crack length

α_k: the phase shift per unit length ($k = p$ for incident P-wave, $k = s$ for incident S-wave)

α_m: the critical dimensionless crack length

α_T, α_N: the tangential and normal coefficient of the visco-elastic boundary

σ_c: the static uniaxial compressive strength

β_k: the wave attenuation parameter ($k = p$ for incident P-wave, $k = s$ for incident S-wave)

γ: the ratio of layer thickness to incident wavelength

$\dot{\gamma}_{xy}, \dot{\gamma}_{yz}, \dot{\gamma}_{zx}$: the shear strain rates

δ_{ij}: the Kronecker delta

ε: the strain

$\varepsilon_1, \varepsilon_2, \varepsilon_3$: the three principal strains

ε_e: the effective strain

$\varepsilon_p, \varepsilon_s$: the normal and shear strains, respectively

ε_{t0}: the tensile strain corresponding to the elastic deformation limit

ε_{tu}: the ultimate tensile strain

$\dot{\varepsilon}_{xx}, \dot{\varepsilon}_{yy}, \dot{\varepsilon}_{zz}$: the strain rates normal to the x, y and z planes

$\dot{\varepsilon}$: the stain rate

$\tilde{\eta}_{BT}, \tilde{\eta}_{BN}$: the proportional coefficient related to the tangential and normal stiffness, respectively

η_{vp}, η_{vs}: the normal and shear viscosities of a rock joint, respectively

θ_j: the angle between the normal direction of the crack surface S at node j and x-axis

λ: the wavelength

λ, μ: the Lame constants

λ_k: the wave length for incident P- or S-wave when $k = p$ or s

λ_p: the P-wave wavelength

v: the particle velocity

v_0: the velocity of the striker

v_1: the velocity at the incident bar end of SHPB

v_2: the velocity at the transmitted bar end of SHPB

v_{Ik}: the incident P- or S-wave ($k = p$ for incident P-wave, $k = s$ for incident S-wave)

v_p, v_s: the particle velocities along normal and shear directions, respectively

$v_{tra}, v_{ref}, v_{inc}$: the particle velocities of transmitted, reflected and incident waves

v_{Tddk}: Transmitted P- or S-wave from DDM when $k = p$ or s

v_{Tek}: Transmitted P- or S-wave from EMM when $k = p$ or s

$v_{Tep,1}$: Transmitted wave from direct wave propagation

$v_{Tep,2}$: Transmitted wave from multiple reflections between joints

$v^+(n, j + 1)$: the particle velocity at time $j + 1$ after the fracture at distance n

$v^-(n, j + 1)$: the particle velocity at time $j + 1$ before the fracture at distance n

ξ: the nondimensional joint spacing (ratio of joint spacing to wavelength)

Π: the artificial viscous pressure

ρ: density

σ: the stress tensor

$\sigma_1, \sigma_2, \sigma_3$: the three principal stresses

$\bar{\dot{\sigma}}$: the average loading rate

σ_{ij}: the stress in direction ij

σ_{max}: the maximum (failure) stress

σ_p, σ_s: the normal and shear stresses, respectively

σ_s: the initial stress

σ_{sc}: the uniaxial compressive strength at quasi-static loading rate

σ_t: the tensile strength

σ_{td}: the dynamic tensile strength

$\dot{\sigma}_{dc}$: the dynamic loading rate

$\dot{\sigma}_{sc}$: the quasi-static loading rate

$\sigma^+(n, j+1)$: the normal stress at time $j+1$ after the fracture at distance n

$\sigma^-(n, j+1)$: the normal stress at time $j+1$ before the fracture at distance n

τ: the fracture process incubation time

τ_j: the contact force vector of node j due to dynamic shear contact stresses between crack surfaces

$|\tau_j|$: the norm of τ_j

τ_p, τ_s: the time of retardation of the Voiget element with normal and shear properties

v: the Poisson's ratio

ϕ: the friction angle

ϕ_j: the weight function of the cover

ψ: the phase angle

ω: wave angular frequency

ω_k: the angular frequency of the incident P- or S-wave when $k = p$ or s

ω_0: the angular frequency of the incident wave

Subject Index

Printed and bound by CPI Group (UK) Ltd, Croydon, CR0 4YY
18/10/2024
01776252-0007